VEGETABLE CROPS

VEGETABLE CROPS
Genetic Resources and Improvement

Editors
DINESH KUMAR SINGH
Professor, Department of Vegetable Science
G.B. Pant University of Agriculture & Technology
Pantnagar, 263145, Uttarakhand
and
HARSHAWARDHAN CHOUDHARY
Senior Scientist
Division of Vegetable Science
Indian Agriculture Research Institute
New Delhi- 110012

2012
New India Publishing Agency
Pitam Pura, New Delhi-110 088

Published by

Sumit Pal Jain for

New India Publishing Agency
101, Vikas Surya Plaza, CU Block, L.S.C. Mkt.,
Pitam Pura, New Delhi-110 088, (India)
Phone : 011-27341616, Fax : 011-27341717
Mobile : 09717133558
E-mail : newindiapublishingagency@gmail.com
Web : www.bookfactoryindia.com

ISBN : 978-93-80235-50-9

Printed at : Jai Bharat Printing Press, Delhi

PREFACE

Genetic resources are often considered as the most important among natural resources on the planet, the others being land, air and water. Plant genetic resources are uniquely placed with in the overall ambit of biodiversity and it plays an important role in providing food, fuel, clothing, medicine and shelter for the whole mankind.Diversegenetic materials are always required to meet the ever changing demands of plant improvement. Plant genetic resources comprising of reservoir of gene and gene complex are basic raw materials for genetic improvement of any crop including vegetables.

The book entitled " Vegetable Crops: Genetic Resources and Improvement" is a compilation of information generated through research work of many scientists in India and abroad. The book contains introductory chapter on various aspects of collection, characterization, conservation and utilization of germplasm in genetic improvement of different vegetable crops highlighting the importance of genetic resource management and their achievement in India. Application of different biotechnological tools and techniques like, tissue culture, molecular markers and bioinformatics in conservation and utilization of plant genetic resources have been included. The issues related to biosafety regulations and IPR have also been discussed. Genetic improvement of different vegetable crops through utilization of genetic resources have been dealt in chapters covering all important vegetables including underutilized vegetables, seed spices and edible mushrooms.

This book is very helpful to the teachers, scientists, students and personals involved in PGR management, who wish to update the knowledge on recent technological advances in genetic resource management and improvement of vegetable crops.

Editors

LIST OF CONTRIBUTORS

1. **A. K. Gaur :** Professor, Department of Molecular Biology and Genetic Engineering, Govind Ballabh Pant University of Agriculture & Technology, Pantnagar-263145, Uttarakhand
2. **A. S. Sidhu :** Director, Indian Institute of Horticulture Research, Hassaraghatta, Bangalore 560089, Karnataka.
3. **Anil Kumar :** Professor & Head, Department of Molecular Biology and Genetic Engineering, Govind Ballabh Pant University of Agriculture & Uttarakhand Technology, Pantnagar-263145, Uttarakhand
4. **B.B. Singh :** Emeritus Professor, Department of Genetics & Plant Breeding, Govind Ballabh Pant University of Agriculture & Tecnology, Pantnagar-263145, Uttarakhand
5. **B. Singh :** Project Co-ordinator, Vegetable Crops, Indian Institute of Vegetable Research, Post Bag No. 01, P.O. Jakhini Shahanshahpur, Varanasi- 221 305, Uttar Pradesh
6. **Durvesh Kumar Singh :** Professor, Department of Vegetable Science, Govind Ballabh Pant University of Agriculture & Technology, Pantnagar-263145, Uttrakhand
7. **D. Roy :** Professor, Department of Genetic & Plant Breeding, Govind Ballabh Pant University of Agriculture & Technology, Pantnagar-263145, Uttarakhand
8. **Dhirendra Singh :** Senior Research Officer, Department of Vegetable Science, Govind Ballabh Pant University of Agriculture & Technology, Pantnagar-263145, Uttarakhand
9. **H. S. Chawla :** Professor, Department of Genetic & Plant Breeding, Govind Ballabh Pant University of Agriculture & Technology, Pantnagar-263145, Uttarakhand
10. **Hari Har Ram :** Chief Research & Development, (Vegetable Seeds), Krishidhan Vegetable Seeds (I) Pvt. Ltd., Sai Capitals V Floor Senapati Bapat Marg, Shivajinagar, Pune-411016, Maharashtra
11. **H. Choudhary :** Senior Scientist, Division of Vegetable Science, Indian Institute of Agricultural Research , PUSA, New Delhi- 110 012
12. **K. K. Gangopadhyay :** Sr. Scientist, Germplasm Evaluation Division,National Bureau of Plant Genetic Resources, PUSA Campus, New Delhi-110012
13. **K. S. Negi :** Officer Incharge, National Bureau of Plant Genetic Resources, Regional Station, Bhowali, Distt. Nainital -263132, Uttarakhand
14. **K. P. Singh :** Principal Scientist, Indian Institute of Vegetable Research, Post Bag No. 01, P.O. Jakhini, Shahanshahpur, Varanasi-221 305, Uttar Pradesh

15. **K. V. Peter :** Retired Professor, Horticulture and Ex-Vice Chancellor, Kerala Agriculturlal University of Thrissur, Kerala,
16. **N. Ahmed :** Director, Central Institute of Temperate Horticulture, PO. Sanatnagar, Srinagar, Jammu and Kashmir-190005, J & K.
17. **N.C. Gautam :** Dean, Post Graduate Studies, Department of Vegetable Crops, Narendra Deva University of Agriculture &Technology, Faizabad- 224 229, Uttar Pradesh
18. **R. K. Yadav :** Sr. Scientist, Division of Vegetable Crops, Indian Agricultural Reserch Institute, PUSA, New Delhi-110012
19. **R. P. Singh :** Emeritus Professor, Department of Plant Pathology, Govind Ballabh Pant University of Agriculture & Technology, Pantnagar-263145, Uttarakhand
20. **Rajeew Kumar :** Junior Research Officer, Department of Agronomy, Govind Ballabh Pant University of Agriculture & Uttarakhand Technology, Pantnagar-263145, Uttarakhand
21. **S.K. Malhotra :** Principal Scientist, Horticulture, Indian Council of Agricultural Research, Krishi Anusandhan Bhawan-II, PUSA, New Delhi-110012
22. **S. K. Mishra :** Ex Head, Germplasm Evaluation Division, National Bureau of Plant Genetic Resources, PUSA Campus, New Delhi-110012
23. **S. K. Verma :** Senior Scientist, National Bureau of Plant Genetic Resources, Regional Station, Bhowali, Distt. Nainital-263132, Uttarakhand
24. **S. R. Sharma :** Retired Head, Indian Agricultural Reserch Institute, Regional Station, Katrain, Kullu Valley, 175129 Himanchal Pradesh.
25. **Soma Marla :** Principal Scientist, National Bureau of Plant Genetic Resources, Indian Agricultural Research Institute, New Delhi-110012,
26. **T. K. Behera :** Scientist, Division of Vegetable Crops, Indian Agricultural Research Institute, New Delhi-110012
27. **Umesh Srivastava :** Assistant Director General, Vegetable Crops, Krishi Anushandhan Bhawan–II, PUSA, New Delhi-110012
28. **Y. V. Singh :** Professor & Head, Department of Vegetable Science, Govind Ballabh Pant University of Agriculture & Technology, Pantnagar-263145, Uttarakhand
29. **I.S. Bisht :** Principal Scientist, National Bureau of Plant Genetic Resources, Pusa Campus, New Delhi- 110012
30. **Ashok Kumar :** Principal Scientist, National Bureau of Plant Genetic Resources, PUSA Campus, New Delhi- 110012
31. **S.K. Yadav :** Senior Scientist, National Bureau of Plant Genetic Resources, PUSA Campus, New Delhi- 110012
32. **K.C. Muneem :** Principal Scientist, National Bureau of Plant Genetic Resources, Regional Station, Bhowali, Distt. Nainital- 263132, Uttarakhand.

33. **Sonu Ambwani :** Asst Professor, Department of Molecular Biology and Genetic Engineering, Govind Ballabh Pant University of Agriculture & Technology, Pantnagar-263145, Uttarakhand.
34. **S.K. Sanwal :** Senior Scientist, Indian Institute of Vetetable Research, Post Bag No. 01, P.O. Jakhini, ShahanshahpurVaranasi -221305 (U.P.)
35. **Gunjeet Kumar :** Senior Scientist, National Bureau of Plant Genetic Resources, PUSA Campus, New Delhi- 110012
36. **K.K. Mishra :** Asst. Professor, Department of Plant Pathology, Govind Ballabh Pant University of Agriculture & Technology, Pantnagar- 263145, Uttarakhand.
37. **Chander Parkash :** Senior Scientist, IARI, Regional Station, Katrain, Kullu Valley, 175129, Himanchal Pradesh.
38. **A. J. Gupta :** Senior Scientist, Directorate of Onion and Garlic Research, Rajgurunagar-410505 Distt-Pune, Maharashtra
39. **K. Hussain :** Central Institute of Temperate Horticulture, Old Air Field, PO. Rangreth, Srinagar,- 190007, Jammu and Kashmir
40. **Shri Dhar :** Principal Scientist, Indian Agricultural Research Institute, PUSA, New Delhi- 110012
41. **B. Bhushan Kumar :** AVRDC-SRTT Project AVRDCRCSA, Birsa Agricultural University, Ranchi-834006, Jharkhand.
42. **P.G. Sadhan Kumar :** Department of Vegetable Crops, Kerala Agricultural University, Thrissur, 680656, Kerala.
43. **S. Nirmala Devi :** Department of Vegetable Crops, Kerala Agricultural University, Thrissur, 680656, Kerala .

CONTENTS

Chapter – 1

Biodiversity of Vegetable Crops: An Overview

D.K. Singh and H. Choudhary

Introduction

The term 'biodiversity' comes from the words biological diversity and quite simply means the variety of all living things, including microbes, plants and animals– from single-celled organisms to the largest mammals and trees. Biodiversity also refers to genetic diversity within a species (essential for evolution) and also the diversity of the woodlands, wetlands and other habitats which provide the food, water and shelter for these species. Human life itself depends upon healthy ecosystems (e.g. wetlands, forests and grasslands) and the biodiversity that they contain. Biodiversity gives us many of the essentials of life – water, oxygen, food, clothing, and medicines, without which we could not survive, and therefore contributes greatly to our economy. Access to nature is also a popular form of relaxation that greatly enriches our lives and helps to keep us healthy. Species are linked in an infinite number of ways via food-webs and the habitats they share. If one species becomes extinct, it may affect many more. If too many species become extinct then whole ecosystems can collapse, with severe consequences for the way we live. Across the world, biodiversity is under threat from human activity such as over-intensive or inappropriate farming, large-scale commercial forestry, forest clearance, mineral extraction, pollution and urban development. Conservation of plant diversity assumes greater importance when the world is facing unprecedented loss of biological diversity.

As per an estimate about 60,000 out of 2, 87, 655 species of plants known in the world are facing the threat of extinction.

Agro-ecosystems are defined as ecological and socio-economic system comprising domesticated plants and/or animals and the people who husband them, intended for the purpose of producing food, fiber, or other agricultural products. Agro-ecosystems are ecological systems transformed and simplified for the purpose of agriculture. In the case of agro-ecosystems productivity, these systems have Holling resilience if, in some state, they are able to maintain productivity and withstand stress or external shocks (e.g., due to lower rainfall and droughts). In many situations, biodiversity provides the link between stress and loss of resilience is a system (Perrings *et al.,* 1995). Genetic variation within species and within population increases the ability to respond to the challenges of environmental stress (Mainwaring, 2001).

Agricultural Biodiversity

Biodiversity is the sum of genetic information that is contained in the genes of individuals, plants, animals and micro organisms. Crop biodiversity, a component of agricultural biodiversity, refers to all diversity within and among wild and domesticated crop species (Qualset *et al.,* 1995; Lenne and Wood, 1999). Crop biodiversity has crucial effects on food production, health and life-support systems. In managed systems, such as agro-ecosystems, crop genetic resources are the raw materials for crop breeding, pest resistance, productivity, stability and future agronomic improvements. Heal (2000) wrote that the "diversity of organisms in an ecosystem is required for that system to function and to provide services to human societies and the removal or addition of even a single type of organism can have far reaching consequences." In a series of plot experiments, plant biomass has been found to be an increasing function of diversity (Tilman and Downing, 1994; Tilman *et al.,* 1996), with higher diversity contributing to increases in the productivity of ecosystems. This is because "multiple species coexistence occurs if there is an interspecific tradeoff such that each species is a superior competitor for a limited range of values of the physical factor and if the physical factor is heterogeneous" (Tilman *et al.,* 2005). Also, genetic variability within and between species confers the potential to resist biotic and abiotic stresses, both in the short and the long term (Giller *et al.,* 1997). Growing more crop species enhance the possibility of producing in years where rainfall regimes or environmental conditions are more challenging. Thus, having functionally similar plants that respond differently to weather and temperature randomness contributes to resilience (Hollings, 1973) and ensures that "whatever the environmental conditions there will be plants of given functional types that thrive under those conditions" (Heal, 2000). Maintaining *in situ* crop biodiversity tends to provide the agro-ecosystem a wider range of productive responses to weather shocks. And, this is particularly important in systems such as agro-ecosystems whose complexity has been simplified and the number of crop species reduced for the purpose of agriculture (Conway, 1993). In such a system crop biodiversity is the most important component of the overall agro biodiversity.

Agricultural biodiversity, as it relates to agricultural crops, can be defined as "the diversity of genetic material contained in traditional varieties and modern cultivars, as well as crop wild relatives and other wild plant species that can be used now or in the future for food and agriculture" (FAO, 1996) There is a richness component to this definition as well as both abundance and temporal components. Agricultural ecosystems exhibit "variety richness" if farmers are growing many varieties that are distant in terms of genetic ancestry. With a rich genetic pool, future breeding efforts are more likely to result in new varieties that improve the human condition. "Variety abundance", on the other hand, refers to the distribution of varieties within a given agricultural ecosystem at a point in time. It decreases as farmers increase the percentage of area planted to the region's most popular variety. With an abundance of varieties, the probability of crop failure decreases since each variety has unique conditions under which it thrives or is susceptible. It is possible, however, to avoid crop failure even with a large percentage of the area planted to a single variety as long as the dominant variety is replaced before its desirable characteristics diminish; that is, before it looses "variety vigor" and becomes susceptible to local crop hazards. In this way, the problems associated with decreasing abundance diversity can be offset by temporal diversity. New varieties and cultivars that are developed in the formal, or scientific, breeding system is referred to as "modern" varieties. They include conventionally bred varieties (Mendelian crosses and hybrids) as well as "biotech" varieties, or those bred through means of biotechnology. Modern varieties and cultivars have tended to be quite similar in genetic ancestry, being the result of manipulating or crossing a few proven elite lines. Farmers' varieties and cultivars, or "landraces", are the product of breeding or selection carried out by farmers. Landraces and wild relatives have widely varied genetic ancestry. (Bellon, 1996; Smale, *et al.*, 1998;Day-Rubenstein and Heisey, 2001).

Farmer's Seed System

Historically, agricultural plant breeders were farmers who nurtured mutations or relied on techniques such as cross-pollination and grafting to develop new varieties and cultivars. Because new varieties were typically bred *in situ*, the genetic pool was continually responding to environmental changes. Very little *ex situ* conservation of agricultural genetic resources occurred - the longer the genetic material remained sequestered, the less adapted it was to prevailing environmental conditions. Farmers saved their seed or took cuttings for the next growing season, traded or sold seed and cuttings to other farmers, and made use of the genetic material in their seed and cuttings to further breeding improvements. New plant varieties and cultivars were thus diffused and adopted via an informal "farmer seed system". The farmer seed system, however, did not encourage continuous development efforts. Farmer-breeders often could not capture the full value of their contribution to enlarging the pool of plant genetic resources. The desirable traits of new varieties and cultivars were encoded in the DNA of the plant's reproductive system and were consistently reproduced in subsequent generations. Once a new variety was distributed among farmers, they simply retained seeds or cuttings for replanting or for exchange with other farmers. The farmer-breeder could control pricing and distribution for first generation progeny (which may or may not have been sufficient to re-coup development costs), but lost all degree of ownership thereafter. Thus there was little economic incentive to develop new varieties and cultivars or to enhance agricultural biodiversity for future breeding efforts.

Importance of Genetic Resources

Genetic resources are sometimes called the "first resource" of the natural resources on this planet - the others being land, air and water. Genes are the link from generation to generation of all living matter. Therefore, attention to genetic resources means attention to the vast diversity among and between species of animals, plants and microorganisms. Within this diversity there is a hierarchy of organization and the term genetic resource has meaning at each level. At one level, genetic resources include all the individuals of a species, particularly if it is threatened with extinction. Genetic resources also include populations, gene pools, or races of a species which possess important attributes not found uniformly throughout the species. Breeding lines and research materials, such as mutant, genetic or chromosomal stocks, are also genetic resources and are important in animal and plant breeding and in all phases of biological research. Finally, genetic resources can refer to genes themselves, maintained in selected individuals or cloned and maintained in plasmids. Genetic resources are the substance of agriculture and food production. Genetic resources must be maintained as an investment for the future.

Strategies for Genetic Resource Conservation

There exist two main approaches to genetic resources conservation: offsite (*ex situ*) conservation, by which is meant the maintenance of the resources in a site or facility which is not their natural or native habitat, and onsite (*in situ*) conservation, by which is meant the preservation of the resources in their native habitats. Four strategy levels for conservation can be distinguished

- Conservation of cloned genes, gametes, embryos, seeds, tissues or whole organisms in a quiescent state.
- Conservation of plants, animals or microorganisms in a confined or controlled environment, such as plantations, gardens, zoological parks, reserves or on host organisms in the case of obligate parasites.
- Conservation of plants, animals or microorganisms in their natural habitats where population size and structure are managed.
- Conservation of plants, animals or microorganisms in their natural habitats without regard to population size or structure.

The successful conservation of any given genetic resource may involve combinations of two or more of these strategy levels, employing both onsite and offsite methods.

Conservation and Use of Vegetable Genetic Resources

Vegetable crops include a large number of species, mainly used as an essential complement to the daily diet, providing vitamins, minerals, fiber, specific amino acids and other active metabolites. Increasing the use of vegetables is considered to offer healthy benefits in all dietary situations. Some vegetables such as tomatoes, cabbages, watermelons and onions are among the most important crops according to total world production. Many others have lower global importance, but their production and use represent relevant nutritional and economic value in specific areas, i.e. asparagus in Europe, traditional leafy vegetables in Africa and Asia and several indigenous cucurbits in India. Several authors have reviewed vegetable genetic resources in the last thirty years (Sloten, 1980; Crisp and Astley, 1985; Kalloo, 1988; Cross, 1998), largely emphasizing the need to accumulate and conserve in gene banks the genetic diversity that is most useful to breeders. One of the primary reasons to sustain conservation of plant genetic resources in gene banks is to prevent the loss of genetic diversity. The great diversity of types in cultivation is also considered a genetic resource itself, to be maintained or increased, since it adds to the diversity of vegetables being grown and consumed and could serve to replace more established but similar crops in case of need. The risk of genetic erosion due to the introduction of single new cultivars was considered especially high by Crisp and Astley (1985) for those vegetables that are built on a narrow genetic base, such as garlic, broccoli, tomato, cucumber, etc. Genetic erosion is difficult to document with solid data. The status of conservation of vegetable crops germplasm has always received less attention than that of the major staple crops such as cereals and legumes. Information on vegetable germplasm can, however, increasingly obtained from online international databases. Maintenance and updating of this information requires a high level of international collaboration.

On a global level, efforts of the International Board for Plant Genetic Resources (IBPGR, now Bioversity International) to conserve vegetable crop germplasm began in 1980, with the definition of a number of priority crops for conservation, according to their importance for rural development and to their economic value for farmers in the tropics (*Abelmoschus esculentus* and related species, *Allium* spp., *Amaranthus* spp., *Brassica* spp., *Capsicum* spp., *Cucurbita* spp., *Lycopersicon esculentum*, *Momordica charantia* and related species, and *Solanum melongena*) (Sloten, 1980). Action on this list of crops included the identification of existing collections, assignment of responsibility to specific gene banks and production of descriptors to facilitate characterization. According to Plucknett *et al.* (1987), a major push to preserve the genetic diversity of vegetables resulted in the doubling in size of vegetable germplasm collections within the five years following IBPGR's intervention in the conservation of vegetable germplasm. They already considered the size of the world collections of landraces of tomato, peppers and amaranths as almost complete in the mid-eighties. According to the data of the FAO World Information and Early Warning System (WIEWS) (http://apps3.fao.org/wiews), in the following fifteen years the number of accessions of major vegetables conserved in gene banks doubled a gain or even tripled (Table 1). The FAO - WIEWS database (Table 2) includes metadata obtained from many different sources, at different times, and offers an approximate indication of the real situation.

In year 1998, FAO estimate regarded vegetable accessions to be about 8 % of a total of around 6 million accessions in ex situ collections. The large majority of these vegetable accessions are represented by the IBPGR priority crops, while minor vegetables (possibly more than 2000 species are used in the world) are numerically much less represented. Cross (1998) considered the comprehensiveness of the collections still inadequate or poor, with the exception of tomato, on the basis of the absolute numbers available and the extent of their provenance from the centers of diversification. It is, however, difficult to make objective statements since a complete analysis of the level of duplication of the world collections and of the actual coverage of genetic diversity of each gene pool is not available.

Table 1. Comparison of world vegetable collections (1987 – 2002).

Crop	Plucknett *et al.* (1987)	FAO – Views (2002)
Tomato	32,000	76,400[1]
Cucurbits	30,000	65,800[2]
Cruciferae	30,000	89,250[3]
Pepper	23,000	59,300
Allium	10,500	26,700
Amaranths	5,000	14,000
Okra	3,600	7,200
Eggplant	3,500	6,850

[1] *Lycopersicon* sp. [2] *Citrullus* sp. and *Cucurbita* sp. [3] *Brassica* sp. and *Raphanus* sp.

Table 2. Total world vegetable collections (Source: FAO -Wiews 2002).

Taxon	Total no. of Accessions
Lycopersicon sp.	76,395
Capsicum sp	59,303
Cucumis sp.	29,869
Cucurbita sp.	27,334
Allium sp.	26,677
Brassica oleracea	17,635
Lactuca sp.	14,126
Citrullus sp.	8,593
Solanum melongena	6,839
Daucus sp.	6,595
Total	**273,366**

The Vavilov Institute, St. Petersburg, Russian Federation is rich in germplasm of all crops and holds the largest collections in Europe of tomato, cabbage, leek and cucurbits. HRI, Wellesbourne, UK holds the largest collections of onions, cauliflowers and carrots. CGN, Wageningen, The Netherlands is very rich in lettuce germplasm. The German gene banks also hold vegetable collections of large sizes

Role of World Vegetable Centre (AVRDC)

World Vegetable Centre is actively involved in conservation of vegetable germplasm. About 10 years back during 1999 its total collection was 45,806 accessions comprising of 75 genera and 191 species (Table 3). July 1999 marked the implementation of a new project "Conservation and Utilization of Indigenous Vegetables." This project was funded by the Asian Development Bank (RETA 5839) and being coordinated by AVRDC. The main objective of the project was to improve the conservation and utilization of indigenous vegetables in South and Southeast Asia. This objective will contribute to the long-term goals of improved human nutrition and reduction in poverty. Five countries are participating in the project: Bangladesh, Indonesia, Philippines, Thailand, and Vietnam. The countries were chosen on the basis of the following criteria: they are within an area of known vegetable genetic diversity and have a rich cultural heritage that includes the use of many traditional vegetable species; they are also suffering genetic erosion at a fast rate due to development. During last decade World vegetable centre has collected and conserved more than 10, 000 new accessions comprising of 85 new genera and 146 new species and now its total collection stands at 56,135 accessions (Table 4).

Table 3. Accessions of vegetable germplasm conserved at GRSU, AVRDC (1999)

Crop	Total no. of Accessions
Glycine	14,142
Capsicum	7341
Lycopersicon	7184
Vigna radiata	5616
Solanum	2341
Brassica	1618
Allium	1071
Sub-total	**39,313**
Other Crops	
Vigna unguiculata	1384
Phaseolus	585
Vigna mungo	481
Luffa	461
Cucumis	387
Abelmoschus	276
Amaranthus	258

Contd.

Cucurbita	247
Vigna unguiculata ssp *sesquipedali*	237
Pisum	214
Lablab	180
Vigna unguiculata ssp *unguiculata*	63
Others	1720
Sub-total	6493
Total	**45,806**
No. of genera	75
No. of species	191
No. of countries	137

Table 4. Accessions of vegetable germplasm conserved at AVRDC (2008)

Crop	Total No. of Accessions
Principal Crops	42,825
Other crops	13,310
Total	**56,135**
No. of genera	160
No. of species	**337**
No. of countries	150

Indian Scenario

Indian National Plant Genetic Resources System (INPGRS) is operated by National Bureau of Plant Genetic Resources (NBPGR) as component of National Agricultural Research System (NARS) for systematic management (collection, characterization, conservation, documentation) and utilization of plant genetic resources. The collaboration of 57 National Active Germplasm Sites (NAGS) for evaluation, maintenance and providing germplasm to the users, is also a vital component of INPGRS. The activities on germplasm evaluation, maintenance and utilization are shared between National Bureau of Plant Genetic Resources and Indian Institute of Vegetable Research (IIVR) as designated as National Active Sites (NAGS) for vegetable crops. More than 10,000 accessions of vegetable crops germplasm have been characterized and evaluated and a number of promising lines have been identified and utilized in improvement programme by various researchers. The information on characterization and evaluation of some of the vegetable crops germplasm namely brinjal (1188), okra (5322), tomato (2980) and fenugreek (171) etc. have been published in the form of catalogues by NBPGR (Thomas *et al.*, 1990;Bisht *et al.*, 1993 & 1995).The present status of germplasm of vegetable crops conserved in the National Gene Bank for long term storage (LTS) and also the same in medium term storage (MTS) have been given in Tables 5 and 6, respectively.

Table 5. Status of vegetable crops germplasm at LTS in the Indian National Gene Bank

Crop	Accessions	Crop	Accessions
Tomato	1609	Bottle gourd	577
Brinjal	3960	Sponge gourd	547
Chilli	2521	Ridge gourd	42
Okra	2292	Bitter gourd	494
Cabbage	67	Kakrol	23
Cauliflower	125	Snake gourd	153
Broccoli	4	Ash gourd	263
Knolkohl	5	Ivy gourd (Kundru)	15
Chinese cabbage	111	Round melon (Tinda)	104
Kale	5	Watermelon	110
Peas	190	Muskmelon	561
Carrot	101	Snap melon	188
Turnip	13	Cucumber	288
Radish	240	Kachri	82
Onion	747	Pumpkin	139
Spinach	115		

Table 6. Status of vegetable crops germplasm at MTS

Crop	Accessions at NBPGR	Accessions at IIVR
Tomato	925 (HQ) + 304 (Hyd)	1052
Brinjal	1860 (HQ) + 611 (Hyd)	350
Chilli	1188 (Bhow) + 3021 (Hyd)	135
Okra	680 (Akola)	373
Bottle gourd	270 (HQ)	36
Sponge gourd	200 (HQ)	-
Ridge gourd	250 (HQ)	-
Bitter gourd	55 (HQ)	176
Ash gourd	20 (HQ)	-
Pointed gourd	-	250
Ivy gourd (Kundru)	-	60
Pumpkin	6 (HQ)	-
Watermelon	-	16
Muskmelon	-	45
Cucumber	-	24
Radish	170 (HQ)	-
Spinach	93 (HQ)	-
Peas	2700	153
Cowpea	2950	127
Onion*	68	60
Garlic*	625	60

* At NRCO&G, Onion (437) and Garlic (269)

Pantnagar Centre for Plant Genetic Resources (PCPGR)

To strengthen the vegetable genetic resource activities, PCPGR was established in 1999 with the financial assistance of World Bank through Diversification Agricultural Support Project-Uttarakhand and Uttar Pradesh as a regional resource centre for collection, maintenance, evaluation and utilization of crop plants with emphasis on horticultural plants specific to the needs of Uttaranchal/Uttar Pradesh. The National Bureau of Plant Genetic Resources, New Delhi is providing the technical assistance to this centre. The project formally became operational on December, 1999. The World Bank funding ended on March, 2004. Considering the importance of the project in overall context of Plant Genetic Resources management in the plant biodiversity rich state of Uttaranchal, the project is being funded by Horticulture Technology Mission-Mini Mission I/Government of Uttaranchal to run the project on a regular basis.

This centre has been able to import a total of 153 elite germplasms lines of soybean, 211 of tomato/cherry tomato, 93 of sweet pepper, 50 of chilli, 16 of amaranth, 5 of eggplant and 3 of mungbean from Asian Vegetable Research and Development Centre, Shanhua, Taiwan. Besides, about 1000 breeding lines/ parental lines/ germplasm of dry bean/ snap bean representing diversity in seed colour and plant growth have been collected from CIAT, Cali, Columbia. A total of 6781 genotypes of different vegetables in collaboration with concerned scientists have been evaluated, documented and conserved under medium term storage (MTS) at PCPGR.

Table 7. Acquisition of elite planting materials and germplasm lines from foreign countries

Source	Crop	No. of accessions
AVRDC, Taiwan	Tomato	211
	Grain/Vegetable Soybean	173
	Mungbean	3
	Sweet pepper	93
	Chilli pepper	50
	Ethiopian eggplant	5
	Bittergourd	7
	Amaranth	16
	subtotal	558
CIAT, Cali, Colombia	French bean	985
IITA, Nigeria	Cowpea	44
	Grand total	**2145**

Segmented leaf type bottle gourd PBOG-54 was identified for the first time in India and was registered in NBPGR meeting held at 18th May 1999. The gene responsible for segmented leaf was found to be dominant in nature.

Pant Sabji Matar-4, Powdery mildew resistant, leafless pea variety was also registered with NBPGR.

Table 8. Germplasm Available at PCPGR in different Vegetable Crops.

S.N.	Crop	No. of accessions	S.N.	Crop	No. of accessions
1.	Amaranthus	219	2.	Arbi	01
3.	Ash gourd	62	4.	Asparagus	02
5.	Broad bean	04	6.	Ban methi	01
7.	Beans	25	8.	Beet leaf	03
9.	Beet leaf	01	10.	Bitter gourd	154
11.	Bottle gourd	217	12.	Brinjal	168
13.	Wild brinjal	03	14.	Broccoli	03
15.	Cabbage	06	16.	Carrot	07
17.	Cauliflower	44	18.	Chenopodium	02
19.	Chilli	1610	20.	Chow chow	01
21.	Cluster bean	10	22.	Cow pea	86
23.	Cucumber	189	24.	*Cucumis sativus var. hardwikii*	08
25.	*Cucubita marrow*	01	26.	Fenugreek	33
27.	Foran	01	28.	French bean	1546
29.	Indian bean	05	30.	Kachari	01
31.	Karam Saag	02	32.	Kasuri methi	01
33.	Kauni	01	34.	Knol khol	02
35.	Kusum Saag	01	36.	Leek	01
37.	Lettuce	07	38.	Long bean	01
39.	Long melon	36	40.	Malabar Spinach	01
41.	Mango melon	01	42.	Meetha karela	01
43.	Mellow	01	44.	Mung bean	03
45.	Musk melon	482	46.	Okra	155
47.	Onion	7	48.	Pehaita	01
49.	Parsley	02	50.	Pea	206
51.	Pointed gourd	02	52.	Pumpkin	174
53.	Radish	40	54.	Red cabbage	01
55.	Ridge gourd	141	56.	Round melon	11
57.	Rye saag	02	58.	Capsicum	96
59.	Snake gourd	24	60.	Snap melon	23
61.	Soy bean	170	62.	Spinach	25
63.	Sponge gourd	134	64.	Taro	02
65.	Tomato	378	66.	Turnip	07
67.	Water melon	224	68.	Wastwahk	01
69.	Wild ivy gourd	02	**Total**		**6781**

Future Thrust

The precise information on the current number of vegetable accessions conserved in the world gene banks is not provided by any global documentation system. vegetables, which have not received the same attention as other crops in the past, it will be especially important that sufficient financial resources continue to be allocated to the maintenance of collections. It is common belief that in several locations, especially in the genetic resource rich but resource poor countries and in the marginal areas, vegetable landraces not represented in the gene banks still exist. It would be important to promote collection of this germplasm and of wild species, together with initiatives to survey and to maintain in situ and on-farm genetic diversity. An integrated and functional documentation system should be developed to increase awareness of the status and distribution of genetic resources.

References

Bellon, M.R. 1996. The dynamics of crop intra-specific diversity: a conceptual freamework at farmer level. Econ. Bot. **50:** 26-39.

Bisht, I.S., Patel, D.P. and Mahajan, R.K. 1997. Classification of genetic diversity in Abelmoschus esculentus germplasm collection using morphometric data. Annals of Plant Biology, **130:** 325-335.

Bisht, I.S., Patel, D.P., Mahajan, R.K., Thomas, T.A. and Rana, R.S. 1995. Catalouge of wild Abelmoschus esculentus germplasm. National Bureau of Plant Genetic Resources, New Delhi. p.51.

Conway. G.R. 1993. Sustainable agriculture: The trade-offs with productivity, stability and equitability, In: Economics and ecology new frontiers and sustainable development, E.D. Barbier (ed.), Chapman, London, 46-65.

Crisp, P. and Astley, D. 1985. Genetic resources in vegetables. In: Progress in Plant Breeding – 1. G.E. Russel(ed.).Butterworths, London. p. 281-310.

Cross, R.J. 1998. Review paper: global genetic resources of vegetables. Plant Varieties and Seeds **11:** 39-60.

Day-Rubenstein, K. and P. Heisey. 2001. Crop Genetic Resources. In: Economics and ecology new frontiers and sustainable development, E.D. Barbier (ed.), Chapman, London, 46-65.

FAO. 1996. The State of the World's Plant Genetic Resources. Rome, Italy.

Giller K. E., Beare , M. H., Lavelle , P., Izac, A. M. N. and Swift, M. J. 1997. Agricultural intensification, soil biodiversity and agro-ecosystem function, Applied Soil Ecology, **6:** 3-16.

Heal, G. 2000. Nature and the Marketplace: Capturing the value of ecosystem services, Island Press, New York.

Holling, C.S. 1973. Resilience and stability of ecological systems, *Annual Review of Ecology* and *Systematics,* **4:** 1-23.

IBPGR. 1992. Report of the Fourth Meeting of the ECP/GR *Allium* Working Group.

IBPGR. 1993. Report of the First Meeting of the ECP/GR *Brassica* Working Group.

Kalloo, G. 1988. Vegetable breeding. Volume III. CRC Press, Inc. Boca Raton, Florida.

Lenne, J. and Wood, D. 1999. Optimizing biodiversity for productive agriculture. In: Agrobiodiversity: Characterization, utilization and management, Wood D. and Lenne J. (eds.), CABI Publishing, Wallingford, UK, 447–470.

Mainwaring, L. 2001. Biodiversity, biocomplexity, and the economics of genetic dissimilarity, *Land Economics,* **77:** 79-93.

Perrings, C., Maler, K.G., Folke, C., Holling C.S. and Jansson, B.O. (eds).1995. Biodiversity loss:Economic and ecological issues, Cambridge: Cambridge University Press.

Plucknett, D.L., Smith, N.J.H., Williams, J.T. and Murthi Anishetty, N. 1987. Gene banks and the world's food. Princeton University Press, Princeton, New Jersey.

Qualset, C.O., McGuire, P.E. and Warburton, M.L. 1995. Agrobiodiversity: key to agricultural productivity, California Agriculture, **49**(6): 45-49.

Smale, M., Hartell, J. ,Heisey, P. and Senauer, B.. 1998. The Contribution of Genetic Resources and Diversity to Wheat Production in the Punjab of Pakistan. *Amer. J. Agr. Econ.*, **80:** 482-493.

Thomas, T.A., Bisht, I.S., Bhalla, S., Saora, R.L. and Rana, R.S. 1990. Catalogue on okra (*Abelmoschus esculentus* (L.) Moench.) Germplasm Part I. National Bureau of Plant Genetic Resources, New Delhi. p.51.

Tilman, D. and Downing, J. A. 1994. Biodiversity and stability in grasslands, Nature, **367:** 363–365.

Tilman, D., Wedin, D. and Knops, J. 1996. Productivity and sustainability influenced by biodiversity in grassland ecosystems, *Nature,* **379:** 718–720.

Tilman, D., Polasky, S. and Lehman, C.2005. Diversity, productivity and temporal stability in the economies of humans and nature, *Journal of Environmental Economics andManagement,* **49**(3): 405-426.

Chapter – 2

Germplasm Utilization in Vegetable Improvement

Umesh Srivastava

Introduction

Vegetable crops are a big group of crop plants, consisting of diverse kinds with differing breeding systems and varying consumer preferences. In India there are native vegetables like eggplant (*Solanum melongena*), beans (*Lablab purpureus*), cucumber (*Cucumis sativus*) and a few gourds, namely smooth gourd (*Luffa cylindrica*), ridge gourd (*Luffa acutangula*), snake gourd (*Trichosanthes anguina*) and pointed gourd (*Trichosanthes dioica*), while there are several introduced crops and which have long history of domestication and adaptation like garden pea (*Pisum sativum*), onion (*Allium cepa*), bottle gourd (*Lagenaria siceraria*), watermelon (*Citrullus lanatus*), cowpea (*Vigna unguiculata),* okra (*Abelmoschus esculentus*) etc. and also some others like tomato (*Lycopersicon esculentum*), cauliflower (*Brassica oleracea* var.*botrytis*), cabbage (*Brassica oleracea* var. *capitata*), chillies, (including *Capsicum*), frenchbean (*Phaseolus vulgaris*) etc, which have been introduced during the last 4 or 5 centuries. With such a diversity of crops, characterization and evaluation of genetic resources in each of them become a stupendous responsibility requiring extensive infrastructure facilities and varying methods of evaluation. The National Bureau of Plant Genetic Resources (NBPGR), New Delhi with its Regional Stations at Shimla, Bhowali, Akola, Thrissur, Jodhpur, Hyderabad, Umiam-Barapani, Base Stations at Ranchi, Cuttack and

Satellite centre at Amravati and Indian Institute of Vegetable Research (IIVR), Varanasi -designated as National Active Germplasm Site (NAGS) for vegetable crops and Indian Institute of Horticultural Research (IIHR), Bangalore, are holding over 40,000 germplasm collections in different vegetable crops.

A large number of lines have been collected from various parts of the country. A number of varieties have developed either through selection from local landraces or through hybridization resulting in development of a large number of open pollinated varieties with resistance to some important diseases, such as wilt. Also, a number of hybrid varieties have been developed both by public and private institutions, however there is a scope to increase further yield and nutritional quality using some of the modern techniques, such as molecular marker aided selection. The promising genetic resources utilized in vegetable improvement work have been described under following three heads:

1) Promising exotic introductions
2) Promising indigenous collections
3) Improved cross-bred and hybrid varieties using conventional methodologies and
4) Improved cross-bred and hybrid varieties using biotechnological tools

I. Vegetable Improvement Using Promising Exotic Introductions

Several exotic germplasms of vegetable crops have been utilized directly as promising varieties in India. The assembled germplasm were evaluated and promising ones identified and further tried with the best local checks available and the outstanding ones were released for direct cultivation. Mentioned below are some promising direct introductions.

A. Solanaceous crops

Tomato: It is an introduced vegetable crop roughly in 18th century and most of the introductions are bred varieties, which have adapted to Indian condition. Over the years, the NBPGR has assembled 2,911 germplasm lines from diverse sources (47 countries). Mostly these are bred varieties with a few allied species. The collected germplasm were evaluated for 41 agro-botanical and economic characters at NBPGR Issapur Experimental Farm, near New Delhi. A wide range of variability was observed for plant height (26-167 cm), number of branches per plant (3-19), number of clusters per plant (4-201), number of flowers per cluster (2-26), number of fruits per cluster (1-22), number of fruits per plant (8-790) and average weight of 10 marketable fruits (30-167 g) (Srivastava *et al.,* 2000). A number of promising accessions for various traits were identified (see table 1). Data on 41 morpho-agronomic traits of 2,021 well- characterized accessions were used for sampling of the core accessions. The standardized data on all the 2,021 accessions were subjected to principal component analysis (PCA). The information on (N x k) scores obtained from PCA were used for extracting the accessions accounting for maximum variability. Higher value of indices (SDI) for majority of the qualitative characters in the selected entries as compared to the whole collection was indicative of better representation of the existent diversity by the selected accessions. The quantitative variability was also fairly represented by the selected accessions for almost all the characters. The accessions sampled from various clusters within a cluster constituted the sample core subset. The accessions from various sample core sub-sets were pooled to form the final sample core set. Srivastava *et al.* (2000) developed a method which first located a point on the relative contribution (RC) curve where the fall in RC stabilizes. After locating the point on the RC curve, the peak point was located on the pooled diversity curve where the pooled diversity (sum or average of SDI -Shannon Diversity Index of all the characters) was the maximum. The method resulted into higher values of SDI/CV (Coefficient of variation) for 36 characters out of 41 characters with lower number of accessions. A sample of 140 accessions (nearly 7% of base collection of 2,021 germplasm lines) was finally selected as 'core collection'.

Wild relatives have also been evaluated for adaptation to low fertility tolerance in soils, drought or excess moisture and resistance to diseases and insect-pests, viz.,*Fusarium* wilt (Sel 1673, EC-37192), TMV (breeding lines B 2247 and Holmes PI) and leaf curl virus *(L. esculentum* lines P- 13, XXXIl-354-A-Silvestra, B 2247 and Sel-498, *L. peruvianum* (LP-6, LP-8), *L. chilense* and *L. hirsutum.* Nematode resistant accessions are Anahu, Florida, Hawaii Cross, Hawaii 55, Nemered, Nematex, Atkinson, VNF 8, 65 N-215-1, 65 N-255-1 and Pusa-120 (Choudhury *et al.,* 1973; Thomas and Umesh Chandra, 1989). Large numbers of germplasm lines including wild accessions have been evaluated at NBPGR (Rai and Gupta, 1995). At IIVR, Varanasi, 436 lines have been characterized and evaluated against TLCV (IIVR, 1997-98).

'Sioux', 'Fireball', 'Rutgers', 'King Humbert', 'Marglobe', 'Best of All' and 'Early Lethbridge' have been found to be promising varieties as direct introductions, 'Roma' (EC 13513 ex. U.S.A.), 'Dwarf Money Maker' (ex Israel); 'la-Bonita' (ex USA) are promising varieties introduced and have pear shaped or oval fruits, prolific in bearing with less seeds, suited to long distant transportation (all from NBPGR).

Varieties HS-101 and HS-102 had the ability to set fruit in April-May when temperature was 35- 42°C in May. EC 130042, *L. cheesmanii* and EC-162935 set fruits at high temperatures due to the stigma exertion of less than 1 mm, whereas other sensitive genotypes produced more than 1 mm of stigma exertion. The most serious problem of high temperature is the reduced size of fruits. Generally there is fruit set in heat tolerant lines but development of such fruit is very slow and poor, with the result that fruits remain smaller. Under Delhi conditions, heat tolerant accessions EC-168070, 130053-1, 165393 recorded to have large fruit (Chandra and Thomas, 1993; Chandra and Gupta, 1994). Germplasm of wild species have also been evaluated which also were remarkable variation for their inherent adaptation to various stress. *L. pennellii, L. cheesmanii and L. esculentum* var. *cerasiforme* are drought tolerant and *L. pimpinellifolium* (PI-205009) require more days to express wilting, thus, showing tolerance to drought conditions. *L. hirsutum* to cold and drought tolerance and EC-130042, EC-65992 and Sel-28, *L. cheesmanii,* K-14, EC-104395, *L. pimpinellifolium* (EC- 65992) and *L. pimpinellifolium*

(Pan American) seem to be the best potential sources for drought tolerance (Seshadri and Srivastava, 2002).

Table 1. Promising tomato accessions identified

Traits	Promising accessions
Early maturing	EC-357833, 357827,362956, 362940, 362944, 362947, 362948, 362950, 99935, 129602
Fresh market type	EC-260636, 232429, 24153, 161245, 7288, 27251, 267729, 67726,11238, 101652
High yield	EC-106285, 118292, 4551, 163683,173856, 170662, 110578, 1127, 129081, 129353, 168084, 50357, 122063, 1301, 251636, 66504, 615481, 315486, 339059, 337827, 339074, 357829, 362940, 357840, 362954, 362957, 361424, 367856, 367857, 257463, 315489, 237288, 114147, 37231, 35401, 15384, 24147, 163602, 362915, 362946, 357839, 357832, 339072
High TSS	EC-31767, 81841
Large fruit	EC-163708, 41340, 267726, 267728, 251751, 168070, 267725, 367857, 361674,362945, 339060, 367856
Late maturing	EC-357829, 357830, 362941, 362942, 362951, 241147, 163594, 212469, 241134
Pear shape fruit	EC-367857-1, 35332, 6192, 326144
Processing type	Bulgarian, EC-942, 1193,8822, 161252,160187, 126676, 4708, 129600, 1154, 2790, 159966, 93739, 155955,24602, 8820,93739,164677, 246028, 89258, 160183
Tolerance to high temperature	EC-numbers: 1127, 4639,11960, 16465, 27910, 31515, 35446, 37226, 37284, 89248, 94181-6, 106265, 110578, 114503, 122063, 125754, 130042, 130053-1, 162598, 162935, 163690, 163704, 164636, 164666, 165393, 165700, 165751, 168064, 168070, 168281, 169308, 170662, 251636, 251674, T-41, NC-57299, PI-205009, *L.cheesmanii. L. pimpinellifolium*-Pan American, Punjab Tropic, Merz, HS-101, HS-102, Mini Rose.
	Saldette and PS 84-58, IIHR – 1124
Tolerance to drought	*L. pennellii,* EC 130042, Sel 28, *L. pimpinellifolium*, PI 205009, *L. cheesmanii,* RFS-1 and RFS -2
Tolerance to salt	*L. pennellii, L.cheesmanii*

Tomato is highly susceptible to a number of fungal, bacterial and viral diseases. Losses up to 95 per cent have been reported due to incidence of tomato diseases. Development of resistant varieties is the most economical method for the control of disease. Tomato is susceptible to several pests also. *L. hirsutum* and *L. hirsutum* f. *glabratum* were generally resistant to numerous pests.

Table 2. Source of resistance to important diseases of tomato*

Diseases/Pathogen	Source of resistance
Fusarium wilt *Fusarium oxysporum* f.sp. *lycopersici*	*L. pimpinellifolium* PI 79532
Race 1	Walter, Columbia, Roma, HS 110, Homestead, Floradade, Sel. 22, and Sel. 30 Race 1 & 2 No. 26915-8 Riustrel Race 2 PI 126915 Race 3 US 629
Fusarium wilt and fruit rot	EC- 160195, 122527, 21626, 103598, 116874, 117008, 115947, 117671, 117163, 128769
Verticillium wilt	*L. pimpinellifolium* Line 64480, Petopride, 10-15-2-2
Early blight (*Alternaria solani*)	NCEBR-1, NCBER-2, *L. peruvianum* var. *dentatum*, *L. pimpinellifolium*, 'Pan American', *L. peruvianum* B 6002177, *L.hirsutum* f. *glabratum*,
	68B 134 South land, AL – 14** and AL 919**
Late blight	*L. hirsutum*, PI 251305, PI 126445, LA 1255, Ottawa 30, 31, Fla P14, WV 38 (*Phytpphthora infestans*) *L. pimpinellifolium*, *L. esculentum*, var. *cerasiforme Septoria* leaf spot *L. hirsutum, L. peruvianum, L. pimpinellifolium, L. glandulosum*, PI 126448Bacterial spot Hawaii 7998 (*Xanthomonas campestris* pv. *vesicatoria)*
Bacterial wilt	*L. pimpinellifolium*, PI 127805A, Venus, Satwon, VC-8-1-2-1 Rodadoe BWR1-1,5, BT-1
Tomato leaf curl virus**	*L. hirsutum* f. typicum, *L. pimpinellifolium* 'A 1921', *L. peruvianum*, *L. hirsutum* f. *glabratum* B 6013, 88±58, 88±31 and 31 x Italy, IIHR 810, IIHR 1949
Tomato yellow leaf curl virus	*L. peruvianum, L. chilense, L. peruvianum, L. cheesmanii,* (sub sp. Min on) LA 1401 TMV (1,2,3&4) *L. chilense*, Ohio MR-9
Fruit borer**	EC-262, 2669, 490, 491, 6200, 7764, 2630, 6987, 9227*L. hirsutum*, *L. hirsutum* f. *Glabratum*, IC 1112064, 101552, 124406, 129599, 129597, 129698, 101552, 124406, 129599, 129597,129698, 129791, 121451, 160191, 161254,119121, 59067, 51/81-3, 14/81-6, 78/81-8, M3/81-3, DM/83-14, DM-3, DM 83-7, 51/81-2
Leaf miner	*L. hirsutum, L. hirsutum f. glabratum*, Pearson, VFL
Resistant to nematode+	Anahu, Florida, Hawaii Cross, Hawaii 55, Nemared, Nematex, Atkinson, VNF, 8, 65 N-215-1, 65 N-255-1 and Pusa-120

*Kalloo, 1993; **NBPGR Annual Reports, IIVR Annual Reports; +Choudhury *et al.,* 1973.

Total soluble solids, sugars, acidity, lycopene and carotene are the main quality components of tomato fruits. The reducing sugars glucose and fructose are the main components of the soluble solids, whereas protein, cellulose, pectic substances and hemicellulose are the main insoluble solids. The flavour depends on the sugar/acid ratio. Varietal differences for sugar content have been reported. A high sugar line has been developed using *L. chmielewskii* as donor parent (Rick and Fobes, 1974). Generally varieties with higher sugar levels have a high solid content. For processing, this is the most important trait because with increased solids, less water is processed and more paste can be manufactured for each ton of tomatoes. The total soluble solids (TSS) in cultivated varieties ranged between 4-6%. However, the wild species have more TSS e.g., *L. chmielewskii* up to 10 per cent (Rick and Fobes, 1974) and *L. cheesmanii* up to 15 per cent. Breeding lines from a cross between cultivated tomato and *L. chmielewskii* have been used to develop varieties with higher TSS (Rick, 1974). It has been estimated that each 0.1 per cent increase in TSS is potentially worth about $ 7,000,000 to US processing industry. At Bichpuri (Agra), evaluation of exotic collections showed that Super Marmande (of France) recorded 6.8 per cent followed by two US accessions viz., EC 2790 and EC 159966 (6.6%) (Chandra, 1997) whereas, at IIVR Varanasi, Microgold and Microtom dwarfing and small fruited varieties recorded TSS 6.0 per cent and 5.8 percent respectively. Generally large leaf area contributes to the soluble solids content, thus selection should be made for enhanced foliage of plants for high TSS. Pericarp thickness is also considered important as it contribute to quality of fruits. Accessions EC 165983, 267725, VF 100, showed consistent contributions towards fruit wall thickness (Chandra, 1997). For processing, pH value is also important and it should not exceed 4.4. *L. pimpinellifolium* is a good source of high acidity. Lycopene and carotene pigments are important quality components. Tomato is a good source of ascorbic acid also. Small fruited varieties and wild species especially *L. peruvianum* and *L. pimpinellifolium* are good sources of ascorbic acid.

Chillies ***(Capsicum*** **spp.*):*** Although recently introduced, probably by Portuguese travelers in 15th -16th century, extensive variability occurs in pungent types with varying fruit shape, size, colour and bearing habit and semi-perennial and perennial types. The NBPGR has assembled a total of 1,014 germplasm accessions comprising both indigenous and exotic at Amravati, Shillong and Bhowali. Nine hundred and three germplasm of chilli have been evaluated for 42 traits at NBPGR Regional Stations at Hyderabad, Thrissur, Shillong, Bhowali and Satellite Centre, Amravati. Accessions that have shown promise are listed in table 5. At IIVR Varanasi, chilli germplasm, 9852- 173 and PDGI, have shown consistently tolerant/resistant against mite and thrips. At IIHR Bangalore, five breeding lines namely 309-1-15, 300-1-5-1, S 118, S 635 and S565, showed consistent tolerance to Thrips. In addition, the studies carried out at Acharya N. G. Ranga Agricultural University, Regional Research Station, Lam (Andhra Pradesh) reveal that Caleapin Red, Chamatkar, NP 46-A, X 1068, X743, X 1047, BG-4, X 226, X 230 and X 233 are tolerant to thrips; LEC 1, Kalyanpur Red, X 1068, X 204 and Goli Kalyanpur are tolerant to mites; LEC 28, LEC 30, LEC 34, Kalyanpur Red and X 1068 are tolerant to aphids. At NBPGR, Hyderabad also, 650 collections (both indigenous and exotic origin) were evaluated. At IIHR Bangalore, chilli lines having resistance to anthracnose, powdery mildew, CMV and PVY have been developed. It is significant that collection from Tezpur (Assam) has been found to have the highest capsaicin content recorded so far anywhere in the world. Also from Assam, a collection has been made which is reported to have fragrance of pure ghee in green fruits.

Table 3. Promising chilli lines identified at NBPGR*

Centres	Year	Identified promising lines
Multiple traits		
NBPGR, New Delhi	1989-90	89-5,89-6, CA-213, CA-282, CA-408, 416, 418, 420, 431, 445, 448, 953
	1996-97	EC-391074, EC-391079-92, EC-399532-581
NBPGR, Amravati	1990- 91	BDJ-224, EC-IO3820, U-20 to 34 1994-95 IC-11960J
	1995- 96	IC-119581, EC-339047, NIC-19991, JBT-22IC-133, IC-119576, NIC-23911, NIC-23309, NIC-20418
	1996- 97	JBT-12/15, JBT-12/17, NIC-20924, EC-382187
NBPGR, Bhowali	1994-95	P-2072, NIC-92250, NIC-72156, NIC-99916, NIC-99912
	1996- 97	P-1893. P-2072. NIC-19943. EC-362918. IC-92123. IC-119241. IC-119787
NBPGR, Shillong	1995- 97	NIC-22146. NIC-22058. NIC-22214. IC-89727,NIC-22076. HM-2804, RS-8. RS-27
NBPGR, Thrissur	1990- 91	IC-88498. IC-88503
Specific Traits		
NBPGR, Hyderabad	EC 399558, 399568, 399578, NIC 19939, IC 147837,	
Early flowering (58 days)	NIC 20416, NIC 20893, EC130820, EC 399581	
Early maturing (101 days)	EC 399588	
High fruit yield	IC 147637, 147719, 18882,119556, EC 246007, NIC 19991, 92153, 92156, 92250, 20901.99912. p 2072 and BDJ 2254	
High fruit set	IC 147537, NIC 20901, IC 147719, IC 147724, IC 119243, IC 119282, IC 119264, IC 119267, IC 92109, IC 119578, IC 119581, IC 119589, IC 119594, IC 119594, IC119598, IC 119696, NIC 2092, IC 110810, IC 119844, IC 119468, IC 119561, IC 119736, IC 119737, IC 119746, IC 119750, JBT 12/20, SA 2407, NIC237779, IC 119688, EC 119592, EC 119659, NIC 20418 and 20905	
PBNV/ Mosaic resistance	IC 119611, EC 121490	

Contd...

PBNV tolerance	EC 1241490-1, IC 147742, IC 119242, IC 119300, IC 9210, IC 119385, IC 119546, IC119592, IC 119561, IC 119430, IC 119756, IC 119758, JBT-12/1, JBT-12/39, JBT-12/96. PBC 6/2
Bacterial wilt resistance	KAU cluster, White Kandu
Resistance to multiple diseases	EC 32333
Anthracnose. Leaf curl virus, TMV and Verticillium wilt	*C. chinense*
Leaf curl virus	*C.. frutescens*
Anthracnose, Phytophthora & CMV	*C. baccatum*

*Gautam and Gangopadhyay, 1998; Singh *et al.*, 2001; IIVR Annual Report; **K S Varaprasad (Personal communication)

In pepper, 'Bell Pepper', 'California Wonder', 'Chinese Giant', 'World Beater' and 'Yolo Wonder' varieties have been found promising as direct introductions. 'Yolo Wonder' is large fruited, 3-4 lobed, medium thick flesh and tolerant to tomato mosaic virus. All these varieties have 3-4 lobes with medium to thick sweet flesh and are heavy yielders. Russian variety R 449 has also given excellent performance in Delhi and Bihar.

Potato *(Solanum tuberosum, L.)*: A large number of potato germplasm of exotic origin have been utilized. 'Craig Defiance' from UK and 'UP-to-date' a selection from Ireland are the most promising varieties introduced as direct introductions.

B. Cucurbits

The NBPGR has initiated the assembling of germplasm in 12 different cucurbits (presently 6,286). Being cross pollinated viny crops, the regeneration, characterization and preliminary evaluation, require extensive land facilities, U.S. breeders who came to India in 1992, undertook exploration in parts of Rajasthan, Madhya Pradesh and Uttaranchal and collected 146 accessions of cucumber. They studied the variation at 21 polymorphic isozyme loci and found them to be, completely different from Chinese cucumber germplasm (Staub *et al.*, 1997) and these were treated as having unique genetic nature.

Indian germplasm in muskmelon (*C. melo*) is unique in that they comprise several kinds of non-dessert (non-sweet) forms of melons which are either used in immature form like cucumber, or cooked. Phoot or snapmelon (*C. melo* var. *momordica),* tar-kakri or long melon or serpent melon (*C. melo.* var *utilissimus*), Vellarikkai of Tamil Nadu - eaten raw like tender cucumber, Budamkaya and Nakka dosa kaya of Andhra Pradesh cooked to a nice delicacy, Chavathakkai and Vellari of Kerala, whose fruits are ripened and cooked and Chibur of Maharashtra (Konkan region) are some of the consciously cultivated and domesticated types. In Rajasthan, Kachri is a partial domesticate with little flesh and small seeds in plenty. All these carry disease resistance characters and bring out the importance of non-dessert kinds of muskmelon as a unique diversification indicating the polymorphic spectrum of the taxon. Similarly in watermelon, wide variability exists in riverbeds and some primitive types of citron like 'Katagolan' whose fruits have long keeping quality without rotting or fermentation. The NBPGR has organized collection of muskmelon and snapmelon totalling to 346 accessions within the country. The Central Institute of Arid Horticulture (CIAH), Bikaner has identified super lines in drought tolerant kachri, viz., AHK 119, AHK 200, in drought tolerant and disease resistant Phoot (snapmelon), viz., AHS 10 and AHS 82 and in watermelon with longer shelf-life Mateera, viz., AHW 19.Tumba (*Citrullus colocynthis)* is a medicinal plant, allied to watermelon (Seshadri and Srivastava, 2002).

At IIVR, Varanasi, preliminary evaluation of some of the cucurbits has been made. In bitter gourd (*Momordica charantia*), 219 collections have been made, besides *M. dioica* and *M. cochinchinensis*. A significant observation in the evaluation studies was a gynoecious segregate, which has been maintained. Nine collections have shown tolerance to CMV. Bottle gourd (*Lagenaria siceraria*) has 58 collections in NBPGR New Delhi. Two accessions (U10-316 and NIC 1225) were found tolerant to red pumpkin beetle, three (U10-316, IC 92362, IC 92418) were tolerant to leaf miner and four (U10-316, IC 92362, Laxmipur and Patna-1) were found to be tolerant to downy mildew. In pointed gourd (*Trichosanthes dioica*), a parthenocarpic line (PG 105) has been identified in IIVR, Varanasi. Root-gall free accessions has also been located and evaluation is in progress in pointed gourd germplasm against heat tolerance, sun-scorching, mites and leaf miner incidence. In watermelon, exotic collections were evaluated for leaf miner incidence and in cucumber against CMV and red pumpkin beetle. In cucurbits, assembling, characterization and documentation have to be systematically done before evaluation. In view of the enormity of the collections in large number of crops and difficulties experienced in regeneration, initial efforts may be confined to a few (6 crops like cucumber, muskmelon, watermelon, pumpkin, bottle gourd and bitter gourd) only.

Some Introductions

Watermelon (*Citrullus lanatus*): 'Asahi Yamato' (Japan) with medium size fruits, having small seeds,'Shin Yamato' (Japan) with yellow, sweet flesh, 'Sugar baby', 'Charleston Gray' and 'New Hampshire Midget' from USA with large number of fruits per plant and palm size-fruits with thin rind and deep red sweet flesh.

Cucumber *(Cucumis sativus, L.):* In cucumber 'Japanese Long Green', and 'Straight Eight' were promising direct introductions. Another cucumber - West Indian Gherkin *(Cucumis anguria)* was also a notable introduction for pickling purpose.

Summer Squash *(Cucurbita paepo, L.)*: 'Australian Green' and 'Zucuni Long' are summer squash varieties found promising and suited to Indian conditions. Another introduction from USA of *C. pepo* with naked seeds was suited for extraction of edible oil.

C. Alliums

This group includes onion, garlic, as well as leek and many varieties from this group have been introduced.

Onion: Onion is an ancient introduced crop into India highly valued for its flavour and medicinal properties. The adaptation of temperate taxon mostly biennial and perennial, to tropical and subtropical conditions of India as an annual is an outstanding example of diversification and adaptation. As many as 1,300 collections in onion and 1,000 collections in garlic have been assembled at NBPGR, Delhi. Besides, 304 collections of onion and 363 of garlic have been collected at National Horticultural Research and Development Foundation (NHRDF), Nasik (Maharashtra). In onion, a peculiar situation is obtaining in this country. There are several improved varieties of onion available for different regions but yet uniformity of produce for exports and processing could not become available mainly because of inadequate and inefficient seed production and delivery systems. Further being a biennial crop, there are handicaps in seed production technology. For processing especially for dehydration, the quality parameters followed in other countries with long day onions cannot be attained with predominantly short day varieties grown under tropical conditions in India. Another difficulty is the non availability of a good resistant source to purple blotch *(Alternaria porri)* and *Stemphylium* among *A. cepa* and its allied species, *A. fistulosum* is promising for *Stemphylium* resistance. At AVRDC Taiwan, virus resistance in garlic has been identified.

'Early Grano' from USA and yellow skinned 'Bermuda Yellow' from Philippines are promising direct introductions with large globular bulbs, which are mildly pungent. 'Pusa Ratnar' is another selection from 'Red Granex' introduced from USA. Its bulbs are large with attractive deep red colour.

Leek (*Allium ampeloprasum*, L.): 'London Flag' is a popular variety introduced from UK with long, thick, white attractive stem, bearing large medium green leaves.

D. Leguminous vegetables

This group includes cowpea used for vegetable, garden pea and French bean. Many direct introductions have been exploited in this group.

Cowpea: Cowpea is a crop with multiple uses, viz., vegetable (tender pods), pulse (dry seeds) and forage as a fodder and green manure or cover crop. The NBPGR maintains 2,350 accessions. Three catalogues have been published (Sardana *et al.*, 2000). Evaluation studies showed that promising accessions were EC 390202, EC 390207, and EC 390211. Pod length (20 to 30 cm) was recorded in EC 392203, EC 390 286, IC 390287 and IC 202821. Vigorous and leafy accessions were EC 390226, EC 390239, EC 202776, EC 209164 and IC 202821. Two accessions (R.17-1-34 and Ala 969-82) were founded to carry resistance to *Cercospora*. Accession EC 297562 was found to be resistant to Bean mosaic virus and Ala Bunch and Dixit to cowpea mosaic virus. Similarly, 702 cowpea accessions were evaluated at New Delhi, Pantnagar, Dharwad and Pusa (Bihar). Seven accessions flowered in less than 40 days, five had matured in less than 65 days, seven had pod length less than 22 cm, six had seeds greater than 15 per pod, nine had pods per peduncle greater than 2.5 and eight had 100 seed weight greater than 209 (Mishra *et al.*, 2000). Two promising exotic introductions of cowpea released under the name 'Pusa Barsati' (ex. Phillipines) and 'Pusa Phalguni' (ex. Canada) have been very popular as direct introductions and have spread widely (Thomas *et al.*, 1983).

Table 4. Promising cowpea accessions identified for resistance against stresses

Name of the Centre/Attributes	Promising accessions identified
IIVR, Varanasi	
Yellow mosaic virus resistance	BC-24074, NC-63382, EC-169371, bc-244009, bc-240770, IC-20215, EC-15567, DCB-253 and IC-39890, IV41209, MS-9804
Resistant to *Pseudocercospora*	VS-389, BC-244002, Arka Garima, BC-240787, RKT -89/743, BC-7096.
	BC-2409-17-A. BC-240770. EC-243948, EC-240750, PDCP-192
Nematode resistant	Colosus-80 Pod borer (*Etiella zinchenella resistance*) Local-l, PDCP-11, PDCP-13, Sel.-2-1, PDCP-16, PDCP-17, EC-4860
NBPGR, New Delhi	Pod borer and Lycaenid butterfly resistant 22 lines were identified
IIHR, Bangalore	Anthracnose resistant 21/141 Moderately susceptible to powdery mildew EC-598. RS-9. IC-322. 2857-A. 2969. 3120.7095.70958 and EC-4326

Garden pea *(Pisum sativum)*: The garden pea is an ancient vegetable crop introduced into India since Harappan days. Early cultivation of pea was for pulse purposes, mostly round seeded varieties grown mainly as a rainfed crop. The garden pea especially, the wrinkled varieties were introductions. Hence germplasm collections of NBPGR are essentially field peas of round seeded varieties and evaluation studies were reported by Sardana and Suneja (2000) on 380 accessions consisting of 215 indigenous and 165 exotic collections. IC 208367 and IC 208364 were leafless tendril type, EC 342007-powdery mildew resistant and EC 342007 and EC 398596 were long podded accessions. Besides *Pisum fulvum* (prostrate growth habit), *Pisum sativum* spp. *abyssinicum* (spreading growth habit) and red smooth seeds and *Pisum sativum* ssp. *elatius* (spreading growth habit) and wrinkled seeds were also maintained. Similarly 700 germplasm lines of indigenous and exotic origin were evaluated for qualitative characters at IARI, New Delhi (Sharma *et al.,* 1998). Powdery mildew resistant lines have been located at different centres such as Jabalpur, Indore, Kanpur, Varanasi, Palampur, Rahuri and Hessarghatta. Paradoxically, it has been found that all the sources of resistance carried the same gene. Multiple disease resistance to powdery mildew, rust and *Fusarium* wilt has been identified at Jabalpur. Among the 140 collections studied for protein content in 2000-01, the variation ranged from 16.7 to 25.5 per cent. Three accessions EC 281866, VKG 2/19 and IC 248997 had more than 23 per cent protein (Seshadri and Srivastava, 2002).

As regards processing, freeze drying and frozen peas are not variety specific and they are very suitable for any cuisine because they retain good texture and colour unlike in canned peas which exclusively meet the needs of defence forces. At IIHR Bangalore, RPC 13-3 has been identified as suitable for canning.

A large number of pea varieties from exotic sources have been found promising as direct introductions. Varieties of peas introduced from USA under the various names such as 'Early to Report', 'Delwiche Commando' and 'Bonneville' have shown promise as direct introductions. Varieties from other countries namely, 'Arkel' from France and 'Meteor', 'Kelveden Wonder', 'Little Marvel' from UK, 'Yorkshire Hero' from Australia have shown promise as direct introductions both as early and main season varieties. Another introduction is from Sweden 'Sylvia' which is edible podded and has proved useful as direct introduction (Thomas *et al.,* 1983).

French bean *(Phaseolus vulgaris,* L.): The National Bureau of Plant Genetic Resources, Regional Station, Shimla is maintaining both indigenous (1,581) and exotic (1,378) collections in French bean. A catalogue of 1,736 collections on13 descriptors was published (Patel *et al.,* 1981). Again these germplasm were evaluated along with new introduced lines (a total of 2,959 accessions) on 26 descriptors (Joshi and Rana, 1995). A core collection of 42 accessions was also established to represent the genetic variability. Varieties like 'Top Crop', and 'Contender' from USA, 'Premier', 'Giant Stringless' BKW-74 (Sweden), 'Wade' and Russian varieties (EC 24940, EC 24958 and EC30021) have been found promising introductions in our country. 'Kentucky Wonder' is another variety introduced from USA which is very good yielder with dwarf habit 'Jampa' another introduction from Mexico is suited from Maharashtra region. 'Watex' is a suitable direct introduction for Nilgiris region.

E. Cruciferous vegetables

These are essentially cruciferous vegetables, namely cauliflower, cabbage, knol khol, broccoli, Brussels sprouts etc. and they have been introduced from the days of East India Company in 14th-15th century when European traders visited this country .The genetic resources in these crops are essentially bred varieties introduced during the last 3-4 centuries and now acclimatization and adaptation have taken place. In cabbage, 'Pride of India' and 'Golden Acre' are the old varieties and the present day strains are acclimatized ones under Indian conditions, probably varying from the original stock. Further, seed production in cabbage is being done in the hills of Kashmir and Himanchal Pradesh. Recently cabbage hybrids of Japanese origin have been introduced for their suitability in mild winter conditions of West Bengal, Maharashtra, Karnataka and Tamil Nadu.

In cauliflower, a distinct group of hot weather or Indian cauliflower varieties has been identified, for their performance during heavy rainfall and high temperature conditions of north Indian plains during August to November. This has made cauliflower a cosmopolitan crop, with tolerance to tropical conditions, a contribution of the farmers and seedsmen of the country made during the 19th and 20th centuries. In fact 'Early Benares' and 'Early Patna' are acclimatized Indian varieties of early 20th century, which found their way to Florida, Hawaii, Philippines etc. Indian cauliflower collections were made at New Delhi (IARI), Pantnagar, Faizabad, Ludhiana and Sabour. Notably self-incompatible lines have been developed in IARI, from this material, besides locating resistance to Black rot and curd and inflorescence blight. Now improved varieties are available for each season, viz., 'Kunwari' (Late August-September), 'Katki' (September-October), 'Aghani' (November-early December), 'Poosi' (mid-late December early January) and 'Maghi' (late January). Total collections maintained at IIVR Varanasi and other centres are estimated to be around 200.

Assembling the collection of cole crops (except tropical cauliflower) has to be done at IARI Regional Station, Katrain (Kullu valley) and Dr YS Parmar University of Horticulture & Forestry, Regional Research Centre, Kalpa (Kinnaur District)., Himachal Pradesh. Evaluation of the collections has to be done in an integrated manner, at a few centres of North Indian plains to evaluate their attributes, along with seed production ability in the hills.

Some promising introductions included

(i) **Cauliflower *(Brassica oleracea* var. *botrytis*):** 'Snowball No. 16' (Holland), 'Improved Japanese' (Israel), 96D (Israel), 'Selandia Osenia' (Denmark), 'AH No. 7' and 'Igloo Osenia' from Denmark. 'Early Nozaki', 'Sakiazume', 'Early Snowall' and 'Takii' series from Japan, 'Erferter' from Netherlands and 'Snowdrift' from Germany are the promising introductions established well in India.

(ii) **Cabbage *(Brassica oleracea*, L. var *capitata*):** 'Golden Acre', 'Sure Head', and 'Drum head' are the promising exotic introductions. 'Pusa Drum Head' is a selection from original variety EC 6774 introduced from Japan long back and is late, flat, large drum-head type.

(iii) **Knol-Khol (*Brassica oleracea* L. var *gonglyoides*):** 'White Vienna' is an early dwarf, globular to flattened, light green, very smooth with creamy white flesh and delicate flavour

(iv) **Radish *(Raphanus sativus*, L.):** 'China Red' is a chinese variety, early type, with attractive red colour and shape; 'White Icicle', 'Rapid Red', '40 days', 'Japanese White' and 'Miya-Shige' are the other exotic promising direct introductions. 'Pusa Chetki' is a selection from variety introduced from Denmark and

(v) **Turnip *(Brassica rapa*, L.):** 'Purple-Top-white-globe', 'Early Milan', 'Snow Ball', 'Golden Ball' are some of the recommended varieties of turnip directly utilized in India (Thomas *et al.*, 1983).

F. Leafy Vegetables

This group includes amaranths, spinach, sarson, lettuce, etc. Amaranth is the most nutritive leafy vegetable, widely distributed in tropical south Asia including India and tropical South Andean region of South America *Amaranthus tricolor* is the best developed species, native to India for vegetable purposes besides *A. blitum, A. dubius, A. viridis, A. hybridus* and *A. spinosus* are the other leafy species. *A. cruentus* and *A. hypochondriacus* are the grain types grown in north Indian hills. Amaranth germplasm has been maintained at NBPGR Regional Station, Shimla (seed amaranth) and National Botanical Research Institute, Lucknow (seed and ornamental amaranth) and at Tamil Nadu Agricultural University, Coimbatore (leafy amaranth). The leafy amaranth collections at Coimbatore total to 450. There are improved varieties, viz., Co-l Co-2 and Co-3 and GKVH-l (from University of Agricultural Sciences, Hebbal and Bangalore). At Kerala Agricultural University, Vellanikkara (Thrissur, Kerala), germplasm collections have been evaluated for bolting habit, oxalate and nitrate content. At Coimbatore, emphasis was laid on leaf/stem ratio in the selection as an index, besides leafiness, to obtain a balance instead of leafiness alone as a criterion.

In sarson – 'Japanese Sarson' is found to be promising introduction from Japan for cultivation in India as leafy mustard. In lettuce, there are three exotic collections, namely, 'Great Lakes', 'Slobolt' and 'Chinese Yellow' have been found to be promising. 'Imperial 847', 'May King' and 'Paris White' are the other exotic introductions of lettuce. 'Virginia Savoy' and 'Early Smooth Leaf' are the recommended varieties of spinach which are direct introductions (Thomas *et al.,* 1983).

G. Okra *(Abelmoschus esculentus)*

In okra, large variability exists in this country such that India is considered a secondary centre of diversity, with a possibility of polyphyletic origin. A case study of in okra comprised 260 germplasm accessions with diverse geographical background for attempting to make a core set. Characterization data on the accessions were documented. Nine distinct quantitative descriptor viz., days to flowering, number of ridges per fruit, plant height, number of internodes, first flowering node, fruits on the main stem, fruit length, fruit width and fruits per plant were used for non-hierarchical cluster analysis and clustering was performed using Euclidean distance on the basis of these quantitative descriptors after standardization. Number of days to flowering and plant height among the quantitative descriptors and pubescence and pigmentation of various plant parts among the qualitative characters were found to be most significant in discriminating accessions and also in the analysis of variability. The accessions from clusters/ sub-clusters were selected using quantitative data through principal component analysis (subjective approach) and qualitative data through SDI-Shannon Diversity Index (Galewey, 1992). Utilizing this methodology, a total of 53 accessions could be sampled as a core set.

In addition, the NBPGR has brought out a series of catalogues with characterization data on okra accessions. Catalogue on okra, Part I comprised 558 accessions with 43 descriptors at Delhi location wherein significant six accessions had flowering within 52 days and fruit maturity within 88 days and fruit length less than 5 cm (for export purposes) were noted in six accessions (Thomas *et al.,* 1990); Part II had characterization data of 432 accessions evaluated at Delhi and Akola locations on 43 descriptors. Twelve accessions at Delhi and 8 at Akola location flowered in < 42 days. Also, fruit length in 5 accessions at Delhi and 8 at Akola measured < 5cm (Thomas *et al.,* 1991). Part III consisted of characterization of 332 and 185 accessions during year 1991 at Delhi and Akola and 153 and 203 accessions during year 1992 at Delhi and Akola locations. Fourteen indigenous accessions (IC) and 9 exotic accessions (EC) flowered in < 45 days (Bisht *et al.,* 1993). Besides, 241 accessions in allied species of *Abelmoschus* were evaluated at Delhi and Akola which comprised *A. ficulneus, A. manihot* var. *tetraphyllus, A. moschatus* and *A. tuberculatus* (Bisht *et al.,* 1995). In all, a total of over 1,500 cultivated and 500 wild collections have been assembled at NBPGR Regional Station, Akola (Maharashtra). The evaluation studies at IIVR Varanasi brought out the following results:

Table 5. Disease resistance in *Abelmoschus* species

Resistant/tolerant	Accessions
YVMV	NIC 9303A NIC 6308 NIC 3322 NIC 9408 NIC 3325 RC 329375 K 4409
Powdery mildew	*A. crinitus A. angulosus*
Cercospara	A. crinitus A. ficulneus A. angulosus
Enation leaf curl	*A. crinitus A. ficulneus A. manihot*
Mites	EC 305694, EC 305695, EC 305714, EC 306731
Jassids	*A. moschatus A. crinitus,* EC 305656 EC 305694. EC 305695. EC 305714. EC 306731
Macrophomina	IC 90186, U43-07, U43-65, U45-79
Multiple resistance	IIHR 68-1, 81-1, 15-1

'Perkin's Long Green' and 'Valvet Green' were recommended for direct use. 'Ghana Red' is another important cultivar for field resistance to yellow vein-mosaic. This is an introduction from Ghana. Besides this, Africa abounds in many important cultivars of okra.

H. Tuber and Root Crops

This includes carrot, beet root and sweet potato, etc.

Carrot *(Daucus carota)* : 'Nantes' (Temperate type), 'Nantes Half Long', 'Chanteney', 'Corelets' and 'Imperator' are promising exotic introductions of carrot utilized for cultivation in India.

Sweet Potato: 'Pusa Saffaid' is white skin selection from Taiwan 'FA 17' and 'Pusa Lal' is red skin selection from American variety, 'Norin'. Both varieties have been found promising under Indian conditions. 'FA 17', or V6 and V.12 or TST white from China and F.8 or FB.4004 from USA are the other varieties found suitable in South India.

Beet root *(Beta vulgaris,* L.): 'Crimson Globe', 'Detroit Dark Red' and 'Crosbys Egyptian' are the promising introductions found suitable for direct cultivation in India (Thomas *et al.,* 1983).

II. Vegetable Improvement Using Promising Indigenous Collections

A large number of vegetable crops have originated in Indian centre of origin of cultivated plants. Therefore, a large variability has been accumulated as a result of selection and also by natural cross pollination and selection. In vegetables, namely brinjal, okra and cucurbitaceous vegetables, very good local cultivars are available which have been utilized in the improvement of vegetable crops. Some of the promising genetic resources of vegetable crops utilized for direct cultivation after initial evaluations are listed below.

A. Solanaceous crops

Tomato: 'Meeruti' and 'Improved Meeruti' have been selected from local cultivars. These varieties are hardy and high yielding. Besides this, HS 101, HS 102 from Hisar, KS 1, KS 2, from Kalianpur (UP) and selection S-11 from Bangalore and Co-1, Co-2 from Coimbatore have been found promising selections. The NBPGR has identified more promising types from the indigenous collection.

Chillies *(Capsicum* spp.): The types NP34, 41, 46 and 51 of pungent chillies have been popularly cultivated by the farmers. A number of selections like 'Kalianpur Red', Kalianpur Sel-1 and 'Kalianpur Yellow' and 'Kalianpur Chaman' have been recommended for growing for condiment and pickle purposes. K-2 of Kavilpatti, (Tamil Nadu), G. and X series of Lam and 'L' series of Pantnagar and CO series of Coimbatore have been released for direct cultivation (Thomas *et al.,*1983).

Brinjal: The genus *Solanum* principally non-tuberiferous species has originated mainly in Asia in the Indo-Myanmar region and has spread all over the world. Therefore, immense diversity of eggplant and its wild relatives exist in various parts of this region. Two African eggplants, the scarlet eggplant *(S. aethiopicum* L) and the gboma eggplant *(S. macrocarpon* L) are extensively cultivated there and at very limited scale in some other parts of world. There are many other allied species (e.g. *S. nigrum, S. nodiflorum* etc.), which are at times cultivated or are semi-cultivated (spontaneously occurring plants are looked after and harvested) in Africa and Asia. Many other species of *Solanum* are used in medicine. An important allied species of medicinal importance (solasodine content) is *S. viarum* (syn *S. khasianum).* Another species *S. torvum* is extensively used in Ayurvedic medicine system (Rai *et al.*, 1995).

Eggplant has immense variability of cultivated varieties especially in fruit shape, colour, size and bearing habit and the wild species numbering around 40. A large number of land races/traditionally grown cultivated varieties of different agro-ecological regions have been collected for testing resistance to biotic and abiotic stresses and adaptation to different environments. Inclusive of collections from neighboring Nepal, Bangladesh and Srilanka, 2,531 collections in S. *melongena* and 337 of related wild species have been assembled by the NBPGR in collaboration with IPGRI.

At NBPGR, 2,419 accessions including wild species have been evaluated for 44 agro- botanical and economic characters and a wide range of variability was recorded for plant height (40.3-96.5 cm), number of primary branches per plant (4.6-10.3), days to 75 per cent flowering after transplanting (23-82), days to horticultural maturity (55-87), number of flowers per cluster (2.2- 9.30), number of fertile flowers per cluster (1.0-5.3), fruit length (4.3-36.7 cm), fruit circumference (6.5-46.3 cm), fruit weight (35- 725 g) and the number of clusters per plant (8.4-35.5) (Seshadri and Srivastava, 2002).

Table 6. Potential source of resistance to diseases and insect-pests and tolerant to abiotic stresses in eggplant*

Disease/insect	Sources (Host resistance / tolerance)
Resistant to *Fusarium* wilt	*S. incanum, S. integrifolium, S.indicum* (single dominant gene), Pusa Bhairav, 119-12-2-1,264+2
F.oxysporum and *Phomopsis vexans*	*S melongena* x *S. indicum,* K 61, K 62, Ghana Local
Resistant to *Verticillium* wilt	S. *melongena* (Harris 468 Special, Hibush), PI 1649, PI 174362 S. *torvum, S. caripense, S. persicum, S.scabrum,* LF 3, 'Florida Market', Harris Hybrid 7763, S. *sisymbrifolium*
Resistant to bacterial wilt	Kopek, Black Beauty, S. *melongena* SM-8l, SM-6, SM 6-1, SM 6-l-M, SM 6-7 SP, PPC, ARU2C, MS-48, SM-56, SM-7l, SM-72, SM-74, H-8, *S. torvum, S. melongena var. insanum, S. xanthocarpum, S. nigrum, S. sisymbriifolium, S. integrifolium,* IIHR 110, 121-2, 181-3 and 85
Resistant to root knot nematode	*S. melongena* (Florida Market), *S. sisymbriifolium, S. torvum Meloidogyne incognita, M. incognita. M. arenaria*
Meloidogyne spp.	*S. torvum,* MM 392, MM 450, *S. aethiopicum* 'Gulla', *Meloidogyne incognita* race 1&2 'Gulla'
Phomopsis fruit rot *(P. vexans)*	*S.gilo, S. integrifolium*
Phomopsis blight resistance	IC-89883, IC 90013, IC 9010, IC 99571, IC 111457, IC 111393, IC 136571, IC 99648, IC 128680, IC 136326, IC 136256, IC 90865, NIC 13021, NIC 13337, BDS-7/331, IC-136326, 127238, NIC-4263, 9420, EC-305069, 316274
Cercospora solani	*S. macrocarpan,* 'Udipigulla', GO-3, Composite 1, IIHR-101, 206, 319 and 434
Little Leaf resistance	*S. integrifolium*
Resistant to shoot and fruit borer *Leucinodes orbonalis*	*S. khasianum, S. sisymbriifolium, S. melongena*, Black Beauty, H-165, H-407, H-408), *S. incanum, S. gilo, S indicum* F1 progenies of *S. melongena* x *S. incanum,* PPC, AM 62, SM 17-4, *S. integrifolium, S. sisymbriifolium, S. xanthocarpum*

Contd...

Jassid *(Amrasea biguttula biguttula)*	S488-2, S 34, S 258, 'Manjari Gota'
Aphis gossypii	AC 49A, S. *sisymbriifolium, S. mammosum*
Tetranycus urticae	*S. macrocarpon*
Tolerant to frost	*S. incanum,* 'Black Tarpedo', 'Long Tom 4', 'Black Beauty'
Tolerant to drought	*S.. macrocarpon* K 2591, 'Supreme', 'Violette Round'

* Kalloo, 1993; Gautam and Gangopadhyay, 1998

Two varieties 'Pusa Purple Long' and 'Pusa Purple Round' were released from Indian Agricultural Research Institute. The former is early, long fruited and high yielding whereas the latter is round, high yielding with dark purple colour. These are selections out of local indigenous stock. 'Arka Sheel' and 'Arka Kusumkar' are selections made by Karnataka state, having long fruits with less seeds. CO.1 and MDU-1 by TNAU, 'Krishnagar Purple round' and 'Green Long' are popularly grown in West Bengal. Besides this, a number of brinjal varieties from different parts of India, have been selected i.e., 'Satputia' and 'Satputia Hara' are selections from Bihar with clustering habit. 'Banarasi Giant' from Varanasi (U.P.). and 'Gola Hara' from Bihar bear very big fruits, 'White Long' (J&K) and 'White Cluster' bear small white egg shaped fruits, 'PB No. 8' is an early prolific variety, round oval in shape, medium size fruit, purple in colour from Punjab. 'Gudiyatham' from Coimbatore' and IC-1855 are reported to be resistant to fruit and shoot borer. 'Surti Gota', 'No. 24-597' and 'Manjri Gota', '28. JG selection' and 'Manjri Local-1', 'S96-2', 'Mukta Keshi', 'White Long' have shown resistance to root-knot nematodes. 'White Long', 'S96-2', 'Round Prickly' and 'Chhangulia' have been reported to show significant field resistance to *Phomopsis* blight. 'Hyderpur Local' has been utilized as direct selection as well as for developing improved varieties. Some wild growing brinjal in Assam has been found to be resistant to *Phomopsis* disease. Many cultivars growing in Potangi (Orissa) region have shown promise as breeding material (Thomas *et al.,* 1983).

Potato: 'Kufri-Red' is a clonal selection from 'Darjeeling Red round' and 'satha' variety from Bihar.

B. Cucurbits

This group includes vegetables such as water -melon, muskmelon, pumpkin, luffa gourd, bottle gourd, pointed gourd, bitter gourd, etc.

Water-melon: "Durgapura Kesar" and "Durgapura Meetha" from Rajasthan, "Farukhabad Giant" from UP are selections out of indigenous collections. Various shapes and sizes are also seen in market which indicates the existing variability available to be exploited.

Muskmelon: 'Arka Rajhans' is selection made by Karnataka with high resistance to disease, 'Arka Jeet' is an improved variety from local 'Bati' strain of Lucknow, 'Sel. No. 445' is a selection from Rajasthan. 'Allhabad Khajira', 'Lucknow Safeda' from UP,

'Durgapura Madhu' from Rajasthan, 'Kutana' and 'Sunheri' from Punjab, 'bathese' from A.P. and 'Sarda' variety of M.P. is promising selections. Besides these, 'Sharbat-E-Anar', 'Hingan' and 'Jalbudata are other varieties under cultivation in A.P. "Hara Madhu' of Punjab is also a popular variety.

Tinda (Round melon): 'Arka Tinda' is a selection made by IIHR from indigenous cultivar. 'Safed' and 'Green tinda' are widely grown and variability also exist in this crop. Selection from Punjab and Annamalai selection of Annamalai University are also popular.

Pumpkin: 'Arka Suryamukhi' is a selection from indigenous collection from Rajasthan. 'Red pumpkin' is also widely grown all over the country having various types of shape and sizes. Co series of Coimbatore are also popular.

Luffa gourd (Sponge gourd and ridge gourd): Variety 'Pusa chikni' of sponge gourd is a selection from a Patna collection. 'Pusa Nasdar' variety of ridge gourd is a selection from Neemuch (MP). There are other *Luffa* also grown near hamlets which are either hermaphrodite or gynomonoecious. These are no doubt small fruited but profuse in bearing.

Bottled gourd: 'Summer Prolific' is a heavy yielding variety grown in UP and around Delhi. These are both long and round. Other shapes of this cultivar are also seen.

Pointed gourd or Parwal: 'Green Oval Green Long Stripped and White Oval are three promising varieiteis of Parwal selected from local cultivars.

Bitter gourd: Variety 'Pusa Do-Mausami' is a selection from local cultivar. It is early with smooth long. 'Pusa Harit' is a selection made by IIHR. Besides this, there are other varieties like 'Coimbatore Long' and 'Coimbatore Green' and 'K. Sona' of Kalianpur and 'MC23' of Vellanikara are popular varieties.

C. Allium spp.

Onion: 'Pusa Red' is a selection from local collection developed by Indian Agricultural Research Institute. 'Line 106',' Line 133', 'Line 137' from Delhi; 'Hisar II' from Haryana, 'S-48' from Punjab, 'Udaipur-101', 'Udaipur-102' from Rajasthan, 'Bellary Red' from Karnataka, 'Co-1', 'Co-2', 'Co-3' and 'Co-4' from TNAU and 'N-53' from Maharashtra are promising selections. Of these 'Pusa Red' and 'N-53' are very popular and have wide area of adaptability. NBPGR has developed selections and have also built up vast collection (about 1600) from different areas of diversity in the country and very good germplasm have been identified (Thomas *et al.,* 1983).

Garlic: 'T-56-4', a selection from Punjab and two local types 'Fawali' and 'Rajalle Gaddi" from Bellary districts are suitable collections from South India. NBPGR has built up about 800 collections from different parts of the country and has identified promising selections which are under study and multiplication.

D. Leguminous crops

Guar: 'Pusa Sadabahar', 'Pusa Navbahar' and 'Pusa Mausami' are the promising guar vegetable varieties released from Indian Agricultural Research Institute. 'Sharadbahar' (IC-110704), 'PIG-850', 'IC-11388' and P. 28-1-1 (selection from Pusa Naubahar) are other promising vegetable varieties developed by the NBPGR. A number of promising selections have been developed now by NBPGR.

Garden Pea: IARI selections 'N.P. 26', 'N.P. 29' and 'selections 18, 30 and 35' are promising collections out of indigenous germplasm. 'UL-2', 'V-3' from Almora (UP) are selections suited to hilly areas. 'G.C. 141', 'G.C. 195' from Gwalior (MP) are suited for drier regions bearing long green pods. 'Darantia Kaip', 'Lucknow Boniya', 'Meerut Boniya', T.17, T.18, T.19, T.61 of UP, P-8, P-35 and 'P-44' of Punjab, 'BR-12' of Bihar have dominated the cultivators field before the introduction of better varieties. 'Gloriosa' of Ooty, 'K.S'. Series of Kalianpur and 'U.D'. Series of Udaipur, are popular varieties of their areas (Thomas *et al.,* 1983).

French bean: 'Black Prince' and 'Black Queen' are the varieties for hills and 'Bigben' of Pantnagar, 'Watex' of Ooty are commonly cultivated and they were selected out of indigenous stocks.

Sem (*Lablab purpureous*): 'Pusa Bunch' a green selection IARI and NBPGR has built up over 1000 collections and identified large number of promising types with very attractive pods, high yields and varying size of pods.

Miscellaneous Legumes: Promising types were identifed in sword bean, winged bean, velvet bean and broad bean by the NBPGR.

E. Cruciferous vegetables

Cabbage: 'Sel.8' from Katrain and 'ARU-Glory' from Almora (UP) are two promising selections of cabbage.

Cauliflower: 'Pusa Katki' Line 294, Line 356, Line 327 and Line 328 of IARI; 'Hisar-I' from Haryana and 'Sel-1', 'Sel-5', 'Sel-9' and 'Sel-12' of IARI are promising selections,. 'Kunwari' of Punjab is also a selection out of indigenous stock. Besides, lot of variability does still exist in the indigenous stock.

Radish: 'Pusa Desi' is a selection from local material with white long roots with green top.

Turnip: 'Pusa Sweti' or 'Sel. No. 6' are selections out of indigenous stock.

F. Leafy vegetables

Palak or Spinach Beet: There are three promising selections of Palak released under the name 'Pusa Jyoti', 'All Green' and 'Jobner Green' which have been under cultivation in India. These are all selections from indigenous stocks.

Methi: 'Kasuri' of Punjab, and 'Pusa Early Bunching' of Hyderabad and 'No. 47' of Maharashtra are promising varieties of methi under cultivation selected out of indigenous collections.

Chakwat: I.C. 3301 is a promising high yielding type developed by NBPGR having broad leaves.

Leafy Amaranth: 'Barichauli' selected by IARI and 'Co-series' from Coimbatore are promising selections.

G. *Malvaceous* crops

Okra: 'Pusa Makhmali' (erstwhile NBPGR), a selection from Bihar, 'Punjab No. 13', 'T-1'and 'T-2' selections from Kanpur (UP), 'Shankar Palli ' and 'Lam Sel-1' from A.P., 'Vaishali Vadhu' and 'Mithila Badhu' from Bihar, 'Co-1' and 'MDU-1' from TNAU and 'Sel.2' from NBPGR are promising selections of okra.

H. Tuberous crops

Sweet potato: 'Musiri Thandal' (SP-1), 'local Musiri' (SP-4), 'local Ratnagar' (SP-8) and Local 'Alangulam Red' (SP-18) have high yielding potential.

Tapioca or Cassava: 'Selection (S-2)' and (S-5) from CTCRI are two promising collections. Some other promising selections of tapioca are '80-OP-14', 'S-82' and 'S-51'.

***Dioscorea* spp:** There are three high yielding selections of *Dioscorea* – 'De-11', 'De-17', and 'De-23', evolved at CTCRI.

III. Improved Cross-Bred and Hybrid Varieties Using Conventional Methodologies

Several exotic and indigenous collections of vegetable crops have been utilized to develop improved cross-bred and hybrid varieties. USA, Canada, Netherlands, Denmark and Japan are the countries where hybrid seeds of vegetables were developed in the beginning and they are still dominating the scene in hybrid seed production. Now, the hybrid seed production of vegetable crops have also become commercialized in several other countries. Many commercial seed firms have taken up hybrid seed production a great deal especially of F_1S and have the monopoly for the potential lines.

Some of the important varieties developed by hybridization in different vegetable crops are described below.

A. Solanaceous crops

Tomato: 'Pusa Ruby' has been produced by crossing 'Sioux' and 'Improved Meeruti', which has uniformly red fruits, prolific bearing and high degree of resistance to virus and water logging. 'Pusa Early Dwarf' is another variety derived from the cross 'Red Cloud' x 'Improved Meeruti'. 'Pusa Red Plum' has been evolved by crossing cultivated tomato with South American wild species, *Lycopersicon pimpinellifolium*. Fruits are small but plants are resistant to virus diseases (all from NBPGR). Another hybrid has been developed at IARI by crossing 'Pusa Ruby' and 'Best of All'. 'Punjab Chuhara' is a selection from the cross 'E.C. 55055' × 'Punjab Tropic'. 'Sweet 72' of Gwalior has the parentage of 'Pusa Red Plum' and 'Sioux' (Thomas *et al.,* 1983).

Chillies (Capsicum spp.): Among pungent chillies, 'Jwala' is a selection evolved from cross 'NP-46 A' and 'Pusa Red'. It has long attractive green fruits. 'NP Hyb'5-1-5' and 'N.P. Hyb. 17-1-1' varieties have been developed (at IARI) by crossing parents with desirable qualities. Bharat (F_1) hybrid released by Indo-American Hybrid Seeds, Bangalore has bell shaped, thick walled, mostly 4-lobed fruit. 'HC-201', 'HC-202' and 'HC-213' are promising chillies selections from Almora, 'Katrain-1' and 'Katrain-2' from HP are other hybrid varieties of capsicum.

Brinjal: 'Pusa Kranti' variety has been bred by combining good qualities of three parents namely 'Pusa Purple Long', 'Hyderpur' and 'Wynad Giant'. It is a midseason variety. 'Pusa Anmol' (F-1 Hybrid) is hybrid between 'Pusa Purple Long' and 'Hyderpur'. 'Pusa Purple cluster' is also promising variety from NBPGR.

Potato: A large number of potato varieties have been released. 'Kufri Chandra Mukhi' is an early maturing variety with attractive oval white tubers. It has good keeping quality. 'Kufri Alankar' is a rapid bulking variety with high yield potential. 'Kufri Lauvkar 'is an early maturing variety having large, round white tubers capable of giving good yield under both seasons in Maharashtra. 'Kufri Chamatkar' is a medium maturing variety of excellent potential plains of Northen India, with round, white and medium sized tubers. 'Kufri Sheetman' is a medium maturing variety having oval, white, medium sized tubers. 'Kufri Deva' is another medium maturing variety with round white and medium sized tubers with frost resistance and excellent keeping and culinary qualities.'Kufri Jyoti' is also a medium maturity variety with large oval, flattened white tubers with high degree of field resistance to late blight. It is immune to wart and has a low rate of degeneration. 'Kufri Muthu' - a late blight resistant variety maturing in 110-120 days. It has a short dormancy period and thus three crops can be taken. 'Kufri Navjot' (SCB/2405 a) is a promising late blight resistant culture for hilly areas with attractive tubers resistant to cracking. 'Kufri Navtal' - another promising culture released as commercial variety for Kumaon hills of Uttaranchal. It has white long tubers and good keeping quality under room storage conditions. 'F3977' and 'P5242' are two medium maturing and medium late varieties possessing a high degree of field resistance to late blight and immunity to wart disease. 'E 3797' is an early maturing selection having round oval, white tubers of large size (Thomas *et al.*, 1983).

B. Cucurbits

Muskmelon: 'Pusa Sharbati' of IARI, developed by hybridization with round fruits with green stripes thick and bright orange coloured flesh and narrow seed cavity and good keeping quality. 'Punjab Hybrid' is a hybrid variety developed at PAU, Ludhiana.

Squash: 'Pusa Alankar' is a hybrid between 'EC 27050' x 'Selection no. IPL. 8'. It is an early, high yielding variety with uniform dark green fruits.

Bottle gourd: 'Pusa Manjari', and 'Pusa Meghdoot' are two F_1 hybrid varieties evolved by using parent (Pusa Summer Prolific Round and Selection-1) and (Pusa Summer Prolific Long and Selection-2) respectively. Both varieties are early and high yielding with green, tender and attractive fruits.

Cucumber: 'Pusa Sanjyog' is F1. Hybrid variety released from IARI with early maturing, long cylindrical and attractive dark green fruits.

Alliums: 'VL-67' (F_1) from Almora is a promising variety.

C. Legumes

Cowpea: 'Pusa Do-fasli' variety is a cross between 'Pusa Phalguni' and 'EC. 21622 (ex. Philippines)'. It is photo insensitive. Grown during spring-summer as well as during rainy season. 'NS24/8-2 (Aseem)' is a cross between 'Pusa Phalguni' X 'EC 21622 (ex. Philippines)'. This is early with yellowish green cylindrical fleshy pods and is also tolerant to wilt and nematodes. 'Brown Seeded Selection (Rituraj)' is photo insensitive selection from 'P 85-2 E 9A' culture. It is very early in maturity, with long pods. 'P 460-1-1' and 'Red Seeded' are other promising varieties of vegetable cowpea evolved by NBPGR (Thomas *et al.,* 1983).

French bean: 'Pusa Parvati' has been evolved from irradiated seeds. It is an early, bush type with green pods, having straight tips, 15 to 18 cm. long, stringless and meaty pods and is high yielding as well as resistant to mosaic and powdery mildew.

D. Crucifers

Cauliflower: 'Pusa Synthetic' and 'Pusa Deepali', 'Early Kunwari','Synthetic Early' are promising varieties.

Radish: 'Pusa Reshmi', 'Pusa Himani'

Turnip: 'Pusa Chandrima' is a selection from across between a 'European type' x 'Japanese White'(an Asiatic type. 'Pusa Kanchan' is a selection froma cross between local 'Red Cloud' x ' Golden Bell'.

E. Miscellaneous crops

Okra: "Pusa Sawani' selected from a cross between 'Pusa Makhmali' and a variety 'Best 1' from West Bengal. MDU-1 is a promising mutant selection developed from irradiated "Pusa Sawani seed. G2 is a selection between 'Pusa Sawani' and 'Ghana Red', both from NBPGR.

Carrot: 'Pusa Kesar' is a selection from a cross between 'Local red' and ''Nantes Half Long'

Improvements for Value Additions

The vegetable production scenario in this country is undergoing several changes and the demand for fresh vegetables is increasing mainly from middle income segment of the population chiefly from urban and semi-urban areas of the country. Such market demands induce farmers to adopt off-season production of vegetables, even under unfavourable conditions and with lower yield but higher market price is the chief attraction. Seasonality in vegetable production, availability and consumption is fast disappearing.

Further improved road transportation and telecommunication facilities ensure long distance usage of fresh vegetables, so that the availability of vegetables is spread over wider and far flung regions of the country and with staggered periods of availability. Hence genetic resources in vegetables have to be sought not only for increased production, disease and pest tolerance but also for tackling abiotic factors like heat or cold tolerance, semiarid conditions (like rainfed cultivation), etc. Besides, there is growing awareness of the advantages of healthy food habits, like increased vegetable consumption and the intake of important nutrients like proteins, vitamins and minerals is facilitated by enhanced vegetable consumption. There is now needed to lay emphasis on the nutrient quality of vegetable crops and their value-added products (Seshadri and Srivastava, 2002).

Nutritional and processing qualities

Protein: In French bean *(Phaseolus vulgaris),* there is a wide range of variability in protein content- especially seed storage protein. French bean protein is high in lysine- a factor of nutritional significance. They are low in total sulphur of amino acid content. Rutgers (1970) screened 343 bean cultivars and found protein content varied from 19 to 31 per cent, with a mean of 24.6 per cent. These data show that cystine and methionine content of bean protein is improved at lower level of protein content. Bonita is a variety identified with high protein content. In garden pea, Tiny is a high protein pea variety. Indian varieties of Faba bean (*Vicia faba)* exhibited higher protein than German varieties. A promise of leaf amaranth is the leaf protein concentrates.

Vitamin A: Vegetables do not contain active vitamin A compound retinol but only carotenoids that are converted into active retinal in the body. Carrot, amaranths, cowpea, beans, sweet potato, spinach etc. are important sources. The carrot variety WC 501 is a rich source of carotene. In tomato, *L. hirsutum*f. *glabratum* is reported to contain 3 to 5 times of β-carotene than the normal variety and 'Caro Red' is an interspecific derivative with high β-carotene content. Vegetable amaranths are also important source of provitamin A (β -carotene).

Even in some cucurbits, carotenoid content is high as in a winter squash (*Cucurbita maxima*) 'Golden Delicious' variety in comparison to low content in common pumpkin (*Cucurbita moschata.*) The total carotene and ascorbic acid content in cantaloupe (*Cucumis melo)* is considerably higher than other green and white fleshed muskmelons. Recently high carotene cucumber germplasm has been identified in pickling cucumber lines developed through introgression from Chinese cucumber having orange flesh *C. sativus* var. *xishuangbanensis* (Simon and Navazio, 1997). In cabbage, the carotene content is very high in Savoy cabbage (995 mg/g) compared to common white cabbage (40 mg/g). Savoy cabbage is a cold tolerant type with dark green wrinkled leaves.

Vitamin C: About 90 per cent vitamin C requirement of human dietary is obtained from fresh fruits and vegetables. Capsicum, cauliflower and bitter gourd are rich sources of vitamin C in capsicum varieties. 'Douxd Alger' has recorded vitamin C 300 mg/ 100g. However more attention has been given to vitamin C in tomato fruits. *L. pimpinellifolium,* a closely related species has been reported to contribute genes for high (ascorbic acid) vitamin C content. It has been reported that as the vitamin C content increased in the fruits, the plant resistance to fruit rot *(Phytophthora parasitica)* is also increased. There existed a relationship viz. small fruited genotypes are richer in ascorbic acid. Cultivar with twice (about 50 mg/100g) the normal amount of vitamin C have been developed eg. Double Rich.

Sugars and Total Soluble Solids: Sugar content is an important quality trait in watermelon, muskmelon, carrot and pea. In processing of tomato and dehydration of onion, total soluble solid content is an important consideration. The free sugars representing 60 per cent of the solids in tomatoes are mainly D-glucose and D-fructose with traces of sucrose, raffinose etc. *L. chmielewski* has been crossed with *L. esculentum* to produce high sugar breeding line (Rick, 1974). *L. pennellii* has been used to improve TSS content in conjunction with high yield (Eshed and Zamir, 1994). In muskmelon and watermelon high soluble solids generally result in higher rating of sweetness of dessert fruit which should be stable to satisfy consumer preference.

Higher solid is not only an indication of nutritive quality but also the processing quality too in tomato. Besides soluble solids, alcohol insoluble solids, which constitute 20 per cent of TSS and consist of pectins, polysaccharides including cellulose and structural proteins, decide the viscosity of fruit juice and are required for products like sauce, ketchup, puree etc. Indian Standards Institution (ISI 1967) has prescribed a minimum 5 per cent TSS by weight in paste and ketchup industry. Low number of locules per fruit is a preferred characteristic of processing varieties. In onion, varieties with high (up to 20 %) dry matter are selected for processing. Dehydrated flakes and powder are usually made from white cultivars with high dry matter content and high pungency. 'Southport White Globe' and 'White Creole' are suitable for dehydration. In the case of eggplant, high dry matter, high protein and low phenol content are suitable for processing specially for dehydration. 'Arka Kusumakar' has been found to be promising. Bitterness in brinjal is traced to glycoalkaloid (as solanin) content.

Flavour: The volatile compound composition in vegetable crops is very complex. The contribution of these many compounds to flavour varies greatly. In some vegetables, a single or a class of compound will have a typical aroma of that vegetable, viz., Nona - 2, 6 dional in cucumber, sulfides in onion and phthalides in celery. In French beans and tomato, no compound which typifies aroma of the vegetable has been found. The concentration of volatile flavour compounds in vegetables is very low, like a fraction of a part per million. Because of extremely complex mixture of compounds, special techniques are required for their analysis. There are also non-volatile compounds like organic acids and carbohydrates contributing to flavour in tomato.

Colour: Colour development is an important character for table tomatoes and for processing. Red fruit of tomato has mostly lycopene pigment, while carotene imparts yellow colour. In normal red fruits, the ratio of lycopene to carotene is 12 to15:1. A proper balance can give a good fruit colour, without sacrificing nutritive value provitamin-

A. Field selection of a tomato that would colour at high temperatures has resulted in the selection for better foliage, but dose not development lycopene at high temperatures.

In carrot, there is a positive correlation between colour intensity and total carotenoids. In watermelon, lycopene content differs among red fleshed watermelon cultivars. The major red colour in capsicum comes from capsanthin from total carotenoids and capsorubin, while the yellow orange colour is from â-carotene and violaxanthin. Capsanthin is more stable of the two. Purple fruit colour in chilli in its immature stage is because of anthocyanin. At Dharwad, a sweet pepper (Paprika capsicum), "Sarpan Madhu" has been evolved with 180 ASTA colour and fruit yield potential of 4.2 to 4.5t/ha. Quality is of much importance in chillies fruit and good pungency, in terms of capsaicin content, bright red crimson colour with colour value in terms of oleoresin concentration. Seed content percentage in the fruit is one of the chief attribute. Two hundred and forty chilli collections were evaluated at NBPGR Regional Station, Bhowali (2001-02).

Carbohydrates: A substantial portion of the carbohydrate found in vegetables, is present as dietary fibres, in the form of cellulose, hemicellulose, pectic substances and lignin. Vegetables are rich resources of dietary fibres ranging from 0.2 g/100g of edible portion of watermelon to 3 g/100g of pointed gourd. Epidemiological studies have shown that fibres found in leafy vegetables like, cabbage, spinach and lettuce etc. are highly useful in preventing some diseases. Nutritional benefit of dietary fibres like protection against certain types of cancer, lowering of blood cholesterol, are now well accepted. Genetic control of the constituents of quality usually depends basically upon the synthesis of enzymes. This condition, for example, in sweet corn, results in a decrease to the synthesis of starch and thus results in the accumulation of sugars. Hence, study of enzymes controlling the constituents of quality may be useful instead of assaying for the constituents at least in some cases (Seshadri and Srivastava, 2002).

References

Bisht, I. S., Patel, D. P., Sapra, R. L., Thomas, T. A., Kopper, M. N. and Rana, R. S. 1993. Catalogue of Okra (*Abelmoschus esculentus* L. Moench) germplasm. Part II. NBPGR,

Bisht, I.S., Patel, D.P., Mahajan, R.K., Kopper, M.N., Thomas, T.A. and Rana, R.S. 1995. Catalogue of wild *Abulmoschus* species germplasm. NBPGR, New Delhi, 51p.

Chandra, Umesh 1997. Studies on genetic divergence of exotic and indigenous germplasm of tomato (*Lycopersicon esculentum* Mill.) under Indian conditions. Ph.D Thesis, Dr BR Ambedkar University (formerly Agra Univ.), Agra, 152p.

Chandra, Umesh and Gupta, P.N. 1994. Evaluation of tomato germplasm adaptable to abiotic stress conditions of Northern India. *Indian J. Pl. Genet. Resources* **7(2)**:465-172.

Chandra, Umesh and Thomas, T.A. 1993. Sources of tomato germplasm adaptable to the hot summer season in India. *In:* C. George Kuo (Ed.) Adaptation of food crops to the temperature and water stress. AVRDC, Taiwan, 13-18 Aug. 1992. p501-502.

Choudhury B., Tiwari, R., Mathai, P. J., Joshi, G.C., Kesavan, V., Chakravarty, A.K., Kalda, T.S. and Sivakami, N. 1973. Breeding for disease resistance in some tropical and sub-tropical vegetable crops. *In:* Breeding Research in Asia & Oceania (S. Ramanujam and R.D. Iyer, Eds), 22-28 Feb. 1973.

Eshed, Y. and Zamir, D. 1994. Introgression from *Lycopersicon pennellii* can improve the soluble solids yield of tomato hybrids. *Theor. Appl. Genetics* **88 (6&7)**: 891-897.

Galewey, N.W. 1992. Verifying and validating the representatives of a core collection. Paper presented in International Workshop on Core collection of Plant germplasm. CENARGEN, Brazil, Sep. 1992.

Gautam, P.L. and Gangopadhyay, K.K. 1998. Biodiversity and genetic resources in vegetable crops. *In*: National Symposium on Emerging Scenario in Vegetable Research and Development, 12-14 December 1998, organised by the Indian Society of Vegetable Science at IIVR, Varanasi, India, p32-42.

IIVR. 1995 to 2000. IIVR Annual Report, 1995-96, 1996-97, 1997-98, 1998-1999 and 1999-2000. Indian Institute of Vegetable Research, Varanasi. U.P.

Joshi, B. D. and Rana, R. S. 1995 French bean in India. NBPGR Sci. Monograph No.5 NBPGR, New Delhi, 168p.

Kalloo, G. 1993. Tomato *In:* Genetic Improvement in Vegetable Crops (Eds. G. Kalloo and B.O. Bergh) Pergamon Press, Oxford, England, p 645-666.

Mishra, S.K., Sharma, B., Tyagi, M.C. and Choudhary, A.B. 2000. Genetic diversity in cowpea germplasm. *In:* International Conference on Managing Natural Resources for Sustainable Agricultural Productions in 21st century. New Delhi, February 2000. Vol. 2, p866-867.

NBPGR. 1995-2001. NBPGR Annual Reports, 1995, 1996, 1997, 1998, 1999, 2000, 2001. National Bureau of Plant Genetic Resources, New Delhi.

Patel, D.P., Thomas, T.A. and Mehra, K.L. 1981 Catalogue on. French bean (*Phaseolus vulgaris* L). germplasm. NBPGR Regional Station, Shimla.

Rai, Mathura and Gupta, P.N. 1995. Genetic resources of solanaceous crops. *In*: Genetic Resources of Vegetable Crops (Eds. R.S. Rana, P.N. Gupta, Mathura Rai and S. Kochhar). NBPGR New Delhi, p91-110.

Rai, Mathura, Gupta, P.N. and Agrawal, R. C. 1995. Catalogue on Eggplant (*Solanum melongena* L.). NBPGR, New Delhi, 207p.

Rick, C. M. 1974. High soluble solids content in large fruited tomato lines derived from a wild green – fruited species. *Hilgardia,* **42:** 493-510.

Rick, C. M. and Fobes, J. F. 1974. Association of an allozyme with nematode resistance. Rep. Tomato Genet. Coop. 25.

Rutgers, J.N. 1970. Variation in protein content and its relationship to other characters in beans. Research Conference, Davis California, **10:** 59-60.

Sardana, S. and Suneja, Poonam 2000. Germplasm and evaluation in pea. *In:* International conference on Managing Natural Resources for Sustainable Agricultural Production in 21st century, Feb 2000. New Delhi Vol. 2, p772-773.

Seshadri, V. S. and Srivastava, Umesh 2002. Evaluation of vegetable genetic resources with special reference to value addition. International Conference on vegetables: Vegetables for sustainable food and nutritional security in the new millennium, organized by Dr Prem Nath Agricultural Science Foundation, Bangalore, Indian society of Vegetable Science New delhi and Indian Institute of Horticultural research Bangalore. 11-14 November, 2002: 41-62.

Sharma, B., Mishra, S.K, Mathur, Neeru and Tyagi, M.C. 1998. Diversity for morphological characters in pea. Paper presented in National Diologue: Issues in Management of Plant Genetic Resources, organized by the Indian Society of Plant Genetic Resources and National Bureau of Plant Genetic Resources at NBPGR, New Delhi, 1-2 December 1998; Abstr. No. P2-3.1, p 119.

Simon, P.W. and Navazio, J. P. 1997. Cucumber Early Orange Mass 400, 402, Late Orange Mass 404: High carotene cucumber germplasm. *Hort. Science* **32(1):** 144-145.

Singh, Mahendra, Kumar, Gunjeet, Abraham, Z., Phogat, B.S., Sharma, B.D. and Patel, D.P. 2001. Germplasm Evaluation. *In:* National Bureau of Plant Genetic Resources: A Compendium of Achievements (Eds. B.S. Dhillon, K.S. Varaprasad, Kalyani Srinivasan, Mahendra Singh, Sunil Archak, Umesh Srivastava and G.D. Sharma). NBPGR, New Delhi, p116-165

Srivastava, Umesh, Sapra, R. L. and Sharma, G. D. 2000. Evaluation studies in world germplasm of tomato and developing a core set, NBPGR, New Delhi, 212p.

Staub, J.E., Serquen, F.C. and McCreight, J.D. 1997. Genetic diversity in cucumber (*C.sativus*) III. An evaluation of Indian germplasm. *Genetic Resources and Crop Evolution* **44(4):** 315-326.

Thomas, T. A., Singh, Ranbir and Prasad, R. 1983. Genetic resources for improvement of vegetable crops. South Indian Hort. 30th Year Commemoration Issue, 1983: 53-73.

Thomas, T.A. and Chandra, Umesh 1989. Genetic resources of tomato in India, their build up, evaluation, maintenance and utilization. Proc. Intl. Symp. On Integrated Management Practices for Tomato and Pepper Production in the Tropics. (Eds. S.K. Green, T.D. Griggs and B.T. McLean). AVRDC, Shanhua, Tainan, Taiwan, 22-25 March, 1988. p22-27.

Thomas, T.A., Bisht, I. S., Bhalla, Shashi, Sapra, R. L. and Rana, R. S. 1990. Catalogue of Okra (*Abelmoschus esculentus* L. Moench) germplasm. NBPGR, New Delhi, 51p.

Thomas, T.A., Bisht, I.S., Patel, D.P., Agarwal, R.C. and Rana, R. S. 1991. Catalogue of okra (*Abelmoschus esculentus*L. Moench)germplasm. Part II, NBPGR, New Delhi, 100p.

❑❑❑

Chapter – 3

Concept and Applications of Core Collections in the Efficient Management of Plant Genetic Resources

S.K. Mishra and I.S. Bisht

Introduction

National Gene Banks are now entering into an era of increased activity and responsibility, particularly for managing their own indigenous germplasm activities extending from collection to actual use of the genetic variation in crop improvement programmes. However, these tasks must be achieved with limited resources in best possible time frame. The core collection offers a potential opportunity to meet these challenges. Recognising that the sheer size of a germplasm collection could deter its use, Frankel (1984) proposed that it could be pruned to a 'core collection', which would represent, "with minimum repetitiveness, the genetic diversity of a crop species and its relatives". The core collection consist of a limited set of accessions derived from a germplasm collection, chosen to represent the genetic spectrum in the whole collection, and including as much as possible of its genetic diversity (Brown, 1995). The remaining accessions would not necessarily be discarded but would be managed as a 'reserve collection'. The proposal, the rationale, purposes and general principles of core collections have been critically examined (Brown, 1989 a and b). As to the size of a core collection, sampling theory of selective neutral alleles in finite populations indicate that about 10 % drawn randomly from the whole collection was relatively efficient in retaining its allelic variation (about 70 % retained). The core collection concept has attracted considerable discussion and debate in the past and has met criticism in many areas. Most of these concerns appear to arise from misuse or misunderstanding of the core approach and a number of core collections have been established during the past decade.

Features of a Core Collection

The main idea behind developing a core collection is that it should represent with minimum redundancy, the genetic diversity of a crop species and its wild relatives. While it is convenient to talk in terms of a core collection of a crop and its wild relatives, in practice, each species is likely to be treated separately in developing any core collection so that differences between species in terms of their biology and the variation they contain can be taken into account. It is essential to realize that core collections are not substitutes for the collections from which they are derived. They are developed to improved accessibility and use of the whole collection. Four elements are basic to the concept of a core collection: (1) the parent whole collection is a large entity (from the standpoint of management or use of many accessions) with taxonomic integrity, (2) the core from this large collection has a restricted size, (3) the core is a representative sample of the collection, and (4) it is diverse (Brown and Spillane, 1999). The following kinds of samples from a collection lack one or more of these four elements. It is, therefore, confusing and inadvisable to apply the term "core collection" to them:

- A highly restricted subsection of a collection (e.g. one type of crop from a limited geographic area).
- A set of accessions that includes the majority of the accessions in the whole collection.
- A series of bulks that together include essentially all the genetic variation in the whole collection.
- A set that is chosen for one specific use or one user.
- The only portion of the collection that is available for distribution.

On the other hand, use of term "core collection" does not require that every part of the whole collection be equally represented. Indeed unequal numbers per class of genetic resources (cultivated vs. wild), per subspecies or per geographic region are to be expected. Nor does it require the absolute maximum possible diversity; otherwise the core would be biased towards large numbers of distant wild relatives. Rather the diversity should be as high as possible; where "possible" entails consistency with the core being a representative genetic resources collection of practical utility for breeders and other researchers.

Core collections have many roles to play in the management and use of genetic resources. This is because most activities in Gene Bank management require the curator to make choice or to set priorities among accessions because of limited resources. The germplasm curator for which a core collection offers distinct advantages is: addition of new accessions, conservation, characterisation, evaluation, germplasm enhancement and germplasm distribution. The general feature of most of these benefits is that the judicious reduction of the number of accessions to be handled in one operation saves resources. These resources are available for a more complete range of activities to be conducted in greater depth and thoroughness because the entries are representative of the collection.

The core collection, by reducing the number of accessions available in the first instance, would foster greater use of genetic resources by plant breeders than large and cumbersome collections. While this had obvious appeal to Gene Bank users and managers, it also implied the need for further refinement to enhance the efficiency of germplasm use. There are thus two primary motives for establishing core collection: assisting germplasm management and enhancing germplasm use. Germplasm specialists seem more concerned with the management motive, while breeders are more concerned with usage. Curators have an interest in both motives but are biased towards Gene Bank management. Establishing a core collection with one of these motives as the main objective does not necessarily serve the other. There are, however, joint benefits: the information obtained using either approach contributes to the success of both applications of the core collection. These two approaches are complementary. The beneficiaries are all those who have a stake in the conservation and judicious exploitation of plant genetic resources (Mackay, 1995). It is worth emphasising herewith that the core collection is not an entity on its own; it is a guide and entry point to the whole collection. Core collection can therefore be expected to increase the value of the whole collection rather than reduce the value of the non-core element. Field Gene Banks is expensive to maintain and are always prone to unavoidable losses. Thus development of "core approach" can guide the choice of accessions for growing in the field and those for developing the use of *in vitro* methods.

Procedure for Developing a Core Collection

The processes involved in developing core collections are as follows:

A. Data assembly

The first stage in the development of any core collection is to assemble the available data on the whole collection. The available passport and characterization data need to be assembled and updated as far as practical. The most flexible approach is to organize the passport, characterization and evaluation data for a hierarchical classification into groups. The attributes are considered in descending order: taxonomy, geographic origin, ecological origin, genetic markers and agronomic data. The material that should be represented by the core i.e. the domain of the core collection- will differ from core to core (Van Hintum, 1999). It will depend on the material available and the purpose of the collection. The domain might be, e.g., the CSIRO collection of perennial *Glycine* (Brown *et al.*, 1987), the complete US germplasm collection of peanut (Hollbrook *et al.*, 1993), the Peruvian quinoa collection (Ortiz *et al.*, 1998), a restricted set of annual *Medicago* (Diwan *et al.*, 1994), the cultivated *Brassica oleracea* in European Collections (Boukema and Van Hintum, 1994), a set of lentil accessions from Chile, Greece and Turkey (Erskine and Muehlbauer, 1991) or the entire genepool of *Hordeum* (Knupffer and Van Hintum, 1995). But it might also include, or even be restricted to, material with specific traits, such as local maize populations with a good combining ability (Radovic and Jelovac, 1994) or *Pisum sativum* germplasm with disease resistance (Matthews and Ambrose, 1994).

B. Grouping

The second step is to assign the accessions to groups, the members of which are likely to be genetically similar (Crossa *et al.*, 1995). In hierarchical approach a divisive procedure is employed, splitting the collection into smaller and smaller groups within groups, as genetically distinct as possible (Van Hintum, 1994 a). The number of groups depends upon the size of the collection, the intended size of the core and the dissimilarity of the groups at the lower level of sorting. The level to which the division of the collection into smaller groups proceeds should depend upon the distinctiveness.

There is considerable evidence (e.g., Peters and Martinelli, 1989) to suggest that country of origin is a reliable unit of grouping and indicator of genetic diversity and this may provide a useful approach to those exploring the development of groups for a particular collection. Where there is a reliable agronomic data, such an approach may be modified to take into account the phenotypic similarity of accessions from different countries (Spagnoletti-Zeuli and Qualset, 1987). Grouping strategies have varied considerably in different core collection programmes, depending on the biology of the crop and the data available. Grouping, however, can be done on the basis of taxonomy, domestication, distribution, breeding history and utilization, or on the basis of molecular marker or other characterization data. An additional approach to creating groups of similar accessions is using multivariate analysis (Crossa *et al.*, 1995).

C. Allocation of entries

Having divided the collection into group, the third step is to select entries from each group. The questions to be addressed here are the number to be chosen from each group and the method of choice within the group. Yonezawa *et al.* (1995) and Schoen and Brown (1995) have discussed the issue in detail. The choice of number of entries to be included in the core is arbitrary and will depend on the purpose of the core collection. Generally, the number of entries chosen is in the 5 to 20 % range. Brown proposed a sample fraction of about 10 % as an appropriate sample size for sampling core entries from a core collection. The optimum sample fraction, however, depends upon various genetic and resource parameters, primarily on the genetic redundancy among accession comprising the whole collection and the amount of resources available for the maintenance of the core entries. The optimum sample fraction is large with a lower redundancy among accessions and a lower initial allelic diversity within accessions, indicating that a larger sample fraction is better in species with a larger selfing rate.

For the allocation of the entries over the groups, a stepwise procedure can be followed. At each division of a group into subgroups, it is decided how the entries in the group should be divided over the subgroups. There are several strategies to use in this process. If no information is available concerning importance of, or diversity in the subgroups, the number of accessions in the subgroups can be used for this decision. In this case the constant (C), proportional (P) and logarithmic (L) strategies are available (Brown 1989b). If quantitative information about the diversity in the subgroups is available the gene diversity (H) or maximization (M) strategy can be used (Schoen and Brown, 1995).

The strategy for allocation of entries over subgroups will depend on the information available on the group and its subgroups, and this can differ by group. For example, if there is no additional information about the accession in a group of wild material that is divided into subgroups by country of origin, one might select the number of entries from each subgroup for the core collection based on the number of accessions in the subgroups. If there is detailed molecular marker information available on the majority of accessions in a group of cultivars, one might use a diversity-based strategy, and if a decision has to be made concerning the division in wild and cultivated subgroups, generally a strategy based on the weight or importance of groups for the core collection objectives will be followed.

D. Choice of accessions

The final choice of specific entries from the groups can be done randomly or on the basis of additional data if available. The objective should be that the diversity in the group should best be represented by the entries. Since it is desirable to have readily available, authentic and well-documented material in the core, practical considerations can also play a role in this choice. This includes factors such as the availability of the seeds and the reliability and quantity of the data on the accessions.

If expensive and complete molecular marker data or other characterization data are available, it is possible to base the choice of entries on the basis of multivariate analysis. This would allow an optimal representation of diversity within the group. Examples of variations of such an approach are Spagnoletti-Zeuli and Qualset (1993), who first clustered durum wheat accessions on the basis of morphological data and then stratified on the basis of country of origin, and Charmet and Balfourier (1995) who clustered their French perennial ryegrass populations on the basis of agronomic traits with the constraint that populations had to be collected in the same area to be clustered together. If the pedigrees of material in a certain group are known, it is possible to use these to maximize the allelic richness in the entries selected from this group (Van Hintum 1994b; Van Hintum and Haalman, 1994).

E. Managing and using core collection

We have yet to acquire experience on how core collections are best managed in gene banks. In many collections, such as those of cereals, the core entries remain within the whole collection. It may be sufficient merely to record in the database a signal that a certain accession is a core entry. In other cases (such as for clonal crops) a spatially separated plantation may be planned, or another conservation method (such as tissue culture, cryogenic set or DNA bank) may be employed. A second question is whether or not each core entry should be made homogeneous. For self pollinating accessions, this would mean making a new pure-line from one in-bred individual. For apomictic species, it would mean propagating one clone as the core entry. However, the cost is a substantial loss of genetic variation.

With the core established, programmes of evaluation can begin. The evaluation could be for finding donors of various characters for use in breeder's crosses, or for agronomic performance or for direct use. The user of a core should sent useful information back to the curator as to the core's validity and value (Galway, 1995).

Issues in the Development of Core Collections

The major concern for those who have expressed doubts about the development of core collection has been the fate of the accessions that are not included in the core. The core is not an entity on its own; it is a guide and an entry point for the whole collection. The core collections are designed to stimulate and improve the use of genetic resources and hence are likely to increase the value of the whole collection rather than decrease the value of the non-core element. However, studies on the links between the core collection and the whole collection are still needed. Another set of questions that have been raised relate to the way in which variation is sampled to create the core collection. This has a number of aspects which include the amount of information needed to create a core collection, the uneven distribution of variation in most collections and the extent to which useful but rare variants will be present in the core.

It has been argued that the creation of reasonably representative core collection would require so much information that is impracticable, and that, once sufficient information has been collected to create such a collection, it would be unnecessary. However, a limited passport data and basic characterization data for major morphological characters can provide an effective grouping strategy for the development of a core collection.

The concern that genetic diversity in collections was so unevenly distributed as to make the creation of a core collection difficult or impossible has itself stimulated some of the most interesting work in this area (e.g. Schoen and Brown, 1995; Perry *et al.*,

1991). Various strategies have been proposed to deal with extremely uneven distributions of diversity and, while these have yet to be fully tested, the available evidence suggests that techniques to take account of uneven distribution are available.

A considerable amount of work remains to be done to improve methods of selecting accessions for inclusion in core collection and, gain experience in their management and use. This will necessarily come from the experience from those who have undertaken the practical work of developing and testing such collections. There is a need to gain experience with a much wider range of crops including self and cross pollinated crops and clonally propagated ones. Developing core collections for this last group of crops raises specific questions of both theoretical and practical nature since core collection theory is largely based on the application of population genetics to seed propagated crops. The extent to which data on quantitative traits can be used needs further study as to the ways in which such data can be combined with that from biochemical or molecular studies. Ways to testing the extent to which the objective of representing diversity has been achieved also need to be improved (Hodgkin, 1994).

The basic concepts that have been developed to identify accessions for a core collection are likely to find other applications. Most obviously, they can be used when new collections are established, to ensure that variation in such collections is maximized and genetic redundancy minimized. The procedures that allow groups to be identified and that ensure effective sampling within groups should also ensure that new collections are not burdened with large numbers of accessions that add little to the overall objective of conserving the maximum possible variation present in a gene pool. In recent years there has been an increase in *situ* conservation, particularly of those wild relatives of crops that constitute the primary, secondary and tertiary gene pool. It is tempting to suggest that a strategy based on using the techniques developed for identifying core collections could also be used to identify populations, which should have highest priority for *in situ* conservation programmes. Herbarium and eco-geographic surveys could be used to identify populations of the target taxa, which could then be grouped according to taxonomic, geographic and ecological criteria. Priority for *in situ* conservation would go to population from different groups whenever practical considerations (such as management, social and political issues) allowed, thus attempting to maximize the conserved diversity (Hodgkin *et al.,* 1995).

Another general issue faced by those forming core collections for use in specific countries or areas of the world is the extent to which accessions that are clearly not adapted to their specific photoperiod response (e.g., sesame, sorghum etc.) with disfavour and avoid them if possible. However, they may well possess characters, such as disease resistance not present in other accessions and their omission might result in excluding a significant fraction of the total variation. Practical considerations will be important in making decisions on such material.

Developing Core Collection: Vegetable Crops

About 70 core collections in different agri-horticultural crops have been developed globally. However, there are very few published reports on use these core collections in even identifying the sources for their use in crop improvement programmes. Studies on establishing core collection of certain indigenous crops have already been taken up at the National Bureau of Plant Genetic Resources. A methodology has been developed to establish the core collection in okra (Bisht *et al.*, 1995; Mahajan *et al.*, 1996). An effort has also been to develop core set of brinjal based on information on about 2400 accessions. The core collections developed in vegetable crops include *Brassicas* (2), *Capsicum*, eggplant, lettuce (2) and okra (2).

Uses of Core Collections

There are many important uses of core collections, which include mainly:

a. Efficient management of plant genetic resources
   * Addition of new accessions
   * Conservation on priority
   * Easy and detailed characterization and evaluation
   * Quick distribution

b. Effective utilization of genetic resources
   * Identification of trait-specific germplasm
   * Easy testing of the core set for combining ability etc.

Future Developments

The 'core selector': The core collection concept works well until users request, for example only half the accessions of the core collection but still desire a representative set, or only material of a specific type but more accessions than are included in the core collection. To meet such requests, the curator has the difficult task to either select a 'core from the core' or to expand all or part of the core collection. To answer these requests, and to allow a much more flexible use of the core collection concept, a system has been devised which is able to generate a representative selection meeting the specific demands. The number of entries no longer has to be fixed and the relative importance of specific parts of the selection can be adjusted to the needs of the user. This system is called the 'core selector' (Van Hintum, 1999). It is a relatively simple system based on a formalization of the normal procedures to create a core collection. It uses information on the structures of the collection using 'path indicators' (Boukema *et al.,* 1997), default importance of specific groups, and finally, a priority for selection of accessions within groups. With this system a request such as '50 accessions representing western European cultivated lettuce, 40 accessions representing the rest of cultivated lettuce material, and 10 accessions representing the wild primary Gene Pool' can be met.

The 'bulk core collection': The increased user-orientation of Gene Banks might urge to make the material available more attractive to the user. One of the options to do this could be the 'bulk core collection', which, and this needs to be stressed, would only be used to serve some users, and not as a tool to conserve overall genetic diversity. The 'bulk core collection' defined as 'a germplasm collection created by bulking accessions in genetically distinct groups', could be used for genetic enhancement. Whereas in a traditional core collection the entries are selected from a group of similar accessions to represent this group, in a bulk core collection these accessions are bulked to form a single entry. These bulked entries might be attractive as a starting point for mass screening for resistance or tolerance to biotic stresses. It would also be feasible to use them in a population breeding programme, exposing them to natural selection or mass selection, or introducing adapted germplasm to the bulk core accessions.

References

Bisht, I.S., Mahajan, R.K. and Rana, R.S. 1995 Genetic diversity in south Asian okra (*Abelmoschus esculentus* (L) Moench) germplasm collection. *Annals of Applied Biology*, **126:**539-550.

Bisht, I.S., Mahajan, R.K., Loknathan, T.R. and Agrawal, R.C. 1998. Diversity in Indian sesame collection and stratification of germplasm accessions in different diversity groups. *Genetic Resources and Crop Evolution*, **45:** 325-335.

Bisht, I.S., Mahajan, R.K. and Kawalkar, T.G. 1998a. Genetic diversity in greengram (*Vigna radiata* L. Wilczek) and its use in crop improvement. *Annals of Applied Biology*, **132:** 301-312.

Bisht, I.S., Mahajan, R.K. and Patel, D.P. 1998b. The use of characterisation data to establish the Indian mungbean core collection and assessment of genetic diversity. *Genetic Resources and Crop Evolution*, **45:** 127-133.

Bisht, I.S., Mahajan, R.K. Loknathan, T.R., Gautam, P.L., Mathur, P.N. and Hodgkin, T. 1999. Assessment of genetic diversity, stratification of germplasm accessions in diversity groups and sampling strategies for establishment of core collection of Indian sesame (*Sesamum indicum* L.) collection. *FAO/IPGRI Plant Genetic Resources Newsletter*, **119 (Supp.):** 34-45.

Boukema, I.W. and Van Hintum, Th. J.L. 1994. *Brassica oleracea*, a case of an integral approach to genetic resources conservation. Pp. 123-129 In: Evaluation and Exploitation of Genetic Resources, Pre-Breeding (Balfourier, F. and Perretant, M.R., eds). *Proc. Genetic Resources Section Meeting of Eucarpia, Clermont-Ferrand.*

Boukema, I.W., Van Hintum, Th.J.L. and Astley, D. 1997. The creation and composition of the *Brassica oleracea* core collection. *Plant Genetic Resources Newsletter*, 111:29-32.

Brown, A.H.D. 1989a. The case for core collections. *In*: The use of plant genetic resources (eds. Brown, A.H.D., Frankel, O.H., Marshal, R.D. and Williams, J.T.). Cambridge, UK: Cambridge University Press.

Brown, A.H.D. 1989b. Core collections: A practical approach to genetic resources management. *Genome*, **31:** 818-824.

Brown, A.H.D., Grace, J.P. and Speer, S.S. 1987. Designation of a 'core' collection of perennial *Glycine. Soybean Genetics Newsletter*, **14:** 59-70.

Brown, A.H.D. and Spillane, C. (1999) Implementing core collections-principles, procedures, progress, problems and promise. *In*: Core collections for today and tomorrow. (eds. Johnson, R.C. and Hodgkin, T.). International Plant Genetic Resources Institute, Rome, Italy. Pp. 1-9

Charmet, G. and Balfourier, F. 1995. The use of geostatistics for sampling a core collection of perennial ryegrass populations. *Genetic Resources and Crop Evolution*, **42:** 303-309.

Crossa, J., Lacy, I.D.De. and Taba, S. 1995. The use of multivariate methods in developing a core collection. *In*: Core collections of plant genetic resources (eds. Hodgkin, T., Brown, A.H.D., Van Hintum, Th.J.L. and Morales, E.A.V.). John Willey and Sons, Chichester, UK.

Diwan, N., Bauchan, G.R. and Mc Intosch, M.S. 1994 A core collection for the United States annual Medicago germplasm collection. *Crop Science*, **34:** 279-285.

Erskine, W. and Muehlbauer, F.J. 1991. Allozyme and morphological variability, outcrossing rate and core collection formation in lentil germplasm. *Theoretical & Applied Genetics*, **83:** 119-125.

Frankel, O.H. 1984. Genetic perspectives of germplasm conservation. *In*: Genetic manipulation: Impact on man and society (eds. Arber, W., Llimensee, K., Peacock, W.T. and Starlinger, P.). Cambridge University Press, Cambridge, UK.

Galwey, N.W. 1995. Verifying and validating the representativeness of a core collection. *In*: Core collections of plant genetic resources (eds. Hodgkin, T., Brown, A.H.D., Van Hintum, Th.J.L. and Morales, E.A.V.). John Willey and Sons, Chichester, UK.

Hodgkin, T. 1994 Core collections and conservation of genetic resources.. *In*: Sesame Biodiversity in Asia-Conservation, Evaluation and Improvement (eds. Arora, R.K. and Riley, K.W.). International Plant Genetic Resources Institute, Office for South Asia, New Delhi. Pp. 41-51

Hodgkin, T., Brown, A.H.D., Van Hintum, Th. J.L. and Morales, E.A.V. 1995. Core collections of plant genetic resources, John Wiley and Sons, Chichester, UK.

Hodgkin, T., Qingyuan, G., Xiurong, Z., Yingzhong, Z., Xiangyun, F., Gautam, P.L., Mahajan, R.K., Bisht, I.S., Loknathan, T.R., Mathur, P.N. and Mingde Z. 1999. Developing sesame core collections in China and India. *In*: Core Collections for Today and Tomorrow (eds. Johnson, R.C. and Hodgkin, T.). International Plant Genetic Resources Institute, Rome, Italy. Pp. 74-81

Hollbrook, C.C., Anderson, W.F. and Pittman, R.N. 1993. Selection of a core collection from the U.S. germplasm collection of peanut. *Crop Science*, **33:** 859-861.

Knupffer, H. and Van Hintum, Th. J.L. 1995. The barley core collection: an international effort. *In*: Core Collection of Plant Genetic Resources (eds. Hodgkin, T., Brown, A.H.D., Van Hintum, Th.J.L. and Morales, E.A.V.). John Willey and Sons, Chichester, UK. Pp. 171-178

Mackay, M.C. 1995. One core collections or many? *In*: Core Collections of Plant Genetic Resources (eds. Hodgkin, T., A.H.D. Brown, Th. J.L. Van Hintum and E.A.V. Morales). John Wiley and Sons, Chichester, UK.

Mahajan, R.K., Bisht, I.S., Agrawal, R.C. and Rana R.S. 1995. Studies on South Asian okra collection: Methodology for establishing a representative core set using characterisation data. *Genetic Resources and Crop Evolution*, **43:** 249-255.

Matthews, P. and Ambrose, M.J. 1994. Development and use of a 'core' collection for the John Innes Pisum collection. In: Evaluation and Exploitation of Genetic Resources, Pre-breeding (eds. Balfourier, F. and Perretant, M.R.). *Proc. Genetic Resources Section Meeting of Eucarpia*, Clermont-Ferrand. Pp. 99-107

Peeters, J.P. and Martenelli, J.A. 1989. Hierarchical cluster analysis as a tool to manage variation in germplasm collections. *Theoretical and Applied Genetics*, **78:** 42-48.

Perry, M.C., Mc Intosch, M.S. and Stoner, A.K. 1991. Geographical patterns of variation in the USDA soybean germplasm collection. II. Allozyme frequencies. *Crop Science*, **31:** 1356-1360.

Ortiz, R., Ruiz-Tapia, E.N. and Mujica-Sanchez, A. 1998. Sampling strategy for a core collection of Peruvian quinoa germplasm. *Theoretical and Applied Genetics*, **96:** 475-483.

Radovic, G. and Jelovac, D. 1994. The possible approach in maize 'core collection' development. *In*: Evaluation and Exploitation of Genetic Resources, Pre-breeding (eds. Balfourier, F. and Perretant, M.R.). *Proc. Genetic Resources Section Meeting of Eucarpia*, Clermont-Ferrand. Pp. 109-115.

Schoen, J. and Brown, A.H.D. 1995. Maximising genetic diversity in core collections of wild relatives of crop species. *In*: Core collections of plant genetic resources (eds. Hodgkin, T., Brown, A.H.D., Van Hintum, Th. J.L. and Morales, E.A.V.). John Willey and Sons, Chichester, UK.

Spagnoletti-Zeuli, P.L. and Qualset, C.O. 1987. Geographical diversity for quantitative spike characters in a world collection of durum wheat. *Crop Science*, **27:** 235-241.

Spagnoletti-Zeuli, P.L. and Qualset, C.O.1993. Evaluation of five strategies for obtaining a core subset from a large genetic resource collection of durum wheat. *Theoretical and Applied Genetics*, **87:** 295-304.

Van Hintum, Th.J.L. 1994a. Hierarchical approaches to the analysis of genetic diversity in crop plants. *In*: Core Collections of Plant Genetic Resources (eds. Hodgkin T., Brown, A.H.D., Van Hintum Th.J.L. and Morales, E.A.V.). John Wiley and Sons, Chichester, UK. Pp. 23-24

Van Hintum, Th.J.L. 1994b. Comparison of marker systems and construction of a core collection in a pedigree of European spring barley. *Theoretical and Applied Genetics*, **89:** 991-997.

Van Hintum, Th. J.L. 1995. Hierarchical approaches to the analysis of genetic diversity in crop plants. *In*: Core Collections of Plant Genetic Resources (eds. Hodgkin T., Brown, A.H.D., Van Hintum Th.J.L. and Morales, E.A.V.). John Wiley and Sons, Chichester, UK.

Van Hintum, Th. J.L. 1999. The Core Selector, a system to generate representative selections of germplasm accessions. *Plant Genetic Resources Newsletter*, **118:** 64-67.

Van Hintum, Th.J.L. and Haalman, D. 1994. Pedigree analysis for composing a core collection of modern cultivars with emphasis on barley (*Hordeum vulgare* L.). *Theoretical and Applied Genetics*, **88:** 70-74.

Yonezawa, K, Nomura, T. and Moroshima, H. 1995. Sampling strategies for use in stratified germplasm collection. *In*: Core Collections of Plant Genetic Resources (eds. Hodgkin, T. Brown, A.H.D., Van Hintum, Th.J.L. and Morales, E.A.V.). John Wiley and Sons, Chichester, UK.

❑❑❑

Chapter – 4

Plant Genetic Resources Management in Vegetable Crops

Hari Har Ram and D.K. Singh

Introduction

Plant Genetic Resources (PGR) are uniquely placed within the overall ambit of biodiversity and play an important role in providing food, fuel, clothing, medicine and shelter for the whole mankind. The value of plant genetic resources or the plant germplasm can be visualized from the fact that Sir Joseph Banks, the Director of Kew Botanical Gardens accompanied Captain Cook on a plant collecting voyage. Plant genetic resources are going to occupy the centre stage in plant breeding and seed industry development in India as will be the case with other countries. However, despite the great importance of plant genetic resources in crop improvement programmes, there is ever increasing threat to the PGR due to degradation of their habitats, changes in ecology, cropping systems, modernization of agriculture, rapid replacement of locally adapted indigenous cultivars by modern high yielding varieties and the effect of massive urbanization. This calls for a sustained and scientific management of plant genetic resources. The issue of PGR management remains basically the same whether we are dealing with field crops, vegetable crops or perennials with slight changes in specific cases. For an effective management of PGR, all the CGIAR funded international agricultural research centres have independent units/divisions on PGR management. IPGRI is coordinating and strengthening the PGR management programmes at the international level. In India, National Bureau of Plant Genetic Resources, New Delhi is the nodal organization on PGR management. This article shall deal with the core components of PGR management in vegetable crops.

Plant Germplasm

Plant germplasm has been defined as the sum total of the hereditary materials in species (Allard, 1960). The word germplasm in context of plants is now referred to as plant genetic resource (PGR) which is a component of the world's biological diversity. Plant germplasm is recognized as an essential resource for successful plant breeding. Plant biodiversity has been used as a foundation for modern plant breeding.

Sources of Plant Germplasm

According to Stoskopf (1993), the common germplasm sources for the plant breeders are:

- Currently grown commercial cultivars
- Obsolete commercial cultivars
- Undomesticated forms of crop plants
- Plant introductions
- Breeding lines and stocks
- Related species
- Mutations
- Inter-specific and inter-generic hybridization

PGR Management in India

The National Bureau of Plant Genetic Resources (NBPGR) established by the Indian Council of Agricultural Research in 1976 with its main campus in New Delhi is the nodal organization with the national mandate to plan, conduct, promote and coordinate all activities concerning plant exploration, collection and also for safe conservation and distribution of both indigenous and introduced variability in crop plants and their wild relatives. The Bureau is also vested with the authority to issue Import Permit (IP) and Phytosanitary Certificate and conduct quarantine checks on all seed materials and plant propagules including transgenic material introduced from abroad or exported for research purpose.

NBPGR has a network of 11 regional stations/base centers representing varied agroecological situations in the country.The regional stations/base centers are located at Shimla, Bhowali, Shillong, Jodhpur, Akola, Hyderabad, Amrawati, Thrissur, Srinagar, Ranchi and Cuttack. NBPGR also has a chain of National Active Germplasm Sites (NAGS) which are responsible for maintenance of active collections. Indian institute of Vegetable Research, Varanasi is the NAGS for the vegetables. National Research Center for onion and garlic, Rajguru Nagar, Pune is the NAGS for onion and garlic. CTCRI, Thiruananthapuram is the recognized NAGS for tuber crops.

Some of the regional stations dealing with vegetable germplasm are as follows:

Akola	:	Winged bean
Bhowali	:	Chilli and French-bean
Hyderabad	:	Chilli, tomato, and brinjal
Shillong	:	Ginger, colocasia, turmeric, chilli
Shimla	:	French-bean, amaranth, chenopod, Meetha Karela
Thrissur	:	Vigna, okra

National Gene Bank

The Indian national gene bank has been established by the ICAR at the NBPGR to conserve national heritage of germplasm collections in the form of seeds, vegetative propagules, tissue/cell culture, embryos, gametes, etc. Four modules (two units of 100 cubic meters each and two units of 176 cubic meters each) have been installed for long term storage of seeds of orthodox species in laminated aluminium foils at – 20 degree C after drying them to 5-7 % moisture content. Vegetatively propagated clonal materials and recalcitrant seed species are maintained under field conditions backed by tissue culture repositories. The new gene bank facility commissioned in 1997 has 13 modules each with a storage capacity of 76,000 samples. One of these modules is used for medium term storage of active germplasm collections and the rest for base collections for long term storage. Its cryo-preservation facility has six liquid nitrogen tanks each containing 1,000 litres of liquid nitrogen. The new gene bank is one of most modern gene banks in the world has a storage capacity of 1.25 million samples. The germplasm holdings of the national gene bank are more than 2 lac accessions of which nearly 20,000 are of different vegetables.

Factors Endangering Genetic Diversity of Cultivated Plants

Dahl and Nanbhan (1992) discussed the issues involved and suggested the factors responsible for genetic erosion of plant genetic resources as given below:

- Introduction of modern varieties and exotic crops
- Loss of seed saving and vegetative propagation skill
- Loss of traditional care takers
- Change in economic base
- Land conversion to industrial agriculture
- Destruction of habitat and farmland
- Herbicide and pesticide impact
- Environmental contamination

- Introduction of exotic pests
- Loss of seeds
- Net reduction in number of farmers
- Inadvertent crossing of varieties

Indian Subcontinent as Gene Center

Indian is one of the 12 centres/regions of diversity of crop plants in the world. As it is, the diverse agro-eco climate of the Hindustan Center, apart from possessing about 166 species of agri-horticultural crop plants, also has rich diversity in wild relatives of crop plants numbering 320 species. In addition, the Indian agriculture has been enriched by a large number of introductions of new crops and their cultivars by man since the ancient times. Prominent vegetable crops having Indian continent as origin are as follows:

Eggplant, ridge gourd, smooth gourd, round gourd or round melon, pointed gourd, colocasia, yams, *Amorphophallus*, kundru and cucumber. Arora (1988) has listed 6 regions in India which exhibit richness in crop plant diversity. These are as follows:

1. Western Himalayas including cold arid tract
2. North eastern region and the eastern Himalayas
3. Eastern peninsular region
4. Western arid/semi-arid region
5. Central tribal region
6. Western peninsular region

The vegetable crops and areas where rich diversity in landraces and primitive cultivars occurs in Indian sub-continent are summarized as follows (Ram and Yadava, 2007).

Eggplant	:	Eastern and western peninsular region, north eastern region
Okra	:	Gangetic and Western arid plains, North-eastern region
Pointed gourd	:	Gangetic plains and North-eastern region
Round gourd	:	Parts of Haryana, Gangetic plains, regions around Aligarh
Bitter gourd	:	Gangetic plains, Central highlands and North-eastern region
Ridge gourd	:	Gangetic plains
Smooth gourd	:	Gangetic plains
Snake gourd	:	Eastern and Western peninsular region, North-eastern region
Cucumber	:	Spradic diversity available all through including the Himalayan region
Leafy types	:	Gangetic plains and North-eastern region

Plant Genetic Resource Conservation

Long term storage

The long term storage conditions of –18° C and 30 ± 3° C are provided mainly for comprehensive base collection. The base collections of germplasm are stored at low temperature for long term secure conservation. From this storage seeds are removed exceptionally only for regeneration when the viability has declined below an acceptable standard or to provide material for an active collection for regeneration. These seeds could also be removed when stocks of accession are no longer available from an active collection. It is thus very clear that such materials are not used for exchange. They are only used for regenerating fresh seed and periodically examined for seed viability level.

Medium term storage

The medium term storage is operated on about 4° C and 40 ±3 % RH. This collection also known as active collection is used for regeneration, multiplication, characterization, evaluation and distribution. This seed will also supplement the base collection in long term storage.

Short term storage

Short term storage means storage at 10 ± 2° C and 40±3 % RH. This collection is known as active collection. This collection is also known as breeder's collection or researcher's collection.

Passport data

Passport data are information about a germplasm sample and the collecting site recorded at the time of collecting or plant germplasm exploration. The passport data sheet as recommended by NBPGR, New Delhi is as follows:

Date..............................Collector's No..Accession No.

Species name..Common name.............................

Cultivar/vernacular name....................................Region explored........................

Village/Block..........................District..............................State.........................

Latitude........Longitude.........Altitude............Temperature.........Rainfall...........

Source	:	Natural wild/ distributed wild/ farmer's field/ threshing yard/ fallow/ farm store/ market/ garden/ institute
Status	:	Wild/weed/landrace/primitive cultivar/breeder's line
Frequency	:	Abundant/frequent/occasional/rare
Material	:	Seeds/fruits/inflorescence/roots/tubers
Sample type	:	Population/pure line/individual plant

Sampling method	:	Bulk/random/selective
Habitat	:	Cultivated/distributed/partly distributed
Disease symptoms	:	Susceptible/ mildly susceptible/ tolerant/ resistant/ immune insect pest/ nematode infection: Mild/moderate/high
Cultural practices	:	Irrigated/ rainfed/ arid/ wet
Season	:	Kharif/ rabi/ spring-summer
Associated crop	:	Sole/ mixed with
Soil colour	:	Black/ yellow/ red/ brown/ other
Soil texture	:	Sandy/ sandy loam/ loam/ silt loam/ clay/ silt
Stoniness	:	Stony/ pulverized
Land aspect	:	Level/ crest/ escarpment/ rounded summit/ upper summit
Slope	:	Terrace/ lower slope/ open depression/ closed depression
Topography	:	Swamp/ plain/ level/ undulating/ hillydissected/ steeply/ dissected/ mountainous / valley
Agronomic scope	:	Very poor/ poor/ average/ good/ very good
Ethnobotanical uses:		
Parts	:	Stem/ leaf/ root/ fruit/ flower/ whole plant/ seed
Kind	:	Food/medicine/fibre/timber/fodder/fuel/ insecticide/pesticide/ other
Informant	:	Local vaidya/house-wife/old-folk/shepherd/other

Farmer/donor name.....................................Ethnic group.............................

Plant characteristics:..

Characterization/Plant Descriptors

Plant descriptors are characterization data or observations collected on traits that are highly heritable, can be seen with eye and are fairly consistently expressed in all environments. Examples of such characters which for some crops are included in the descriptor list published by International Plant Genetic Resources Institute are flower colour, pod size, leaf shape, time to flowering, etc. For characterization related traits in specific vegetable crops, the book by Ram (2005) may be referred.

Plant Quarantine

Plant quarantine is a government endeavor enforced through legislative measures to regulate the introduction of planting materials, plant product, soil pathogens, and weeds harmful for agriculture of a country/state/region, and if introduced, to prevent their establishment and further spread. Various methods of pest disease control are exclusion, eradication, protection therapy, resistance and biological control. Exclusion or keeping out is the fundamental to the concept of plant quarantine while eradication methods are employed to eliminate a newly established pest/pathogen. Plant quarantine, may therefore be defined as –rules and regulations promulgated by governments to regulate the introduction of plants, planting materials, plant products, soil living organisms, etc. with a view to prevent inadvertent introduction of exotic pests, weeds and pathogens harmful to the agriculture or the environment of a country/region, and if introduced, to prevent their establishment and further spread.

Main features of the existing plant quarantine regulations in India are as follows:

1. No consignment of seeds/planting materials shall be imported into India without a valid Import Permit which is to be issued by a competent authority to be notified by the central government from time to time in the official gazette. Presently NBPGR has been given this authority for seeds for research purpose.
2. No consignment of seeds/planting materials shall be imported into India unless accompanied by a Phytosanitary Certificate issued by the official Plant Quarantine Service of the source country.
3. All consignments of plants and seeds for sowing/propagation/planting purposes shall be imported into India land custom stations, seaport, airport at Amritsar, Bombay, Calcutta, Delhi and Madras and as such other entry points as may be specifically notified by the central government from time to time, where these shall be inspected and if necessary fumigated/disinfected/disinfested by the authorized plant quarantine official before quarantine clearance.
4. Seeds/planting materials for isolation growing under detention shall be grown in post-entry quarantine facility approved and certified by the Designated Inspection Authority (DIA) to conform to the conditions laid down by Plant Protection Advisor to the Government of India.
5. Hay/straw/or any other material of plant origin shall not be used as packing material.
6. Import of soil/earth/sand/compost/plant debris accompanying seeds/planting materials shall not be permitted. However, soil can be imported for research purposes under a special permit issued by the Plant Protection Advisor to the Government of India.
7. No transgenic material is permitted for experimentation in open environment without prior authorization from the Government of India.
8. Supplier of the transgenic material shall certify that the transgenic has the genes as has been described in the permission.
9. The transgenic supplier shall also certify that the transgenic material does not contain terminator gene.

International Treaty on Plant Genetic Resources for Food and Agriculture

The FAO Treaty on Plant Genetic Resources for Food and Agriculture came into force on 29 June, 2004. This legally binding instrument provides framework for national, regional and international efforts to conserve and sustainably use plant genetic resources for food and agriculture and sharing the benefits equitably in line with Convention on Biological Diversity (CBD). It recognizes the enormous contributions made by farmers in all regions of the world towards the conservation and development of PGR and identifies ways of protecting and promoting Farmers' Rights. It also establishes a multilateral system of access and benefit sharing.

Biological Diversity Act-2002 Enacted by Govt. of India

The biological diversity act-2002 enacted by Govt. of India came into force on 5/2/2003. According to the provisions of this act, the following persons shall be required to have the previous approval of National Biodiversity Authority (headquartered in Chennai) for obtaining any biological resource occurring in India or knowledge associated therefore research or commercial utilization or for bio-survey and bio-utilization.

(i) A person who is not citizen of India
(ii) A citizen of India who is non-resident
(iii) A body corporate/association/organization not incorporated or registered in India

The results of research are not to be transferred to certain persons without the approval of NBA. However, transfer does not include publication of research papers or dissemination of knowledge in any seminar or workshop. Applications for IPR are not to be made without approval of NBA. No person who is a citizen of India or a body corporate/organization/association which is registered in India shall obtain any biological resource for commercial utilization or bio-survey and bio-utilization except after giving prior intimation to State Biodiversity Board concerned. However, this provision shall not apply to the local people and communities of the area including growers and cultivators of biodiversity and *vaids* and *hakims* who have been practicing indigenous medicines. For details Ram and Yadava (2007) may be referred.

References

Allard, R. W. 1960. Principles of Plant Breeding. John Wiley and Sons. Inc., New York and London. pp 485.

Arora, R. K. 1988. The Indian gene centre-Priorities and prospects for collection. *In:* Plant Genetic Resources: Indian Perspective. (eds.) R. S. Paroda, R. K. Arora and K. P. S. Chandel. NBPGR, New Delhi. pp 66-75.

Dahl, K. and Nanbhan, G. P. 1992. Conservation of Plant Genetic Resources: Grassroots Efforts in North America. ACTS Press, Nairobi.

Ram, Hari Har 2005. Vegetable Breeding: Principles and Practices. Kalyani Publishers, Ludhiana. pp 514.

Ram, Hari Har and Yadava, Rakesh 2007. Genetic Resources and Seed Enterprises: Management and Policies. New India Publishing Agency, New Delhi. pp 63-71.

Stoskopf, N. C. 1993. Plant Breeding: Theory and Practice. West View Press, Inc. pp 531.

Chapter – 5

Collection, Characterization and Conservation of Indigenous Germplasm of Vegetable Crops

S.K. Mishra, Ashok Kumar, S.K. Yadav and K.K. Gangopadhyay

Introduction

Plant genetic resources represent a huge reservoir of genes of economic importance, which acts as buffer against environment fluctuations. These genes are largely spread in the land races and traditional cultivars, which have been conserved by virtue of their specific merits. The human and natural selection interventions have resulted in establishment, improvement and differentiation of numerous types, which represent genetic wealth of the crop species and are commonly referred as plant genetic resources (PGR).

Genetic Erosion and Need for Conservation

In view of the availability of high yielding varieties/hybrids, the traditional cultivars/ land races have been disregarded by the farmers due to their low economic potential. Sustaining the green revolution, changing cultivation practices and intensification of land use pattern have been the potential threats for rapid erosion of plant genetic diversity. The erosion of these resources has resulted into the extinction of several valuable genetic materials. Realizing the consequences of genetic erosion and importance of plant genetic resources, more diversity needs to be conserved for future. Prior to conservation the collection and characterization of germplasm through morphological and molecular markers for establishing the identity of genotype is essentially required. For effective utilization of PGR in vegetable improvement programme, the proper characterization and evaluation of genotypes for breeder driven traits including biotic and abiotic stresses in target environments is urgently needed. In addition, value addition is an important area of concern especially in vegetable crops. After implementation of recommendations of Convention on Biodiversity (CBD), the major thrust has been given on conservation of bio-resources. In view of the globalization and emerging IPR regimes (UPOV, TRIPS, ITPGFRA etc.), there is a need for having suitable mechanisms for the protection of bio-resources.

Management of Vegetable Genetic Resources

Global diversity of vegetable crops constitutes about 400 species, which are distributed in eight eco-geographic regions. In the tropical Asian region, both India and China hold maximum diversity (Arora, 1995). Around, 80 species of major/minor vegetables, apart from wild types occur in this region. The Indian gene center is the primary center of origin of several vegetable crops like, brinjal, cucumber, ridge gourd, sponge gourd, smooth gourd, amaranthus, basella, sword bean, winged bean, kundru, dolichos, Indian lettuce and drumstick. In addition, this center is also secondary center of diversity for many vegetable crops like cowpea, okra, chillies, pumpkin, and brassicas. Interestingly, India is also an important centre of diversity for a large number of vegetable crops some of which include wax gourd, bottle gourd, parsley and chow chow.

Indian National Plant Genetic Resources System (INPGRS) is operated by National Bureau of Plant Genetic Resources (NBPGR) as component of National Agricultural Research System (NARS) for systematic management (collection, characterization, conservation, documentation) and utilization of plant genetic resources. The collaboration of 57 National Active Germplasm Sites (NAGS) for evaluation, maintenance and providing germplasm to the users, is also a vital component of INPGRS. The activities on germplasm evaluation, maintenance and utilization are shared between National Bureau of Plant Genetic Resources and Indian Institute of Vegetable Research (IIVR) as designated as National Active Sites (NAGS) for vegetable crops.

In addition, Bioversity International (earlier IPGRI) has active collaboration for organizing joint exploration and collection, and also assisting in developing programme relevant to management and conservation of plant genetic resources in this region. Considering the importance of distribution of diversity and assessing the facilities available, the IPGRI (now Bioversity International) has been assigned, the global (eggplant, okra and lablab bean) and Asian (chilies and radish) responsibility for the conservation of vegetable genetic resources.

Exploration and Collection

The Indian centre has rich diversity in vegetable crops (Singh and Pandey, 2005). Broadly, two kinds of explorations, based on the priority of the crops and areas to be surveyed are undertaken as follows:

1. Multi crop specific (region specific)
2. Crop specific/trait-specific (gene pool specific)

A total 48,438 accessions of different vegetable crops have been collected from diverse agro-climatic zones across the country during 1976-2005. (Malik and Srivastava, 2006). A wide range of diversity has been observed in different regions (Table 1).

Table1. Distribution of vegetable crops diversity in different agro-ecological regions of India

Agroecological regions	Geographical Ranges	Crops
Humid Western Himalayan Region	J&K, H.P. and parts of U.P.	Cucurbits, radish, carrot, turnip, peas, cowpea, chillies, brinjal, okra, spinach, fenugreek, amaranth, *Solanum khasianum, S. hirsutum, Sechium edule, Basella rubra*
Humid Bengal / Assam Basin	W.B. and Assam	Cucurbits, radish, cowpea, chillies, brinjal, okra, spinach beet, *Abelmoschus manihot ssp. manihot*, amaranth, *Solanum indicum, S. khasianum, S. surattense , Cucumis sativus var. vikkimensis, Edgeria dargelingensis, Melothria assamica, Momordica cochinchinenssis, Sechium edule, Tuladiantha coordifolia, Basella rubra*
Humid Eastern Himalayan Region and Bay Lands	Arunachal Pradesh, Nagaland, Manipur, Mizoram, Tripura, Meghalaya, Andaman & Nicobar islands	Cucurbits, radish, peas, cowpea, chillies, brinjal, okra, spinach, amaranth, *Abelmoschus manihot ssp. tetraphyllus, Solanum khasianum, S. torvum, S. sisymbrifolium, S. ferox, S. verbasifolium, Cucumis hystrix, Luffa echinata, Sechium edule*
Sub humid Sutlej Ganga Alluvial Plains	Punjab, U.P., Bihar	Cucurbits, radish, peas, brinjal, okra, pinach beet, fenugreek, onion, garlic, *Abelmoschus manihot ssp. Tetraphyllus var. pungens, A. tuberculatus, Solanum indicum, S. khasianum, S. torvum S. surattense, S. hispidum, Cucumis hardwickii, C. trigonus*
Humid Eastern and South Eastern Uplands	East M.P., Orissa, and A.P.	Cucurbits, radish, carrot, cowpea, chillies, brinjal, okra, spinach, amaranth, garlic, *Abelmoschus manihot ssp. manihot, Solanum surattense, S. torvum.*
Arid Western Plains	Haryana, Rajasthan and Gujarat	Cucurbits, cauliflower, radish, carrot, peas, cowpea, chillies, brinjal, okra, spinach beet, fenugreek, onion, garlic, amaranth, *Abelmoschus tuberculatus, A. ficulens, manihot ssp. tetraphyllus , A. tuberculatus, Solanum torvum, S. nigrum, Citrullus colocynthes*

Contd...

Semi arid Lava plateau and Central Highlands	Maharasthra and West M.P.	Cucurbits, cauliflower, radish, carrot, cowpeas, chillies, brinjal, okra, spinach beet, fenugreek, amaranth, onion, *Solanum surattense, S. torvum, S. nigrum, S. khasianum, Cucumis setosus, Luffa acutangula var. acutangula*
Humid to semi arid Western Ghats and Karnataka plateau	Karnataka, Tamilnadu, Kerala, and Lakshadweep Islands	Cucurbits, chillies, brinjal, okra, *Abelmoschus crinitus, A. angulosus, A. ficulens, A. moschatus, A. manihot var. tetraphyllus, Solanum triobatum, S. indicum, S. insanum, S. pubescens, S. surattense, S. torvum, Luffa acutangula var. acutangula, Melothria angulata, Basella rubra*

However, there are several areas which need to be explored for augmentation of un-tapped variability. Some of the potential areas are as follows:

Crop	Area/region to be explored
Cucumber	Indo-Gangetic plains, sub-Himalayan tract, Western Ghats and eastern peninsular tract
Cucumis sp.	J.K., H.P. U.P. Rajasthan, A.P. and Karnataka
Pointed gourd	Bihar, Bengal and Assam
Ivy gourd	Eastern Madhya Pradesh, West Bengal, North Eastern hill region, Uttar Pradesh, Bihar
Bottle gourd	Indo-Gangetic plains, Tarai region and North Eastern plains
Sponge and Ridge gourd	Indo-Gangetic plains, Tarai region and North Eastern plains
Snake gourd	Southern peninsular tract and Kerala
Bitter gourd	Uttar Pradesh, Bihar, Kerala and Karnataka
Pumpkin and Chow	North Eastern hill region
Lab lab bean	Bihar, Orissa, Gujarat, Maharashtra, Goa and Eastern peninsula
Sweet potato	Bihar, Orissa, Uttar Pradesh and West Bengal
Taro	Kerala, Andhra Pradesh, Orissa, Bihar, Uttar Pradesh, West Bengal, North East region, Nilgiri and Anamalai hills
Yam	Andaman and Nicobar Islands, Bihar, North East hill region
Jack fruit	Eastern Uttar Pradesh, Bihar, Orissa and Southern region

Characterization and Evaluation

Plant Genetic Resources (PGR) comprises landraces, traditional varieties, wild, weedy and other related species constituting primary, secondary and tertiary gene pools. Plant Genetic Resources (PGR) is the basic raw material for any crop improvement activity. Indian subcontinent being one of the Vavilovian centres of origin of crop plants possesses a lot of diversity in several important vegetable crops. After the green revolution, India became self sufficient in major food grain crops that met the food requirement of the country. But, the vegetables production was low due to lack of improved varieties for different agro-ecological zones. Vegetables are an important and essential component of balanced diet, which provides carbohydrate, protein, fiber, mineral and essential vitamins in most economic way. The nutritional security can be achieved by making vegetable at everyone's doorstep. In this endeavour proper assessment of the worth of the vegetable crops germplasm for different yield, nutrient, biotic and abiotic stress parameters assumes greater significance.

Characterization essentially means the recording of the observations on highly heritable traits, can be observed and expressed across the environment. Besides morphological characterization, germplasm is also characterized for biochemical (Protein/ Isozyme) and molecular (DNA based) markers.

In vegetable crops, germplasm of brinjal, okra and onion have been characterized biochemically. The evaluation of germplasm is pre-requisite for their effective utilization in vegetable improvement programmes. Most of the desirable traits are polygenic in nature and are influenced by the environments. Therefore, it is better to evaluate the germplasm at multilocations to draw unbiased conclusion about their genetic potential. In vegetable crops, a large number of germplasm have been evaluated for different agro morphological traits and have been documented in the form of catalogues.

In okra, range and pattern of variation have been studied using morpho- agronomic traits and representative core set has also been established for efficient management and effective utilization of assembled germplasm (Bisht *et al.*, 1997) For detailed evaluation of vegetable crops, the following parameters need attention for effective utilization:

(a) Evaluation of vegetable germplasm for biotic and abiotic stresses, value addition and shelf life etc.

(b) Characterization of promising germplasm through biochemical and molecular markers.

(c) Documentation of vegetable germplasm using agro morphological and biochemical data

(d) Pre breeding/germplasm enhancement in indigenous vegetables using collection, of primary, secondary and tertiary genepools.

More than 10,000 accessions of vegetable crops germplasm have been characterized and evaluated and a number of promising lines have been identified and utilized in improvement programme by various researchers. The information on characterization and evaluation of some of the vegetable crops germplasm namely Brinjal (1188), Okra (5322), Tomato (2980) and Fenugreek (171) etc. have been published in the form of catalogues by NBPGR (Bisht *et al.* 1993, 1995, Thomas *et al.* 1990a&b). Recently multilocation evaluation of vegetable crops germplasm in tomato, brinjal, okra and cowpea has been initiated in collaboration with AICRP (Vegetable Crops). Based on evaluation data, promising accessions have been identified in different crops (Table 2).

Table 2. Promising Accessions Identified for Various Traits in Different Vegetable Crops

Crop	Economic Trait	Accessions
Cucumber	Extended shelf life	IC203838, IC203839
	Early & determinate	EC398030
	Gynoecious line	EC382739
	High yield	IC203838, EC237658
	Anthracnose resistance	PI197087, Poinsett
	Downy mildew resistance	PI197087
	Powdery mildew resistance	Poinsett, Yomaki, PI179376
	Cucumber Scab resistance	Wisconsin SMR9
	Angular leaf spot resistance	Poinsett,MSU9402,PI169400
	Bacterial wilt resistance	PI200815, PI200818
Musk melon	High yield	EC303044
	Downy mildew resistance	Seminole, Edisto
	Powdery mildew resistance	Edisto, PMR45
Snap melon	High yield	EC303049, IC92305
Water melon	High yield	EC28435, K3228, U8-246
	Anthracnose resistance	Charleston Gray, Crimson Sweet
	Powdery mildew resistance	Arka Manik
Bottle gourd	Summer Type	BDJ628, IC1612, U10-315
	Kharif type	EC176860, IC92904
	High yield	IC29404, IC92429, IC110389
Pumpkin	High yield	IC90512, IC92499, IC92571
Sponge gourd	Early	IC92604, IC92779, IC93445
	High yield	NIC23288, NIC23291, NIC23292
	Long and heavy fruit	EC305688,IC92475,IC201217
	Ridge gourd	Early fruiting NIC20402
	High yielding	IC93399, IC12136, NIC20213
	Cluster Bearing	NIC10222, NIC10232, NIC10288
Bitter gourd	Long fruit duration (>150 days)	IC256226
Tomato	Early	EC99935, EC362940, EC362947
	High TSS (>6^{0}B)	EC162519, EC113820, EC103460
	Pear shaped	EC35332, EC367857-1
	Processing type	Bulgarian, EC246028
	Tolerance to high temperature	EC251674, HS-101
	High yield (> 5 Kg)	EC35346, EC129154, EC13903
Brinjal	High yield (> 3Kg /plant)	IC89870, IC90842 , IC111068
	Early flowering (<60 days)	IC89953, IC99676, IC137770
	Dwarf (<40 cm)	IC89964, IC112588, IC112692
	Cluster bearing	EC304951, IC99747, NIC5904
	Long fruit (>35 cm)	IC90102, IC90157, IC11384

Contd...

	Large round fruit	EC305096, NIC6393
	Phomopsis blight resistance	IC90013, IC128680, IC136326
Chilli	High fruit yield	IC119556, IC147637, IC147719
	Long fruit (>10cm)	IC136061, IC136252, IC136266
	Drought tolerance	NIC23797, NIC23837
	PBNV/Mosaic resistance	IC119611, EC121490
Okra	YVMV resistance	EC305616, IC13999, IC43181
	Long fruit	EC89895, EC90075, EC90077
	High yield	IC116970, IC116983, EC169341
	Pod borer resistance	IC27821, IC27831, EC169401
Onion	Attractive bulb	IC34540, IC42900,IC48205
	High TSS (>15^0 B)	IC35549, IC35758, IC42911
	High pungency	IC42911, IC42992, IC49130
	High Tillering (seed crop)	IC48320, IC49207, IC48336
Garlic	Early maturity	NIC81, NIC82, NIC2369
	High bulb yield	IC39382, IC49373
	Compact bulb with bold clove	EC97537, EC172362, EC182916
	Purple Blotch resistance	IC48988, IC3222
Radish	High yield	IC143928, IC143937, EC589790
Chinese cabbage	Early maturity	EC182742, EC182743, EC182754
	Late maturity	EC182751, EC182758
Spinach	Foliage type	NIC14420, EC31894

Recently a catalogue on Chilli germplasm has been published based on evaluation data of 549 accessions (Verma *et al.,* 2007). Detailed information on germplasm characterization, evaluation and utilization for the last hundred years has been compiled by Mishra *et al.* (2006).

Conservation

Conservation of plant genetic resources includes their preservation or protection in particular state for their sustainable use. Both *in situ* and *ex situ* approaches of conservation are used depending on the objectives and priorities. *In situ* conservation, allows for continued adaptation of plants to the environment. The distribution pattern of wild genetic resources in different agroclimatic regions, where rich diversity occurs are of special significance for *in situ* conservation. In vegetable crops, species of interest for *in situ* conservation, include parennial types which do not set seed including wild relatives. The land races of vegetable crops, have specific environment for their proper genotypic expression e.g. root development of radish (in Jaunpur), aroma and quality development in water melon (in Nagour) and muskmelon (in Tonk) need attention.

Ex-situ conservation approach requires maintenance of germplasm outside their original habitats, however, it devoid of natural evolution process, that in always operative in nature. The *ex-situ* conservation includes the following:

Field gene bank

The plant germplasm which is difficult to manage through seeds and may be perennial in nature can be maintained through field gene banks. In vegetable, the germplasm of bulbous and rhizomatous crops are maintained. In addition, the perennial germplasm of under exploited vegetables like jackfruit and drumstick, were also maintained in field gene banks in different agro ecological regions.

Seed gene bank

There is global emphasis on conservation of germplasm (orthodox seed) in gene banks. Two types of collections are being maintained in gene bank as active collections, germplasm which is conserved under medium term conditions (4°C). From active collections, seed samples are drawn for evaluation, multiplication and distribution. The base collections are conserved under -20°C for long term storage. The germplasm is monitored periodically for viability. Such collections are not available to users/curators, frequently. The present status of germplasm of vegetable crops conserved in the National Gene Bank for long term storage (LTS) and also the same in medium term storage (MTS) have been given in Tables 3 and 4, respectively.

Table 3. Status of vegetable crops germplasm at LTS in the National Gene Bank

Crop	Accessions	Crop	Accessions
Tomato	1609	Bottle gourd	577
Brinjal	3960	Sponge gourd	547
Chilli	2521	Ridge gourd	42
Okra	2292	Bitter gourd	494
Cabbage	67	Kakrol	23
Cauliflower	125	Snake gourd	153
Broccoli	4	Ash gourd	263
Knolkohl	5	Ivy gourd (Kundru)	15
Chinese cabbage	111	Round melon (Tinda)	104
Kale	5	Watermelon	110
Peas	190	Muskmelon	561
Carrot	101	Snap melon	188
Turnip	13	Cucumber	288
Radish	240	Kachri	82
Onion	747	Pumpkin	139
Spinach	115		

Table 4. Status of vegetable crops germplasm at MTS

Crop	Accessions at NBPGR	Accessions at IIVR
Tomato	925 (HQ) + 304 (Hyd)	1052
Brinjal	1860 (HQ) + 611 (Hyd)	350
Chilli	1188 (Bhow) + 3021 (Hyd)	135
Okra	680 (Akola)	373
Bottle gourd	270 (HQ)	36
Sponge gourd	200 (HQ)	
Ridge gourd	250 (HQ)	
Bitter gourd	55 (HQ)	176
Ash gourd	20 (HQ)	
Pointed gourd	-	250
Ivy gourd (Kundru)	-	60
Pumpkin	6 (HQ)	-
Watermelon	-	16
Muskmelon	-	45
Cucumber	-	24
Radish	170 (HQ)	-
Spinach	93 (HQ)	-
Peas	2700	153
Cowpea	2950	127
Onion*	68	60
Garlic*	625	60

* At NRCO&G, Onion (437) and Garlic (269)

Cryo bank (Cryo preservation)

The long term conservation of vegetatively propagated species, orthodox and recalcitrant seeds can be maintained through crop preservation. The reproductive materials, pollen, embryo tissue and seed material can be preserved in liquid nitrogen (-196°C) and vapour phase (–150°C). More than 200 accessions of various vegetable crops have been maintained in NBPGR.

In vitro bank (In vitro conservation)

In vitro technology is quite important for the maintenance of vegetatively propagated species and may be an alternative to field gene bank. Through tissue culture technology, germplasm is maintained in disease free conditions. In vegetable crops, the germplasm of rhizomatous crops, bulbous crop (*Garlic*) and wild Alliums (*Allium* sp) are being conserved.

Under-exploited vegetable crops

The native plants are used as traditional vegetables in different parts of the country and these need collection and further improvement. Some of such materials are as under:

Common name	Botanical Name	Areas of diversity
Jack fruit	*Artocarpus heterophyllus*	Eastern Uttar Pradesh, Bihar Orissa and Southern Parts
Drum Stick	*Moringa oleifera*	Madhya Pradesh, Orissa and Bengal
Ivy gourd	*Coccinia indica*	Madhya Pradesh, Orissa and Bengal
Round gourd	*Citrullus vulgaris var. fistulosus*	North Western plains, Indo-Gangetic plains, Bihar, etc.
Snake gourd	*Trichosanthes spp*	Eastern Madhya Pradesh and Western ghats
Kankoda	*Momordica dioica*	North Maharashtra, Madhya Pradesh, Bihar, Orissa
Parslane	*Portulaca oleracea*	Himalayan region
Poly	*Basella rubra*	North Eastern region, Orissa, Bengal
Chaulai	*Amaranthus sp.*	Himalayan region, North Eastern Region, Eastern and Southern region
Chenopod	*Chenopodium sp.*	Sub tropical and tropical region

Landraces and wild relatives

In ridge gourd, several morphotypes of satputiya which are hermaphrodite in nature and bear fruits in cluster possess breeding/commercial potential. Similarly, Jethua Belgian a land race of eggplant is suitable for off season (summer) and has good potential. Wild relatives of vegetable crops possess resistant genes to various biotic and abiotic stresses. In okra, *Abelmoschus tuberculatus*, a wild relative, has low susceptibility to fruit and stem borer, a major pest of cultivated okra. Similarly, *Cucumis hardwickii* a progenitor of cucumber possess genes for multiple disease resistance and especially to downy mildew. *Allium fistulosum* have resistance to *Stemphyllium blight*. In future, the resistance potential of wild species can be exploited through biotechnological approaches.

Registration of germplasm

Registration of germplasm has become as an important activity of management and conservation of genetic stocks. At present 35 genetic stocks have been registered with the N.B.P.G.R. (Table 5). These materials can be shared with crop breeders on specific request.

Table 5. List of Registered Germplasm Conserved at NBPGR

Crop	Accs.	Character(s)
Tomato	3	Resistance to RKN, TLCV, BW
Brinjal	2	BW resistance, Better ratooning and semi- spreading
Chilli	7	CMS, High pungency, CGMS, Fertility restorer, Early and High yielding
Pea	4	PM resistance, MS line, Yellow wrinkled seed
Sword bean	1	Drought tolerant
Amaranth	2	Thick stem, White rust resistance
Garlic	2	Compact bulb, Big cloves
Bottle gourd	2	Andromonoecious, Segmented leaf type
Bitter gourd	1	Gynoecious line with high yield and attractive fruits,
Round gourd (Tinda)	1	Intermediate, Tolerant to DM and root rot wilt complex
Water melon	3	Drought hardy, Yellow pulp, Unlobed leaf
Snap melon	2	Drought hardy, High yielding
Cucumber	2	Long fruit, Small fruit
Kachri	2	Salad type, pickle type
Pointed gourd (Parwal)	1	Obligate parthenocarpic and seedless fruit

Future Thrusts

India has rich diversity in case of vegetable crops. Although systematic efforts have been made on collection, characterization and conservation of germplasm of vegetable crops, there is still need for collection in un-explored areas to conserve the depleting diversity. Now emphasis should be given on trait-specific collection. Likewise, evaluation is also an important activity of PGR management through which the true genetic potential of the germplasm can be established. Therefore, efforts are needed to characterize and evaluate the existing germplasm on priority basis in a network mode involving multi-disciplinary approach. The wild relatives are undoubtedly carriers of important genes/gene complex, which need to be exploited with suitable tools and techniques. There are several under-utilized vegetable crops, which are used locally. They need to be collected and improved for their sustainable production and consumption. Evaluation of vegetable germplasm for shelf life, quality traits and resistance to biotic and abiotic stresses needs to be given priority. Germplasm enhancement/pre-breeding in indigenous vegetables (okra, brinjal and cucurbits) involving collections from primary, secondary and tertiary gene pools needs to be initiated in a network mode.

References

Arora, R. K. 1995. Ethnobotanical studies in plant genetic resources: National efforts and concern. *Ethnobotany*, 7:125-136.

Bisht, I. S., Patel, D. P. and Mahajan, R. K. 1997. Classification of genetic diversity in *Abelmoschus esculentus* germplasm collection using morphometric data. *Annals of Plant Biology*, **130**: 325-335.

Bisht, I. S., Patel, D. P., Mahajan, R. K., Thomas, T. A. and Rana, R. S. 1995. Catalouge of wild *Abelmoschus* species germplasm. National Bureau of Plant Genetic Resources, New Delhi. p. 51.

Bisht, I. S., Patel, D. P., Sapra, R. L., Thomas, T. A., Kopar, M. N. and Rana, R. S. 1993. Catalogue of okra [*Abelmoschus esculentus* (L.) Moench.] Germplasm Part III. National Bureau of Plant Genetic Resources, New Delhi. p. 77.

Malik, S. S. and Srivastava, Umesh. 2006. Exploration and collection of genetic diversity in crops. *In: Hundred Years of Plant Genetic Resources Management in India* (eds. A. K. Singh, K. Srinivasan, S. Saxena and B. S. Dhillon). National Bureau of Plant Genetic Resouces, New Delhi: pp. 109-132.

Mishra, S. K., Singh, M., Kumar Ashok and Rana, J.C. 2006. Germplasm characterization, evaluation and utilization. *In: Hundred Years of Plant Genetic Resources Management in India* (eds. A. K. Singh, K. Srinivasan, S. Saxena and B. S. Dhillon). National Bureau of Plant Genetic Resouces, New Delhi. pp. 133-170.

Singh, A. K. and Pandey, Chitra. 2005. Vegetable germplasm diversity in Asia and its use in India. *In: Crop Improvement and Production Technologies of Horticultural Crops Vol I* (eds. K. L. Chadha, B. S. Ahluwalia, K. V. Prasad and S. K. Singh). Malhotra Publishing House, New Delhi. pp. 62-81.

Thomas, T. A., Bisht, I. S., Bhalla, Shashi, Sapra, R. L. and Rana, R. S. 1990a. Catalogue on okra [*Abelmoschus esculentus* (L.) Moench.] Germplasm Part I. National Bureau of Plant Genetic Resources, New Delhi. p. 51.

Thomas, T. A., Bisht, I. S., Patel, D. P., Agarwal, R. C. and Rana, R. S. 1990b. Catalogue on okra [*Abelmoschus esculentus* (L.) Moench.] Germplasm Part II. National Bureau of Plant Genetic Resources, New Delhi. p. 100.

Verma, S. K., Mandal, S., Bhandari, D. C., Muneem, K. C. and Mahajan, R. K. 2007. Catalogue on Chilli (*Capsicum* spp.) Germplasm Evaluation. National Bureau of Plant Genetic Resources, New Delhi. p. 134.

❑❑❑

Chapter – 6

In-Situ Conservation Approach for Germplasm of Vegetable Crops

S.K. Verma, K.S. Negi and K.C. Muneem

Introduction

Vegetable crops include a large number of species, mainly used as an essential complement to the daily diet, providing vitamins, minerals, fiber, specific amino-acids and other active metabolites. Largely emphasizing the need to accumulate and conserve in gene banks the genetic diversity that is most useful to breeders. One of the primary reasons to sustain conservation of plant genetic resources in gene banks is to prevent the loss of genetic diversity. The great diversity of types in cultivation is also considered a genetic resource itself, to be maintained or increased, since it adds to the diversity of vegetables being grown and consumed and could serve to replace more established but similar crops in case of need (Crisp and Astley, 1985). The risk of genetic erosion due to the introduction of single new cultivars was considered especially high by Crisp and Astley (1985) for those vegetables that are built on a narrow genetic base, such as garlic, broccoli, tomato, cucumber, etc. Genetic erosion is difficult to document with solid data available in vegetable crops. On a global level, efforts of the International Board for Plant Genetic Resources (IBPGR, now IPGRI) to conserve vegetable crop germplasm began in 1980, with the definition of a number of priority crops for conservation, according to their importance for rural development and to their economic value for farmers in the tropics (*Abelmoschus esculentus* and related species, *Allium* spp., *Amaranthus* spp.,

Brassica spp., *Capsicum* spp., *Cucurbita* spp., *Lycopersicon esculentum*, *Momordica charantia* and related species, and *Solanum melongena*). During last six–seven decades, germplasm of promising and profitable varieties have been built up and developed by different agencies, initially assembling local types and selecting from them and by introducing promising types from other parts of the world. The introduction and acclimatization of vegetable germplasm from within the country and all over the world is important for exploitation and incorporation of useful characters such as early maturity, high yielding capacity, resistance to diseases and pests and tolerance to different climatic hazards, viz., drought, heat, cold and flood situations.

Indian National Scenario of Vegetables

There are about 25-30 vegetable crops which are commercially important and these include both the indigenous and exotic species variability; the introductions being well adapted to local needs and agro-ecology. So far research has confined on 30 vegetables (prominent are tomato, brinjal, chilli, okra, cauliflower, melons, onion, peas etc.). Native tribes use another 521 species as early vegetables /green vegetables. In India there are native vegetables like eggplant (*Solanum melongena*), beans (*Lablab purpureus*), cucumber (*Cucumis sativus*) and a few gourds, namely smooth gourd (*Luffa cylindrica*), ridge gourd (*Luffa acutangula*), snake gourd (*Trichosanthes anguina*) and pointed gourd (*Trichosanthes dioica*), while there are several introduced crops and which have long history of domestication and adaptation like garden pea (*Pisum sativum*), onion (*Allium cepa*), bottle gourd (*Lagenaria siceraria*), watermelon (*Citrullus lanatus*), cowpea (*Vigna unguiculata*), okra (*Abelmoschus esculentus*) etc. and also some others like tomato (*Lycopersicon esculentum*), cauliflower (*Brassica oleracea* var. *botrytis*), cabbage (*Brassica oleracea* var. *capitata*), chilli, (including *Capsicum*), frenchbean (*Phaseolus vulgaris*) etc, which have been introduced during the last 4th or 5th centuries. With such a diversity of crops, characterization and evaluation of genetic resources in each of them become a stupendous responsibility requiring extensive infrastructure facilities and varying methods of evaluation. The National Bureau of Plant Genetic Resources (NBPGR), New Delhi with its Regional Stations at Shimla, Bhowali, Akola, Thrissur, Jodhpur, Hyderabad, Umiam-Barapani, Base Stations at Ranchi and Cuttack and Indian Institute of Vegetable Research (IIVR), Varanasi and Indian Institute of Horticultural Research (IIHR), Bangalore, designated as National Active Germplasm Site (NAGS) for vegetable crops and are holding over 40,000 germplasm collections in different vegetable crops. Diversity zones in India are narrated below:

Hindustani Centre of Origin/Diversity of Cultivated Plants

The Indian gene centre holds a prominent position among the eight Vavilovian Centres of crop plants origin. Out of 2, 40, 000 economic plant species recounted by Zeven and deWet (1982) distributed in 12 mega gene centres of diversity, 160 species are enumerated in the Hindustani centre, distributed in eight diverse phyto-geographical/

agro-ecological regions of India. This diversity exhibits preponderance of variable landrace forms/primitive types belonging to different crops viz., cereals, millets, legumes, vegetables, fruits, forages, fibres, oilseeds, sugar yielding crops, spices, condiments, medicinal and aromatic forms etc. The Indian sub-continent is one of the centers of origin and /or diversity in vegetable crops. Around 80 species of major and minor vegetables, apart from several wild/gathered kinds occur (Choudhury, 1967; Sheshadri, 1987). India is the primary center of diversity for crops such as eggplant, sponge gourd, ridge gourd and cucumber and secondary center for cowpea, okra, chilli, pumpkin and several Brassicae.

Distribution of Diversity

Majority of the tropical vegetable crops are grown throughout the country as suitable agro-climates are available. However, temperate vegetables are grown mainly in the Himalayan ranges and to some extent in higher ranges of the Western and Eastern Ghats, which are also well adapted as "Rabi" crops in northern/northwestern plains and in Tarai foot hills. Important areas of diversity in cultivated crops are as follows:

(i) **The Northwestern and Eastern Himalayan Region:** Temperate crops predominate. Enormous diversity occurs in *Allium* species - leak, shallot and other introductions including *A. sativum* and *A. cepa*. Sporadic diversity also occur in asparagus, spinach, chenopods, amaranth, *Beta vulgaris*, leafy and other vegetables, *Brassicae*, squash, *Cucurbita* spp., cucumber, chilli, bell pepper, pea, faba bean, french bean, cowpea, horse radish, artichoke, potato, colocasia, tomato, parsley, coriander, ginger, *Mentha* spp. and in *Sechium edule* as well as in *Cyclanthera pedata* (meetha karela).

(ii) **The North-Eastern Region:** The above mentioned crops predominate in this region also but more diversity occurs in leafy vegetables, amaranth, *Brassicae* spp., *Malva verticillata* and others; chilli, tomato, eggplant, okra, ginger, taros and yams. Several wild species of cucurbits are growing in this region. Among cultivated ones the diversity occurs in cucumber, pointed gourd, chayote or chow-chow, bitter gourd, *Momordica dioica* and *Cyclanthera pedata*.

(iii) **The Northern Plains including Tarai Region:** This is one of the richest pockets of diversity in vegetable crops. The vegetables can be grown during rainy, winter and summer seasons, more diversity prevails here and it is fitted well into the cropping patterns. Relatively, traditional genetic resources are less and improved cultivars cover vast areas. Rich variability exists in cucurbits, eggplant, tomato, chilli, okra, leafy vegetables, Brassicae, carrot, radish, onion, garlic, tuber and root crops etc.

(iv) **The North-Western Plains:** The diversity occurring in Northern plains extends to North-Western plains though the extent of diversity varies, particularly in cucurbits such as *Cucumis*, *Momordica* and *Citrullus* spp., eggplant, amaranth, *Chenopodium*, okra, chilli and garlic.

(v) **The Central Region/Plateau:** Much of the diversity grown in Northern India occurs in Central India but its nature and extent varies. More diversity is exhibited in *Cucurbita* spp., different kind of squashes, cucumber, ash gourd, round gourd, bitter gourd, pointed gourd and ridge gourd, okra, eggplant, chilli, tomato and in root and bulbous crops and sporadically in leafy vegetables. More indigenous diversity occurs in cucurbits, eggplant, okra and chilli.

(vi) **The Western and Eastern Peninsular Region:** This is an extremely rich region. Here rich diversity occurs in eggplant, okra and particularly in chilli, both annual and perennial types. In cucurbits, apart from cucumber, bitter gourd, bottle gourd and squashes are important. More landrace diversity occurs in snake gourd in the western and for *Luffa* spp., and eggplant in the eastern region. Leafy vegetables, amaranth, Brassicae, Chenopods, spinach and *Beta vulgaris* are grown sporadically. Three to four categories of genetic resources predominate and the promising diversity in vegetable crops represented in each category is given below.

(a) **Introduced germplasm:** Tomato, peas, french bean, scarlet bean, lima bean, winged bean, faba bean, Brassicae (cauliflower, cabbage, knol-khol, turnip, leafy mustard etc.), radish, carrot, *Beta*, capsicum, chayote, potato, *Cyclanthera pedata*, *Allium* spp. (onion, leek, shallot, garlic), asparagus, artichoke and parsley.

(b) **Indigenous germplasm:** Eggplant, sponge gourd, ridge gourd, snake gourd, round gourd, *Cucumis melo*, phunt, kakri, cucumber, leafy vegetables viz., *Basella, Chenopodium, Amaranthus, Brassica*, spinach, rumex/ sorrel; taro, yam, elephant foot yam, ginger, *Allium* spp. and bitter gourd.etc.

(c) **Germplasm of exotic origin for which India is a centre of diversity:** Okra, pumpkin, chilli, cowpea, *Brassicae*, chyote and coriander.

(d) **Under utilized / under exploited vegetables:** In tribal areas of India many minor vegetables are grown in their field as well as in kitchen garden. Such crops are consumed locally. The genetic resources of such minor vegetable crops include over 50 species (Arora, 1995). Some important ones are: *Allium rubellum*, *Amaranthus* spp., *Apium graveolens*, *Asparagus officinalis*, *Basella alba/ rubra*, *Bamboo* spp., *Fagopyrum cymosum*, *Hibiscus sabdariffa*, *Ipomoea aquatica*, *Lactuca sativa*, *L. indica*, *Malva* spp., *Moringa oleifera*, *Ocimum* spp., *Parkia roxburghii*, *Polygonum* spp., *Pilea* spp., *Portulaca oleracea*, *Sesbania grandiflora*, *Solanum torvum*, *Trichosanthes* spp., *Trigonella* spp., *Vigna* spp., *Zizania latifilia*, several aroids and yams.

(e) **Wild relatives and related species of vegetable crops:** The wild relatives and related species of vegetable crops belongs to the categories, legumes–31 spp., vegetables–54 spp., spices and condiments–27 spp.

(Arora and Nayar, 1984) and these are distributed in the western and eastern Himalayas, Northeastern region, Gangetic plains, Indus plains and in the western and eastern peninsular regions. Botanically these can be classified as follows (Arora, 1995):

Distribution of Wild Relatives of Vegetable Crops in Different Parts of India (Area wise)

Area	Wild species of vegetable crops
Western Himalayas	*Abelmoschus manihot (tetraphyllus), Cucumis hardwickii, C. trigonus, Luffa echinata, L. graveolens, Solanum incanum, S. indicum, Trichosanthes multiloba, T. himalensis.*
Eastern Himalayas	*Abelmoschus manihot, Cucumis trigonus, Luffa graveolens.*
North-eastern region	*Abelmoschus manihot (pungens* forms*), Alocasia macrorhiza, Amorphophallus bulbifer, Colocasia esculenta, Cucumis hystrix, C. trigonus, Dioscorea alata, Luffa graveolens, Momordica dioica, M. cochinchinensis, M. macrophylla, M. subangulata, Solanum indicum, Trichosanthes cucumerina, T. dioica, T. dicaelosperma, T. khasiana, T. ovata, T. truncata,.*
Gangetic plains	*Abelmoschus tuberculatus, A. manihot, (tetraphyllus* form*), Luffa echinata, Momordica cymbalaria, M. dioica, M. cochinchinesis, Solanum incanum, S.indicum.*
Indus plains	*Citrullus colocynthis, Cucumis prophetarum, Momordica balsamina,.*
Western peninsular tract	*Abelmoschus angulosus, A. moschatus, A. manihot (pungens* form*), A. ficulneus, Cucumis setosus, C. trigonus, Luffa graveolens, Momordica cachinchinensis, M. subangulata, Solanum indicum, Trichosanthes anamalaiensis, T. bracteata, T. cuspidate, T. perottitiana, T. nerifolia, T. villosa..*
Eastern peninsular tract	*Amorphophallus campanulatus, Abelmoschus manihot, A. moschatus, Colacasia antiquorum, Cucumis hystrix, C. setosus, Luffa acutangula var. amara, L. graveoens, L. umbellate, Momordica cymbalaria, M. denticulate, M. dioica, M. cochinchinesis, M. subangulata, Solanum indicum, S. melongena (insanum* types*), Trichosanthes bracteata, T. cordata, T. lepiniana, T. himalensis, T. multiloba..*

Distribution of Wild Relatives of Vegetable Crops in Different Parts of India (species wise)

Family	Genera
Cruciferae/Brassicaceae	Brassica, Lepidium
Malvaceae	Abelmoschus, Hibiscus, Malva
Leguminosae	Canavalia, Cicer, Dolichos, Moghania, Trigonella, Vigna
Cucurbitaceae	Citrullus, Coccinea, Luffa, Momordica, Neoluffa, Trichosanthes
Umbelliferac	Carum
Compositae	Chicorium
Solanaceae	Solanum
Amaranthaceae	Amaranthus
Chenopodiaceae	Chenopodium
Polygonaceae	Fagopyrum, Polygonum, Rumex
Zingiberaceae	Alpinia, Curcuma, Zingiber
Dioscoreaceae	Dioscorea
Lliliaceae/Amaryllidaceae	Allium
Araceae	Alocasia, Amorphophallus, Colocasia

In situ and *ex situ* Conservation

The status of conservation of vegetable crops germplasm has always received less attention than that of the major staple crops such as cereals, legumes and other horticultural crops. Cost- effective and reliable *ex situ* conservation remains a challenge that can benefit from sharing responsibilities within crop networks. In these cases, the discussion of common problems (long-term storage, safety-duplication, regeneration) can lead to effective collaborative solutions. The agriculture related indigenous knowledge possessed by the elderly people, mostly womenfolk, is facing series of challenges due to variety of factors and out-migration of younger generation to urban areas. The documentation of traditional landrace-based indigenous knowledge system would play significant role in making value addition of the traditional crops and their better management on-farm. *In situ* conservation and crop improvement can complement one another in marginal areas. Breeding programmes that evaluate landraces and use them in local improvement efforts are expected to produce material of direct value for marginal agroclimatic zones as well as achieving significant local conservation (McNeely 1988; Harlan 1992; Brush 1999; Smale and Bellon 1999; Almekinders and Elings 2001). The databases, holding mainly passport data, can be analyzed for the identification of duplications and gaps among collections. At the same time, the management of collections can be based on better knowledge of the diversity in stock. The enhancement of the links between germplasm conservation and use will continue to depend, inter alia, on easy access to the genetic material. Further, locally common alleles are more important for conservation and interesting to users.

The emphasis on *in situ* and *ex situ* approaches will, however, depend on the conservation context- the object, aims and location of conservation (Iwanaga, 1995; Jarvis and Hodgkin, 1998). Iltis (1974) proposed the *in situ* conservation model for the first time. *In situ* conservation is concerned with maintaining species populations in habitats in which they occur and has been reviewed in greater details by several authors (Oldfield and Alcorn 1987; Maxted *et al.* 1997; Brush and Meng 1998). Conservation *in situ*, with local communities and farmers ensures that resources remain directly in the hands of the primary users. *In situ* conservation means preserving the varieties / germplasm in their original agro-ecosystem cultivated by farmers using their own selection methods and criteria. There is however lot of literature available now on various activities involved and generalized models for the conservation of genetic diversity on-farm (Altieri *et al.* 1987; Bellon 1996, 1997; Brush 1989, 1991, 1995, 1999; Worede 1992, 1993, 1997; Worede and Hailu 1993; Eyzaguirre and Iwanaga 1996; Bellon *et al.* 1997; Maxted *et al.* 1997, 2002; Jarvis and Hodgkin 1998; Jarvis *et al.* 2000 and Long *et al.* (2000). *In situ* (on-farm) conservation will be most effective when targeted to specific areas with significant plant genetic resources and with communities who are willing to participate in conservation programmes. Strategies to enhance local participation through marketing and education must be identified.

Planning of *In situ* Conservation – A Methodological Approach

Identification of hot spot: *In situ* conservation involves the knowledge of traditional agriculture and growing zones/areas where genetic diversity of the species could be concentrated. In this methodological approach we have to consider the family/ community as primary level with micro-climatic and micro-geographical region in a different hierarchical level. Maintenance of biodiversity should be done in a holistic context that includes socio-cultural, economic and agro-ecological aspects.

The work places have been determined considering the conservation of wild and old landraces/ species. This is a geographical area with ecological conditions, agricultural systems and cultural patterns that make possible the survibility and use of the biodiversity in this area. We have to focus the actual geographical concentration of cultivated and genetically (varieties, clones, genotypes, etc.) different species. This concentration of variability could have ethnical and cultural origin besides biological evolution. It is an assert that the farmers family is able to identify genetic variations based on phenotypic elements, growing period, soil adaptation, colour, shaped, flavour and quality. One person should also get trained with officials to know the agricultural system phenotypically/ linguistically feasible with the specific area. Time is a variable factor to define. We begin over the assert that the *in situ* conservation work is done with the farmers and take place in areas where the species occurs with *in situ* cultures and we observe it for long periods of time to follow the results. However we should also see the gene pool of the specific crop.

Gene pool classification: It is a proposition of some guidelines for classification or grouping of genetic diversity based on cross-compatibility relationships that can simplify use of genetic diversity. Harlen (1992) developed '*gene pool*' concept by assigning the constituent taxa to primary, secondary and tertiary gene pools. At the intra-specific level, cultivars are grouped into races and sub-races.

Primary gene pool: Here those species/germplasm falls which are easily crossable; hybrids are fertile with good chromosome pairing; normal gene segregation and gene transfer is generally easy. It includes spontaneous races (wild and/or weedy) as well as cultivated races. It can be divided into two subspecies: subspecies A to include the cultivated races and subspecies B to include the spontaneous races.

Secondary gene pool: Includes species which are crossable within the crop species. Gene transfer is possible, but with barriers (poorly or not at all). This gene pool is available for use; however, the plant breeder or geneticist will have to put an extra effort to overcome the cross-ability barriers with application various possible cyto-genetic manipulations to establish a fertile hybrid.

Tertiary gene pool: Refers to wild species that produce hybrids with crop species which are lethal or completely sterile. Gene transfer is either not possible with known techniques, requires embryo culture or grafting to obtain hybrids, doubling chromosome number or using bridging species or biotechnological techniques. It is the outer limits of potential genetic reach. It is rather ill defined.

Survey of hot spot and characterization: We have to make the survey and hot spots could be identified previously therefore to array the data and then analyze it under historical, geographical, agro-ecological, cultural, social and economic points of view. Variables habitat, number and cultivated area in each zone, zone importance, crop and wild species concentration, uses, customs, accessibility and relation between hot spot is the prime factors. The growers constitute the primary source of information data. It is important to strengthen the relations with the growers to get first hand information. The base of this relation is mutual respect and the responsibility to assume commitments based on friendship.

Handling the habitate conservation: Here we should keep constant watch on the system activities. A household develops its activities in different hierarchical levels closely with its productive system. The most important characterization points are (i) Utilization of natural resources with its existing soil and cropping patterns. (ii) Physio-graphical and agro-ecological levels (iii) number of habitants/tribes / community involved and its essentialities (iv) Local technologies used by the habitates (v) Agricultural system followed (vi) Farmers level (small/marginal/large) (vii) Other related limiting and favourable factors.

Participation of the Growers and their families: It is very necessary to have a constant dialogue with the growers and their families to transfer the information data they possess. The people and the local organizations should participate in different steps of the process.

Monitoring the local effectiveness: The household (family), which is the "owner" of the biodiversity and takes determination about how to handle, use and conserve the resources becomes in analysis unit at this level. The community has the highest hierarchy. *In situ* conservation dynamic is based on genetic material movements through the time. There are movements of germplasm produced by the growers from the plot, household, and passing through local fairs to local markets. Exchange, gifts or pay for job occurs between households of the community. There is also exchange of seeds which should be taken into account as percentage of gene mixing. Thus, *in situ* conservation begins with sowing, harvest, sowing that produce conditions to be registered and quantified and monitored time to time. There is flow among all the family (very close relatives). It is interesting to know about preferences to exchange the variety and the frequency. Secondly communities located at a particular region visit different size fairs/ local markets to give or take genetic material to sow in their pieces of land. They are selecting the propagation material or the material to be used as seed for the next production campaign.

Scientific and Institutional challenges

(i) **Production cycle:** Factors to be monitorized are frequency of use, frequency of selling, kind of varieties – mainly landraces / folk varieties / primitive form of crop plants / farmer's selections etc, which give information about the production cycles.

(ii) **Elite material identification:** Landraces / cultivars to be identified and to be collected in farmland to improve *ex – situ* collections.

(iii) **Monitoring the out flow:** Seed fairs local markets, mutual exchange etc. Developing niche markets for traditional crops and promoting agro-processing industries for local produce including the wild edibles. There is need of land consolidation and development of village marketing cooperatives through appropriate policies to avoid exploitation through middlemen to providing maximum returns.

(iv) **Incorporation of basic technology:** Improving seed management of local landraces and access of farmers to crop genetic diversity for increased use by identifying elite germplasm with the potential for use in food industry and multiplication of these seed types for both local and urban consumption. Increased use of local crop resources for value addition based on traditional knowledge and farmer participation in crop breeding.

(v) **Commercialized production:** Production of seed of the commonest cultivars to commercialize- Intervention to a consolidated hot spot to increase productivity, money incomes and well being for the community. Empowerment of women and benefit sharing by the gender which is the main conserver and manager of agro-biodiversity.

(vi) **Inventorisation of crop plants:** Suitable inventory of crops plants, their varieties and other plants may be prepared for maximum utilization.

(vii) **Seed storage:** Development of community seed bank and linking the community conservation with *ex situ* conservation.

(viii) **Other factors:**

(a) **Land tenure issues:** The complexity in land ownership and tenurial rights makes it difficult for survey, demarcation and consolidation of land. Unequal distribution of land resources is responsible for increasing dependence on forests by certain sections of the society leading to diversity degradation.

(b) **Inter-departmental coordination:** Needs a close inter-departmental coordination to the sustainable management of vegetable resources in the region.

(c) **Insufficient basic infrastructure:** Insufficient basic infrastructure of R & D works and shortage of skilled manpower and improper institutional arrangements and limited role played by financial institutions in diversity conservation. Regeneration and cultural practices for many species need to be researched and standardized for their cultivation.

Possibilities in Uttarakhand

The Uttarakhand has its own unique combination of living species, habitats and ecosystems which all together make up it a diversity rich state. In all living organisms, the species is the single most useful unit to use in diversity assessment. Species richness and the relative abundance of different species is another criteria to measure the degree of diversity The number of endemic species also reflects into account while assessing the richness of diversity (Negi *et al*., 2003).

Potentialities of the region: Central Himalayan Region is popularly known as Uttarakhand, located between latitudes 29^0 5'-31^0 25' N and longitude 77^0 45'-81^0 E and covering an area of 51124 Km^2. Uttarakhand being the mountainous region is the core area of global biodiversity. It is located in the center of Western Himalaya comprising of 13 districts of Garhwal and Kumaon regions namely Almora, Bageshwar, Chamoli, Champawat, Dehradun, Haridwar, Nainital, Pauri Garhwal, Pithoragarh, Rudrapryag , Tehri, Udham Singh Nagar, Uttarkashi inhabited by several communities dwelling in remote and inaccessible areas. Great variations in climatic, topographic, socio-economic and cultural aspects in the Himalayan region led to the evolution of diverse traditional farming systems. This Western Central Himalayan Region is one of the richest reservoir of genetic variability and diversity of different vegetable crops. Wild and domesticated vegetable crops are important sources of range of nutritional factors in the human diet and yet are often considered secondary to the staple crops during discussions of crop diversity and genetic erosion (Rana *et al*., 1985).

The inhabitants of this region had been consuming a large number of plants from their natural habitats and even some grown in their backyards. The diversity for vegetable crops of this region has mainly been managed by local farmers, often women. Considerable diversity exists among the regional vegetable species including variation in plant type, morphological and physiological characteristics, reactions to diseases and pests, adaptability and distribution. Apart from the nutritional value, many regional vegetable crops are used for medicinal purposes, income generating and poverty alleviation programmes in the rural areas. In addition, the region is well known for its richness in variety of land races/primitive cultivars / local cultivars of several agricultural, horticultural crops besides a huge floristic wealth of great economic importance including medicinal and aromatic plants (Maikhuri *et al.,* 1996; Nautiyal *et al.,* 2002; Swarup, 1993; Verma *et al.,* 2000). A wide range of vegetable crops grown in this region. It includes solanaceous vegetables, cucurbitaceous, okra, various kinds of beans, tubers & roots crops, spices, Cole crops as well as some varieties of leafy vegetables etc. Occurrences of their wild relatives are also a subject of potential exploitation, towards the crop improvement. Some underutilized crops like tree tomato, chow-chow, drum stick, etc. required attention.

Conclusion

In situ and *ex situ* conservation of biodiversity for agriculture and forestry are complementary approaches. Genetic diversity continues to meet farmers' needs and plays an important role in traditional agro-ecosystems. New materials were also incorporated into existing landraces, permitting the agricultural system to evolve without total replacement. The local landraces are also being replaced with high yielding landraces from neighboring areas in case of common food crops like wheat, paddy and brassica.

Perilla, a minor oilseed crop, is almost replaced by amaranths. It has now been well realized that in many areas farmers have developed distinct systems of classification and description of local landraces (Boster, 1985). The agriculture related indigenous knowledge possessed by the elderly people, mostly womenfolk, is facing series of challenges due to variety of factors and out-migration of younger generation to urban areas.

The Uttarakhand Himalaya is one where people are still practicing traditional landrace-based cultivation. Farmers' dependence on varietal mixtures, multiple crops, intercropping, growing genetically diverse varieties of individual crops fits with high variability in their edaphic and biological environments and their limited access or inability to acquire purchased inputs. Gradual reduction in area of several traditional crops and farmers preferences for certain other traditional and introduced crops is induced by the economic and socio-cultural factors. The market forces are creating new preferences. It suggests that the hill agro ecosystems with traditional crops are ecologically and economically viable and still have the potential to support the food requirements in the Himalayan region. *In situ* (on-farm) conservation will be most effective when targeted to specific areas with significant plant genetic resources and with communities who are willing to participate in conservation programmes. Strategies to enhance local participation through marketing and education must be identified. By including decentralized breeding as part of *in situ* programme, farmers and crop biologists can become partners in local crop improvement efforts for marginal agrocilmatic zones and for crops without national breeding programmes. This 'grassroot breeding' can build upon existing knowledge and skills of farmers and link farmers from different regions through the exchange of information and landraces (Iwanaga 1995; Brush and Meng 1998; Berthaud *et al.* 2001; Cooper *et al.* 2001). In situ (on-farm) conservation can also be seen as a conservation strategy that is complementary to ex situ conservation (Visser and Engels 2000; Almekinders *et al.* 2000; Almekinders and Elings 2001).

References

Almekinders, C., Boef, W. de and Engels, J. 2000. Synthesis between crop conservation and development. *In*: Encouraging Diversity the Conservation and Development of Plant Genetic Resources (Almekinders C. and Boef W. de (eds). Intermediate Technology Publication, London, pp. 330–338.

Almekinders, C. J. M. and Elings, A. 2001. Collaboration of farmers and breeders: participatory crop improvement in perspective. *Euphytica,* **122:** 425–438.

Altieri, M., Anderson, K. and Merrick, L.C. 1987. Peasanat agriculture and the conservation of crop and wild resources. *Conserv. Biol.,* **1:** 49–58.

Arora, R. K. 1995. Genetic resources of vegetable crops in India: Their diversity and conservation. *In:* Genetic Resources of Vegetable Crops (eds.), R.S. Paroda, P.N. Gupta, Mathura Rai and S. Kochhar, NBPGR, New Delhi. pp. 29-39.

Arora, R. K. and Nayar, E. R. 1984. Wild relatives of crop plants in India, NBPGR Science Monograph No. 7. National Bureau of Plant Genetic Resources (NBPGR), New Delhi.

Bellon M. R. 1996. The dynamics of crop intraspecific diversity: a conceptual framework at the farmer level. *Econ. Bot.,* **50:** 29–36.

Bellon, M.R., Pham, J.L. and Jackson, M.T. 1997. Genetic conservation: a role for farmers. *In:* Plant Genetic Conservation: The in Situ Approach. (eds) Maxted N., Ford-Lloyd B.V. and Hawkes J.G. Chapman and Hall, London, pp. 263– 289.

Berthaud, J., Clement, J. C., Emperaire, L., Louettee, D., Pinton, F., Sanou, J. and Second, G. 2001. The role of local-level geneflow in enhancing and maintaining genetic diversity. *In:* Broadening the Genetic Base of Crop Production. (eds), Cooper H.D., Spillane C. and Hodgkin T. IPGRI/FAO, Rome, pp. 81–103.

Boster, J.S. 1985. Selection for perpetual distinctiveness: evidence from Aguaruna cultivars of *Manihot esculenta*. *Econ. Bot.*, **39:** 310–325.

Brush, S. B., 1989. Rethinking crop genetic resources conservation. *Conservation Biology,* **3:** 19-29.

Brush, S.B. 1991. A farmer based-approach to conserving crop germplasm. *Econ. Bot.,* **45:** 153–165.

Brush, S.B. 1995. In situ conservation of landraces in centres of crop diversity. *Crop Sci.,* **35:** 346–354.

Brush, S.B. 1999. Genes in the Field: On-farm Conservation of Crop Diversity. Lewis Publishers, Boca Raton, Florida, USA.

Brush, S.B. and Meng E. 1998. Farmers' valuation and conservation of crop genetic resources. *Genet. Resour. Crop Evol.,* **45:** 139–150.

Choudhury, B. 1967. Vegetables, National Book Trust, India.

Cooper, H.D., Spillane, C. and Hodgkin, T. 2001. Broadening the genetic base of crops: an overview. *In:* Broadening the Genetic Base of Crop Production. (eds.) Cooper H.D., Spillane C. and Hodgkin T., IPGRI/FAO, Rome, pp. 1–23.

Crisp, P. and Astley, D. 1985. Genetic resources in vegetables. *In:* Progress in Plant Breeding –1 (ed.) G.E. Russel Butterworths, London. p. 281-310.

Dobriyal, R.M.G.S., Singh, K.S. and Saxena, K.G. 1997. Medicinal plants resources in Chhakinal watershed: Traditional knowledge, economy and conservation. Journal of Herbs, Spices and Medicinal Plants **5:** 15–27.

Eyzaguirre, P. and Iwanaga, M. 1996. Farmers' contribution to maintaining genetic diversity in crops and its role within total genetic resources systems. *In:* Participatory Plant Breeding. (eds). Eyzaguirre P. and Iwanaga M. IPGRI, Rome, pp. 9–18.

Harlan, J. 1992. Crops and Man. 2nd ed. American Society of Agronomy, Madison,WI, pp. 284.

Iltis, H.H. 1974. Freezing the genetic landscape. Maize Genet.Coop. *News Lett.*, **48:** 199–200.

Iwanaga, M. 1995. IPGRI strategy for in situ conservation of agricultural biodiversity. *In: In situ* Conservation and Sustainable Use of Plant Genetic Resources for Food and Agriculture in Developing Countries. (ed.) Engels J.M.M., IPGRI/DSE, Rome, pp. 13–26.

Jarvis, D. and Hodgkin, T. 1998. Strengthening the scientific basis of *in situ* conservation of agricultural biodiversity on farm. Options for data collection and analysis. Proceedings of a workshop to develop tools and procedures for *in situ* conservation on-farm, 25–29 August 1997, IPGRI, Rome.

Jarvis, D. I., Myer, L., Klemick, H., Guarino, L., Smale, L., Brown, A. H. D., Sadiki, M., Sthapit, B. and Hodgkin, T. 2000. A Training Guide for In situ Conservation On-farm. Version 1. International Plant Genetic Resources Institute, Rome, Italy. Jarvis D., Staphit B. and Sears L. 2000b. Conserving agricultural biodiversity in situ: a scientific basis for sustainable agriculture. IPGRI, Rome.

Long, J., Cromwell, E. and Gold, K. 2000. On-farm Management of Crop Diversity: An Introductory Bibliography. Overseas Development Institute for ITDG, London, pp. 42.

Maikhuri, R.K., Nautiyal, S., Rao, K. S. and Saxena, K.G. 1998. Role of medicinal plants in traditional health care system: A case study from Nanda Devi Biosphere Reserve, Himalaya. *Curr. Sci.* **75:** 152–157.

Maikhuri, R. K., Rao, K. S. and Saxena, K. G. 1996. Traditional crop diversity for sustainable development of Central Himalayan agroecosystems. *Int. J. Sust. World Ecol.* **3:** 8–31.

Maxted, N., Ford-Lloyd, B. V. and Hawkes, J. G. 1997. Complementary conservation strategies. In: Maxted N., Ford-Lloyd B.V. and Hawkes J.G. (eds), Plant Genetic Conservation: The In Situ Approach. Chapman and Hall, London, pp. 15–40.

Maxted, N., Guarino, L., Myer, L. and Chiwona, E.A. 2002. Towards a methodology for on-farm conservation of plant genetic resources. *Genet. Res. Crop Evol.* **49:** 31–16.

McNeely, J. A. 1988. Economic and biological diversity: developing and using economic incentives to conserve biological resources. International Union for Conservation of Nature and Natural Resources, Gland.

Nautiyal, S., Maikhuri, R. K., Rao, K. S., Semwal, R. L. and Saxena, K.G. 2002. Agroecosystem function around a Himalayan Biosphere Reserve. *J. Environ. Syst.* **29:** 71–100.

Negi, C. S., Nautiyal, S., Dasila, L., Rao, K. S. and Maikhuri, R. K. 2003. Ethnomedicinal plant uses in a small tribal community of Central Himalaya, *India. J. Hum. Ecol.* **14:** 23–31.

Oldfield, M. L. and Alcorn, J. B. 1987. Conservation of traditional agroecosystems. *Bioscience*, **37:** 199–209.

Rana, R.S., Gupta P.N., Rai Mathura and Kocchar S. 1985. Genetic resources of vegetable crops, management, conservation and utilization. NBPGR, New Delhi, pp1-19.

Sheshadri, V.S. 1987. Genetic resources and their utilization in vegetable crops, *In*: Plant Genetic Resources: Indian Perspective (eds.), R.S. Paroda, R.K. Arora and K.P.S. Chandel, NBPGR, New Delhi. pp. 335- 343.

Smale, M. and Bellon, M. R. 1999. A conceptual framework for valuing on-farm genetic resources. *In:* Agrobiodiversity: Characterisation. Utilization and Management, (eds) Wood D. and Lenne J, CAB International, Wallingford, pp. 387–408.

Swarup, R. 1993. Agricultural economy of Himalayan region with special to Garhwal Himalaya. Vol. II. Gyanodaya Prakashan, Nainital, India, pp. 288.

Verma, S. K., Muneem, K. C., Verma, V. D., Negi, K. S., Gautam, P. L., Mishra, K. K., Parakh, D. B., Gupta, S., Singh, R.V., Arya, R. R. and Pandey, G. P. 2000. Diversity in horticultural wealth of UP Hills. *In*: international Conference on "Managing Natural Resources for Sustainable Agricultural Production in the 21st Century" at IARI, New Delhi –12 w.e.f. 14 – 18 Feb. 2000, appeared in Agri-biodiversity Extended Summary, **2:** 827-830.

Visser, B. and Engels, J. 2000. Synthesis: the common goal of conservation of genetic resources. *In:* Encouraging Diversity. The Conservation and Development of Plant Genetic Resources. (eds) Almekinders C. and Boef W. de. Intermediate Technology Publication, London, pp. 145–153.

Worede M. 1992. Ethiopia: a genebank working with farmers. *In:* Growing Diversity: Genetic Resources and Local Food Security. (eds) Cooper D., Vellve R. and Hobbelink H., Intermediate Technology Publications, London, pp. 78–94.

Worede M. 1993. The role of Ethiopian farmers on the conservation and utilization of crop genetic resources. Int. Crop Sci. Soc. Am. **1:** 395–399.

Worede M. 1997. Ethiopian in situ conservation. *In:* Maxted Plant Genetic Conservation: The *In situ* Approach. (eds) N., Ford-Lloyed B.V. and Hawkes J.G. Chapman and Hall, London, pp. 290–301.

Worede M. and Hailu M. 1993. Linking genetic resources conservation to farmers in Ethiopia. *In:* Cultivating Knowledge: Genetic Diversity Farmer Experimentation and Crop Research. (ed.) de Boef D. Intermediate Technology Publications, London, pp. 78–84.

Zeven, A. L. and deWet, J. M. I. 1982. Dictionary of cultivated plants and their regions of diversity. Centre of Agricultural Publicity and Documentation, Wageningen. The Netherlands.

❑❑❑

Chapter – 7

Molecular Markers and Genetic Resource Management of Vegetable Crops

H. Choudhary and D.K. Singh

Introduction

Plant genetic resources management essentially consists of two phases; its conservation and utilization. Germplasm conservation includes acquisition, in which germplasm is safeguarded *in situ* (by establishing reserves) or *ex situ* (by assembling collections through exchange or exploration). It also involves maintenance, protecting germplasm *in situ* in reserves or storing it *ex situ* under controlled conditions, propagating it while preserving its original genetic profile with maximum fidelity, monitoring its viability and health in storage or *in situ* and maintaining associated passport information and other data. Germplasm conservation also involves characterization, assessing highly heritable morphological and molecular traits for taxonomic, genetic, quality assurance and other management purposes. The second phase of germplasm management, encouraging utilization, includes evaluation, assessing agronomically or horticulturally important traits with relatively low heritability and high components of environmental variance e.g. yield adaptation and host-plant resistance to certain abiotic/biotic stress. It also includes genetic enhancement, making particular genes more accessible and usable to breeders by adopting exotic germplasm to local environments without losing its essential exotic genetic profile and/or introgressing high value traits from exotic germplasm into adapted varieties. Genetic enhancement can be subdivided into introgression i.e. backcrossing a few genes controlling desired characters into adapted stocks and incorporation i.e. the large scale development of locally adapted populations good enough to enter the adapted genetic bases of the crops concerned.

What is a marker: Marker is that which marks a tag for identification.

Genetic marker: Marker is a piece of genetic material that bears or produces distinctive feature. A marker is usually a mutant gene.

Features of genetic markers

1. Highly polymorphic: Many alleles per locus
2. Highly heritable: Phenotypic expression is relatively unaffected by environmental variation or by G X E interaction.
3. Simple inheritance: Ideally, they should be single Mendelian genes with co dominant alleles.
4. Different loci should be well dispersed throughout the genome.
5. It should not differentially affect a plant's fitness such as some albino mutations do.
6. It should be scored rapidly and early in the plant's life cycle.
7. Assays should be inexpensive and innocuous to humans.

Type of genetic markers

1. Morphological marker
2. Karyology (Chromosome no and morphology)
3. Secondary metabolites (anthocyanin, flavonoid)
4. Protein (Seed protein, Isozyme)
5. DNA

Development of markers: There has been five eras in evolution of different genetic markers:

(a) Morphological & Cytological markers (1950's)
(b) Protein & Allozyme markers (1960-70's)
(c) RFLP & Minisatellites markers (1970-85)
(d) RAPD & AFLP (1986-95)
(e) Complete DNA sequence with known and unknown function (1996 onwards)

Morphological markers: Morphological traits are the oldest and most widely used genetic markers and they may still be optimal for certain germplasm management applications. Their main advantages are simplicity and rapid, inexpensive assays. The linkage between seed size and coloration in bean was shown for the first time in plants by Sax during early 1920s. Walker(1923) reported association of pigmented bulb with the resistance to onion smudge and that of non pigmented bulbs with susceptibility. White bulb onions were found to be susceptible against *Lasiodiplodia theobromae*, while the yellow and red bulbs onions were resistant (Cramer and Havey,1999).

Limitations of morphological markers

1. Less in number
2. Confer indistinguishable phenotype
3. Influenced by the environment
4. Influenced by the genetic background
5. Influenced by the ontogeny
6. Most of the markers produce lethal effects
7. No stable inheritance

Karyology: Chromosome number and cytomorphological traits have also served as genetic markers, especially in polyploid crop species. Chromosome morphology has similarly elucidated the evolution and systematic relationships of domesticated and wild *Capsicum* peppers (Pickersgill, 1971 and 1981).

Secondary metabolites: Anthocyanin and flavonoid pigments possess some of the cost/ time efficiency advantages of morphological markers. But they too may not be interpretable by locus/allele models, nor are they necessarily selective neutral. The genetic basis of pigment polymorphism is well known in *Pisum* and amaranthus.

Protein: Seed protein and isozyme variants that migrate at different rates under electrophoresis have been most widely employed genetic markers during last quarter of 20 the century. Isozymes are generally fractionated by starch gel electrophoresis in studies of genetic diversity and divergence, whereas seed proteins are generally analyzed via polyacrylamide gels. Protein electrophoretic migration rates are generally highly heritable and ample polymorphisms are available for many germplasm management purposes.

Molecular marker: A molecular marker is a DNA sequence that is readily detected and whose inheritance can easily be monitored.

Salient features of molecular markers

1. It must be polymorphic.
2. Its inheritance should be codominant.
3. It should be frequently and evenly distributed throughout the genome.
4. It should be easy, fast and cheap to detect.
5. It should be reproducible.

Advantages of molecular markers

- High variability
- Minimal environment effect
- No genetic interaction
- Generally codominant in nature
- Unambiguous data scoring
- Small tissue sample required
- Tissue, growth and season independent

Introduction to molecular markers

All living organisms are made up of cells that are programmed by genetic material called DNA. This molecule is made up of a long chain of nitrogen-containing bases (there are four different bases–adenine [A], cytosine [C], guanine [G] and thymine [T]). Only a small fraction of the DNA sequence typically makes up genes, i.e. that code for proteins, while the remaining and major share of the DNA represents non-coding sequences, the role of which is not yet clearly understood.

The genetic material is organized into sets of chromosomes and the entire set is called the genome. In a diploid individual (i.e. where chromosomes are organized in pairs), there are two alleles of every gene–one from each parent. Molecular markers should not be considered as normal genes as they usually do not have any biological effect. Instead, they can be thought of as constant landmarks in the genome. They are identifiable DNA sequences, found at specific locations of the genome, and transmitted by the standard laws of inheritance from one generation to the next. They rely on a DNA assay, in contrast to morphological markers that are based on visible traits and biochemical markers that are based on proteins produced by genes.

Different kinds of molecular markers exist, such as Restriction Fragment Length Polymorphisms (RFLPs), Random Amplified Polymorphic DNA (RAPDs) markers, Amplified Fragment Length Polymorphisms (AFLPs), Microsatellites and Single Nucleotide Polymorphisms (SNPs). They may differ in a variety of ways such as, their technical requirements (e.g. whether they can be automated or require use of radioactivity), the amount of time, money and labour needed, the number of genetic markers that can be detected throughout the genome and the amount of genetic variation found at each marker in a given population. The information provided to the breeder by the markers varies depending on the type of marker system used. Each has its advantages and disadvantages and, in the future, other systems are likely to be developed. Some popular molecular markers have beeen compared for different application aspects in Table 1.

Table 1. Comparison of different molecular marker techniques

Features	RFLP	RAPD	STMS	AFLP	PCR-seq
Development costs	Medium	Low	High	Low	High
Level of polymorphism	Low-medium	Medium	High	Medium	Medium
Reliability	High	Low	High	Medium	High
Level of skill required	Medium	Low	Low-high	Medium	High
Automation cost	High	Medium	High	High	High
Samples/day	20	50	50	50	20

Restriction fragment length polymorphisms (RFLP)

Restriction fragment length polymorphisms (RFLPs) were the first DNA-based molecular markers. An application of Southern analysis, RFLPs exploit the ability of single stranded DNA to bind (hybridize) to DNA with a complementary sequence. RFLP markers detect variation in DNA sequences at the same loci in different individuals or accessions. Technically, RFLP technology involves the hybridization of cloned DNA to restriction fragments of differing molecular weights from restriction enzyme-digested genomic DNA. The digested DNA fragments are size-separated on agarose gels by electrophoresis and transferred as denatured (single stranded) arrays of fragments to filters through capillary action. The filters are then incubated with specific labelled probes (genes or anonymous fragments of single stranded DNA), washed and exposed to x-ray film. To identify polymorphisms between individuals or accessions, the genomic DNA extracted from each individual is digested with a series of restriction enzymes to find enzymes that produce fragments (bands) that differ in molecular weight between accessions and can be distinguished by hybridization with a given probe. To ensure that probes hybridize to single fragments on a gel, the DNA used as a probe should be from a single or low copy (non-repetitive) region of the genome. Probes may represent genes (i.e. derived from complementary DNA [cDNA]) or they may represent anonymous sequences derived from genomic DNA. The polymorphisms detected by RFLPs may result from single base changes causing a loss of restriction sites or a gain of new restriction sites, or from insertions and deletions (indels) between restriction sites.

PCR-based markers

Many advances in molecular marker technology have come through applications of the PCR method. In PCR, a thermo-stable DNA polymerase enzyme makes copies of a target sequence beginning from two small pieces of synthetically produced DNA (primers) that are complementary to sequences bracketing the target. Through iterations of the process with heating to separate the double stranded DNA molecules and cooling to allow the primers to re-anneal, the target sequence is exponentially amplified. Polymerase chain reaction-based markers require much less DNA per assay than RFLPs and are more compatible with automated high-throughput genotyping (i.e. the ability to process large numbers of samples quickly and efficiently).

Randomly amplified polymorphic DNA markers (RAPD)

Randomly amplified polymorphic DNA markers (RAPDs) use PCR to amplify stretches of DNA between single primers of arbitrary sequence. Amplification occurs only where sequences complementary to the primers are in close enough proximity for successful PCR. The typical oligonucleotide used for RAPDs is ten bases long and will amplify many loci simultaneously, allowing multiple markers to be assayed in a single PCR reaction and a single lane on an agarose gel. As the primers are arbitrary, RAPD technology can be applied directly to any species with no prior sequence knowledge. This technology is particularly useful when there is a need to assay loci across the entire genome. The polymorphisms are detected only as the presence or absence of a band of a particular molecular weight, and it is not possible to differentiate between homozygous and heterozygous markers. RAPDs are notoriously unreliable because, aside from sequence differences, the amplification or failure of amplification of any band may be sensitive to any number of factors, including DNA template quality, PCR conditions, reagents and equipment.

Amplified fragment length polymorphisms (AFLP)

Amplified fragment length polymorphisms (AFLPs) are molecular markers derived from the selective amplification of restriction fragments. Genomic DNA is digested with a pair of restriction enzymes and oligonucleotide adaptors are ligated to the ends of each restriction fragment. The fragments are amplified using primers that anneal to the adaptor sequence and extend into the restriction fragment. Only a portion of restriction fragments will be within the range of sizes than can be amplified by PCR and visualized on polyacylamide gels (between 50 and 350 bp).

For large genomes, additional selective bases can be added to the primers to reduce the number of co-amplified bands. AFLPs have many of the advantages of RAPDs, but have much better reproducibility. AFLP technology requires greater technical skill than RAPDs and, because AFLPs run on polyacrylamide gels instead of agarose, they also require a larger investment in equipment than RAPDs. Using manual gels, AFLP bands are detectable using silver stain, or by labelling of the primers with a radioactive isotope. Alternatively, for higher throughput, AFLPs can be detected with an automated DNA sequencer by using fluorescently labelled primers.

Developing molecular markers with DNA sequence information

When the DNA sequence is available, it is possible to design primers to amplify across a specific locus. However not all loci will be polymorphic. Targeting highly variable sequence features increases the likelihood of detecting polymorphism. These highly variable features include tandem repeats such as microsatellites and dispersed complex repeats such as transposable elements.

Microsatellites

Simple sequence length polymorphisms (SSLPs) also known as simple sequence repeats (SSRs), or microsatellites, consist of tandemly repeated di-, tri- or tetra-nucleotide motifs and are a common feature of most eukaryotic genomes. The number of repeats is highly variable because slipped strand mis-pairing causes frequent gain or loss of repeat units. With their high level of allelic diversity, microsatellites are valuable as molecular markers, particularly for studies of closely related individuals.

PCR-based markers are designed to amplify fragments that contain a microsatellite using primers complementary to unique sequences surrounding the repeat motif. Differences in the number of tandem repeats are readily assayed by measuring the molecular weight of the resulting PCR fragments. As the differences may be as small as two base pairs, the fragments are separated by electrophoresis on polyacrylamide gels or using capillary DNA sequencers that provide sufficient resolution.

Without prior sequence knowledge, microsatellites can be discovered by screening libraries of clones. Clones containing the repeat motif must be sequenced to find unique sites for primer design flanking the repeats. Microsatellite marker development from pre-existing sequence is far more direct. Microsatellites discovered in non-coding sequence often have a higher rate of polymorphism than microsatellites discovered in genes.

Microsatellite markers have several advantages. They are co-dominant; the heterozygous state can be discerned from the homozygous state. The markers are easily automated using florescent primers on an automated sequencer and it is possible to multiplex (combine) several markers with non-overlapping size ranges on a single electrophoresis run. The results are highly reproducible, and the markers are easily shared among researchers simply by distributing primer sequences. Although SSRs are abundant in most eukaryotic genomes, their genomic distribution may vary. Uneven distributions of microsatellites limit their usefulness in some species. Inter-SSRs (ISSRs) are another type of molecular marker that makes use of microsatellite sequences. ISSRs use PCR primers anchored in the termini of the repeats extending into the flanking sequence by several nucleotides. PCR products are produced for each pair of microsatellites that are in sufficient proximity for PCR to occur, or may be generated by anchoring one primer in the SSR motif and using a second "universal" primer corresponding to a sequence that has been ligated onto the ends of restriction fragments. Markers at multiple loci are assayed as the presence or absence of bands of particular sizes. ISSRs can be visualized on agarose gels, on silver stained polyacrylamide gels or fluorescently labelled for detection with an automated DNA sequencer.

Transposable element-based markers

Transposable elements (TEs) are another rapidly changing feature of the genome that can be exploited as a source of variability for molecular markers. Discovery of TE sequences is a prerequisite for their use as markers. While TEs may be discovered as mutations in alleles of genes conferring mutant phenotypes, they have also been discovered directly in genomic sequence. Transposon display is a modified AFLP procedure that differs only in that one of the two primers is designed within the consensus sequence of a TE family so that amplification depends on the presence of a TE insertion within a restriction fragment. Using this approach, the presence or absence of a TE can be assayed simultaneously at many loci throughout the genome. To assay for a TE insertion at a specific locus, single copy "anchor markers" can be designed with primers located in unique sequences flanking the region of interest. A size polymorphism indicates the presence or absence of the TE in that particular location. Anchor markers are advantageous because they are co-dominant, can be run on a simple agarose gel system and are biologically informative in that they provide evidence of both complete, or incomplete, insertion or excision events. This methodology can also be applied to any known indel feature regardless of whether or not it is derived from a TE.

Single nucleotide polymorphisms

Single nucleotide polymorphisms (SNPs) are an abundant source of sequence variants that can be targeted for molecular marker development. Of all the molecular marker technologies available today, SNPs provide the greatest marker density. SNPs are often the only option for finding markers very near or within a gene of interest, and can even be used to detect a known functional nucleotide polymorphism (FNP). Discovery of SNPs requires obtaining an initial DNA sequence in a reference individual followed by some form of re-sequencing in other varieties to find variable base pairs. There is a myriad of other SNP assay technologies in development and to date no single method stands out as superior to the others. The benefits of SNP assays include increased speed of genotyping, lower cost and the parallel assay of multiple SNP.

Single feature polymorphisms and microarray-based genotyping

Indel polymorphisms, also known as single feature polymorphisms (SFPs), are particularly amenable to microarray-based genotyping. These assays are done by labelling genomic DNA (target) and hybridizing to arrayed oligonucleotide probes that are complementary to indel loci. Each SFP is scored by the presence or absence of a hybridization signal with its corresponding oligonucleotide probe on the array. The SFPs can be discovered through sequence alignments or by hybridization of genomic DNA with whole genome microarrays. The advantage of microarray platforms for genotyping is that they are highly parallel, and they are well suited for applications such as quantitative trait loci (QTL) analysis, where whole genome coverage with many markers is desirable.

Special considerations for diversity studies and germplasm evaluation

The interpretation of molecular marker data for germplasm classification and diversity can be confounded by uncertainty about the underlying sources of the polymorphisms and by homoplasy (false homology). The ratio of indels to base changes is important for diversity studies because, when molecular markers are used to estimate nucleotide divergence, the divergence will be overestimated if indel-derived polymorphisms are common. The greatest certainty of the underlying polymorphism comes from SNP technologies that directly assay for single base changes. For SSR markers among closely related individuals, most polymorphism should be caused by expansion or contraction of the number of repeat units. However, as genetic distance between the varieties increases, there is an increasing chance that indel events will cause additional size polymorphism. Thus, the use of stepwise SSR mutation models would be inappropriate for highly diverged populations. Homoplasy is also a problem in SSR markers because the hyper-variability leads to some shared allele sizes through parallelism, convergence and reversion. Homoplasy from reversions can affect transposonbased markers or any markers with polymorphisms potentially derived from Class II DNA transposable elements. This class of TEs has a cut and paste mechanism of transposition, so a TE may insert onto a locus and later excise. In RAPDs, ISSRs and AFLPs, homoplasy can occur when two or more loci produce PCR fragments of similar molecular weight. Although it is desirable to have high numbers of bands to maximize the amount of information per lane, this must be balanced against the increasing risk of homoplasy as more loci are represented.

Reproducibility of molecular marker data

For orphan species, clearly there is a huge value to the anonymous primer approaches (AFLP, DArTs, ISSRs and RAPDs) that do not require sequence information or much up-front investment. However, the data can be difficult to score, and reproducibility requires a lot of technical skill. Technologies that depend on the presence or absence of PCR amplified bands are susceptible to changes in PCR conditions and the quality of sample DNA, and the data from separate experiments may differ. Further, in any method that depends on accurate measurement of molecular weight differences between bands (e.g. SSRs), the exact molecular weights assigned to each allele may be different in each analysis because of differences in labelling of PCR products, rounding of allele molecular weight estimates and binding of alleles. Without controls for each allele encountered, it is difficult or impossible to merge separate sets of data. Despite discrepancies in the exact data derived from molecular markers, the results and conclusions should be consistent within independent experiments. For reliability in making inferences across independent data-sets, SNP markers are preferred. SNP data-sets can be easily integrated based on sequence, and SNPs have properties (such as a low mutation rate) that are particularly valuable for evolutionary inference .

Application of molecular markers

Plant genetic resources conservaton can be distinguished into 3 main activities (Hodgkin, *et al.*, 2001)

(a) locating and describing the available diversity

(b) developing effective conservation procedure and

(c) identifying materials and methods that are needed by users

Genetic information plays a significant part in determining the effectiveness of work in all of these areas and molecular markers now provide an excellant ways of obtaining large amounts of genetic data to inform the conservation process. Molecular studies have provided information on crop taxonomy and evolution, on geographic and ecological aspect of the extent and distribution of genetic diversity and on the processes that have given rise to observed patterns of variation. The different molecular methods have been used in widely differing way to support work on the conservation and use of plant genetic resources. They each have specific advantages and disadvantages and the appropriateness of individual marker systems may vary depending on the nature and objective of the investigation and the properties of the species (Karp *et al.,* 1997 and Powell *et al.,* 1996). The application of molecular markers for germplasm management can be broadly grouped into 3 major objectives:

(a) To know the extent and distribution of diversity.

(b) Ex situ conservation and management of plant genetic resources.

(c) Utilization of plant genetic resources in crop improvement.

(a) Extent and distribution of diversity

Molecular markers have significant advantages in the analysis of genetic diversity. They do not vary with the environment, many mrkers can be detected and, in many of the systems developed, hetrozygotes and homozygotes can be distinguished. They have proved to be useful in providing estimates of the amount of genetic diversity, in determining its distribution and its relationship with geographic or other variables, and in understanding the natural process of evolution, migration and selection. They have also allowed us to obtain a deeper insignt into the biological properties of species with respect to aspects such as taxonomic relationships between species, mating behavior and breeding systems. Knowing the amount of diversity in different taxa and its distribution in different regions or environments in which the taxa occur is a fundamental element in the development of adequate and cost effective conservation strategies. The information allows us to choose how much of which populations should be maintained *ex situ* or managed *in situ* in order to achieve identitified conservation objectives. It also greatly assists users in identifying which material to explore and to find specific useful traits.

Molecular makers have been used to provide information on the quantity of variation in different taxa or populations and have provided graphic evidence of the differences in amounts of variation that can be found in wild relatives and crops. For example, Miller & Tanksley (1990) analysed the level of polymorphism in the genus *Lycopersicon* using RFLPs and found that there were large differences in the amount of polymorphism found in different species. Cultivated *L. esculentum* had very little variation compared with wild and weedy relatives and the three self- incompatible species studied contained nearly three times as much diversity as the four self-compatible species combind.The application of different molecular markers for assessment of genetic diversity in vegetable crops has been listed in the following Table2.

Table-2. Assessment of genetic diversity in vegetable crops.

S.N.	Crop	Technique	References
1.	Brinjal	RAPD, STMS	Karihaloo *et al.*,1995; Behera *et al.*, 2006
2.	Tomato	Microsatellites, RAPD, RFLP	Rus Kortekaas *et al.*,1994; Villand *et al.*, 1998; Archak *et al.*, 2002.
3.	Pepper	AFLP,RAPD	Paran et al., 1998; Rodriguez *et al.*, 1999
4.	Potato	AFLP, Microsatellites, ISSR, RAPD	Mc Gregor *et al.*, 1998; Ashkenazi *et al.*, 2001.
5.	Cucumber	ISSR, Microsatellites, RAPD, RFLP	Staub *et al.*, 1997; Stachel *et al.*, 1998; Horejsi and Staub, 1999; Danin *et al.*, 2001; Dijkhuizen *et al.*, 1996
6.	Vegetable Brassica	Microsatellites, RAPD	Margale *et al.*,1995; Cansian and Echeverrigaray, 2000
7.	Melons	RAPD	Garcia *et al.*, 1998
8.	Bitterground	ISSR, RAPD	Singh *et al.*, 2007; Dey *et al.*, 2006
9.	Ashground	RAPD	Sureja *et al.*, 2006
10.	Pumpkin	RAPD	Stachel *et al.*,1998; Gwanama *et al.*, 2000
11.	Pea	RAPD	Samec and Nasinec,1996
12.	Beans	RAPD, RFLP	Stockton and Gepts, 1994; Skroch and Nienhuis,1995
13.	Allium Spp.	AFLP, Microsateliite, RAPD	Maass and Klass, 1995; Bradly *et al.*, 1996; Al Zahim *et al.*, 1997; Fischer and Bachmann, 2000
14.	Sweet potato	RAPD	He *et al.*, 1995
15.	Carrot	AFLP	Shim and Jorgensen, 2000
16.	Radish	RAPD	Rabbani *et al.*, 1998
17.	Spinach	Microsatellites	Groben and Wrickle,1998
18.	Lettuce	AFLP, Microsatellites	Hill *et al.*, 1996; Witsenboer *et al.*, 1997
19.	Amaranth	RAPD	Ranade *et al.*, 1997
20.	Asparagus	RAPD	Khandka *et al.*,1996; Roose & Stone, 1996
21.	Artichoke	RAPD	Tivang *et al.*, 1996

Source: Karihaloo and Bhatt, 2002 (modified)

Germplasm of narrow genetic base is obviously unlikely to harbour novel genes e.g. those conferring resistance to biotic and abiotic stresses. RAPD analysis of pepper breeding lines (Las Heras Vezquez *et al.*, 1996) revealed very narrow genetic base with more than 50% of the DNA bands being common among all the lines. In an assessment of the world collections of tomato, Villand *et al.* (1998) found South American accessions to have greater diversity than old world accessions. Horejsi and Staub (1996) using RAPD analysis recorded limited genetic diversity in 118 cultivars and elite lines of cucumber. Shim and Jorgensen (2000) carried out AFLP analysis in wild and cultivated carrots and found that the old varieties released between 1974-76 were more heterogeneous than newly developed F1 hybrid varieties. Archak *et al.* (2002), using RAPD markers in tomato, found old introductions and locally developed varieties of 1970s exhibiting significantly greater variation than the ones released in 1990s.

Clusters of crops accessions and cultivars generated on the basis of molecular diversity analysis often possess some common agronomic or physiological features. In such cases diversity analysis can be used to predict the performance or adaptability of accessions without actually screening for such traits. In lettuce, Hill *et al.* (1996) found positive linear relationship between distance estimates based on AFLP data and kinship coefficient calculated from pedigree data. Bradley *et al.* (1996) using DNA fingerprinting, found that the Australian garlic cultivars classified into bolting and nonbolting types and into groups related to growing seasons. Similarly, Al Zahim *et al.* (1997) could group 27 garlic cultivars into bolting and nonbolting types on the basis of RAPD analysis. RAPD analysis was also useful in grouping onion accessions according to their geographical locations (D' La Thierry *et al.*, 1997). Molecular analysis of melon breeding lines revealed that RAPD and agronomic traits were highly concordant (Garcia *et al.*, 1998). RAPDs were better than agronomic traits in predicting genetic distance and hybrid performance.

Establishing centres of diversity

VNTR analysis has been used effectively to describe the process of domestication, centres of domestication and path of dissemination of cultivated beans (Sonnante *et al.*, 1994). Two distinct bean gene pools, one Middle American and other Andean were identified (Becerra-Velasquez and Gepts, 1993).

(b) *Ex situ* conservation and management of PGR

There are a number of aspects of *in situ* and *ex situ* management where molecular techniques should be useful. However, to date, there is rather little evidence of their extensive application. Molecular markers have been used to study the efficiency of maintaining large numbers of very similar types of varieties. Phippen *et al.* (1997) performed a RAPD assay on *Brassica oleracea* var. *capitata* accessions stored in a genebank to quantify and partition the molecular varition between them. The results indicated that entries of Golden Acre" cabbage- a group of varieties known to be extremly similar morphologically could be reduced to as four groups with a potential loss of variation of only 4.6% of the absolute current genetic variation in the group as a whole. This would allow an approximate saving of 70% of the cost at each regeneration cycle. However. it might be argued that the 4.6% of diversity lost reflected the differences between the varieties and was, perhaps, the most important part of the diversity to try and retain.

There has been increasing interest in the development of core collections (Brown and Spillance, 1999) as aids to management and use of *ex situ* collections. Molecular markers have been used in this process to help in identifying groups from which core collection accessions can be selected or to monitor the effectiveness of other selection strategies in capturing the genetic diversity found in the whole collection. Molecular markers have been used by Chavarria-Aguirre *et al.*(1999) to analyse genetic diversity and identity redundancy in a core collection of cassava. In an interesting investigation of diversity found in a core collection, Skroch *et al.* (1998) used RAPDs to compare two sets of 90 accessions of *Phaseolus vulgaris*, one from the core collection and the other chosen randomly from the whole collection. They indicated that selection of the core had not led to any reduction in variation and that the methods used to develop the core had not been any more efficient in capturing diversity than a random sampling procedure.

(c) Utilization of plant genetic resources for crop improvment

Molecular genetic techniques are already playing a significant part in many aspects of crop improvement and in the development of improved crop varieties. These are the detection of new useful genes, the analysis of genome synteny between related species, an improved understanding of important phenomena such as heterosis and the analysis, at the molecular level, of key traits such as diseas resistance. The use of different molecular markers for tagging of resistance genes in vegetable crops have been listed in Table 3.

Table 3. Molecular markers linked to major resistant genes in vegetable crops (Kumar *et al.*, 2003)

Crop	Pathogen	Gene	Marker(s)	Reference
Tomato	*Meloidogyne incognita*	Mi	RAPD	Williamson *et al.*, 1994
	Meloidogyne incognita/javanica	Mi3	RAPD, RFLP	Yaghoobi *et al.*, 1995
	Cladosporium fulvum	Cf2	RFLP	Dixon *et al.*, 1995
	Phytophthora infestans	Ph2	RFLP	Moreau *et al.*, 1998
	Verticillium dahliae	Ve	RAPD	Kawchuk *et al.*, 1994
	Verticillium dahliae	Ve	SCAR	Kawchuk *et al.*, 1998
	Verticillium dahliae	Ve	RFLP	Diwan *et al.*, 1999
	F. oxysporum f. sp. radicislycopersici	Fr2	RAPD	Fazio *et al.*, 1999
	F. oxysporum f. lycopersici	sp.12	RFLP	Sarfatti *et al.*, 989
	Cucumber mosaic virus	cmr	RFLP	Stamova and Chetelat, 2000
	Yellow leaf curl virus	Ty1	RFLP	Zamir *et al.*, 1994
	Yellow leaf curl virus	Ty2	RFLP	Hanson *et al.*, 2000
	Tomato spotted wilt virus	Sw-5	RFLP	Brommonschenkel and Tanksley, 1997
	Tomato spotted wilt virus	Sw-5	RAPD	Chaque *et al.*, 1996
	Tomato mosaic virus	Tm2nv	RAPD	Tian *et al.*, 2000
	Tomato mosaic virus	Tm2a	RAPD	Daxet *al.*,1994
	Tomato mosaic virus	Tm1	RAPD/ SCAR	Ohmori *et al.*,1996
	Tomato mosaic virus	Tm22	SCAR	Dax *et al.*, 1998

Contd....

	Tomato mosaic virus	Tm2	SCAR	Sobir *et al.*, 2000
	Pyrenochaeta lycopersici	py	RAPD/ RFLP	Doganlar *et al.*, 1998
Pepper	Tomato spotted wilt virus	Tsw	RAPD	Jahn *et al.*, 2000
	Tomato spotted wilt virus	Tsw	CAPS	Moury *et al.*, 2000
	Xanthomonas vesicatoria	Bs2	AFLP	Tai *et al.*, 1999
	Leveillua taurica	Lv	RFLP	Chunwongse *et al.*, 1997
Bean	Common bean mosaic virus	I	RAPD	Melotto *et al.*, 1996
	Uromyces appendiculatus	Up2	RAPD	Miklar *et al.*, 1993
Pea	Pea common mosaic virus	Mo	RFLP	Dirlewanger *et al.*, 1994
	Erysiphe polygone	Er	RAPD	Dirlewanger *et al.*, 1994
	Fusarium oxysporum	Fw	ISAT	Dirlewanger *et al.*, 1994
Cucumber	*F. oxysporum f. sp. melonis*	Fom2	SSP	Wechter *et al.*, 1998
Melon	*F. oxysporum f. melonis*	Fom2	RAPD	Wechter *et al.*, 1995

Increasingly detailed studies of particular genes are providing important new insights into the nature and extent of divesity around useful loci. These studies are likely to play an important part in the search for, and detection of, new useful variants in germplasm collections and in wild relatives still occuring naturally. one exmnple where both map and sequence knowledge have been used to analyse divesity has been described by Sicard *et al.* (1999) for resistance to downy mildew in lettuce. Resistance gene candidates encoding nuclear binding sites and leucine rich regions have been idnetified and one family of over 20 memebers was found to be localized in the major disease resistance cluster. Three markers were developed which could be used to screen genotypes of cultivated and wild species. The markers were able to discriminate between accessions that had previously showed to be resistant to all known isolates of *Bremia lactucae*. Similarly, Kelly & Miklas (1998) reviewed many studies in which molecular markers (RAPD or SCAR) were found to be linked to 15 different genes conferring disease resistance to pathogens in *Phaseolus vulgaris*. As our knowledge of the molecular architechure of resistance genes becomes more detailed and additional sequence data is obtained, it is increasingly likely that markers can be identified that are closely correlated with resistance in a wide range of materials.

Each plant species has a wide variety of defense mechanisms to protect itself against pathogens and pests. When describing the genetics of resistance we consider only the genes that are responsible for the genetic variation, like genes that switch the processes on or off that are involved in resistance. Many more genes may be involved in the resistance mechanisms, but generally these do not cause a distinct variation and are usually not a target for plant breeding. Polygenic resistance is not associated with a typical mechanism of resistance, but only refers to the number of genes involved in resistance. These genes are designated 'quantitative trait loci (QTLs)'. Each QTL is supposed to have an additive effect to the resistance. As a consequence polygenic resistance is characterised by quantitative differences in the level of resistance. Monogenic race specific resistance usually gives clear phenotypes with a qualitative effect like no sporulation, necrotic spots or complete lack of symptoms. However, 'weak' HR-genes may give intermediate phenotypes like 'incompletely resistant'. For example, the *Cf-1* gene in tomato gives resistance to *Cladosporium fulvum* but the fungus is still able to grow and sporulate and the phenotype can be described as 'incompletely resistant' or 'intermediately resistant' (Patrick *et al*., 1971). Quantitative resistance is often described as 'field resistance', 'horizontal resistance', 'tolerance' (a confusing term as it is also used to indicate the amount of damage that is caused by a pathogen that infects a plant), 'partial resistance', etc. (Van der and Plank, 1963). In natural populations quantitative resistance is often found. Breeders and gene banks maintain germplasm collections that represent the biodiversity of cultivated species and their wild relatives The genetics of quantitative resistance are hard to study, as the effect of each gene is small and often influenced by the environment or by the interaction with other genes (epistasis). For genetic studies, progenies are obtained by crossing genotypes that contrast for the level of resistance. The segregation of quantitative resistance in these populations enables to estimate the number of genes involved in the resistance. These estimations are based on the assumptions that each genotype has the same environmental variation (experimental error) and that the quantitative effects of each gene are similar and additive. As these assumptions are debatable and often not true these estimations have limited value. The advent of DNA markers has enabled to identify loci on the genome that are involved in quantitative traits (quantitative trait locus, QTL). QTL mapping is accomplished by estimating the chance that a QTL is present at a position on the genome over the chance that a QTL is absent. The loci are characterised by DNA markers and for each locus on the genome this chance is expressed as the logarithm of this ratio. Peaks in the 'LOD profile' indicate those locations on the chromosomes where the QTLs are most likely present. The accuracy of QTL mapping is dependent on the density of markers, the accuracy of the trait analysis, the size of the population used and of the size of the explained variation of the QTL. The different disease resistance QTLs detected in vegetable crops are described in Table 4.

Table 4. Detection of disease resistance QTLs in vegetable crops (Michelmore, 1995)

Host/Pathogen	No. of markers	Population size/Method of analysis
Tomato		
Pseudomonas solanacearum	67 RFLP, 12 RAPD	71 F2, MMQTL
A. solani	141 RFLP,23 AGA	145, SIMQTL324
Clavibacter michiganensis	51 RFLP, RAPD, SCAR	324
Chilli		
Cucumber mosaic virus	93 AFLP, 51 RFLP, 38 RAPD	78, CIMQTL
Cucumber mosaic virus	84 RAPD, 51 RFLP	94DH, CIMQTL
Cole crops		
Plasmodiophora brassicae	198 RFLP,	90 F_2/F_3 , MMQTL
Plasmodiophora brassiceae	99 RAPD, 21 RFLP	138 F_2

Contd....

French bean		
Xanthomonas campestris	152 RFLP	70 F_2/F_3 , MMQTL
Soyabean		
Heterodera glycines	36 RFLP, 7 RAPD	56 F_2/F_3 , AV
Pea		
Ascochyta pisci	56 RFLP, 6 others	174 F2 , MMQTL

Tanksley and McCouch (1997) described ways in which molecular genetic studies could lead to the identification and introduction of useful traits from a wider range of plant materials than had hitherto been possible. They suggested that such methods would be particularly useful in the identification and tansfer of QTLs associated with useful traits that occured in crop wild relatives. Thus, De Vicente and Tanksley (1993) carried out an RFLP analysis on an interspecific F_2 population from wild tomato (*Solanum pennellii)* x cultivated tomato. Markers were identified to be associated with a number of QTLs for useful traits such as yield which showed a transgressive segregation. It was shown that marker assisted selection might be useful to impove traits of agronomic importance including those present in wild species for which appear phenotypically inferior to their cultivated counter parts. At Cornell University (Ithaca, USA) BIL populations have been generated from crosses of the cultivated tomato with several wild *Lycopersicon* species (Monforte and Tanksley, 2000). Such population of BILs can be reliably and accurately screened for quantitative traits including resistance. By doing so QTLs for fruit size, plant vigour and yield have been identified (Grandillo *et al.,* 1999).

These approaches provide powerful ways of identifying specific useful genes in crop wild relatives or other materials where the genes may not otherwise be easily detected and of transferring them to improved cultivars. However they can still only be used on a few acessions at a time since each analysis requires its own crossing programme and its own analysis of progeny using molecular markers and analysis of traits expression. The advent of technologies, that permit genetic mapping using molecular markers has provided maps of great detail and accuracy. From such mapping studies it has become clear that there are very substantial similarities in gene order along substantial lengths of chromosome in related species. Genome synteny provided opportunities for using information gained from detailed studies of one species in work on a related species in the same genus or even family. This has particularly important implications for horticultural crops and for negelected and underutilized crop generally. Such crops could never command sufficient resources for detailed molecular studies on their own account but could well benefit from the development of markers, probes and data obtained for other species or from collaborative integrated genus level programmes of work. Prince *et al.* (1993) working with pepper and tomato mapped two quantative collinear loci associated with number of flowers per node and were able to use comparative mapping to identify other loci associated with viral disease resistance. The first QTL mapping in cucumber was reported on fruit related traits in F3 families by Kennard and Havey in 1995. Later, Serquen et al. (1997) and Fazio *et al.(*2003) identified QTLs for a number of agronomic traits in cucumber such as lateral branch number, number of fruits per plant, and days to anthesisin the F3 and recombinant inbred lines (RILs) from the same cross (G421X H-19) , respectively. Sakata et al. (2005) made a QTL analysis of powdery mildew resistance in cucumber under two temperature (200C and 260C). QTL mapping for some other important agronomic traits (flower sex, fruit shapeindex, parthenocarpy etc.) were also reported (Dijkuizen and Staub, 2002; Wang *et al.*, 2005; Sun *et al.*, 2006; Yuan *et al.*, 2008). An advance backcross QTL study was performed (Rao *et al.*, 2003) in pepper using a cross between the cultivated species *Capsicum annuum* cv. Maor and the wild *C. frutescens* BG 2816 accession and 58 QTLs were reported for 10 yield contributing traits. Six QTLs controlling capsaicinoid content were detected in analysing F2 populationin an inter specific cross between *Capsicum annuum* and *C. frutescens* with the use of 728 molecular markers. The QTL mapping results such as the number and loci and their explained phenotypic variances have been shown affected by many factors, including genotype, population size, environment and trait itself.

Molecular markers contributed to a better understanding of the biological basis of the phenomenon of heterosis and to improved ways of identifying potentially heterotic combinations. Bonierbale *et al.* (1993) used RFLPs to examine the association between maximum level of heterozygosity and several components of yield in three populations of tetraploid potatoes. The results indicated that the value of maximum heterozygosity was dependent on the genetic background of the material under evaluation. There are a number of other characters of wide general significance in the production of new varieties where molecular analysis is making a substantial contribution. These include the identification and manipulation of male sterility in the production of F_1 hybrids, and the control of flowering and of photoperiod response.

Future Prospects

Molecular markers have improved our knowledge of the distribution of genetic diversity in vegetable crops and their wild relatives and started to make useful contributions to efficient conservation methods. There are a number of areas such as monitoring the identity or purity of accessions and testing for the effectiveness of regeneration procedures where molecular methods could make a much more substantial contribution.

References

Al Zahim, M., Newbury, H.J., Ford Lloyd and B.V. 1997. Classification of genetic variation in garlic (*Allium sativum* L.) revealed by RAPD. *HortScience,* **32:** 1102-1104.

Archak, S., Karihaloo, J.L., Jain, A. 2002. RAPD markers reveal narrowing genetic base of Indian tomato cultivars. *Current Science,* **82:** 1139-1143.

Arnon, Ben-Chaim, Borovsky Yelena, Falise, Mathew, Mazourek, Michael, Cheorl kang, Byoung, , Paran, Iian and Jahn Molly-=-. 2006. QTL analysis for capsaicinoid content in *Capsicum. Theor. Appl. Genet.,* **113:** 1481-1490.

Ashkenazi, V., Chani, E., Lavi, U., Levy, D., Hillel, J. And Veilleux, R. E. 2001. Development of microsatellite markers in potato and their use in phylogenetic and fingerprinting analyses. *Genome.,* **44:** 50-62.

Becerra-Velasquez, V.L. and Gepts P. (993). Characterization of the genetic diversity in common beans (*Phaseolus vulgaris* L.) using RFLP markers. Phaseolus Beans Advanced Biotechnology Research Network. Cali (Colombia). Centro Internacional de Agricultura Tropical, Cali (Colombia). 7-10 Sep. 1993.

Behera, T. K., Sharma, P., Singh, B. K., Kumar, G., Kumar, R., Mohapatra, T. and Singh, N. K. (2006). Assessment of genetic diversity and using species relationship in eggplant (*solanum melongena* L.) using STMS markers. *Scientia Horticulture,* **107:** 352-357

Bonierbale, M. W., Plaisted, R. L. and Tanksley, S. D. 1993. A test of the maximun heterozigosity hypothesis using mlecular markers in tetraploid potatoes. *Theoretical and Applied Genetics,* **86:** 481-91.

Bradley, K. F., Rieger, M. A. and Collins, G. G. 1996. Classification of Australian garlic cultivars by DNA fingerprinting. *Australian Journal of Experimental Agriculture,* **36:** 613-618.

Bromonschenkel, S.H. and Tanksley, S.D. 1997. Map based cloning of the tomato genomic region that spans the 100-seed weight-5 tospovirus resistance gene in tomato. *Mol. Gen. Genet.,* **256:** 121- 126.

Brown, A.H.D. and Spillane, C. 1999. Implementing core collections-principles, procedures, progress, problems and promise. *In:* Core collections for today and tommorrow, Johnson. R.C., Hodgkin, T. (eds.), 1-9. IPGRI, Rome, Italy.

Cansian, R.L. and Echeverrigaray, S. 2000. Discrimination among cultivars of cabbage using randomly amplified polymorphic DNA markers. *Hort. Science* **35:** 1155-1158.

Carmer, C.S. and Havey, M.J. 1999. Morphological, biochemical and molecular markers in onion. *Hort science,* **34:** 589-593.

Chaque, V., Mercier, J.C., Guenard, M., Courcel, A.D., Vedel, F. and De Courcel, A. 1996. Identification and mapping on chromosome 9 of RAPD markers linked to *Sw-5* in tomato by bulked segregant analysis. *Theor. Appl. Genet.,* **92:** 1045- 1051

Chavarriaga-Aguirre, P., Maya, M.M., Tohne, J., Duque, M.C., Iglesias, C., Bonierbale, M. W., Kresovich, S., Kochert, S. and Kochert G. 1999. Using microsatellites, isozymes and AFLPs to evaluate genetic diversity and redundancy in the cassava core collection and to assess the usefulness of DNA- based markers to maintain germplasm collections. *Molecular Breeding,* **53:** 2634-273.

Chunwongse, J., Doganlar, S., Crossman, C., Jiang, J. and Tanksley, S.D. 1997. High-resolution genetic map of the *Lv* resistance locus in tomato. *Theor. Appl. Genet.,* **95:** 220-223.

D' La Thierry, E.M.I., Panaud, O., Robert, T. and Ricroch, A. 1997. Assessment of genetic relationships among sexual and asexual forms of *Allium cepa* using morphological traits and RAPD markers. *Heredity,* **78:** 403-409.

Danin, P.Y., Reis, N., Tzuri, G. and Katzir, N. 2001. Development and characterization of microsatellite markers in *Cucumis. Theoretical and Applied Genetics,* **102:** 61-72.

Dax, E., Livneh, Edelbaum, Kedar, N., Gavish, N., Karchi, H., Milo, J., sela, I. and Rabinowitch, H.D. 1994. A random amplified polymorphic DNA (RAPD) marker for the *Tm-2a* gene in tomato. *Euphytica,* **74:** 159-163.

Dax, E., Livneh, O., Aliskevicius, E., Edelbaum, O., Kedar, N., Gavish, N., Milo, Geffen F, Blumenthal A, Rabinowitch HD and Sela, I. 1998. A SCAR marker linked to the ToMV resistance gene, *Tm22*, in tomato. *Euphytica,* **101:** 73-77.

De Vicente, M.C. and Tanksley, S.D. 1993. QTL analysis of transgressive segregation in an interspecific tomato cross. *Genetics,* **134:** 585-96.

Dey, S.S., Singh, A.K., Chandel, D. and Behera, T. K. 2006. Genetic diversity of bitter gourd (*Momordica charantia* L.) genotypes revealed by RAPD markers and agronomic traits. *Scientia Horticulture,* **109:** 21-28

Dijkhuizen, Arian, Kennard, Wayne C., Havey, Michael, J. and Staub, Jack, E. 1996. RFLP variation and genetic relationships in cultivated cucumber. *Euphytica,* **90:** 79-87.

Dijkuizen A, Staub JE. 2002. QTL conditioning yield and fruit quality traits in cucumber (*Cucumis sativus*

L.): effects of environment and genetic background. *J New Seeds,* **4:** 1-30.

Dirlewanger E, Issac PG, Ranade S, Belajouza M, Causin R, and de Viene D 1994. Restriction fragment length polymorphism analysis of loci associated with disease resistance genes and developmental traits in *Pisum sativum* L. *Theor. Appl. Genet.,* **88:** 17-27.

Diwan, N., Fluhr, R., Eshed, Y., Zamir, D. and Tanksley, S.D. 1999. Mapping of Ve in tomato: a conferring resistance to the broad spectrum pathogen, *Verlicillium dehliae* race 1. *Theor. Appl. Genet.,* **98:** 315-319.

Dixon, M.S., Jones, D.A., Hatzixanthis, K., Ganal, M.W., Tanksley, S.D. and Jones, J.D. 1995. High resolution mapping of the physical location of the tomato *Cf-2* gene. *Mol. Plant Microbe Interact.,* **8:** 200-206.

Doganlar, S., Dodson, J., Gabor, B., Beck-Bunn, T., Crossman, C. and Tanksley, S.D. 1998. Molecular mapping of the *py-l* gene for resistance to corky root rot (*Pyrenochaeta lycopersici*) in to- mato. *Theor. Appl. Genet.,* **97:** 784-788.

Edwards, J. D. And Mccouch, S.R. 2007. Molecular marker for use in plant molecular breeding and germplasm evaluation. *In* Marker-Assisted Selection: Current Status and Future Perspectives in Crops, Livestiock, Forestry and Fish. FAO, UN, Rome 29-50 pp.

Fazio G, Staub JE, Stevens Mr. 2003a. Genetic mapping and QTL analysis of horticultural traits in cucumber (*Cucumis sativus* L.) using recombinant inbred lines.*Theor Appl Genet.,* **107:**864-874.

Fazio, G., Stevens, M.R. and Scott, J.W. 1999. Identification of RAPD markers linked to fusarium crown and root rot resistance (*Frl*) in tomato. *Euphytica,* **105:** 205-210.

Fischer, D. and Bachmann K, 2000. Onion microsatellites for germplasm analysis and their use in assessing intra and inter specific relatedness within the subgenus *Rhizirideum. Theoretical and Applied Genetics,* **101:** 153-164.

Foolad, M.R., Zhang, L..P., Khan, A.A., Nino, L.D. and Lin, G.Y. 2002. Identification of QTLs for the early blight (*Alternaria solanai*) resistance in tomato using back cross populations of a *Lycopersicon esculentum x L. hirsutum* Cross. *Theor. Appl. Genet.,* **104:** 945-959.

Garcia, E., Jamilena, M., Alvarez, J.I., Arnedo, T., Oliver, J.L. and Lozano, R. 1998. Genetic relationships among melon breeding lines revealed by RAPD markers and agronomic traits. *Theoretical and Applied Genetics,* **96:** 878-885.

Grandillo, S., Ku, H.M. and Tanksley, S.D. 1999. Identifying the loci responsible for natural variation in fruit size and shape in tomato. *Theor. Appl. Genet.,* **99:** 978–987.

Groben, R. and Wricke, G. 1998. Occurrence of microsatellites in spinach sequences from computer databases and development of polymorphic SSR markers. *Plant Breeding,* **117:** 271-274.

Gwanama, C., Labuschagne, M.T. and Botha, A.M. 2000. Analysis of genetic variation in *Cucurbita moschata* by random amplified polymorphic DNA (RAPD) markers. *Euphytica,* **113:** 19-24.

Hanson, P.M., Bernacchi, D., Green, S., Tanksley, S., Muniappa, V., Padmja, A.S., Kuo, G., Fang, D. and Chen, J. 2000. Mapping of a wild tomato introgression associated with tomato yellow leaf curl virus resistance in a cultivated tomato line. *J. Amer. Soc. Hortic. Sci.* **125:** 15-20.

He, G.H., Prakash, C.S. and Jarret, R.L. 1995. Analysis of genetic diversity in a sweetpotato (*Ipomoea batatas*) germplasm collection using DNA amplification fingerprinting. *Genome* **38:** 938-945.

Hill, M., Witsenboer, H., Zabeau, M., Vos, P., Kesseli, R. and Michelmore, R. 1996. PCR-based fingerprinting using AFLPs as a tool for studying genetic relationships in *Lactuca* spp. *Theoretical and Applied Genetics* **93:** 1202-1210.

Hodgkin, T., Roviglioni R., Vicente, M. C., De and Dudnik, N. 2001. Molecular methods in the conservation and use of plant genetic resources. Methods D' Analyse moleculaire pour la conservation et L'Utilization Des resources gentiques vegetables. *Acta Horticulture,* 546

Horejsi, T., and Staub JE. 1996. Genetic variation in cucumber (*Cucumis sativus* L.) as assessed by random amplified polymorphic DNA. *Genetic Resources and Crop Evolution,* **46:** 337-350.

Jahn, M., Paran, I., Hoffmann, K., Radwanski, E.R,, Livingstone, K.D., Grube, R.C., Aftergoot, E., Lapidot, M. and Moyer, J. 2000. Genetic mapping of the Tsw locus for resistance to the Tospovirus tomato spotted wilt virus in *Capsicum spp.* and its relationship to the Sw-5 gene for resistance to the same pathogen in tomato. *Mol. Plant Microbe Interact.,* **13:** 673-682.

Karihaloo, J. L. and Bhat, K.V. 2002. Molecular markers for enhanced efficiency of breeding vegetables. In: Proceding of *International Conference on Vegetables,* November, 11-14, Bangalore. 162-171.

Karihaloo, J.L., Brauner, S. and Gottlieb, L.D. 1995. Random amplified polymorphic DNA variation in the eggplant, *Solanum melongena* L. (Solanaceae). *Theoretical and Applied Genetics* **90:** 767-770.

Karp, A., Issar, P.G. and Ingram, D.S. (eds.) 1998. Molecular tools for screening biodiversity. Chapman & Hall, London.

Karp, A., Kresovich, S., Bhat, K.V., Ayad, W.G. and Hodgkin, T., 1997a. Molecular tools in plant genetics resources conservation: a guide to the technologies. IPGRI Technical Bulletin No. 2, IPGRI, Rome, Italy.

Kawchuk, L.M., Hachey, J. and Lynch, D.R. 1998. Development of sequence characterized DNA markers linked to a dominant Verticillium wilt resistance gene in tomato. *Genome* **41:** 91-95.

Kawchuk, L.M., Lynch, D.R., Hachey, J., Bains, P.S. and Kulcsar, F. 1994. Identification of a codominant amplified polymorphic DNA marker linked to the Verticillium wilt resistance gene in tomato. *Theor. Appl. Genet.* **89:** 661-664.

Kelly, J.D. and Miklas, P.N., 1998. The role of RAPD markers in breeding for disease resistance in common bean. *Molecular Breeding,* **4:**1-11

Kennard, W.C. and Havey, M.J. 1995. Quantitative trait analysis of fruit quality in cucumber: QTL detection, confirmation, and comparision with mating-design variation. *Theor Appl Genet.* **91:** 53-61.

Khandka, D.K., Nejidat, A., Golan and Goldhirsh A. 1996. Polymorphism and DNA markers for asparagus cultivars identified by random amplified polymorphic DNA. *Euphytica* **87:** 39-44.

Kulakow, P.A., Hauptli, H. And Jain, S.K. 1985. Genetics of grain amaranthus. I. Mendelian analysis of six colour characteristics. *J. Hered.* **76:** 27-30.

Kumar, S., Kumar, S., Singh, M. And Rai, M. 2003. Marker aided selection for disease resistance in vegetable crops. *Vegetable Sciencne.* **30(1):** 10-20.

Las, Heras Vazquez F.J., Jimenez J.M.C., Vico, F.R. 1996 RAPD fingerprinting of pepper (*Capsicum annuum* L.) breeding lines. *Capsicum and Eggplant Newsletter* **15:** 37-40.

Maass, H.I. and Klaas, M. 1995. Infraspecific differentiation of garlic (*Allium sativum* L.) by isozyme and RAPD markers. *Theoretical and Applied Genetics* **91:** 89-97.

Margale, E., Herve Y., Hu, J. and Quiros, C.F. 1995. Determination of genetic variability by RAPD markers in cauliflower, cabbage and kale local cultivars from France. *Genetic Resources and Crop Evolution* **42:** 281-289.

Marker, C.L. Ellis, T.H.N. and Coen, E.S. 1990. Identification and genetic regulation of the chalcone synthase multigene family in pea. *Plant Cell* **2:** 184-193.

McGregor, C.E., Lambert, C.A., Greyling, M.M., Louw, J.H. and Warnich, L. 2000. A comparative assessment of DNA fingerprinting techniques (RAPD, ISSR, AFLP and SSR) in tetraploid potato (*Solanum tuberosum* L.) germplasm. *Euphytica* **113**: 135-144.

Melotto, M., Afanador, L. and Kelly, J.D. 1996. Development of a SCAR marker linked to the I gene in common bean. *Genome* **39:** 1216-1219.

Michelmore, R.W. 1995. Molecular approaches to manipulation of disease resistance genes. *Ann. Rev. Phytopath.* **33:** 393-427.

Miklar, P.N., Afanador, L. and Kelly, J.D. 1993. Identification and potential use of molecular marker for rust resistance in common bean. *Theor. Appl.Genet.* **85:** 745-749.

Miller, J. C. and Tanksley, S. D. 199). RFLP analysis of phylogenetic relationships and genetic variation in the genus *Lycopersicon. Theoretical and applied genetics,* **80:** 437-48

Monforte, A.J. and Tanksley, S.D. 2000. Development of a set of near isogenic and backcross lines containing most of the *Lycopersicon hirsutum* genome in a *L. esculentum* genetic background: A tool for gene mapping and discovery. *Genome* **43:** 803–813.

Moreau, P., Thoquet, P., Olivier, J., Laterrot, H. and Grimsley, N. 1998. Genetic mapping of Ph-2, a single locus controlling partial resistance to *Phytophthora infestans* in tomato. *Mol. Plant Microbe Interact.* **11:** 259-269.

Moury, B., Pflieger, S., Blattes, A., Lefebvre, V. and Palloix, A. 2000. A CAPS marker to assist selection of tomato spotted wilt virus (TSWV) resistance in pepper. *Genome* **43:** 137-142.

Paran, I., Aftergoot, E. and Shifriss, C. 1998. Variation in *Capsicum annuum* revealed by RAPD and AFLP markers. *Euphytica* **99:** 167- 173.

Patrick, Z.A., Kerr, E.A. and Bailey, D.L. 1971. Two races of *Cladosporium fulvum* new to Ontario and further studies of *Cf-1* resistance in tomato cultivars. *Can J Bot* **49:** 189–193.

Phippen, W.B., Kresovich, S., Candelas, F.G. and McFerson, J.R. 1997. Molecular characterization can quantify and partition variation among genebank holdings: a case study with phenotipically similar accessions of *Brassica oleracea* var. *capitata* L. (cabbage) 'Golden Acre'. *Theoretical and Applied Genetics*, **94:** 227-34.

Pickersgill, B. 1971. relationships between weedy and cultivated forms in some species of chili pepers (genus Capsicum). *Evoluation* **25:** 683-691.

Pickersgill, B. 1981. Biosystematrics of crop-weed complexes. *Kulturpflanze* **29:** 377-388.

Powell, W., Morgante, M., Doyle, J. J., McNicol A. W., Tingey S. V. and Rafalski A.J., 1996a. The comparison of RFLP, RAPD, AFLP and SSR (Micro satellite) markers for germplasm analysis. *Molecular Breeding*, **2:** 225-38.

Prince J.P., Pochard E. and Tanksley S.D., 1993. Construction of a molecular linkage map of pepper and a comparison of synteny with tomato. *Genome,* **36:** 404-17.

Rabbani, M.A., Murakami, Y., Kuginuki, Y. and Takayanagi K, 1998. Genetic variation in radish (*Raphanus sativus* L.) germplasm from Pakistan using morphological traits and RAPDs. *Genetic Resources and Crop Evolution* **45:** 307-316.

Ranade, S.A., Kumar, A., Goswami, M., Farooqui, N. and Sane, P.V. 1997. Genome analysis of amaranths: determination of inter- and intra-species variations. *Journal of Biosciences* **22:** 457-464.

Rao, G.U., Ben Chaim, A., Borovsky, Y. And Paran, I. 2003. Mapping of yield-related QTLs in pepper in an interspecific cross of *Capsicum annuum* and *C. Frutescens*. *Theor Appl Genet.* **106:** 1457-1460.

Rodriguez, J.M., Berke, T., Engle, L. and Nienhuis, J. 1999. Variation among and within Capsicum species revealed by RAPD markers. *Theoretical and Applied Genetics* **99:** 147-156.

Roose, M.L. and Stone, N.K. 1996. Development of genetic markers to identify two asparagus cultivars. In: Nichols M, Swain D (Eds.) Proceedings of the VIII International asparagus symposium. 1993 Nov 21, Palmerston North, New Zealand. *Acta Horticulturae* No. **415:** 129-135.

Rus, Kortekaas, W., Smulders, M.J.M., Arens, P. and Vosman, B. 1994. Direct comparison of levels of genetic variation in tomato detected by a GACA-containing microsatellite probe and by random amplified polymorphic DNA. *Genome* **37:** 375-381.

Sakata, Y., Kubo, N., Morishita, M., Kitadani, E., Sugiyama, M. and Hirat, M. 2006. QTL analysis of powdery mildew resistance in cucumber (*Cucumis sativus* L.) *Theor Appl Genet.* **112:** 243-250

Samec, P. and Nasinec, V. 1996. The use of RAPD technique for the identification and classification of *Pisum sativum* L. genotypes. *Euphytica* **89:** 229-234.

Sarfatti, M., Katan, J., Fluhr, R. and Zamir, D. 1989. An RFLP marker in tomato linked to the *Fusarium oxysporum* resistance gene 12. *Theor. Appl. Genet.* **78:** 755-759.

Serquen, F. C., Bacher, J. and Staub, J. E. 1997. Mapping and QTL analysis of horticultural traits in a narrow cross in cucumber (*Cucumis sativus* L.) using random-amplified polymorphic DNA markers. Mol Breed **3:** 257-268.

Shim, S.I. and Jorgensen, R.B. 2000. Genetic structure in cultivated and wild carrots (*Daucus carota* L.) revealed by AFLP analysis. *Theoretical and Applied Genetics* **101:** 227-233.

Sicard, D., Woo, S.S., Arroyo-Garcia, R., Ochaoa, O., Nguyen, D., Korol, A., Nevo, E. and Michelmore, R. 1999. Molecular diversity at the major cluster of disease resistance genes in cultivated and wild *Lactuca* spp. *Theoretical and applied genetics,* **99:** 405-18.

Singh, A.K., Behera, T.K., Chandel, D., Sharma, P., and Singh, N.K. 2007 Assessing genetic relationships among bitter ground (*Momordica charanntia* L.) accesssions using inter-simple sequence repeat (ISSR) markers. *Journal of Horticulture Science & Biotechnology* **82 (2):** 217-222.

Skroch, P.W. and Nienhuis, J. 1995. Qualitative and quantitative characterization of RAPD variation among snap bean (*Phaseolus vulgaris*) genotypes. *Theoretical and Applied Genetics* **91:** 1078-1085.

Skroch, P.W., Nienhuis, J., Beebe, S., Tohme, J. and Pedrazza, F. 1998. Comparison of Mexican common bean (*Phaseolus vulgaris* L.) core and reserve germplasm collection. *Crop Science*. **38:** 488-96.

Sobir, O.T. , Murata, M. and Motoyoshi, F. 2000. Molecular charceterization of the SCAR markers tightly linked to the *Tm-2* locus of genus *Lycopersicon. Theor. Appl. Genet.* **101:** 64-69.

Sonnante, G., Stockton, T., Nodari, R.O., Becerra-Velasquez, V.L. and Gepts P. 1994. Evolution of genetic diversity during the domestication of common-bean (*Phaseolus vulgaris* L.) *Theoretical and Applied Genetics* **89:** 629-635.

Stachel, M., Csanadi, G., Vollmann, J. and Lelley, T. 1998. Genetic diversity in pumpkins (*Cucurbita pepo* L.) as revealed in inbred lines using RAPD markers. *Report Cucurbit Genetics Cooperative* **21: 48-50.**

Stamova, B.S. and Chetelat, R.T. 2000. Inheritance and genetic mapping of cucumber mosaic virus resistance introgressed from *Lycopersicon chilense* into tomato. *Theor. Appl. Genet.* **105:** 527-537

Staub, J.E., Box, J., Meglic, V., Horejsi, T.F. and McCreight, J.D. 1997. Comparison of isozyme and random amplified polymorphic DNA data for determining intraspecific variation in *Cucumis. Genetic Resources and Crop Evolution* **44:** 257-269.

Staub, J.E., Serquen, F.C. and Gupta, M. 1996. Genetic markers, map construction, and their application in plant breeding. *HortScience* **31:** 729-741.

Stockton, T. and Gepts. P. 1994. Identification of DNA probes that reveal polymorphisms among closely related *Phaseolus vulgaris* lines. *Euphytica* **76:** 177-183.

Sun, Z., Staub, J.E., Chung, S.M. and Lower, R. L. 2006. Idnetificiation and comparatiive analysis of quantitative trait loci associated with parthenocarpy in processing cucumber. *Plant Breed* **125:** 281-287.

Sureja, A.K., Sirohi, P.S., Behera, T.K. and Mohapatra, T. 2006. Molecular diversity and its relationship with hybrid performance and heterosis in ash ground [Benincasa hispida (Thunb.) Cogn.]. *Journal of Horticulture Science & Biotechnology* **81 (1): 33-38.**

Tai, T., Dahlbeck, D., Stall, R.E., Peleman, J. and Staskawicz, B.J. 1999. High-resolution genetic and physical mapping of the region containing the *Bs2* resistance gene of pepper. *Theor. Appl. Genet.* **99:** 1201-1206.

Tankley, S.D. and McCouch, S.R. 1997. Seed Banks and Molecular Maps Unlocking Genetic Potential from the wild. *Science,* **227:** 1063-6.

Tian, M.Y., Feng, L.X., Yang, C.R. and Gong, H.Z. 2000. Identification of a molecular marker linked to ToMV resistance gene *Tm2nv* in tomato using randomly amplified polymorphic DNA. *Acta Phytopathologica Sinica* **30:** 158-161.

Tivang, J., Skroch, P.W., Nienhuis, J. and De, Vos N. 1996. Randomly amplified polymorphic DNA (RAPD) variation among and within artichoke (*Cynara scolymus* L.) cultivars and breeding populations. *Journal of the American Society for Horticultural Science* **121:** 783-788.

Van, der and Plank, J.E. 1963. Plant diseases: epidemics and control. Academic Press, New York/London, 349 pp.

Villand, J., Skroch, P.W., Lai, T., Hanson, P., Kuo, C.G. and Nienhuis, J. 1998. Genetic variation among tomato accessions from primary and secondary centers of diversity. *Crop Science* **38:** 1339-1347.

Walker, J.C. 1923. Disease resistance to onion smudge. *J. Agric. Res.* **24:** 1019-1040.

Wang, G,, Pan, J.S., Li, X.Z., He, H.L. and Cai, R. 2005. Construction of a cucumber genetic linkagemap with SRAP markers and location of the genes for lateral branch traits. *Sci China* C., **48:** 213-220.

Wechter, W.P., Whitehead, M.P., Thomas, C.E. and Dean, A. 1995. Identification of a randomly amplified polymorphic DNA marker linked to the *Fom 2* Fusarium wilt resistance gene in muskmelon MR-1. *Phytopath.,* **85:** 1245-1249.

Williamson, V.M. 1998. Root-knot nematode resistance genes in tomato and their potential for future use. *Ann. Rev. Phytopath.*, **36:** 277-293.

Williamson, V.M., Ho, J.Y., Wu, F.F., Miller, N. and Kaloshian, I. 1994. A PCR based marker tightly linked to the nematode resistance gene, *Mi*, in tomato. *Theor. Appl. Genet.* **87:** 757-763.

Yaghoobi, J., Kaloshian, I., Wen, Y. and Williamson, V.M. 1995. Mapping a new nematode resistance locus in *Lycopersicon peruvianum*. *Theor. Appl. Genet.*, **91:** 457-464.

Yuan, X. J., Pan, J. S., Cai, R., Guan, Y., Liu, L. Z., Zhang, W.W., Li, Z., He, H. L., Zhang, C., Si, L.T. and Zhu, L. H. 2008. Genetic mapping and QTL analysis of fruit and flower related traits in cucumber (*Cucumis sativus* L.) using recombinant inbred lines. *Euphytica* (doi: 10.1007/s10681-008-9722-5)

Zamir, D., Ekstein-Michelson, I., Zakay, Y., Navot, N., Zeidan, M., Sarfatti, M,. Eshed, Y., Harel, E., Pleban, T., Oss, H., Van, Kedar, N., Rabinowitch, H.D. and Czoshek, H. 1994. Mapping and introgression of a tomato yellow leaf curl virus tolerance gene, *Ty-l. Theor. Appl. Genet.,* **88:** 141-146.

❑❑❑

Chapter – 8

Intellectual Property Rights in Relation to Germplasm Management

H.S. Chawla

Introduction

The WTO was established on 1st January 1995 and is responsible for making and enforcing rules for trade between nations. WTO marks a major change in global trade rules. As an organization, it replaces the General Agreement on Tariffs and Trades (GATT), which had been in existence since 1947. The Eighth Round of Multilateral Trade Negotiations under GATT, which started in Uruguay in 1986, was concluded in 1994, leading to the creation of WTO as the new permanent international trade organization. The role of WTO is much more extensive than that of GATT, which dealt with trade in goods. Apart from goods, the two other broad areas that WTO covers are services and intellectual property, which previously belonged to the domestic domain. Accordingly, WTO administers not only the Multilateral Trade Agreements (MTAs) in goods but also the General Agreement on Trade in Services (GATS) and the Agreement on Trade Related Aspects of Intellectual Property Rights (TRIPS), which came into existence with WTO. All the agreements annexed to the Agreement establishing the WTO were signed as part of a package deal. Member countries did not have the option of choosing some and rejecting others. Another important difference with the erstwhile GATT is that WTO has a stronger compliance mechanism (Chawla, 2007a).

As one of the WTO agreements, TRIPS is binding on all member countries of WTO. TRIPS aims at establishing strong minimum standards for intellectual property rights (IPRs). Apart from patents, intellectual property includes copyrights, trademarks, geographical indications, industrial designs, integrated circuits and trade secrets. The protection of IPRs is binding and legally enforceable. Intellectual property (IP) is a product of the mind. Intellectual property is intangible in contrast to real property (land) or physical property, which one can see, feel and use. With any type of property there are property rights. When IPs are expressed in a tangible form, they can also be protected. Intellectual property rights (IPRs) have been created to protect the right of individuals to enjoy their creations and discoveries. In fact, IPRs can be traced back to the fourteenth century, when European monarchs granted proprietary rights to writers for their literary works.

IPRs have been created to ensure protection against unfair trade practice. Owners of IP are granted protection by a state and/or country under varying conditions and periods of time. This protection includes the right to:

(i) defend their rights to the property they have created;
(ii) prevent others from taking advantage of their ingenuity;
(iii) encourage their continuing innovativeness and creativity; and
(iv) assure the world a flow of useful, informative and intellectual works.

Forms of Protection

IPRs can be defined as the rights given to people over the creation of their minds. They usually give the creator an exclusive right over the use of his/her creation for a certain period of time. Intellectual property includes patents, copyrights, trademarks, geographical indications, industrial designs, integrated circuits and trade secrets (Chawla and Singh, 2005). The protection of IPRs is binding and legally enforceable.

Patents

A patent is a government granted exclusive right to an inventor to prevent others from practicing i.e. making, using or selling his invention. A patent is a personal property, which can be licensed or sold like any other property. The purpose of a patent is to encourage and develop new innovations. The Patent Law recognizes the exclusive right of a patentee to gain commercial advantage out of his invention. There are three criteria to issue a patent for the innovation (Chawla and Singh, 2007).

(i) **Novelty:** The inventor must establish that the invention is new or novel. The novelty requirement refers to the prior existence of an invention. If an invention is identical to an already patented invention, the novelty requirement is not met, so a patent cannot be issued.
(ii) **Inventiveness (Non-obviousness):** It is an invention and not merely discovery. It is non obvious to one skilled in the field. The non-obvious requirement refers to the level of difficulty required to invent the technology.

If an invention is so obvious that anyone having an ordinary skill would have thought of it, then it does not meet this requirement.

(iii) **Usefulness (Industrial application):** It has a utility or is useful for the society. The useful requirement refers to the practical use of invention. If an invention provides a product that is required or needed in some manner, then it meets this requirement.

In the patent adequate disclosure should be made so that others can also work on it. It should have the features like: i) be a written description; ii) enables other persons to follow; iii) adequate and iv) deposit mechanism.

The present law, Patents Act 1970, amendment 2005 is effective from January 1, 2005. Product patents on all items including food, agro-chemical and pharmaceuticals have also been allowed making the Patents Act fully TRIPS compliant.

The patent system was developed as a means to reward inventions which would be useful to the society. However, in order to ensure the interests of society, as per the Indian Patents Act, certain things have been excluded from the purview of patentability. The sections relevant to plant material and agriculture which are excluded from patentability are:

Section 3(h): a method of agriculture and horticulture.

Section 3(i): any process for medicinal, surgical, curative, prophylactic (diagnostic therapeutic) or other treatment of human beings or any process for a similar treatment of animals to render them free of disease or to increase their economic value or that of their products.

Section 3(j): plants and animals in whole or any part thereof other than microorganisms but including seeds, varieties and species and essentially biological processes for production or propagation of plants and animals.

Section 3(p): an invention which in effect, is traditional knowledge or which is an aggregation or duplication of known properties of traditionally known component or components.

Further the mere discovery of any new property or new use for a known substance or the mere use of a known process, machine or apparatus unless such known process results in a new product or employ at least one new reactant is not patentable. Also a patent claim for a substance obtained by merely mixing ingredients resulting only in the aggregation of the properties of the components is not a patentable invention.

Microorganisms *per se* can be claimed for protection provided they are not mere discovery of organisms. It is mandatory to deposit the biological material in International Depositary Authority (IDA). In India, Institute of Microbial Technology (IMTECH), Chandigarh is a recognized international depositary for some category of micro-organisms. If an applicant mentions a biological material in the patent specification then disclosure requirements prescribed for biological materials have been notified in the list of the Central Government or for indicating its source and geographical origin [Section:10.4(d)].

However, in India, method for rendering plants free of diseases or to increase their economic value or that of their products can be claimed for patent protection.

The purpose of a patent is to promote the progress of science and useful arts. The patent law promotes this progress by giving the inventor the right of exclusion. In exchange for this right to exclude others, the inventor must disclose all details describing the invention, so that when the patent period expires, the public may have the opportunity to develop and profit from the use of invention. A patent is enforced in the country which issues it, meaning thereby territorial in nature. For each country a separate application is to be filed in that country where protection is sought.

Microorganism Patents

The first patent on living organism was granted to a micro-organism *Pseudomonas*. The first example was the classical judgment in the Diamond *vs*. Chakrabarty case in 1980. In the Chakrabarty case USPTO rejected the patent application on the ground of product of nature but the US Supreme Court decided that a microorganism was not precluded from patentability solely because it was alive. Thus a *Pseudomonas* bacterium manipulated to contain more than one plasmid (four plasmids were present) controlling the breakdown of hydrocarbons (therefore more useful in dispersing oil slicks than the natural organism containing only one such plasmid) was "a new bacterium with markedly different characteristics from any found in nature" and hence not nature's handiwork but that of inventor. The "product of nature" objection therefore failed and the modified organisms were held patentable. This precedent is being followed even today to define the patentability of microorganisms (Chawla, 2007 b).

Plant Patents

Plant patents are obtainable in US, Europe and Japan. The US Plant Patent Act of 1930 (PPA) granted property rights for privately developed plant varieties of asexually reproducing plants. These rights were extended to new and distinct asexual varieties for a period of seventeen years. Advances in breeding technology provided the momentum for the 1970 Plant Variety Protection Act (PVPA). The PVPA provided protection for sexually reproducing plants, including seed germination. In 1980 Diamond vs. Chakrabarty case set in motion the trend towards the legal acceptance of the commodification of germplasm. The court held that a live, man made bacterium was patentable under the PPA and the 'product of nature' objection therefore failed and the modified organisms were held patentable. In the Hibberd case (1985), involving a tryptophan-overproducing mutant, the patent office ruled that plants could be patented and there is no distinction between asexually and sexually propagated plants. Following the principle established in the Chakrabarty case, it was decided that normal US utility patents could be granted for other types of plant e.g. genetically modified plants. Plant patents have been granted by European Patent Office (EPO) from 1989. But in 1995, EPO severely restricted the scope of Plant Genetic Systems (Belgium) patent on herbicide resistant plants and allowed claims only on the herbicide resistant gene and the process used in the generation of plants. In Japan, plant patents are allowed, but there are some disputes over territorial rights. Life forms of plants and animals except microorganisms are not patentable in India. In pursuance to the TRIPS agreement, India has enacted "Protection of Plant Varieties and Farmers' Rights" (PPV&FR) Act, 2001, a *sui generis* system of plant variety protection which has been described in detail separately.

Plant Variety Protection in India

As stated India is signatory to WTO agreements and it has to abide by the TRIPS regulations. As per article 27.3(b) of the TRIPS which demand that member countries should protect their plant varieties either by patent, or an effective system of *sui generis* protection, or a combination of these two. In this context India chose a *sui generis* system for protection of plant varieties. An Act named as Protection of Plant Varieties and Farmers' Rights (PPV&FR) Act 2001 has been passed and Rules have been framed. PPV&FR Authority has been constituted with its Head Office located at Delhi. The PPV&FR Act is TRIPS compliant and compatible with UPOV system of plant variety protection (Anonymous, 2003).

The PPV&FR Act 2001 provides protection to following types of plant varieties (Anonymous, 2003):

(i) Newly bred varieties.

(ii) Extant varieties – The varieties which were released under Indian Seeds Act, 1966 and have not completed 15 years as on the date of application for their protection.

(iii) Farmer's varieties – The varieties which have been traditionally cultivated, including landraces and their wild relatives which are in common knowledge, as well as those evolved by farmers.

(iv) Essentially derived varieties.

(v) Transgenic varieties.

An application for registration can be made by any person claiming to be the breeder of the variety, successor of the breeder, assignee, any farmer or group of community of farmers, any person authorized for the above mentioned categories or any University or publicly funded agricultural institution claiming to be the breeder of the variety. It is pertinent to note that the Act recognizes the farmer as a cultivator, conserver and breeder. This embraces all farmers, landed or landless, male and female. To qualify for registration under the act, a new variety has to conform to the criteria of novelty (N), distinctiveness (D), uniformity (U) and stability (S). Besides, a denomination has to be given for the registration of variety. Denomination refers to the label or title of the variety. It is the denomination that is registered. For extant and farmers' varieties which are in public domain the DUS features will be considered while the novelty feature will not be taken because these varieties are not new and are in public domain. In this act a special clause has been put which states that any variety with terminator gene sequences will not be registered. Thus any transgenic material with genetic use restriction technology (GURT) sequences will not be registered.

All the varieties will be registered with PPV&FR Authority. DUS guidelines for 35 crops have been prepared by ICAR while guidelines for 12 crop species have been notified by PPV&FR Authority in the gazette. PPV&FR Authority has established testing centres for each and every crop species. In the first phase which has started in May, 2007, the registration of varieties will be done for 12 crop species of cereals and legumes. The registration will then extend to 35 crops which include cereals, pulses, oilseeds, vegetable and two flower species. DUS guidelines are also being prepared for medicinal and aromatic plants, spices, ornamentals and forest trees for which task forces have been constituted by the PPV&FR Authority.

Indian PPV&FR Act allows farmers to save, use, sow, resow, exchange, share or sell his farm produce including seed of a variety protected under this Act, but it prohibits that the farmer shall not be entitled to sell branded seed of a variety protected under the Act [Sec. 39, 1(iv)]. The farmers have been given the right to register farmers varieties themselves [Sec. 39,1(i)], right to claim compensation for under performance of a protected variety from the promised level [Sec. 39(2)], benefit sharing for use of biodiversity conserved by farming community [Sec. 41]. According to the concept of benefit sharing, whenever a variety submitted for protection is bred with the possible use of a landrace, extant variety or farmer's variety, a claim can be referred either on behalf of the local community or institution for a share of the royalty [Sec. 41(1)] (Anonymous, 2003). In the Act a provision of compulsory license has also been put. According to this, after the expiry of three years from the date of issue of certificate of registration of a variety, any person interested can claim in an application to the authority alleging that reasonable requirements of the public for seeds or other propagating material have not been satisfied or that the seed or other propagating material is not available to the public at a reasonable price and pray for the grant of a compulsory license to undertake production, distribution and sale of the seed or other propagating material of that variety [Sec. 47(1)] (Anonymous, 2003).

The Act had laid down the norms for registration of plant varieties, fee structure, provisions of opposition, DUS testing of material, etc. If any farmer or association of farmers is applying for registration of a plant variety then this category is not required to pay any fee for either registration or DUS testing. Once the variety has been tested for its features then the Registrar of the Authority will issue the certificate of registration. It shall have the validity of nine years initially in case of trees and vines with renewal up to a period of 18 years. For other crops certificate of registration will be issued for six years initially with renewal up to 15 years. In case of extant varieties the validity period is 15 years from the date of notification of that variety by the Central Government under section 5 of the Seeds Act 1966.

References

Anonymous, 2003. The Protection of Plant Varieties and Farmers' Rights Act, 2001 and Rules, Universal Law Publishing Co., Delhi, 2003.

Chawla, H. S., 2007a. Managing intellectual property rights for better transfer and commercialization of agricultural technologies. *J Intellectual Property Rights,* **12:** 330 – 340

Chawla, H. S., 2007b. Intellectual Property Rights, *J. Eco-friendly Agriculture,* **2(2):** 103-112

Chawla, H. S. and Singh, A. K., 2005. Intellectual Property Rights. Vol II: Copyrights, Trade Marks, Trade Secrets and Geographical Indications. Pantnagar University Press, 75p.

Chawla, H. S. and Singh, A. K., 2007. Intellectual Property Rights: Patents, Plant Variety Protection and Biodiversity, Published by Intellectual Property Management Centre, G.B. Pant Univ. of Agric & Tech., Pantnagar, 54 p.

Chapter – 9

Impact of Genetic Modified Crops on Biodiversity and Environment: Biosafety Issues and Regulations

Anil Kumar and Sonu Ambwani

Introduction

Indian subcontinent has a rich and varied heritage of bioresources, encompassing a wide spectrum of habitats from tropical rain forests to alpine vegetation and from temperate forests to coastal wetlands. It is one of the eight centres of origin and one of the 12 mega centres of the world. It possess 11.9 % of world flora and about 33% of the country's recorded flora are endemic to the region and are concentrated mainly in the North-East, Western Ghats, North-West Himalayas and the Andaman and Nicobar Islands. Of the 49,219 higher plant species, 5,725 are endemic and belong to 141 genera under 47 families. Of theses 3,500 are found in the Himalayas and adjoining regions and 1,600 in the Western Ghats alone. India has 26 recognized endemic centres that are homeland to—

- One-third of all the flowering plants identified and described to date
- 167 cultivated species and 329 wild relatives of the crop plants
- There are about 583 cultivated crops in India
- It has about 30,000-50,000 landraces of rice, pigeonpea, mango, turmeric, ginger, sugar cane, gooseberries etc.

- Nearly 9,500 plants species of ethano-botanical usage are reported from the country

Preserving biodiversity in the face of a variety of well-documented encroachments is more than an aesthetic or strictly environmental concern. There are two basic ways to conserve bioresources/genetic diversity: *in-situ* (in the natural environment, as in national park lands or on farms) and *ex-situ* (removed from the natural environment, as in gene banks). With *in-situ* conservation, dynamic evolutionary processes continue to operate, including the possibility of mutation and the threat of extinction. With *ex-situ* conservation, the long-term safety and integrity of genetic resources is maintained by collecting and preserving seeds, living plants, cuttings and tissue cultures. If a plant becomes endangered *in-situ*, it can be moved to a gene bank for *ex-situ* conservation. The functions of a gene bank include maintenance and expansion of germplasm collections, long term conservation, including multiplication and regeneration, characterization and evaluation of samples & data management of bioresources, exchange of germplasm among researchers and the promotion of germplasm use to enhance crop productivity. *Ex situ* conservation in gene banks is a safe and cost-effective method of preserving the genetic diversity of crops and wild species of plants, as long as the seeds can tolerate desiccation and storage at low temperatures. Biodiversity plays an important economic, social, and cultural role in the lives of many people, particularly indigenous and local communities.

Biotechnology in Crop Improvement

From the late nineteenth century, plant breeders began using hybridization and cross breeding aggressively as a means of improving agricultural crops employing available gene pools/bioresources. The advances in conventional plant breeding led to the development of more productive varieties of food crops with better nutritional quality. From the 1980's, scientists began using recombinant DNA (r-DNA) techniques as a new tool for developing crops with specific beneficial traits. Recombinant DNA techniques allow scientists to isolate certain desired gene sequences from various organisms and introduce them into other organisms. The scene for the growth and expansion of biotechnology in India was firmly set in place by the establishment of the National Biotechnology Board in 1982 by the Government of India Act, which was upgraded four years later to a separate **Department of Biotechnology (DBT).** The vision statement of DBT is **"Attaining new heights in Biotechnology research, shaping it into a premier precision tool of the future for creation of wealth and ensuring social justice-especially for the welfare of the poor"**. DBT is regularly advised by a distinguished Scientific Advisory Committee (SAC-DBT) and the Standing Advisory Committee (overseas) SAC- 0 about new and emerging areas. In addition there are Biotechnology based programmes related to food and nutritional security.

Agriculture is expected to feed an increasing population, forecasted to reach 8 billion by 2020, out of whom 6.7 billion will be in developing countries where the carrying capacity of agricultural lands will soon be reached. Fifteen years ago, plant biotechnology comprised only a few applications of tissue culture, recombinant DNA technology and monoclonal antibodies. Today, genetic transformation and marker aided selection and breeding are just a few of the examples of the applications in crop improvement. Plant biotechnology applications must respond to increasing demands in terms of food security, socio-economic development and promote the conservation, diversification and sustainable use of plant genetic resources as basic inputs for the future agriculture of the region.

Food security is defined by FAO as the access by all people at all times to the food needed for a healthy and active life. The concept means the achievement of the food self-sufficiency and guarantees that this condition will be sustained in the future. Food security implies reaching productive growth and the preservation of the environment. Agricultural biotechnology is intended to address some of the critical challenges faced by farmers, which include threats to crop yield from diseases, pests, weeds or nutrients and water deficiency. In the recent past, there has been rapid increase in research and development of Genetically Modified Organisms (GMOs) / Living Modified Organisms (LMOs) and their exploitation for commercial purposes.

Thereafter adoption of GM crops has been at a very fast pace. The estimated global area of all GM crops was more than 114 million hectares in 2007 which is nearly 100 times the GM crops planted in 1996. Realizing that GM crops can accelerate future global food security and produce foods that are higher in nutrient content, taste better and have higher yields with less chemical use, and also cost less in cultivation. It has triggered international market for GMOs at unprecedented rate.

Genetically Modified Crops

Transgenic technology has opened new vistas in agriculture. The economic policy of globalization in India is going to give tremendous boost to agricultural productivity using tools of genetic engineering for providing necessary competitive edge in productivity with custom build quality. Thus, transgenic technology as evident from the various on-going and prospective applications, hold considerable promise to meet the challenges for accelerated and sustainable agricultural production. Transgenic plants have been deliberately developed for a variety of reasons: longer shelf life, disease resistance, herbicide resistance and pest resistance, non-biological stress resistances, such as to drought or nitrogen starvation and nutritional improvement. The first modern transgenic crop approved for sale in the US, in 1994, was the FlavrSavr tomato, which was intended to have a longer shelf life.

Genetic engineering has wide ramification not only in diversification but also by genetic restructuring of plants. This offers new opportunities for the development of (i) transgenic crops for important agronomic traits, resistance to diseases, pests and abiotic stresses, improvement of quality traits and manipulations of physiological attributes for greater productivity and sustainability, (ii) transformation of plant cells which acts like biofactories for the production of biologically important molecules used in diagnostics and therapeutics such as immunoglobulins, interferons, growth factors and safe, effective and less expensive immunoprophylaxis such as recombinant vaccines and edible vaccines.

Although, theoretically there is no boundary of the application of transgenic technology for restructuring of plants and development of designer crops, however, imagination, planning and priority setting mechanisms are needed for harnessing the new potential. Some of these designer crops of superior traits are listed in Table 1.

Table 1. Designer crops of superior traits

- Agronomic Traits
  - Biotic Stress
    - Insect Resistance (i) Toxins (bt-toxin) (ii) Antifeedant (proteinase inhibitor) (iii) Lectins (iv) Sesqui-terpenoids (v) Pheromones
    - Disease Resistance: Viral, Bacterial, Fungal, Nematode
      (i) Enzymes : β-glucanase and chitinase, cystatin (ii) coat protein: viruses (iii) SAR: signal transduction pathways in plants (Figure 3)
    - Weed, herbicide tolerance, parasitic weeds
  - Abiotic Stress
    - Drought, Cold, Heat, Salinity, Poor soils
- Reproduction: sex barriers, male sterility , seedless ness
  (a) Hybrid; (i) Dihaploids (ii)CMS (iii) TGMS
- Quality Traits
  - Yield- Nitrogen Assimilation, Starch Biosynthesis, O_2 Assimilation
  - Processing
  - Prevention of post-harvest spoilage
  - Preservation (shelf life): (i) Ethylene pathway-slow ripening (ii) polygalacturonase softening (iii) antimicorbial protection:rottening
  - Processsing (i) breadmaking : glutenin subunits (ii) oil extraction; lipase (iii) Rancidity ; lipase free and (iv) strach quality : amylose/ amylopectin
  - Value addition: (i) Colour (anthocyanin) flowers and cotton fibres, (ii.) Aroma: in rice
  - Nutrients (Nutraceuticals)
    - Macro: Protein, Carbohydrates, Fats, Fiber
    - Micro: Vitamins (vitamin A, β –carotene), Minerals (Iron, Zinc and iodine), Antioxidatnts, Isoflavonoids, Phytoestrogens, Condensed tannins
  - Anti-nutrients: oxalic acid, neuro-lathyrotoxin, Goitergenic, Phytase, Allergen and Toxin removal
  - Taste
  - Architecture
  - Ornamentals: color, shelf-life, morphology, fragrance

- Novel Crop Products
  - Oils: PUFA
  - Proteins: amino acid imbalance, nutraceuticals
  - Polymers: Carbohydrate, starch, PHB
  - Renewable Resources, Biofuels, feed stocks for synthetics
  - Molecular farming through bioreactor plants

High quality biological proteins like plantibodies (antibodies from plants), recombinant or edible vaccines, interferon, interleukins, herbal medicines, hormones and growth factors.

Summary of Transgenic Research in India

Target Crops/Vegetables

Cotton, Cron, Mustard, Rice, Soybean, Potato, Tabacco, Coffee, Tomato, Brinjal, Cauliflower, Pea, Cabbage, Banana, Muskmelon, Pigeonpea, Chickpea, Bell-pepper, Blackgram, Chilli, Watermelon etc.

Transgences Employed

Bt. Toxin genes, Harbicide tolerant genes (CP4 EPSPS, Bar gene), Xa21, cetx-B and tep of V.cholera, Chitinase, Glucanase, ACC synthase, RIP, Protease Inhibitor, Lectin, Ama-1, OXDC gene, Rabies glycoprotein gene, Bar, Barnase, Barstar, GNA gene, Vip-3 gene, Bacterial Blight Resistance gene, Osmotin etc.

Insect resistant GM crops in India

Nine transgenic crops with seven genes under field testing

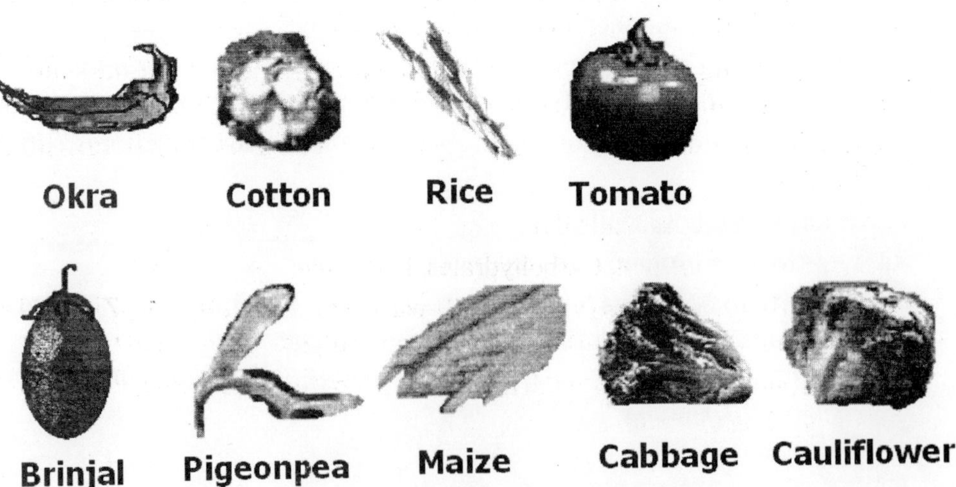

Genes: Cry1Aa, Cry1Ab, Cry1Ac, Cry1F, Cry1B, Cry2Ab & Vip-3A

Other traits under development

- Biotic stress resistance (Disease resistance)
- Enhancement of quality
- Abiotic stress resistance (water, salinity, temperature)
- Post Harvest attributes
- Edible vaccines

Concerns in Context to GM Crops

Proponents of agricultural biotechnology point to the fast-growing global population, arguing that most of the planet's arable farmland is already being used and that the use of genetically modified (GM) crops is the best hope of alleviating widespread hunger and starvation. Commonly cited benefits of food biotechnology are increased yields (without overtaxing soils); hardier varieties that can withstand heat and drought; enhanced nutritional—and medicinal—value; improved storability; and built-in pest resistance that may decrease the need for chemicals. There is also considerable concern about ensuring safety in research on GMOs, particularly with respect to their handling, containment and the impact on environment and biodiversity on their eventual release.

Despite of the several proven advantages of biotechnology in agriculture, there are several concerns surrounding the commercial use of genetically modified crops. The concerns broadly linger on two main questions; are the GM plants safe to eat and do they pose dangers if they spontaneously breed with wild relatives or conventional crops. There is also a widespread apprehension as what could happen if transgenic plants accidentally crossbreed with their wild and domesticated relatives. The issue revolves around the extent to which GM plants may spread their novel traits into surrounding crops and ecosystems via pollen or seed and just what sort of problems one might encounter from this "gene flow." However, numerous studies conducted over the past 20-25 years on the safety of genetically modified organisms have concluded that there is no direct negative impact either on environment or on the wild species or wild relatives of the crop plants. Thousands of field trials conducted on transgenic crop plants over the last two decades have proved the safety of the introduced transgenic plants in the environment. In fact it is difficult to say with certainty that crops and food designed by traditional plant breeding will not lead to health or environmental problems which GM foods and crops can do.

Indian government in January 1999 announced that it would encourage use of GM (Genetically Modified) seeds in agriculture and give high priority to transgenic research. Till date only transgenic variety which is commercialized in India is Bt cotton which occupies a total area of 3.6 Mha. While there is large consensus that agricultural research and development (including biotechnological applications) need to be pursued vigorously in order to meet future food demands of the increased world population, there are concerns over the safety of GM crops. These concerns relate to consumer safety, as well as, potential adverse effects on ecosystems which are considered essential for sustainable agriculture and for maintaining agricultural biological diversity. Some of the perceived threats include:

- Threat of "Introgression" or gene flow or gene dispersal, between genetically engineered organisms and related species
- Non-target effects or ecological effects
- Invasiveness or weediness of transgenics
- Creation of 'super-weeds' and super-viruses
- Development of resistance in target populations against engineered bio - pesticides
- Long-term consequences on human health such as toxicity, allergenicity, carcinogenicity, food intolerance and nutritional value.
- Expression of undesirable phenotypic traits
- Erosion of biological diversity

In this regard, a science based evaluation system that is requiring the benefits and risks of each individual GMO product. This calls for a cautious case-by-case approach to address legitimate concerns for the biosafety of each product prior to its release into the environment or placed on the market. The possible effects on biodiversity, the environment and food safety needs to be evaluated and the extents to which the benefits of the product or process outweigh assessed risks are subjected to scientific assessment.

The objections to the use of genetic modification in the production of agricultural crops are several. The main concerns over the use of GM crops centre on their implications for food safety and human health, environmental effects and the general ethics of transferring genes, particularly between non-related species. They revolve around issues of: who benefits? What is their effect on the environment and biodiversity? What is the effect on human health when they are consumed? Many of the GM crops now being grown around the world have been engineered for traits that benefit the economic return from production – such as Bt maize, which is resistant to the corn borer pest, or the 'Round-up Ready' crops, which are tolerant to the herbicide 'glyphosate'. Unlike other applications of transgenic technology, these benefits are perceived to be more limited and mainly to the advantage of the grower, while the advantages to wider society are perceived as being less clear.

Gene flow is not something peculiar to transgenic plants. In the world of living creatures, gene flow is as old as life itself. It happens naturally in ecosystem and one organism breeds with a related species, thus passing along their combined DNA to the offspring. There is a continuous interaction between the environment and the living organisms, which make the biosphere a dynamic interactive space. The plant interacts dynamically with biosphere and with the rhizosphere. In biospheric interaction plant and their related species and animals co-exists in an eco-system. Plant interacts with a biotic component of the environment including the microbes for the mutual existence. Plant-to-plant interaction is through gene transfer. Plants normally exchange genes within their own species and some of the related species depending upon their evolutionary development and domestication.

Most of the research and discussions on environmental biosafety are mainly focused on GM crops. Some of identified risks inherent to GM crops are discussed here:

- **Allergens:** One of the biggest concerns related to transgenic food is that an allergen (a protein that causes an allergic reaction) could be introduced into a food product. It is extremely unlikely that it will be introduced into GM food. Even so, allergenically screening is important part of safety looking before; crop can enter into food market. A variety of questions and tools must be considered to remove this apprehension. Science based risk assessment of GM food is the only way to address public concerns.
- **Antibiotic resistance:** Concerns have been raised that these markers genes could move from GM crops to the microorganisms that normally reside in a person's gut and lead to an increased the antibiotic resistance. Although likelihood of antibiotic genes moving from GM crops to any other organism is extremely remote. It should be noted that the widespread use of antibiotics or food additives for animals and prescribed medicines for human, causes a greater risk for creating antibiotic resistant bacteria than transfer of marker genes from GM plants. Nevertheless, in response to public concerns, the transgenic developers should continue to more rapidly use alternative marker strategies for the selection of new varieties. Alternative strategies are new being followed for transgenic development and selection.
- **Potential of the introduced genes to outcrosses to weed relatives:** A major environmental concern associated with GM crops is their potential to create new super weeds out-crossing with wild relatives or supply persisting in the wild themselves. That a herbicide tolerant gene may escape through pollen into nearby farms and fields, to another GM cultivars or to a wild and weedy relative is a possibility. However, the possibility of these escaped gene may lead to super weed, seems very unlikely. The genes may escape through pollen into the nearby farm to another non-GM cultivar or wild weedy relatives. It is therefore important to evaluate individual GM crops on a case by case basis.
- **Direct effects on non-target organisms:** Some scientists however urged caution over interpretation of the study because it reflects a different situation than that in the environment, and it would therefore be inappropriate to draw any conclusions about the risk of Monarch population in the field solely on these initial results. Furthermore, a collaborative research effort in USA has concluded that in most commercial hybrids, Bt expression in pollen is low and field studies show no acute toxic effects to non-target wildlife.

- **Development of insect resistance:** Another concern with Bt crops is that it will lead to the development of insect resistance to Bt. This is not a possibility but a reality. Scientists have to develop strategies like pyramiding of insecticidal genes with divergent mode of action to overcome this problem. Additional resistance management practices are also need to be developed by scientists to overcome this problem.
- **Risks to biodiversity:** Monoculture and the rapid and extensive adoption of new, improved varieties have caused the reduction in the variety of crops plants being planted in farmers fields. It has been apprehended that the rapid adoption of transgenic crops will also cause loss in biodiversity in a similar manner.
- **Genetic contamination (Crop-to-crop gene flow):** An important issue, which is frequently discussed in the risks of gene flow from GM species, is the likelihood that transgenic plants will move "off the farm". It is generally expected that if a GM variety of a crop is grown near non-modified varieties, gene flow by natural means, such as birds, insects, and wind will be a fairly common occurrence. There are plenty of documented cases of this kind of spontaneous "crop-to-crop" gene flow occurring between transgenic and conventional varieties of corn and canola. Therefore not surprisingly many scientists are concerned about the widespread release of genetically modified organisms.

Crop-to-wild relatives gene flow It is the fact that gene flow between domesticated crops and wild relatives has been relatively common in the history of agriculture. Negative consequences have been relatively uncommon, but there have been instances in which such gene flow has done damage. For example, breeding between wild and domesticated sugar beets has been economically devastating to European sugar producers because the unproductive hybrids can take over a crop and render it worthless. Hybrids created by spontaneous breeding between wild rice and domesticated rice are blamed for the extinction of Taiwanese wild rice.

- **Factors affecting pollen dispersal and cross-pollination:** The extent of cross-pollination between field crops or between crops and wild plant populations is largely dependent on the scale of pollen emission and dispersal. Pollen produced by some crops can be dispersed over considerable distances by both wind and insects. The weather is known to affect the behavior of pollinating insects on the crop resulting in significant variation in amount of cross- pollination crop-to-crop and day-to-day. The numbers and even species of natural pollinating insects can vary considerably crop to crop affecting successful pollination. Pollen dispersal is influenced by the weather and changes in temperature, humidity and light, as well as wind and rain: Wind strength can also have an important role in distributing pollen grains over significant distances within their viability periods. Immediate local environment such as the nature of the plant canopy, surrounding vegetation and topography may also influence patterns of pollen dispersal.

Biological factors influencing successful pollination begin with the ability of the donor plant to produce viable pollen and the length of time the pollen grain retains its potential for pollination. Pollen viability can vary greatly between species but is also dependent on environmental variables such as temperature and humidity. The amount of out-breeding in the crop is an important aspect to consider apart from morphology and pollen characteristics. There must be some overlap in flowering times between the pollen donor and the receptor plant. In its broad sense 'hybridization' can be defined as the cross-breeding of genetically dissimilar individuals. Gene flow can be defined as 'the incorporation of genes into the gene pool of one population from one or more populations. Such gene movement is a major determinant of genetic structure in natural populations. Gene flow is strongly influenced by the biology of the species and is likely to vary with different breeding systems, life histories and modes of pollination.

- **Risks Associated with the Transfer of Genes that add Fitness to the Environment:** It is feared that newly created "transgenes" will spread unintentionally from target crops to related weed species. The ecological harm caused by crop-to-wild relative gene flow is also routinely discussed. The genetic trait passed on to the new hybrid may give the plant a "fitness" advantage over the wild plant. A new hybrid could become established and may even replace the wild plants, grow in new places, and change ecological balances. Another one is the possibility of the transgenic crop becoming a weed, that is, the newly acquired gene or trans gene has increased the ability of the crop to survive out of cultivation, gain weedy characteristics or become invasive. The discovery of a canola variety growing on the roadside as a weed or "volunteer" possessing resistance to three different herbicides is cited as an example. Two of the resistance traits were acquired by the inadvertent cross breeding between two genetically engineered canola plants while one was acquired from an herbicide resistance canola produced through conventional breeding. The second report is of the Starlink *Cry9C* allele appearing in a variety of supposedly non-engineered com. Although unintentional mixing of seeds during transport or storage may explain the contamination of the traditional variety, inter- varietal crossing between seed production fields could be just as likely. Without careful checking, there are plenty of opportunities for them to move from variety to variety. The field release of "third generation" transgenic crops that are grown to produce pharmaceutical and other industrial biochemicals will pose special challenges for containment.
- **Possible Cascading Adverse Effect on the Ecosystem:** A scenario that has followed the introduction of transgenics in the environment is the unintended effect that could change the components of an ecosystem. This risk is associated with genes producing products toxic to pests. One scenario is the loss of an insect species that serve as food of another species, which could also become extinct because of the loss of its food source. However, it has been known that

predators in their native habitats do not often rely on a single species as food. Predators usually have their preferred food but also prey on other species when their preferred food is not available.

Biosafety Regulations

With the development of transgenics, the issues of biosafety, food safety, the environmental impact and their risk assessment have become one of the great concerns of consumers around the world. Therefore, it is of crucial importance that transparent and science based approaches are used to adopt to conduct safety assessments on biotechnology products before their commercialization. It is important that correct information are regarding the new technology should be communicated properly and effectively to the general public and stakeholders. It is imperative 'to build public confidence through proper communication system so that the public at large is well informed about the benefits and risk if any, of the GM foods. Of the various genetically modified organisms developed, GM crops are the major cause of concern unlike transgenic farm animals and recombinant bacteria because of its reproductive structures, such as seeds and pollen, which could be freely dispersed whereas the reproduction of farm animals can be sufficiently controlled. Several regulations are put forth in relations to recombinant DNA technology and transgenic research (Table 2).

Since 1989, the Ministry of Environment & Forests, New Delhi under Environmental Protection act expressed concern at the inadequacy of the appraisal procedures for GM products and requested the Department of Biotechnology (DBT) to develop an enhanced risk evaluation procedure to examine each GM seed, food, feed or crop prior to their being placed on the market in India. The procedure was designed to address the food safety and environmental concerns, taking a precautionary, science based, case-by-case approach to regulatory approval.

Table 2. Genetically Modified Organisms (GMOs) and R-DNA Products Governed by

- Environment (Protection) Act, 1986 - Rules, 1989
- Industries (Development & Regulation) Act, 1951
  - New Industrial Policy & Procedures, 1991
- Drugs & Cosmetics Act, 1940
  - Drugs (Price Control) Order - 1995
  - Drug Policy-1986 & Modification in September, 1994 & February, 1999.
- Seeds Act, 1966
- Seeds Rules, 1968
- Seeds (Control) Order, 1983
- Seeds Policy, 1988
- Protection of Plant Varieties and Farmers' Right Act, 2001

The bioethics committee of UNESCO established in 1993 has evolved guidelines for ethical issues associated with the use of modern biotechnology. Biosafety guidelines for Genetically Improved Organisms (GIOs) need to be strictly followed to prevent harm to human health or the environment. The Recombinant DNA Safety Guidelines are based on a 3-tier system, involving the Institutional Biosafety Committee (IBSC), the Review Committee on Genetic Manipulation (RCGM) and the Genetic Engineering Approval Committee (GEAC). The IBSC is established at every institution engaged in research on genetically engineered organisms. The RCGM and GEAC closely monitor the field experiments involving transgenics, before their commercialization is contemplated, as well as granting permission for commercial use of the products containing GMOs and/or recombinant DNA after making necessary biosafety evaluation. Based on the recommendations of the RCGM, trial permits are issued by the DBT. The RCGM can approve applications for generating research information on transgenic crops, using contained greenhouses as well as small plots. The small experimental field trials are limited to a total area of 20 acres in multi-locations in one crop season. In one location where the trial is to be carried out on transgenic plants, the land use should not be more than one acre. The trial design requires approval from the RCGM. Environmental safety of the transgenic lines, including animal and animal health, gene flow etc, are required to be investigated for each transgenic, besides the economic advantages over the existing non-transgenic varieties Once the integration and confirmation of transgene in the seeds is confirmed using molecular tools, there are number of steps involved in the commercialization of transgenic seeds.

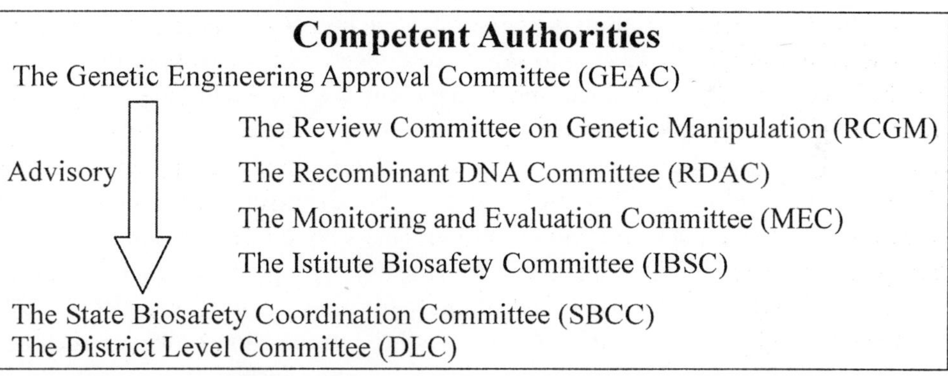

All genetically engineered crops/varieties will be tested for environment and biosafety before their commercial release, as per the regulations and guidelines of the Environment Protection Act (EPA), 1986. The EPA, 1986, read with the Rules, 1989 would adequately address the safety aspects of transgenic seeds/planting materials. A list will be generated from Indian experience of transgenic cultivars that could be rated as environmentally safe. Seeds of transgenic plant varieties for research purposes will be imported only through the National Bureau of Plant Genetic Resources (NBPGR) as per the EPA, 1986.

Transgenic crops/varieties will be tested to determine their agronomic value for at least two seasons under the All India Coordinated Project Trials of ICAR, in coordination with the tests for environment and bio-safety clearance as per the EPA before any variety is commercially released in the market. After the transgenic plant variety is commercially released, its seed will be registered and marketed in the country as per the provisions of the Seeds Act. After commercial release of a transgenic plant variety, its performance in the field will be monitored for at least 3 to 5 years by the Ministry of Agriculture and State Departments of Agriculture. Transgenic varieties can be protected under the PVP legislation in the same manner as non-transgenic varieties after their release for commercial cultivation. All seeds imported into the country will be required to be accompanied by a certificate from the Competent Authority of the exporting country regarding their transgenic character or otherwise. If the seed or planting material is a product of transgenic manipulation, it will be allowed to be imported only with the approval of the Genetic Engineering Approval Committee (GEAC), set up under the EPA, 1986. Packages containing transgenic seeds/planting materials, if and when placed on sale, will carry a label indicating their transgenic nature. The specific characteristics including the agronomic/yield benefits, names of the transgenes and any relevant information shall also be indicated on the label. Emphasis will be placed on the development of infrastructure for the testing, identification and evaluation of transgenic planting materials in the country.

Regulatory Issues

In the beginning of transgenics development, there were many unknown about transgenic crops and about their possible effect on humans and the environment. There was a need to establish detection to safety the concerns of general public as well as the scientific community. The open and transparent system established experience and information gathered about safety of transgenic crops, it is time for our regulatory agencies to consider whether some or all the regulatory requirements can be gradually relaxed. The molecular and genetic principles of plant biotechnology and plant breeding and selections are the same. No compelling evidences have been presented to show that transgenic crops are innately different from the non transgenic product of breeding and selection. What then is the basis and rationale for many restrictions still placed on the field plating and human use of transgenic foods? The decision however must be based on science and facts, and not upon political or ideological considerations. Genuine concerns about gene flow, development of resistance to antibiotics and pests or pathogens can be met adequately with currently available and emerging technologies.

In view *of* above said risks and necessity; a governing body is needed which takes" care of all the rules and protocols for the introduction of these recombinant organisms in the environment. This task is well accomplished by first and only international law on genetic engineering/ genetically modified organisms known as **Cartagena Protocol on Biosafety** (CPB). It is a protocol under the UN convention on biological diversity and came in to force on 11 September 2003 currently, there are around 120 Parties. Most of them are developing countries, with the majority from Africa and Small Island developing states. The Biosafety Protocol makes clear that products from new technologies must be based on the precautionary principle and allow developing nations to balance public health against economic benefits. Objective of the Protocol In accordance with the precautionary approach, contained in Principle 15 of the Rio Declaration on Environment and Development, the objective of the Protocol is to contribute to ensuring an adequate level of protection in the field of the safe transfer, handling and use of 'living modified organisms resulting from modern biotechnology' that may have adverse effects on the conservation and sustainable use of biological diversity, taking also into account risks to human health, and specifically focusing on trans-boundary movements (Article 1 of the Protocol, SCBD 2000).

Biosafety concerns have prompted both developing as well as developed countries to implement Biosafety guidelines governing testing, safe use and handling of GMO's. A Cartagena Protocol on Biosafety regulations has been published. The Department of Biotechnology (DBT) and the Ministry of Environment & Forests (MoEF) are responsible for implementing Biosafety regulations within the country. The MoEF notified the legal frame work for the same in 1989 and the DNA safety guidelines were framed by the DBT in 1990. Subsequently, DBT framed guidelines for transgenic plants and recombinant clinical products in 1998 and 1999 respectively.

It is important to give a clear explanation of the new biotechnologies to the public to allay their fears. New models of cooperation and partnership have to be established to ensure close linkages among research scientists, extension workers, industry, the farming community, and consumers. Gene transformation is done worldwide with four broad objectives:

(a) to develop products with new characteristics;

(b) to develop pest and disease resistance;

(c) to improve nutritional value; and

(d) to modify fruit ripening to obtain longer shelf life. Thus the aims and objectives are laudable and the tools are available.

The new technology does, however, call for a cautious approach following appropriate biosafety guidelines.

About 25,000 field trials of genetically modified crops have been conducted worldwide. The anticipated benefits are better planting material, savings on inputs, and genes of different varieties can be introduced in the gene pool of crop species for their improvement. The potential risks include weediness, transgene flow to nontarget plants, and the possibility of new viruses developing with wider host range and their effects on unprotected species. For crops such as corn and cotton with single gene introductions, there is very little problem expected. When multiple genes are involved scientists have to be more cautious. The time has arrived for a serious look at ethical and biosafety aspects of biotechnology. Researchers, policymakers, NGOs, progressive farmers, industrialists, government representatives, and all concerned players need to come together and share a platform to address the biosafety, ethical, environmental, social and economical issues. To achieve the goal of nutritional and economical self-reliance, India will require a strong educational and scientific base; clear public understanding of the value of new biotechnologies, and involvement of society in many of these biological ventures. India has a large research and educational infrastructure comprising 29 agriculture universities, 204 central and state universities, and more than 500 national laboratories and research institutions. It should therefore be possible to develop capabilities and programs so that these institutions act as regional hubs for the farming community, where they can get direct feedback about new technological interventions. It will be equally important to establish strong partnerships and linkages with industry, from the time a research lead has emerged until the packaging of the technology and commercialization are achieved.

References

Bansal K.C., Singh N.K., Prasanna B.M., Tiwari S.P. and Dhillon B.S. 2004. Transgenic crops: Status and Prospects in Indian Agriculture. *In:* Proceedings of the National Conference on Transgenics in Indian Agriculture, New Delhi (March 9-10, 2004) pp. 17-32

Crawley M.J., Brown, S.L., Hails, R.S., Kohn, D.D. and Rees, M. 2001. Biotechnology: Transgenic crops in natural habitats. *Nature,* **409:** 682-683.

Daniell, H. 1999. GM crops: public perception and scientific solutions. *Trends in Plant Science,* **4:** 467-469.

Eastham, K. and Sweet, J. 2002. Genetically modified organisms (GMOs): The significance of gene flow through pollen transfer. Environment issue report No. 28. European Environment Agency.

Gressel, J. 1999. Tandem constructs: preventing the rise of superweeds. *TIBTECH.,* **17:** 61-366.

Hoyle, B. 1999. Canadian farmers seek compensation for genetic pollution. *Nat. Biotechnol,* **17:** 747-748.

James, C. 1998. Global Review of Transgenic Crops: 1998. ISAAA Brief No. 8. Ithaca, N.Y.: International Servicefor the Acquisition of Agri-biotech Applications.

Losey, J.E., Rayor, L.S. and Carter, M.E. 1999. Transgenic pollen harms monarch larvae *Nature,* **399:** 214.

Palm C.J., Seidler, R.J., Donegan, K.K. and Dharris. 1993. Transgenic plant pesticides: fate and persistence in soil. *Plant Physiol. Suppl,.* **102:** 166.

Raghuwanshi A, Singh, R.K. & Jain, R.K. 2004. The Golden Rice: Five Years After. *Physiol. Mol. Biol. Plants* **10(1):** 1-5

Saxena. D., Flores, S. and Stotsky, G. 1999. Transgenic plants: Insecticidal toxin in root exudates from Bt corn. *Nature,* **402:** 480.

Sharma, M. 2003. India: Biotechnology Research and Development. *In:* The Business of Biotechnology. ASSOCHAM's Publication for the Knowledge Millennium III (21-23 March, 2003), New Delhi, pp 47-53; The Associated Chambers of Commerce & Industry of India, New Delhi.

Chapter – 10

Role of Bioinformatics in Crop Diversity Analysis

Soma Marla

Introduction

Nature provides a wealth of genetic variation in both wild and cultivated plants. In the advanced selections conducted in developing superior crop varieties cultivated today agriculture represent only a small portion of the available gene pool. However wild species remain a reservoir of valuable genetic variation and represents a valuable source for exploitation of unexplored genetic variation for improvement of various agronomic traits. Most success has come from introgression of specific single gene controlled traits largely for disease resistance (Tomleyson, 2000 and Sergio *et al.,* 2000). The identification of genes and co-adapted gene complexes that control complex agronomic traits such as adaptation to specific environmental stresses, crop quality, yield, plant architecture, flowering time (especially important for vegetable crops) has been less effectively addressed so far and remains a major and immediate challenge. Also, existence of low genetic diversity in breeding lines of many species used in various plant breeding programs is often resulting in development of insignificant improvement of the trait under investigation (Christopher, 2000). Understanding the spectrum, frequency and distribution of allelic variation in genes controlling such multiple genes controlled complex traits will improve our knowledge of exploitation of natural variation and promotes effective management and exploitation of functional biodiversity. In India efforts have been initiated in the exploitation, conservation of crop germplasm and technologies that are needed to fully exploit this natural diversity available in our major crop plants and their wild relatives. For example, centers that maintain diverse germplasm collections (both by ICAR and State Agricultural Universities nation wild are well established.

However application of genomics and bioinformatics to understand and exploit the natural biodiversity present in the accessions of wild and cultivated plants via traditional or biotechnological means will be one of the major biological and technological challenges of the twenty first century. So far research conducted in molecular biology laboratories centered on gene by gene approach (i.e. one gene at a time) by analyzing, identification, cloning and sequencing of individual genes responsible for manifestation of a trait. Whereas most of the agronomically important plant traits such as grain yield, flowering time, product quality, resistance to stress (both abiotic as well as biotic) are governed by multiple sets of genes & proteins and require systematic functional analysis to be conducted at organism level. Genomic research offers valuable laboratory techniques to address these traits enabling identification & analysis of multiple genes controlling expression of a trait (Alpert *et al.,* 1995). Also, a large amount of valuable data generated from field and laboratory experimentation. This valuable experimental data requires to be integrated with sequenced model organism (Rice and *Arabidopsis thaliana)* genomics generated data so as to make meaningful inferences.

Biodiversity and Agri Informatics

Agri Informatics can be defined as application of Bioinformatics tools to understand the phenotypic trait expression, genetic relationships exhibited by related genes and proteins at various levels of cell organization. Employing various bioinformatics tools, genomics facilities for the discovery of natural genetic diversity through a rigorous analysis of molecular polymorphisms at the DNA sequence level.

The primary objectives of such analysis can be listed as:

1. In silico identification of genes governing the trait under investigation,
2. Validate the role of a candidate gene (emerged from related or unrelated species) in a given trait,
3. Identify a range of alleles of potential value in crop or germplasm enhancement,
4. Understand the mechanisms and process which drive the generation of biodiversity (to generate rational strategies for subsequent exploitation by tapping evolutionary the advanced relationships),
5. Identify candidate genes via patterns of DNA sequence and /or haplotype diversity
6. Inference of both structure and properties of the gene/protein under study.

While molecular analyses of monogenic traits are now relatively straight forward that of complex traits is compounded by their polygenic control and large interaction with environmental factors. Its study these requires quantitative trait locus (QTL-segments of related DNA) analysis and the cloning of QTL loci. By functional genomics approaches or exploitation of biodiversity and QTL analysis it is possible to identify performing alleles of genes involved in disease tolerance, drought adaptation, carbon fixation, or even in the quality of our foods. The latter has recently been show to be feasible using map based cloning strategies for species where large molecule genetic resources are available (e.g. *Arabidopsis*, *Medicago*, rice and tomato). The relevance of research in model plants to other plants is being increasingly demonstrated - for instance the genetic control of flowering time is controlled by related genes is both *Arabidopsis* and rice. In analysis of polygenic traits involving QTL & chromosomal gene mapping, linkage & correlation studies is facilitated by bioinformatics tools like QTL maker and MAPMAKER (Bee, 2006).

Crop Diversity Databases

A database is a structured collection of information and a sort of an electronic filing cabinet that allows storing information and retrieving it with ease. The purpose of crop databases is not merely to collect and organize data, but mainly to allow advanced data retrieval relating to plant phenotype, gene & protein information, their functions, relationships, all crop related published literature, nutritional information, cultivated varieties, agronomic package & practices etc.

Example Query

Find all genes in varieties of field bean vegetables which where resistance to downy mildew disease is presently cultivated in Tarai region.

Building Evolutionary Relationships

In a collection of germplasm diversity analysis can be conducted and existing evolutionary relationships found using bioinformatics tools. Sequences used for organism phylogenies.

- DNA
- Protein
- rRNA
- Mitochondrion DNA sequence
- RAPD bands
- AFLP bands
- Other

Finding Orthologs, Paralogs and Discovery of New Genes

Through course of evolution and speciation it is interesting to note several genes controlling cellular metabolism and responsible for some agronomics traits have been conserved on the chromosomes at specific points of location. Employing comparative genomic techniques and bioinformatics tools, it has now become possible to locate orthologs and paralogs associated with functions of genes of our interest. On line sequence information of genes & proteins is now available from *Arabidopsis* and rice is being actively exploited in locating and identification of similar genes controlling traits of our interest. Bioinformatics tools like blast (basics local alignment tool), cogs (conserved orthologous genes) are some examples:

1. To compare geneomes of different organisms, detecting homologs and identification of new genes/proteins.
2. To estimate the level of variation present in genes/proteins.
3. To estimate the genetic and molecular basis for variation in these traits within and among species.

Gene Sequence Databases

Experimental data generated from field, biotechnology research on important genes/ proteins of our interest is deposited in online database in NCBI such as:

1. Gene bank (Gene Sequences)
2. Swiss Port (Protein sequences)

These are the repositories of sequences of genes from different organisms that are sequenced and deposited in on-line gene banks. These data bases allow free for public download of gene sequences:

They are: DNA Data Bank of Japan (DDBJ), EMBL Nucleotide Sequence Database and Gene Bank (USA). Once a genome is assembled, genic regions are annotated to give information about gene functionality. COG (Cluster of Orthologhous Group), gene discovery databases. (GeneScan and GRAIL). Similarly there are repositories of sequences of proteins from different organisms that are sequenced and deposited in on-line gene banks. These databases allow free for public download of protein sequences.

They are: SWISS-PROT/TREMBL curated protein sequences: http://www.expasy.ch/sprot,

InterPro: Protein families and domains http://www.ebi.ac.uk.interpro,

Protein Information Resource (PIR): http://pir.georegetown.edu/,

UniPort: http://www.expasy.uniport.org/index.shtml and

Interpro: Protein families and domains http://www.ebi.ac.uk/interpro/

References

Alpert, K. B., Grandillo, S. and Tanksley, S. D. 1995. A major QTL controlling fruit weight is common to both red- and green-fruited tomato species. Theoretical and Applied Genetics, **91** (6-7): 994-1000.

Bee Lian Ong, 2006. Molecular approaches to overcome abiotic stress in vegetable crops. *Annals of Applied Biology,* **110(3):** 666-68.

Christopher S. Cramer. 2000. Breeding and genetics of Fusarium basal rot resistance in onion, *Euphytica,* **115:** 3.

Sergio H. Brommonschenkel, Amy Frary, Anne Frary, and Steven D. Tanksley. 2000. The Broad-Spectrum Tospovirus Resistance Gene *Sw-5* of Tomato Is a Homolog of the Root-Knot Nematode Resistance Gene *Mi, Molecular Microbe Interactions,* **13(10)**: 1130-1138

Tomleyson, J. A. 2000. Epidemiology and control of virus diseases of vegetables Perspectives Publ. in Perspectives in ASEAN cooperation in vegetable research and development, http://www.avrdc.org/pdf/aarnet_proc/006.pdf

Chapter – 11

Application of Cell and Tissue Culture Techniques for Plant Genetic Resources Management

H.S. Chawla

Introduction

Cell and tissue culture techniques are becoming increasingly popular as alternative means of plant vegetative propagation. Plant tissue culture involves asexual methods of propagation and its primary goal is crop improvement. The success of many *in vitro* selection and genetic manipulation techniques in higher plants depends on the success of *in vitro* plant regeneration. Crop breeding and rDNA technology require the widespread use of reliable true to type propagation and better regeneration methods. Plant regeneration can be grouped in to different categories of enhanced release of axillary bud proliferation, organogenesis and embryogenesis. The oldest practical application of tissue culture techniques is the propagation of disease free plants from meristem tip culture. Meristem tip consisting of dome and one or two subjacent leaf primordial with a size up to 0.25 mm are aseptically cultured in appropriate nutrition media. The resulting plants are disease free because the pathogen is not generally present in the meristematic cells.

The aim of germplasm conservation is to ensure the availability of useful germplasm at any time. Breeding programs rely heavily on locally adapted ancient plant varieties and

and their wild relatives as sources of germplasm. Further, the continuing search for high yielding varieties of crop plants with resistance to pathogens and pests warrants the availability and maintenance of large collections of germplasm. In species which are propagated through seeds, it is economical to preserve the seeds. These seeds are dried to water content of 5 - 8% and then stored at a temperature of -18° C or lower and low humidity. In case of vegetatively propagated species, preservation of germplasm heavily taxes manpower and land resources. As *in vitro* techniques are becoming more important in crop improvement through the use of somatic cell genetics, the material produced *in vitro* will have to be conserved *in vitro*. The *in vitro* system is extremely suitable for storage of plant material, since in principle it can be stored on a small scale, disease free and under conditions that limit growth. Germplasm storage *in vitro* is crucial for the future development and safety of agriculture.

The conservation of plants *in vitro* has a number of advantages over *in vivo* conservation. These are as follows:

(i) *In vitro* culture enables plant species that are in danger of being extinct to be conserved.

(ii) *In vitro* storage of vegetatively propagated plants can result in great savings in storage space and time.

(iii) Sterile plants that cannot be reproduced generatively can be maintained *in vitro*.

(iv) In *in vitro* culture it is possible to efficiently reduce growth, which decreases the number of subcultures necessary.

(v) If a sterile culture is obtained, often with great difficulty, then subculture *in vitro* is the only safe way of ensuring that it remains sterile.

There are two main approaches for *in vitro* storage of germplasm

Cry Preservation

Cry preservation (Greek, Kryos = frost) means "preservation in the frozen state". In practice, this is generally meant to be storage at very low temperatures, e.g. over solid carbon dioxide (–79°C), in low temperature deep freezers (–80°C or above), in vapor phase of nitrogen (–150°C) or in liquid nitrogen (–196°C). Generally the plant material is frozen and maintained at the temperature of liquid nitrogen (–196°C). At this temperature, the cells stay in a completely inactive state. The theoretical basis of freeze preservation is the transfer of water present in the cells to the solid state. While pure water becomes ice at 0° C, the cell water needs a much lower temperature because of freezing point depression by salts or organic molecules. The lowest temperature at which water has been demonstrated to be liquid is –68°C. Cryopreservation of plant cells, tissues and organs received widespread importance mainly due to the emphasis placed in recent years on *in vitro* manipulations with cultured cells. Storage at very low temperature considerably slows down or halts metabolic processes and biological deterioration. Two major considerations have to be taken into account for freezing of specimens.

(1) Susceptibility or degree of freeze tolerance displayed by a given genotype to reduced temperatures.

(2) The formation of ice crystals within the cells. The cryopreservation of plant cell culture and seeds and eventual regeneration of plants from them involves the following steps:

(i) Raising sterile tissue cultures

(ii) Addition of cryoprotectants and pretreatment

(iii) Freezing

(iv) Storage

(v) Thawing

(vi) Determination of survival/viability

(vii) Plant growth and regeneration

Raising Sterile Tissue Cultures: The morphological and physiological conditions of the plant material influence the ability of an explant to survive freezing at -196° C. Different types of tissues can be used for freezing such as the apical and lateral meristems, plant organs (embryos, endosperm, ovules, anther/pollen), seeds, cultured plant cells, somatic embryos, protoplasts, calluses, etc. In general, small, richly cytoplasmic, meristematic cells survive better than the larger, highly vacuolated cells. Cell suspensions have been successfully frozen in a number of species. These should be in the late lag phase or exponential phase. Callus tissues, particularly among tropical species, are more resistant to freezing damage and have also been used. A rapidly growing stage of callus shortly after one or more weeks of transplantation to the medium is best for cryopreservation. Old cells at the top of callus and blackened area should be avoided. Cultured cells are not the ideal system for cryopreservation. Instead, organized structures such as shoot apices, embryos or young plantlets are preferred. Plant meristems have been quite often used for preservation. Plant meristems when aseptically isolated are incubated for few days in growth media before proceeding to the next step. Water content of cells or tissues for cryopreservation should be low because with low freezable water, tissues can withstand extremely low temperatures.

Addition of Cryoprotectants and Pretreatment: Seeds are one of the most preferred explants for storage due to ease of handling. Studies have shown that the orthodox seeds can be successfully stored at -20^{0} C after desiccating to 5 - 7% moisture content without loss of viability using conventional storage methods. However, many crops and spices, which are indigenous to the tropics and belong mainly to the category of recalcitrant seeds, do not produce orthodox seeds and hence are not amenable to the conventional storage methods e.g. Arecanut (*Areca catechu* L.), Black Pepper (*Piper nigrum* L), Cardamom (*Elettaria cardamomum* Maton), Coffee (*Coffea arabica* L.,

C. canephora Pierre., *C. liberica* Bull.), Coconut (*Cocos nucifera* L.), Oilpalm (*Elaeis guineensis* Jacq.), Rubber *(Hevea brasiliensis* Muell-Arg.), Tea (*Camellia Sinensis* L.O. Kuntze). These cannot be desiccated to low moisture content without substantial loss of viability and hence are difficult to conserve even for short periods at low temperatures. There are various pretreatment desiccation procedures *viz.* Air Desiccation-Freezing, Pregrowth-Desiccation, Encapsulation-Dehydration, Vitrification for desiccation before freezing which can be used.

Addition of Cryoprotectants: The tissue is prepared for freezing by treating with various cryoprotectants to withstand the freezing conditions of –196°C. The effect of temperature on plants depends equally on genotype and environment, as well as physiological conditions. While genotype for practical purposes is an unchangeable factor, the other two factors can be altered according to circumstances. The freeze tolerance of a culture can be improved by timing the harvest of cells (in exponential phase) or by hardening process (growing the plants for shorter duration of one week at a lower temperature). Culturing of shoot apices for a brief period under optimum conditions before freezing has proved beneficial. Growing plants at 4°C for 3-7 days before taking explant also increases the survival of specimens.

Addition of cryoprotectants controls the appearance of ice crystals in cells and protects the cells from the toxic solution effect. A large number of heterogeneous groups of compounds have been shown to possess cryoprotective properties with different efficiencies. These are glycerol, dimethyl sulfoxide (DMSO), glycols (ethylene, diethylene, propylene), acetamides, sucrose, mannose, ribose, glucose, polyvinylpyrrolidone, proline, etc. Most frequently used are DMSO, sucrose, glycerol and proline. Cryoprotectants depress both the freezing point and the supercooling point of water i.e. the temperature at which the homogeneous nucleation of ice occurs thus retarding the growth of ice crystal formation.

In practice, the material suspended in the culture medium and treated with a suitable cryoprotectant is transferred to sterile polypropylene cryovials or ampoules with a screw cap. A dilute solution of the cryoprotectant (e.g. 5 – 10 % DMSO) is added gradually or at intervals to the ampoules containing the tissue material to prevent plasmolysis and protects cells against osmotic shocks. After the last addition of cryoprotectant, but before cooling, there should be an interval of 20 – 30 min. The ampoules containing the cryoprotectants are then frozen.

Freezing: The type of crystal water within stored cells is very important for survival of the tissue. For this reason, three different types of freezing procedures have been developed. Various types of cryostats and freezing units are available by which different cooling rates can be easily regulated.

(a) Rapid freezing: The plant material is placed in vials and plunged into liquid nitrogen, and a decrease of -300 to -1000°C/min or more occurs. This method is technically simple and easy to handle. The quicker the freezing is done, the smaller the intracellular ice crystals are. A somewhat slower temperature decrease is achieved when the vial containing plant material is put in the atmosphere over liquid nitrogen (-10 to -70°C/min). Dry ice (CO_2) instead of nitrogen can also be used in a similar manner. It is also possible that ultra-rapid cooling may prevent the growth of intra-cellular ice crystals by rapidly passing the cells through the temperature zone in which lethal ice crystal growth occurs. Rapid freezing has been employed to cryopreserve shoot tips of carnation, potato, strawberry and *Brassica napus* and somatic embryos.

(b) Slow freezing: In this method, the tissue is slowly frozen with a temperature decrease of 0.1 - 10° C/min from 0° C to –100° C and then transferring to liquid nitrogen. Survival of cells frozen at slow freezing rates of -0.1 to -10° C/min may involve some beneficial effects of dehydration, which minimizes the amount of water that freezes intracellularly. Slow cooling permits the flow of water from the cells to the outside, thereby promoting extracellular ice formation instead of lethal intracellular freezing. It is generally agreed that upon extracellular freezing the cytoplasm will be effectively concentrated and plant cells will survive better when adequately dehydrated. Slow freezing is generally employed to freeze the tissues for which computer-programmed freezers have been developed. Successful examples of cryopreservation of meristems include diverse species as peas, strawberry, potato, cassava, etc.

(c) Stepwise freezing: This method combines the advantages of both rapid and slow methods. A slow freezing procedure down to -20 to - 40° C, a stop for a period of time (approximately 30 min) and then additional rapid freezing to -196° C is done by plunging in liquid nitrogen. Slow freezing procedure initially to -20 to -40° C permits protective dehydration of the cells. An additional rapid freezing in liquid nitrogen prevents the growing of big ice crystals in the biochemically important structures. The stepwise freezing method gives excellent results in strawberry and with suspension cultures.

Storage: Storage of frozen material at the correct temperature is as important as freezing. The frozen cells/tissues are immediately kept for storage at temperature ranging from –70°C to –196°C. A liquid nitrogen refrigerator running at –150°C in the vapor phase or –196°C in the liquid phase is ideal for this purpose. The temperature should be sufficiently low for long-term storage of cells to stop all metabolic activities and prevent biochemical injury. Long-term storage is best done at –196°C. Thus as long as regular supply of liquid nitrogen is ensured in liquid nitrogen refrigerator, it is possible to maintain the frozen material with little further care. The injury caused to cells could be due to freezing or storage. The latter occurs primarily when the cells are not stored at sufficiently low temperatures.

Thawing: Thawing is usually carried out by putting the ampoule containing the sample in a warm water bath (35 – 45°C). Frozen tips of the sample in tubes or ampoules are plunged into the warm water with a vigorous swirling wrist action just to the point of ice disappearance. It is important for survival of the tissue that the tubes should not be left in the warm bath after the ice melts. Just at the point of thawing, quickly transfer the tubes to a water bath maintained at room temperature (20 – 25°C) and continue the swirling action for 15 sec to cool the warm walls of the tubes. Then continue with washing and culture transfer procedures. It is necessary to avoid excessive damage to the fragile thawed tissues or cells by minimizing the amount of handling at pre-growth stage.

Determination of Survival /Viability: Regrowth of plants from stored tissues or cells is the only realistic test of survival of plant materials. However, at any stage, the viability of frozen cells can be determined. It should be noted that viability of cells based on staining reaction alone would be insufficient and misleading and cannot replace the plant regrowth experiments.

Plant Growth and Regeneration: Major biophysical changes occur in the cell during freezing and thawing. The freshly thawed cell is strongly prone to further damage and requires appropriate nursing. Regrowth of cryopreserved materials would depend upon the manner in which specimens are handled, such as plating the cells without washing away the cryoprotectants, incorporation of special additives in regrowth medium and the physical environmental conditions the cells are exposed to during the early stages of regrowth.

Slow Growth Method

Cells or tissues can be stored at non-freezing conditions in a slow growth state rather than at the optimum level of growth. This technique is comparable to the traditional technique of Japanese gardeners known as Bonsai, in which trees are kept alive in small pots over hundreds of years. Growing processes are reduced to a minimum by limitation of a combination of factors like temperature, nutrition medium, hormones, etc. By this method, water in the tissue is maintained in the liquid condition, but all biochemical processes are delayed. All cultures in slow growth are static and therefore require only shelf space in suitable environments. This approach can be applied to tissue cultures mainly by following methods:

(a) **Temperature:** The tissues (shoot tips) can be cultured at low temperatures of 1 - 4°C. The explants such as callus, shoot tip cultures are established and maintained for 3 - 4 weeks under standard growth conditions. The explants are then transferred to lower temperatures of 1 - 4°C for cultures normally grown at 20 - 25°C and to 4 - 10°C for cultures normally grown at 30°C. A 16 hr photoperiod with reduced lighting (~50 lux) may be beneficial.

(b) **Low oxygen pressure:** The oxygen pressure can be regulated by atmospheric or partial pressure.

(c) **Hormones:** It is possible to use plant growth regulators like succinic acid, 2, 2-dimethyl hydrazide (B-9), cycocel (CCC) and abscisic acid (ABA) to delay culture growth. These chemicals cause reduced growth, mainly shorter internodes of the plants.

(d) **Osmotic inhibitors:** Cultures can be grown on a medium supplemented with 3 – 6 % (w/v) mannitol or high levels of sucrose as an osmotic inhibitor which slows down the growth. By this method, virus free strawberry plants were maintained for 6 years at 4° C. Grape plants have been stored for 15 years at 9° C by yearly transfer to fresh medium. This method holds great promise in the industry employing micro propagation techniques. During the period of low demand for a particular genotype the cultures may be simply shelved in refrigerator to avoid frequent sub-culturing.

References

Chawla, H.S. 2002. Introduction to Plant Biotechnology (2nd edition). Oxford & IBH Publishers, Delhi, India, p 538.

❑❑❑

Chapter - 12

Exploitation of Germplasm for Okra Improvement in India

N.C. Gautam

Introduction

Okra, *Abelmoschus esculentus* (L.) Moench, is annual herb grown in tropical and subtropical parts of world for green tender immature fruits and consumed as cooked mostly fresh but sometimes sun dried or frozen also used as vegetable. Some parts of plant and roots are used for medicinal as well as purification of sugarcane juice. Green fruit contains vitamins, calcium whereas seeds possess 13-22 per cent edible oil and 20-24 per cent edible protein. India is leading producer of okra and exporting fresh as well as processed fruits to other countries and also ranked first in area (0.48 million ha.), production (3.5m tones) and productivity (9.8 t/ha) during 2005 against the world area (0.78 million ha.), production (4.99 m tones) and productivity (6.4t/ha).

Origin

Okra is old world origin and probably from tropical Asia and Africa. Domestication started in Ethiopian region (Vavilov, 1936) and West Africa (Joshi *et al.*, 1974). The cultigens *Abelmoschus esculentus* might have originated in Asia (Joshi and Hardas, 1976) and India is also considered one of the centre of origin (Zeven and Zhukovsky, 1975 and Zeven and DeWet, 1982). Van Borssum – Waalkes (1966) advocated South Asia as centre of diversity of okra. Available diversity of *A. ficulneus* and along with *A. tuberculates* in India further support that Asia is the centre of origin of *Abelmoschus esculentus*.

Taxonomy and Classification

The genus name was changed from *Hibiscus* to *Abelmoschus* on the distinguishing features of calyx: spathulate with five short teeth, connate to corolla and caducous after flowering in latter. About fifty species of *Abelmoschus* have been described (Kalloo,1986) but six - eight species belong to this genus has been accepted now and recent classification developed by Van Borssum- Waalkes (1966) and its continuation by Bates(1968) and adopted by okra workshop IBPGR (1991) is as follows:

Table-1. Classification of *Abelmoschus* species

Classification developed by van Borssum-Waalkes (1966)	**Classification adopted by International Okra Workshop (IBPGR,1991)**
A. moschatus Medikus	*A. moschatus* Medikus
–ssp. *Moschatus*	–ssp. *Moschatus*
var. *moschatus*	var. *moschatus*
–ssp. *moschatus*	*–ssp. Moschatus*
var. *betulifolius* (Mast.) Hochr.	var. *betulifolius* (Mast.) Hochr.
–ssp. *biakensis* (Hochr,) Borss.	–ssp. *biakensis* (Hochr.) Borss.
–ssp. *tuberosus* (Span.) Borss.	–ssp. *tuberosus* (Span.) Borss.
A.manihot (L.) Medikus	*A. manihot* (L.) Medikus
–ssp. *Manihot*	
–ssp. *tetraphyllus*	–ssp. *tetraphyllus*
(Roxb.ex. Hornem.)Borss.	(Roxb. ex Hornem.)Borss
var. *tetraphyllus*	var. *tetraphyllus*
–ssp. *tetraphylls*	var. *pungens*
var. *pungens (*Roxb.)Hochr.	
A. esculentus (L.)Moench.	*A. esculentus* (L.) Moench. *A. tuberculatus*
(including *A. tuberculatus*)	
A. ficulneus (L.)W.&A.ex. Wight	*A. ficulneus* (L.)W.& A.ex Wight
A. crinitus Wall.	*A. crinitus* Wall.
A. angulosus Wall. ex. W.&.A.	*A. angulosus* Wall.ex.W.& A.
	A. caillei (A.Chev.)Stevels

Cytology

The most detailed studies have been done on *A. esculentus*. The chromosomes are short, mostly median or submedian primary constrictions and few have secondary constrictions. This genus appears as a regular series of polypoids with x = 12 with eight types (A to H) of chromosomes (Datta and Naugh, 1968) and basic genomes are respectively called (T) *A. tuberculatus,* n = 29; (M) *A. moschatus*, n = 36; (F) *A. ficulneus*, n = 36. The cultivated okra (*A. esculentus*) is considered to be polyploidy origin E = T' + Y where T' is slightly different from T and Y and accepted as T'Y genome which is closely homologous to two Indian wild species (T' F), (Joshi and Hardas, 1956). The chromosome number (2n) of *A. esculentus* varied from 56 to 196 (Siemonsma, 1982) but species with 2n = 130 belongs to *esculentus* types. Ten species /subspecies found in India carry the varied number of chromosomes *viz. A. angulosus* (2n = 56) *A. tuberculatus* (2n = 58), *A. manihot* (2n = 66), *A. moschatus* (2n = 72), *A. ficulneus* (2n = 72), *A. esculentus* (2n = 130), *A. tetraphyllus* var. *tetraphyllus* (2n=138), *A. tetraphyllus* var. *pungens* (2n = 138) and *A. cirnitius* (2n = 138) and *A. callei* (2n = 196).

Diversity

Zeven and Zhukovsky (1975) and Zeven and De wet (1982) have claimed *A. esculentus* is of Indian origin with maximum variation for different traits as compared to other species. India and Turkey have maximum diversity as compared to other countries of South Asia (Pakistan, Afganistan, Bagladesh) presently existing for economic traits *viz.* number of fruits/ plant, fruit length, number of ridges, length width ratio, leaf shape, branching habit, pigmentation and hairiness on fruit and other attributes.

Distribution and Adaptation in India

A. esculeutus is widely distributed througout the country while *A. moschatus* is present in wild form in Kerla, Tamil Nadu, Maharastra, Karnataka, Uttar Pradesh, Uttarakhand and Bihar. *A. tetraphyllus* is found in natural form in the forest land as well as barren land of Uttar Pradesh, Bihar and Madhya Pradesh where, its roots are also used in purification of sugarcane juice for the preparation Gur. The remaining species are fully wild texa and are being used for others purposes of academic and applied research.

Table 2. Distribution of wild *Abelmoschus* species in different phytogeographical regions of India. (Ghose and Kalloo, 1995)

Species	Phytogeographical regions							
	Indus Plains	Ganges Plains	Western Plains sular Region	Eastern Plains sular Region	North western Himala-yas	Eastern Himala-layas	North Eastern regions	Indian Ocean Island
	1	2	3	4	5	6	7	8
A. angulosus	–	–	+	+	–	–	+	–
A. crinitus	–	+	+	+	+	+	+	–
A. ficulneus	+	+	+	+	+	–	+	–
A. manihotb	–	+	+	+	+	+	+	+
A. moschatus	+	+	+	+	–	–	+	–
A. tuberculatus	+	+	–	–	–	–	–	–

1. North western semi-arid region;
2. Sub humid northern plains, Tarai belt and parts extending to the east;
3. Western ghats and adjoining tracts;
4. Eastern ghoats and adjoining tracts;

5. Western Himalayas, Himachal Ladak;
6. North Bengal, Sikkim and parts of Arunachal Pradesh;
7. Assam and adjoining states;
8. Andaman, Nicobar Island; Lakshadweeep.

Two wild species *A. crinitus* and *A. angulosus* are found in wild form in India and other Asian countries. *A. tetraphyllus* is being grown in western region of India and about 18 old landraces have been developed in this species. *A. crinitus* cultivated at low altitudes in regions with a marked dry season, being fire resistant and *A. angulosus* cultivated at altitudes between 750 and 2000 m in Nilgiri Hills.

Exploration and Collection

In India, okra germplasm exploration and collection programme was started just before independence (1946) at Indian (Imperial) Agriculture Research Institute, New Delhi and still continue by NBPGR, New Delhi with collaboration of IPGRI (IBPGR). A large number of accessions have been collected from different parts of India and neighboring countries. About 1200 accessions of cultivated okra (*A. esculentus*) and more than 600 accessions of wild species of *Abelmoschus* have been added to gene bank located at NBPGR, New Delhi. Location specific variations were observed in cultigens/landraces of okra found at different places of India. Multi-edged types tolerant to insect-pests and diseases were found in Rajasthan and adjoining areas while primitive landraces, having field resistance to biotic stress, were collected from eastern parts of India.

Existence of Different Species in India

The existence of different species of *Abelmoschus* at various places of India summarized by Dhankhar *et al*., (2005) is as follows:

(i) ***Abelmoschus crinitus***: Tehri, chamoli Bagheswar, Pauri-Garwal, Chhindwara and Sagar

(ii) ***A. cancellatus***: Uttrakhand (Dehradun Almora, Chamoli, Bering) and Uttar Pradesh (Saharanpur), Himachal Pradesh (Sirmour), Punjab, (Manguwal, Bankhandi, Dhangadi and Hosiarpur and Orissa (Jaspur Estate).

(iii) ***A. manihot***: Uttar Pradesh (Behat and Mahoba), Uttrakhand (Corbet-Park), Erunelli range (Kerala), Tinemghat (Goa), Moti, Kherawal range (Daman and Diu)

(iv) ***A. manihot*** ssp, ***tetraphyllus*** var. ***pungens (A. pungens):*** Himachal Pradesh, Jammu, Uttrakhand (Chamoli, Dehradun, Tehri), Orissa (Jaspur), Andaman (Khasihills)

(v) ***A. moschatus***: Uttarakhand (Almora, Pauri), Andman, Orrisa (Khasi hills)

(vi) ***A. moschatus ssp. tuberosus***: Uttarakhand (Uttar Kashi-Kot Bangla).

(vii) ***A. tuberculatus***: Uttar Pradesh (Saharanpur) and Rajasthan (Sirohi and Banswara).

(viii) ***A. ficulneus***: Jammu, Rajasthan, Madhya Pradesh, Maharasthra, Sawrashtra Tamil Nadu (Ganjam), Andhra Pradesh (Guntur)

(ix) ***A. angulosus***: Tamil Nadu (Nilgirihills) and Kerala (Westrn Ghat).

(x) ***A. manihot ssp. tetraphyllus var. tetraphyllus***: Rajasthan (Banswara), Uttarakhand (Pithoragarh). Uttar Pradesh (Rampur)

(xi) ***A. esculentus***: Gujarat (Gir and Dwarka), Punjab (Bathinda) and Uttarakhand (Dehradun), Uttar Pradesh.

North eastern hill regions and forest ranges of India are still under active exploration mission through different projects of AICRP (VS) and NBPGR in collaboration with SAU's and NGO's.

Introduction (Exotic collections)

Number of imported accessions found suitable to agro-climatic condition of India was promoted for cultivation at initial stage by the Plant Introduction Section, Genetics Division of I.A.R.I, New Delhi under the able leadership of Late Dr. Har Bhajan Singh. Important entries introduced through USDA were evaluated and recommended for cultivation were Clemson's spineless, Perkins Long Green, Smooth long, Valvet Green and White Velvet. Collection form Ghana was observed to have high degree of resistance to Yellow Mosaic Vein Virus and were extensively used by the breeders to develop resistant varieties. More than 150 exotic germplasm from Nigeria and Brazil each and 177 from the World Okra Care Collection through IPGRI were obtained by NBPGR and are being maintained and conserved under long term storage conditions at NBPGR and PGPR centres located at SAU's.

Germplasm Characterization and Evaluation

1. Morphological Characterization:

Almost all accessions were evaluated and characterized scientifically for more than 40 traits and their mean and range of expression of some important horticultural traits (Table 3, 4, 5) were recorded by various scientists (Thomas *et al.*, 1991 and Bisht *et al.*, 1993).

Table 3. Horticultural characters

Characters	Species
High number of fruits/plant	*A. tetraphyllus, A. mainhot ssp. manihot, A. caillei*
High number of branches/plant	*A. tetraphyllus*
Dark green fruit and extended fruiting period	*A. manihot ssp. Manihot*
Attractive smooth green fruits	EC187251, stock 510171

Diversity for a range of morphological characters namely, epicalyx segments, shape, size, persistence, petal colour and blotch, stem and fruit pubescence, pigmentation of various plant parts, days to flowering, plant height and fruit yield/ plant were important components of variation in germplasm of the important four species (*A. ficulneus, A. manihot* ssp. *tetraphyllus, A. moschatus and A. tuberculatus*) were characterized at NBPGR

Table 4. Range, Mean and Coefficients variant (Source: Bisht *et* al., 1995b).

Character	Min.	Max.	Mean	CV (%)
Ridges/fruit (no.)	5.0	8.0	5.6	14.2
Plant height (cm)	5.7	202.5	96.0	33.4
Internodes (no.)	5.0	33.5	16.5	23.6
Days to flowering (no.)	40.0	98.0	54.0	24.1
First fruiting node (no.)	2.0	24.0	8.6	46.5
Fruits on main stem(no.)	1.0	22.0	6.5	63.1
Fruits length (cm)	3.2	23.3	12.3	33.6
Fruits width (cm)	1.2	4.7	3.5	17.1
Fruits/plant (no.)	1.0	134.5	14.3	93.6
Seed/fruit (no.)	10.0	102.0	45.5	41.1

A large number of accessions of cultivated and related wild species of okra have been characterized for various traits including resistance/ tolerance to biotic and abiotic stresses.

Table 5. Resistance/tolerance to biotic stress

Traits	Genotype
A. Disease	
Fusarium wilt (oxysporum *Fusarium* f.*vasinfectum*)	*A. manihot*, PI379584
Powdery mildew (*Erysiphe cichoacearim*)	*A. tetraphyllus*, *A. manihot*, *A. manihot* ssp. *manihot*, *A. moschatus* (immune), *A. esculentus* cv. Nigeria, *A. angulosus*.
Cercospora and Alternaria blight	*A. crinitus*, *A. angulosus*, *A. moshchatus*, cv. 71, Round selction, cv. Nigeria, EC32598, IC1542, IC8248, ICI10238
Yellow Vein Mosaic Virus	EC305616 (tolerant), *A. angulosus*, *A. manihot*, *A. manihot ssp. manihot*, *A. manihot ssp. manihot* var. ghana, *A. tetraphyllus*, *A. crinitus* (immune), *A. pungens* (immune), Vaishali Vadhu and Red Wonder (symptoms suppressed), EC3299375, NIC3322, NIC3325, NIC6308, NIC9303A, NIC9408
Enation leaf curl virus	*A. crinitus*, *A. ficulneus*. *A. manihot*.
Microphomina	IC-90186, U-4307,U-4365, U-4579
B. Insect-pests	
Jassids (*Amrasca bigutulla*)	IC7194, IC8889(*A. esculentus*), *A. manihot* ssp. *manihot* var. ghana, *A. moschatus*, *A. crinitus*, EC.305656, EC305694, EC305695, EC305714, EC306731
Shoot & fruit borer (*Earias* ssp.)	*A. tuberculatus*, *A. caillei* cv. Narnaul Special
Mites (*Tetranychus* ssp.)	EC305656, EC305664, EC305696
C. Resistance/tolerance to abiotic stress	
Low temperature and frost	*A. angulosus*
Salinigy	Pusa Sawani (moderate tolerance)
Photo-sensitivity	Pusa Sawani (moderate tolerance)

2. Molecular Characterizations

A large numbers of accessions of *A. esculentus* along with its related species *viz. A. ficulneus, A. manihot ssp. tetraphyllus, A. tuberculats, A. manihot,* collected from different parts of India, south Asia and Africa were studied at NBPGR using isozyme electrophoresis and random amplified polymorphic DNA (RAPD) analyses. The study covered allelic variation at 13 isozyme loci and 189 amplification products obtained by RAPD using 22 random primers of 10 nucleotides length. Moderate genetic diversity was observed within *A. ficulneus and A. moschatus* by both isozyme and RAPD analyses. Diversity within *A. manihot ssp. Tetraphyllus, A. tuberculatus* and *A. escculentus* was narrow while genome of *A. moshatus* was observed to be quite distinct from that of other four species. Large similarities were recorded among above species by using the techniques of parsimony analysis (Dhankhar *et al.,* 2005).

3. Germplasm Utilization

Under the leadership of Late Dr. Har Bhajan Singh, varietal improvement programme was initiated during 1950's at Plant Introduction (section), Division of IARI and as a result of Pusa Sawami was developed from intevarietal cross between IC-1542 and Pusa Makhmali as resistant variety against Yellow Mosaic Vein Virus. The research for resistance to YVMV lad identification of an exotic collection from Ghana, which belongs to A. *manihot ssp. manihot* and mostly this introduction (Ghana) was used purely in developing number of elite genotypes namely, G-2,G-24 (NBPGR), Panjab Padmini and P-7 (PAU) Parbhani Kranti (MAU), Pusa A-4 (IARI), Varsha, Hisar Uphar and Hisar Unnant while another species, *A. manihot ssp. tetraphyllus* was used to developed Arkra Anamika and Arka Abhaya (IIHR).

Table-6. Breeding procedures and varieties developed in India

Breeding procedure	Variety developed
Pure line selection	Pusa Makhmali
Intervarietal hybridization	Pusa Sawani, Sel-2, HRB-9-2, HRB-55, NDO-10, VRO-6, JNDOL-O3-1
Interspecific hybridization	Punjab Padmini P-7, Parbhani Kranti, Arka Abhya, Arka Aramika, Hisar Unnat, Hisar Upar.
Heterosis breeding	DVR-1, DVR-2, DOH-1, DOH-2, GOH-3, GOH-4, HYB-7, HYB-8, BO-1, BO-2, Vishal, Vijay, Varsha, SOH-101, NBH-180

Interspcific hybridization has led to large scale variation in Okra. Amphidiploids using wild relative were also developed while studying inter relationship between different *Abelmoschus* species (*A. esculentus x A. mainhot; A. esculentus x A. tuberculatus; A. mainhot x A. tuberculatus; A. esculentus x A. tetraphyllus; A. ficulneus x A. tuberculatus, A. esculentus*.x A. *manihot* ssp. *Manihot)* but these were not stable (Jambhle and Nerkar, 1981). These F_1s were largely sterile but fertility could be restored to varying degree through polyploidization for backcrossing. Existence of natural amphidiploids of *A. esculentus and A. mainhot* has also been reported (Siemonsma, 1982) but Kuwada (1966) could able to produce artificial amphidiploids (2n = 192) of *A. esculentus* (2n = 124) × *A. manihot* (2n = 68).

It is difficult to maintain wild species due to differences in their adaptability and seed dormancy. The F_1s involving certain partially crossable forms are highly sterile and genes carrying resistance to stresses in wild relative are linked with undesirable genes that slow down the progress in breeding programmes, therefore new approaches of biotechnological research has to be applied for obtaining fertile and potential genotypes of worth use to mankind.

Germplasm Conservation

Technology has been standardized to the well dried (5 per cent moisture contents) under ambient conditions at 4°C (short term) and at –20°C (long term) in heat sealed aluminum foils at NBPGR regional stations, ICAR's institutes and SAU's. As per IPGRI germplasm database more than 46 institutions in different countries maintaining about 11,000 accessions of cultivated and wild species of okra.

References

Bates, D. M. 1968. Notes on the cultivated malvaceae. *Abelmoschus*, Baileya, **16:** 99-112

Bisht, I. S., Patel, D. P., Mahajan, R. K., Thomas, T. . and Rana, R. S. 1995b. Catalogue of wild *Abelmoschus* Species Germplasm. Nation Bureau of Plant Genetic Resources, New Delhi, India

Bisht, I. S., Patel, D. P., Sapra, R. L., Thomas, T. A., Kapoor, M. N. and Rana R. S. 1993. Catalogue of okra [*Abelmoschus esculentus* (L.) Moench.] Germplam, PartIII. National Bureau of Plant Genetic Resources, New Delhi, 77p.

Datta, P. C. and A. Naugh. 1968. A few strains of *Abelmoschus esculentus L.* Moench. Their caryological study and relation to phylogeny and organ development. *Beitr. Biol. Pflan,* **45:** 113-126.

Dhankhar, B. S., Mishra, J. P and Bisht, I. S. 2005. Okra. *In:* Plant Genetic Resources: Horticultural crops, Narosa Publishing House Pvt. Ltd., New Delhi, India. pp. 59-74.

Ghose, S. P. and Kalloo, G. K. 1995. Okra: Genetic resources of Indigenous Vegetables and their uses in South Asia, IIVR, Varanasi.

IBPGR. 1991. International crop network series 5- report of on International workshop on Okra Genetic Resources, International Board for Plant Genetic Resources, Rome, Italy.

Jambhale, N. D. and Nerkar, Y. S. 1976. Occurrence of spontaneous amphidiploids in an interspecific cross between *A. esculcutus* and *A. tetraphyllus*. *J. Maharastra Agric. Univ.* **6:** 167.

Joshi, A. B. and Hardas M. W. 1976. Okra *In:* Evolution plants. N.W. Simmonds (ed.) Longmans, London. U.K. pp. 194-195.

Joshi, A. B. and Hardas, M. W. 1956. Alloploid nature of okra *(Abelmoschus esculentus). Nature,* **178:** 1190

Joshi, A. B., Godwal, V. R. and Hardas, M. W. 1974. Okra. *In:* Evolutionary studies in world crops. Diversity and change in the Indian sub-continent. J.B. Hutchinson (ed.) Cambridge University Press, PP99-105

Kalloo . G. 1986. Okra: Genetic resources vegetables crops, Vegetable Breeding, CRC Press.

Kuwada, H. 1966. The new amphidiploid plant named *Abelmoschus tuberculatus esculentus,* Obtained from the progeny of the reciprocal crossing between *A. tuberculatus* and *A. esculeutus. Jap. J. Breeding,* **16:** 21-30

Siemonsma, J.S. 1982. West African Okra. Morphhological and cytological indications for the existence of a natural amphidiploid of Abelmoschus esculentus L. (Moench) and A. manihot. Midikus. *Euphytica* , **31:** 214-252

Thomas, T. A., Bisht, I. S., Patel, D. P., Agarwal, R.C. and Rana, R. S. 1991. Catalogue on okra (*Abelmoschus esculentus* [L.]Moench.) *Germplasm II.* National Bureau Plant Genetic Resources,New Delhi, India,100p.

Van Borssum-Waalks J. 1966. Malasian Malvaceae-revised. *Blumea.,* **14:** 1-251.

Vavilov, N.I. (1936). Studies on the origin of cultivated plants. *Bull. Appl. Bot.,* **26:** 1-248.

Zeven, A. C. and De Wet, J. M. J. 1982. Dictionary of cultivated plants and their centres of diversity. Centre of Agricultural Publishing and documentation Wageningen, Netherlands p 259.

Zeven, A. C. and Zhukovsky, P. M. 1975. Dictionary of cultivated plants and their centres of diversity .Centre of Agricultural Publishing and documentation, Wageningen, Netherlands p 299.

Chapter – 13

Genetic Diversity of Indigenous Underutilized Cucurbits

S.K. Verma, K.S. Negi, K.C. Muneem and R.R. Arya

Introduction

Genetic resources are the building blocks for new varieties. The Indian gene centre is floristically extremely rich. Most of the tropical and sub-tropical fruits believed to have originated from Hindustan and Indo Malaya-Region. Hindustani gene centre possess a rich diversity of 152 crop plants distributed in different agro-ecological regions of India (Arora, 1985). The Indian sub-continent is one of the centers of origin and diversity of vegetables crops. Around 80 species of major and minor vegetables occur in wild forms (Choudhury, 1967, Seshadri 1987). Among the 118 genera belonging to the Cucurbitaceae family *Cucurbita*, *Cucumis*, *Citrullus* and *Lagenaria* genera are of great importance. There is tremendous genetic diversity within the family, genera and even within species level. The adaptation range varies from tropical and subtropical regions to extreme arid desert and in temperate regions from Tarai belts to higher altitudes. Extreme germplasm collections are maintained at different centers of the world which represents valuable resources for breeding even in wild form, primitive cultivars, landraces/ folk varieties/ farmer's selection in a much localized pockets. Now we have very good strains for highly nutritive value, multiple disease resistant and extended shelf lives from our local material through intensive breeding.

Indigenous plant germplasm are adapted component of plant biodiversity. Such adapted components are more valuable resources either as a cultivar or as a parental line in hybridization programme. It might be also a putative material for biotechnological treatment for tolerance to biotic, abiotic factors or nutritional and quality components. Indigenous germplasm can be basically of two types, one which has been domesticated and subjected to human intervention, the other one may be called wild plants or wild relatives and still in the process of evolution without the human intervention. Biochemically the cucurbits are characterized by bitter principles, called cucurbitacins. Great variety of cucurbits contains bitter principle in portion of the plant at some stage of development. Cucurbitacins are tetracyclic triterpenes having extensive oxidation level. Bitter principals found in roots differ from those in fruits. Seedlings specially radicals contains cucurbitacins which are considered primary. The pollen also carries bitter principles and hence when bitter pollen fertilized non-bitter ovule the resulting fruit will be bitter known as meta-xenia effect. Cucurbits are consumed in various forms i.e. salad (cucumber, gherkins, longmelon), sweet (ash gourd, pointed gourd), pickles (gherkins), deserts (melons) and culinary purpose. Some of them (e.g. bitter gourd) are well known for their unique medicinal properties. In recent years, *abortifacient proteins with ribosome-inhibiting properties* have been isolated from several cucurbit species, which include *momorcharin* (from *Momordica charantia),* luffaculin (from *Luffa operculata),* trichosanthin (from *Trichosanthes kirilowif)* and beta-trichosanthin (from *Trichosanthes cucumeroides).* Among these, Trichosanthin is of particular interest because its ribosome-inhibiting properties have been found to be effective in inhibiting the replication of human immunodeficiency virus (HIV), indicating its potential as a therapeutic agent for AIDS. In cucurbits three conditions are found (i) fruit and vegetative parts bitter (ii) fruit not bitter but vegetative parts bitter (iii) bitterness absent in both vegetative parts and fruits. Hence some kind of co-evolutionary relationship between cucurbits found and they are very helpful during evaluation of germplasm.

Indigenous Cucurbit Diversity in India

Almost all the cucurbits are of tropical origin, mostly in Africa, Tropical America and South East Asia

(Gupta *et al.*, 1995, Gupta and Rai, 1995). Warm climate and well-drained soils are suitable for their proper growth but are sensitive to water logging and freezing temperatures. The country produced 82.7 million tones of vegetables from 5.98 million hectares and contributed 14.4% of the total world vegetable production of 574 million tones from 42 million ha during the year 2000. In group of under utilized vegetables, a number of less common vegetables are included because they are grown on very limited scale. The under utilized vegetables fall under different prominent culinary group like minor brassicas, minor root crops, minor leafy vegetables and some minor cucurbits etc. Recently much attention is being paid for rare vegetables because their good demand in multi star hotels and export values. Under utilized vegetables are also much preferred by amateur gardeners because they are particularly suited to small gardens and easy to grow.

Biological Diversity

Biological diversity is defined as the variability among living organisms from all sources including, inter alias, terrestrial, marine and other aquatic ecosystem and the ecological complex in which they survive, this include diversity within species, between species and of ecosystem. Thus in short, biodiversity refers to the quality, range or extent of differences between the biological entities in a given situation. Here all forms of a given species even released varieties, wild form, primitive cultivars, landraces/folk varieties/farmer's selection in a very localized pockets are considered. The status of conservation of vegetable crop germplasm has always received less attention than that of the major staple crops such as cereals, legumes and other horticultural crops. The documentation of traditional landrace-based indigenous knowledge system would play significant role in making value addition of the traditional crops and their better management on-farm. *In situ* conservation and crop improvement can complement one another in marginal areas. Breeding programmes that evaluate landraces and use them in local improvement efforts are expected to produce material of direct value for marginal agro-climatic zones as well as achieving significant local conservation (McNeely 1988; Harlan 1992; Brush 1999; Smale and Bellon 1999; Almekinders and Elings 2001). The databases, holding mainly passport data, can be analyzed for the identification of duplications and gaps among collections. At the same time, the management of collections can be based on better knowledge of the diversity in stock. The enhancement of the links between germplasm conservation and use will continue to depend, inter alias, on easy access to the genetic material. It is also important to test how variable are common varieties than less common varieties. Further, locally common alleles are more important for conservation and interesting to users. Important areas of diversity in cultivated cucurbitaceous vegetable crops (Arora, 1995 and Ram, 1997) are given as follows:

(i) The Northwestern and Eastern Himalayan Region

Temperate crops predominate in this region. Rich diversity were found mainly in squash, cucumber and chow-chow. Other under-utilized cucurbits are *Cucumis hardwickii, C. trogonus, Cucurbita* spp., *Cyclanthera pedata. Luffa graveolens, Trichosanthes multiloba* and *T. himalensis.*

(ii) The Northeastern Region

This region has a good diversity in cucumber, pointed gourd or parwal, chow-chow and bitter gourd. Other minor cucurbits are *Cucumis hystrix,C. trigonus, Cyclanthera pedata, Luffa graveolens, Momordica cochinchinensis, M. dioica , M. macrophylla, M. subangulata, Trichosanthes cucumerina, T. dioica, T. dicaelosperma, T. khasiana, T. ovata* and *T. truncata.*

(iii) The Northern Plains including Tarai Region

This is one of the richest pockets of diversity in vegetable crops. As vegetables can be grown during the rainy, winter and also in summer season, hence more diversity prevails and it is fitted well into the cropping patterns. Rich variability occurs in cucurbits - *Cucumis* spp. (cucumber, Phut, Kakari), *Luffa* spp (ridge gourd, smooth gourd),

Cucurbita spp. (squashes), *Benincasa hispida* (ash gourd), round gourd, bottle gourd, bitter gourd, *Trichosenthes dioica* (pointed gourd), *Luffa echinata, Momordica cymbalaria, Momordica dioica, Momordica cochinchinensis.*

(iv) The North Western Plains

The type of diversity particularily in *Citrullus* spp. (*Citrullus colocynthis*), *Cucumis* spp. (*Cucumis prophetarum*) and *Momordica* spp. (*Momordica balsamina*) are found.

(v) The Central Region/Plateau

More diversity occurs in *Cucurbita* spp., different kind of squashes, cucumber, ash gourd, round guard, bitter gourd, pointed gourd and ridge gourd. Other under- utilized less known are *Momordica balsamina, Citrullus colocynthis* and *Cucumis prophetarum.*

(vi) The Western and Eastern Peninsular Region

This is extremely rich region in cucumber, bitter gourd, bottle guard, and squashes. Other are *Cucumis hystrix, C. setosus, C. trigonus, Luffa acutangula* var. *amara, L. graveolens, L. umbellata, Momordica cochinchinensis, M. cymbalaria, M. denticulata, M. dioica, M. subangulata, Trichosanthes anamalaiensis, T. bracteata, T. cordata, T. cuspidata, T. himalensis, T. horsfieldii, T. lepiniana, T. multiloba, T. nerifocia, T. perottitiana* and *T. villosa*

Specific Distribution of Cucurbits in India

Among the cucurbits, cucumber originated in India and wax gourd *(Benincasa hispida)* is from Southeast Asia. Considerable amount of genetic diversity in wild and cultivated species of *Citrullus, Cucumis, Coccinia, Cucurbita, Luffa, Momordica,* and *Trichosanthes* is found. *Citrullus colocynthis* exhibits much variation in North-Western plains. In *Cucumis, C. hardwickii* (supposed to be progenitor of cultivated cucumber) and *C. trigonus* are distributed in Himalayan belt with considerable diversity. The distribution of C. *setosus* is restricted to Eastern plains while diversity in *C. hystrix* ranges from Eastern plains to North-Eastern hills of Assam, Tura range and Mishimi hills of Megbaiaya. In *Luffa,* most of the species are naturally growing on road side, near railway tract, forest openings etc. *Luffa acutangula* var. *amara* is usually found in Peninsular India and *L. echinata* in the Western Himalayan belt, central India and upper Gangetic plains. Another important species, *L. graveolens* (considered to be wild progenitor of *L. hermaphrodita)* can be spotted in Bihar, Sikkim, Tamil Nadu and *L. umbellate* is confined to eastern coast. Similarly, *Momordica balsamina* extensively occur in semi-dry Northwestern plains and sporadically in upper Gangetic regions and in the Northern parts of Western and Eastern ghats. *M. dioica and M. cochinchinensis* occurs in wild, semi-wild forms in the Gangetic plains. The distribution of *Momordica cymbalaria* is restricted to the Western ghats, Maharashtra with sporadic occurrence in the Eastern Peninsular region. *M. subangulata* and *M. macrophylla* occur largely in the North-Eastern region and exhibits sporadic distribution in Deccan plateau, extending to the Eastern Ghats. *M. deudata* is largely confined to Eastern Peninsular tract. *Trichosanthes* has 21 species in India and the major zones of its concentration distributed along the Malabar Coast in Western-Ghats and low and medium elevation zones (up to 1500 m) in Eastern Ghats and North-Eastern region. A widely distributed species, *T. bracteata,* occurs in Eastern India, extending to the South and Himalayan belt (1500 m). *T. cordaia* (related to *T. anguina)* is distributed in peninsular region extending to North-Eastern plains and hills. There are several cucurbits, which have adaptability in a particular region of India, where their expression with respect to yield and quality is better than other regions. Chow-chow *(Sechium edule)* has specific adaptation in Mizorum and *M. chochinchinensis* in Tripura, Assam and West Bengal, *T. dioica* in Eastern Uttar Pradesh, Bihar and West Bengal.

Genetic Diversity

The genetic diversity in cucurbits extends to both vegetative and reproductive characteristics and considerable range in the monoploid *(x)* chromosome number including 7 *(Cucumis sativus),* 11 *(Citrullus* spp., *Lagenaria* spp., *Momordica charantia* and *Trichosanthus* spp.), 12 *(Benincasa hispida, Coccinia cordifolia, Cucumis melo, C. melo* var *utilissimus* (Kakari), *C. vulgaris* var *fistulosus* syn *Praecitrullus fistulosus* (Tinda), *C. melo* var *momordica* (Phoot), *C. callosus* (Kachari) and *Sechium* spp.), 13 *(Luffa* spp.), 14 (*Momordica dioica* (sweet gourd), *M. cochinchinensis* (spine gourd) and 20 *(Cucurbita* spp.).

Some Important Underutilized Cucurbits are:

Ash gourd *(Benincasa hispida)* (Petha)

Annual plant growing upto 6 m. It produces leaves from June to October and flowers from July to September. The flowers are monoecious (individual flowers are either male or female, but both sexes can be found on the same plant) and are pollinated by Bees. The plant is self-fertile. We rate it 2 out of 5 for usefulness. The plant prefers light (sandy), medium (loamy) and heavy (clay) soils and requires well-drained soil. The plant prefers acid, neutral and basic (alkaline) soils. It cannot grow in the shade. It can tolerate drought and because of its waxy coating, it can be stored for several months, sometimes as long as a year.

Balsam pear *(Momordica balsamina)*

Glabrous to slightly hairy perennial herb with a tuberous rootstock. Plant smells like the common thorn apple or *Datura stramonium,* more so when bruised. Stem mostly annual, prostrate or climbing, about 5 m long, cut twigs exude clear sap. Tendrils simple, leaves waxy, lower surface paler than upper, deeply palmately 5-7 lobed, about 12 cm long, margins toothed and stalked. Flowers solitary, male and female flowers occur on the same plant (monoecious). Male flowers prominently bracteate (subtended by a leaflet), bract ± ovate, to 18 mm long, pallid, green-veined, calyx green to purplish-black, corolla white to yellow, apricot or orange, green-veined, with grey, brownish or black spots near the bases of the three inner petals, 10-20 mm long, anthers orange. Female flowers inconspicuously bracteate, corolla rather smaller than males.

Fruit spindle shaped, dark green with 9 or 10 regular or irregular rows of cream or yellowish short blunt spines, changes its colour from bright orange to red on ripening. Seeds ovate in outline, rather compressed, up to 11 mm long, light brown, surface sculptured encased in a sticky scarlet red fleshy covering that is edible and sweet, tasting like watermelon. The balsam pear grows in white, yellow, red and grey sandy soil, also loam, clay, alluvial, gravelly and calcareous soil. It thrives in full sun and semi-shade in grassland, savanna, woodland, forest margins, coastal dune forests and in river bank vegetation as well as disturbed areas. The leaves and green fruits are cooked and eaten as spinach, sometimes with groundnuts, or simply mixed with porridge. The young leaves contain vitamin C. The raw ripe fruits are also eaten. The fruit is eaten by birds, ants and also by humans.

Chow chow *(Sechium edule)*

Chow chow is mostly grown in area of higher altitudes like in NEH region in North India, Bangalore region of Karnataka and known as Askas in Uttarakhand. Plant is monoecious and develops perennial vine. Its immature fruits are used for vegetable purpose. Unlike the other many seeded cucurbit fruit, chow chow has a single large seed. It is a predominantly cross-pollinated, but is self-compatible. It has been grown to a limited extent mainly grown in door yards/ kitchen garden / back yards as a neglected crop in mountainous region of Sikkim Himalayas to Central Himalayas from 800 m to 1800 m altitude. It is a native to Mexico and Central America - Guatemala and also known as Vegetable Pear, Mango Squash, Chayote, Spanish Citrayota, Gayto and Askas. Chow-chow is a tender, perennial-rooted cucurbit with climbing vines, light green pear shaped fruit. The whole fruit is planted as a seed. Each fruit has a single large seed that sprouts soon as the fruit matures (viviparous in nature). Some trellis or support for climbing vines is required. Both male and female flowers occur on the same vine. Fruits mature about 35 days after insect pollination. This Station is maintaining 11 accessions of chow-chow germplasm in Field Gene Bank collected from East and West Garo Khasi hills, Kumaon and Garhwal Region of Himalayas (Table 1).

Table 1. Mean value of various characters in different germplasm in Chow - Chow

S.N.	Genotypes	Days to Sprouting	Fruit dia. (cm)	Fruit length (cm)	No. of fruit/plant	Days to 50% fruit setting	Fruit weight (g)
1.	IC - 340574	24.33	9.38	14.47	32.33	182.66	637.33
2.	IC – 340575	27.00	10.25	14.60	43.33	173.66	471.33
3.	IC – 340576	28.66	9.16	15.93	16.33	180.00	442.33

Contd...

4.	IC – 340577	32.66	9.19	11.67	39.66	174.66	315.83
5.	IC – 340578	19.66	9.11	13.67	13.50	175.33	537.33
6.	IC – 340579	23.00	9.04	11.95	17.33	185.00	553.33
7.	IC – 340580	24.66	9.54	13.85	12.00	185.00	512.83
8.	IC – 340581	24.66	10.32	12.96	11.66	188.33	458.23
9.	IC – 340582	24.66	9.78	14.60	18.00	176.66	434.00
10.	IC – 340583	25.00	8.13	13.79	14.33	175.33	389.00
11.	IC - 340584	23.66	8.61	13.90	15.00	188.66	340.66
	Range	23.00 - 32.66	8.13 - 10.32	11.67- 15.93	11.66 - 43.33	173.66 - 188.66	315.83 - 637.33
	SEM	1.49	0.66	0.84	8.40	2.87	52.92

The most significant morphological differences between cultivated and wild Chow-Chow are the difference in size of the vegetative and reproductive structures. Wild plants are more robust and their leaves, flowers and staminate inflorescence are bigger than those found in cultivated plants and they are different in their chromosome number. They do differ in the size of the fruit and the pedicels of the staminate flowers, as well as in the outline of the leaves and the shape of the lobules and have smaller prickles only at the base. Collections produce fruits with such diverse morphology (round white, long white, pointed green, broad green, oval green) mainly with regard to form, size (vary in size and taste, and generally they are partially or entirely covered with spines), surface texture (pyriform fruit) and flesh flavour (soft and pleasant) and, to a lesser degree, colour (light green or white, smooth or unarmed, approximately 15 cm long and weighing approximately 637 g), in presentation (no signs of premature germination, physical damage or marks produced by pathogens) and in texture and flesh flavour (soft and pleasant). All are having endocarpic and precocious germination hence all are maintained in Field Gene Bank. All these 11 accessions were evaluated for 20 agro-botanical characters under Bhowali condition (Singh *et.al.* 2002) .

Ivy gourd (Kundru or tondli) *(Coccinia grandis)*

Coccinia grandis, also called tindora (tindori, tindoori), giloda, kundri, kundru, kowai, kovai, tindla. Its young fruits are used for vegetable purpose and said to be good for the diabetic people. Fruits come in the market during rainy season. Crop is common both in the Eastern and Southern states of India. Propagation is done by stem cutting for which 12-15 cm cuttings of pencil thickness having 5-6 leaves are taken and planted in basins of 60 cm diameter dug 175 cm apart.

It is a semi-perennial crop of 4-5 years, yielding fruits in summer and rainy season. Stem cuttings, 25 to 30 cm long, 1.5 cm to 2 cm thick are used for planting. Thicker cuttings sprout earlier. They are planted in basins of 60 cm diameter, spaced at 2 meters and cuttings are planted at 3 cm depth in July or February. About 10 per cent male plants have to be planted to ensure good fruit set. Some clones produce parthenocarpic fruits and some bitter fruits too. Bowers are erected at about 1.5 metre height. Fruiting starts in 10 to 12 weeks after planting. At least once a year, the pits should be manured. Ivy gourd prefers warm and humid climate and it remains dormant during winter season. Plant requires perfect drainage and is very susceptible to water logging. In South and Central India, fruiting is round the year, while in North India, fruiting terminates when the temperature comes down in November. A vine can yield 200-350 fruits, weighing 3 to 4 kg going up to 8 kg which works out to about 120 quintals per hectare.

Kadawi kakari (*Cucumis sativus* var. *hardwickii*)

Annual herbaceous vine; stems angular and rough; leaves rough, 5-10 cm long, middle lobe sometimes ovate, flowers monoecious, solitary, fruit size of small orange, green and yellow variegated becoming yellow when ripe, easily broken, light yellowish-orange to pale yellow; intensely bitter; seeds numerous, ovoid, compressed, smooth, borne on parietal placenta. Produces flowers in summer, it is distributed from 200 to 2200 msl from Northern plains to Himalayan region. Twenty six accessions were evaluated at this Station for 20 different agro-botanical characters. Variability among the accessions is presented in Table 2.

Table 2. Range in different characters in *C. sativus* var. *hardwickii*

Characters	Range
Days to First Flowering	78 – 112
Days to 80% Maturity	86 – 120
Node to 1st Fruiting	3.3 – 7.0
No. of Fruits / Plant	8.5 – 24.7
Five Fruits weight (g)	103.3 – 666.7
Fruit Length (cm)	3.7 – 9.9
Fruit Diameter (cm)	2.2 – 5.4
Yield / Plant (g)	550 – 3760
No. of Seeds / Fruit	141 – 250.7
100 Seed Weight (g)	2.6 – 8.1

Kheksa and Kakoda (*Momordica* spp.)

It is also known as sweet gourd. There are two species of *Momordica,* viz., *cochinschinensis* Spreng and *dioica* L. which are grown in a limited way in some parts of India, but are often considered synonymous because of confusing vernacular names like *kakado, keksa, kakrol,* etc., signifying either of the two species. *M. cochinchinensis* is called *gol kakoda* or sweet gourd, grown in North Bengal, N.E. India while *M. dioica* is grown in tribal areas of Bihar, Orissa and Thane district of Maharastra. Both are dioecious perennials, tuberous rooted, but differ in fruit and seed characters. Fruits of *M. cochinchinensis* are ovate 10-15 cm, pointed, orange or red, densely covered with conical spines, very fleshy orange – red pulp containing numerous seeds, ovoid and brownish black in colour. It grows in warm humid weather and tuberous roots are planted in pits spaced 120 cm apart. The vines are trained in bowers and 5-10 percent of male plants are provided for good fruit set. Flowering starts in April and fruiting ends in October-November. The plants remain dormant in winter. The yield is 30-50 fruits per plant weighing around 4 kg. The tubers are left *in situ* and they over winter. The cultivation of *M. dioica* is somewhat similar. It is called kartoli in Thane district of Maharastra, cultivated by tribal populations as in Bihar and Orissa. Tuberous roots are planted here also. Fruits are smaller, 2.5-5 cm long, ovoid ellipsoidal, densely echinate with soft spines, seeds slightly compressed 3.0-7.0 mm long.

It is a dioecious and perennial plant, bears oval shaped green in colour fruit. Fruits are not bitter and used for vegetable purpose. Kokoda (*M. dioica*) is also a dioecious and perennial plant but develops small size fruit as compared to *M. cochinchinensis*. *M. dioica* closely confused with it. *M. dioica* has estipulate fruits while in *M. subangulata,* the fruits are longitudinally date. Kakoda and kheksa both are mainly propagated by their tuberous roots. Since both the dioecious, tubers from male and female plant should be planted together, keeping the population of 8-10 per cent male plants to ensure proper pollination.

Meetha Karela (*Cyclanthera pedata*)

This is an annual cucurbit introduced from S. America, called *meetha karela* (*Cyclanthera pedata*) grown in Western Himalayan regions. It is grown on arches, pandals or bowers in the hills of U.P., up to about 1600 m above mean sea level. It is primarily grown for local consumption. The plant prefers light (sandy), medium (loamy) and heavy (clay) soils and requires well-drained soil. The plant prefers acid, neutral and basic (alkaline) soils. It can grow in semi-shade (light woodland) conditions. It is a monoecious cucurbit with smooth fruits, oblong with narrow base, green when young, white when mature and containing 8 to 10 seeds in each fruit. Four to six green pickings are possible.

Pointed gourd (parwal) (*Trichosanthes dioica L.*)

In North Bihar and Eastern Uttar Pradsh, it is extensively grown in 'diara' lands and in Assam, Bengal, Orrisa, etc. It is also grown in loamy soils and in places with hot and humid climate. The main precaution is to prevent water logging. Being a vegetatively propagated crop, there are several clones with distinct characters found in different States. Seed propagation is avoided because a) germination is poor, b) about 50 per cent plants male and non-bearing and c) flowering takes nearly two years.

Cuttings are planted in two systems. In one system, cuttings 60 cm long or more from one year old plants are taken in October when the plants complete fruiting and vines are mature. These cuttings are coiled in the shape of a ring and planted directly in the hills of prepared land or nursery. Here a number of nodes are covered by the soil and rooted cuttings are planted in February-March in permanent places. In the other system, furrows are dug 30 cm deep and after filling with organic manure, 60 cm long cuttings are planted 15 cm deep keeping both ends exposed. In root sucker propagation, the smaller root suckers at the nodes of creepers are uprooted in October and planted in hills. At the time of planting it is to be ensured that 10 to 15 per cent of the cuttings should have male plants for adequate fruiting.

In some areas like in Eastern U.P., growers do staking. During summer, frequent irrigation may be necessary. In garden land conditions, the crop is grown as a perennial for three to four years with ratooning practiced in late October. However, in river beds, the crop is planted every year. The crop starts fruiting from February in Bengal and later in other states like Bihar. The fruiting continues up to South-West monsoon when another flush of flowering begins, it may continue until October. The fruits are harvested when they are green and tender. In the first year, the yield would be around 75-80 quintals per hectare, while in the second year it may increase to 140-150 quintals.

Tumba (*Citrullus colocynthis* (L.) *Schrad.*)

Annual or perennial (in wild) herbaceous vine; stems angular and rough; leaves rough, 3 to 7 lobed, 5-10 cm long, middle lobe sometimes ovate, sinuses open; flowers monoecious, solitary, peduncled, axillary, corollas 5-lobed; ovary villous; fruit is a pepo type, nearly globular, 4-10 cm in diameter with somewhat elliptical fissures, about size of small orange, green and yellow variegated becoming yellow when ripe, with hard rind, pulp light in weight, spongy, easily broken, light yellowish-orange to pale yellow; intensely bitter; seeds numerous, ovoid, compressed, smooth, dark brown to light yellowish-orange, borne on parietal placenta. Flowers in summer.

Dried pulp of unripe fruit is used medicinally for its drastic purgative and hydragogue cathartic action on the intestinal tract. The fruit is used to repel moths from wool. The vine is planted as a sand binder. Seed often removed from the poisonous pulp and eaten. Fruits considered cathartic, ecbolic, emmenagogue, febrifuge, hydragogue, purgative and vermifugal. The colocynth is used for amenorrhea, ascites, bilious disorders, cancer, fever, jaundice, leukemia, rheumatism, snakebite, tumors (especially of the abdomen), and urogenital disorders. Roots may also be used as purgative against ascites, for jaundice, urinary diseases, rheumatism and for snake-poison.

Endangered Cucurbits of India

There are several species, which are being extinct day by day in India, especially because of monoculture, adoption of improved' open pollinated varieties and hybrids, urbanization, fragmentation of habitats and deforestation. As a result several cucurbits in India are at the verge of extinction and considered as endangered (Rai *et al.*, 2005) Table 3.

Table 3. Rare and endangered cucurbitaceous species in India

Cucurbitaceous species	Biogeographic zones
Corallocarpus gracillipes (Naud.) Cogn.	Western Ghats
Gomphogyne macrocarpa Cogn.	Eastern Himalaya
Indogevillea khasiana Chatterjee	North East India
Luffa umbellata (Kleir) Roem.	Western Ghats
Melothria amplexicaulis Cogn.	Deccan Peninsula
Momordica sub angulata Bl.	Deccan Peninsula, Western Ghats
Trichosanthes lepiniana (Naud.) Cogn.	Deccan Peninsula, Western Ghats
Trichosanthes perrotteliana Cogn.	Western Ghats
Trichosanthes villosula Cogn.	Deccan Peninsula, Western Ghats

Cultivated and wild relatives: Important wild and wild relatives cucurbits are *Cucumis sativus* var. *hardwickii, C trigonus, C. prophetarum, C. setosus, C.hystrix, Luffa graveolens, L.echinata, L.acutangula var. amara,* L. umbellate, *Trichosenthes multiloba, T. himalensis, T. cucumerina, T. dioica, T. dicaelosperma, T. khasiana, T. ovata, T. anamalaiensis, T. bracteata, T. cuspidate, T. horsfieldii, T. perottitiana, T. nerifolia, T. villosa, T. truncata, T. cordata, Momordica cochinchinensis, M, microphylla, M. subangulata, M. cymbalaria, M. diocia, M. balsamina, M.denticulate, and Citrullus colocynthis,* Others wild species are given in Table -4

Table 4. Cultivated and wild relative – cucurbits

Crops	Species
Benincasa spp.	*Benincasa hispida*
Citrullus spp.	*Citrullus lanatus, C. lanatus* var. *lanatus, C. lanatus* var. *spermus, C. lanatus* var. *vulgaris, C. lanatus* var. *fistulosus, Citrullus colocynthis (L.), C. ecirrhosus, C.s naudinianus*
Coccinia spp.	*Coccinia grandis*
Corallocarpus spp.	*Corallocarpus conocarpus, C. epigalus, C. gracilipes, C. palmatus*
Cucurbita spp.	*Cucurbita moschata, C. maxima, C. pepo, C. ficifolia, C. mixta or C. argyrosperma*
Group I	*Cucurbita digitata, C. palmata, C. californica, C. cylindrica , C. cordata*
Group II	*Cucurbita. lundelliana, C. okeechobensis, C. martinezii, C. mixta, C. palineri, C. sororia, C. grassicolor*
Cone specific group	*Cucurbita pepo, C. texana*
Group –5	*Cucurbita maxima, C. andrina, C. pedatifolia, C. moschata, C. foetidissima, C. ficifolia, C. ecuadorensis*
Cyclanthera spp.	*Cyclanthera pedata*
Dactyliandra	*Dactyliandra welwitchii*
Dicoelospermum	*Diocoelospermum ritchiei (monotypic genus)*
Diploycyclos	*Diplocyclos palmatus (L)*
Edgaria	*Edgaria darjeelingenesis*

Contd..

Luffa spp.	*Luffa acutangula, L. cylindrica , L. echinata (dioecious), L. graveolens, L. hermaphrodita, L. tuberosa, L. umbellata*
Momordica spp.	*Momordica charantia L.* var. *charantia, M. balsamina, M. charantia* var. *muricata, M. cochinchinensis, M. denundata, M. dioica, M. macrophylla, M. subangulata*
Sechium	*Sechium edule*
Solena	*Solena heterophylla or S. amplexicaulis or Melothria heterophylla*
Squash melon	*Praecitrullos fistuloses*
Zanonia	*Zanonia indica L.* var. *indica, Z. indica* var. *pubescens*
Zehuria	*Zehuria indica, Z. mysorensis*

Plant Genetic Resource Management in Cucurbitaceous Crops

The Indian National Plant Genetic Resource System (IN-PGRS) under the aegis of the Indian Council of Agricultural Research with National Bureau of Plant Genetic Resources (NBPGR) as a apex body, is a very strong system and holds a prominent place among the global gene bank. National Bureau of Plant Genetic Resources (NBPGR) is a nodal agency for importing elite fruit germplasm for enhancement and diversification of old traditional agricultural land use pattern. It has a national mandate to plan, conduct, promote and co-ordinate all activities concerning plant exploration and collection, germplasm import and exchange, plant quarantine, germplasm multiplication and distribution, evaluation and characterization, documentation and conservation of both indigenous and introduced genetic variability in cultivated plants and their wild relatives. At conceptual level, the need for establishment of an agency for "Organized Plant Introduction" in India was expressed by late Dr. B.P. Pal and "Crops and Soil Wing" of the Board of Agriculture and Animal Husbandry in India way back in 1935. Historically, Plant Genetic Resources (PGR) activities were started as early as in 1946 with the initiation of Plant Introduction Scheme in Botany Division of Indian Agricultural Research Institute, New Delhi. This scheme was further strengthened into a full-fledged Plant Introduction and Exploration Organization in 1956 and was upgraded into an independent Division of Plant Introduction of the Indian Agricultural Research Institute in 1961 and from 1976 it's become an independent institute National Bureau of Plant Genetic Resources (NBPGR). The NBPGR Regional Station, Bhowali (29° 20' N latitude and 70° 30' E longitude), Nainital, Uttarakhand has been entrusted the responsibility for the collection, characterization, documentation, maintenance, conservation and sustainable utilization of agro-biodiversity of Central Himalayan Region for today as well as for posterity. The PGR of this region will be of immense value for their utilization in crop improvement programmes in developmental activities as well as for increasing over all farming production. The climate of this site is sub-temperate with minimum and maximum temperature ranging between –7°C to 33°C and average annual rainfall around 1600 mm. From 1990 onwards several crop specific explorations were undertaken.

Different Plant Germplasm Exploration Activities

Modern plant improvement programmes are highly influenced by the characteristics of present day crops, which were significantly influenced by the natural selection, the fundamental basis of adaptation. The availability of specific phenotypes with useful gains and high combining ability has been instrumental in breeding for desirable high yielding types in various crops. Initially such germplasm accessions for known traits can be collected from the agro-climatic niche earmarked in the coarse grid surveys and then these can be utilized for developing improved varieties for the respective climatic conditions and prevalent cropping patterns (Verma *et al.,* 2000). The main objectives of the plant explorations are to sample maximum amount of viable variability. Emphasis is being given to collect representative diversity of the area however, off routes, far flanged and less explored or unexplored areas were given due attention during survey and collection of plant germplasm. In addition need based germplasm of different vegetable crops are also being collected from other institutes/ parts of the country so that in addition of enrichment of vegetable gene pool, their gene or constellation of gene could be in fluxed into adaptive bases of central Himalaya. Few points to be taken into consideration before proceeding for exploration of horticultural crops. They are (i) Collection need, (ii) Agro-ecology of the region/plant distribution, (iii) Local contacts in the area of survey, (iv) Sampling, (v) Transportation of material and (vi) After exploration care (Verma, 2005).

Exchange of Plant Genetic Resources

The National Bureau of Plant Genetic Resources (NBPGR), New Delhi has been functioning as a national research cum service organization for introduction and exchange of plant genetic resources of agri-horticultural and agri-silvicultural plants in small quantities for research purposes. It assists various ICAR crop based Institutes, All India Crop Improvement programmes, Agricultural Universities/Private seed industries with research base in the country by way of providing native variability of agri-horticultural plants as well as exotic introductions procured under phyto-sanitary conditions. The Bureau has close linkages for germplasm exchange with more than 85 countries through its cooperating Plant Introduction Agencies / Organizations, Universities, Botanical Gardens as well as with International Crop Research Institutes, as well as other centers like AVRDC (Taiwan) and WARDA (Libera) besides IBPGR/FAO.

Procedure for Exchange of Germplasm

The exchange of plant material on world basis has been carried out with well defined procedures by countries which have well established plant introduction organizations, like USA, Russia, Australia, Canada, Brazil, Japan and other for channelising import and export of plant, material under quarantine control. To do justice to its cause, plant introduction agency has to follow a set of certain procedures for introduction, collection of germplasm viz., either through 'plant exploration' or 'collection through contacts by correspondence'.

Germplasm Evaluation and Utilization

It is difficult to detect particular micro-evolutionary features of accessions in limited screening, characterization or evaluation due to huge size of collections, paucity of designs of experiments and ensuing statistical analysis. Thus the selection of accessions for initial breeding material becomes a limiting factor. There is a close relationship between plants behavior patterns or response to environment and its physiological conditions. The influence of individual environmental factors such as temperature, moisture, light etc. and their interrelations in the whole eco-system has been significant in shaping a variable trait from which would be evident from specific genotype environment interactions. Success of a breeding programme for yield increase along with specific adaptability of the improved cultivars needs appropriate base material and combining ability. The concept of adaptability at the improved cultivar level is narrowed down from the corresponding situations at the level of domesticated crop species / landraces / primitive cultivar.

Characterization

Characterization and preliminary evaluation will be the main responsibility of curator, while further characterization and evaluation should be carried out by the user (plant breeder). Characterization is an account of plant morphology, either through out its developments or only at maturity (Rai *et al.*, 1994). Hence it is important to checklist the steps taken at various stages of germplasm characterization and evaluation of effective documentation.

a. Criteria for selecting germplasm

Germplasm may be added through new collections, exotic introductions, including elite material or rejuvenation of old germplasm.

b. Package of agronomical practices

Optimal package and practices for the crop suitable under local condition should be followed.

c. Diseases / pest control

There should be an optimal balance between the screening for natural incidence of diseases or pest but we should have our more attention towards for saving the germplasm.

d. Data recording

Data recording should judiciously planned so that it is recorded over a period of crop growth without crowding only at harvest.

(i) **Passport data:** Passport data refers to the origin of the sample or its known history. It provides a means for classifying germplasm and for studying pattern of variability and hence is essential in managing collections to reduce redundancy.

(ii) **Characterization:** Consists of highly heritable characters, which can be expressed in all environments.

(iii) **Preliminary evaluation:** Covers a limited number of traits. Minimal descriptors should be used viz. site data, plant data- vegetative, flower, fruit, stress susceptibility etc. Some crops specific characters must be recorded which has direct correlation with yield and quality parameters.

Germplasm Conservation

The status of conservation of vegetable crops germplasm has always received less attention than that of the major staple crops such as cereals, legumes and other horticultural crops. The databases, holding mainly passport data, can be analyzed for the identification of duplications and gaps among collections. At the same time, the management of collections can be based on better knowledge of the diversity in stock. The enhancement of the links between germplasm conservation and use will continue to depend, inter alias, on easy access to the genetic material. Further, locally common alleles are more important for conservation and interesting to users. The emphasis on *in situ* and *ex situ* approaches will, however, depend on the conservation context- the object, aims and location of conservation. *In situ* conservation means preserving the varieties / germplasm in their original agro-ecosystem cultivated by farmers using their own selection methods and criteria.

Due to open pollinated nature of Cucurbitaceae, controlled pollination is necessary during seed increase to avoid contamination of accessions. It also requires a consideration of the use of populations large enough to preserve segregating alleles within accession samples. Potentially valuable alleles may occur at very low frequencies in accession populations, especially in wild material and landraces. Many of the traditional agro-ecosystems across the world including those of hilly agro-ecosystem still found constitute major *in situ* repositories of both indigenous crops and wild plant germplasm. Many traditional agro-ecosystems are located in centres of diversity, and thus contain populations of variable and adapted landraces as well as wild and weedy relatives of crops (Harlan, 1992).

Establishment of planting propagules

In perennial cucurbits, due to its long juvenile period and depending on the degree of threats to erosion, adequateness of germplasm and socio-economic need, collection, propagation and establishment of germplasm is decided. Four major types of establishment may be followed;

(a) **Seed repository:** In case of wild types and at rare species level collected seeds must be maintained with careful attention in cold storage at -20^0C.

(b) **Tissue culture repository:** Disease free tissue cultures of recalcitrant species/ viviparous (chow-chow etc.) developed from stable explants, will be maintained using appropriate growth limiting media and by sub-culturing at suitable intervals.

(c) **Cryo-preservation:** Seed samples of selected species are being preserved using Cryo-preservation techniques. Cryo-preservation of meristem tips, shoots and buds (if they will be collected in appropriate physiological state) is also useful as its provides a cheap, space saving method for long-term storage (chow chow).

(d) **Clonal repository:** Many collections are found in botanic gardens, gene sanctuaries and arboreta and wild forms in forests. The germplasm is satisfactorily maintained in these prohibited areas (*In situ*). Clones usually are maintained *in vivo* in green houses or field gene banks (chow-chow). *In vivo* maintenance requires more space and high cost of aftercare. For preparing collections to transfer into field gene banks they must be pre-established in controlled environment/ suitable conditions. Not only these field plantings are vulnerable to losses, screen houses or green houses are needed to maintain a duplicate set of clones (chow-chow).

Maintenance of germplasm: Good quality seed is one of the major factors that determine the success of a crop. The unavailability of good vegetable seed germplasm is due to inadequate technology for seed production, poor quality control, post harvest, seed harvest, seed handling etc. In most of the vegetable crops, the potential yield is evaluated at the marketable green fruit stage of the crop. The fruits left for seed collection are sometimes those which bear at the late stage of the plant growth, and very often the seeds produced are of poor quality. The plants are weak at the end of the production season thus, the left over fruits are not physically and physiologically fully mature. Positive selections are to be carried out to choose the best plant and fruits for seed collection. The correct isolation distance is not usually followed to prevent cross-pollination. The plant population then becomes genetically mixed is succeeding plantings. If the same germplasm is grown continuously, disease infection particularly by virus may become a problem. The weekened infected plants produce poor seeds and if the virus is seed-born, it will be carried in to the next generation. The vegetable seeds produced during *Kharif* crop (wet season) sometimes exposed to fungal and bacterial disease infection. Again this allows seed borne pathogens to infect the next crop. Moreover, the seeds are not properly extracted, processed and stored. All these contribute to production of inferior quality seeds.

The best plants in terms of growth and yield and the best fruits should be selected for seed production for Gene Bank storage. Any plant or fruit with suspected symptoms of disease and pest attack should be excluded. Sufficient fruits in each accession should then be allowed to mature fully, either on the plant on during post-harvest storage before seed extraction. Off types in each germplasm accession should be rouged or pulled out. Any disease and insect infection should be controlled more rigorously. For some biennial vegetables (radish and cabbage), special treatments have to be done and/ or planting times has to be adjusted to induce flowering and seed setting. These vegetables need lower temperature and /or longer days for flower initiation.

Maintenance of accessions/varieties

To prevent accession/varietal mixture, it is important to harvest and process each accession/variety separately, lot by lot. All the seed cleaning, drying and processing articles and seed containers should be cleaned and free from seed of another accession/variety. The seed containers should have proper labels with accession numbers and date of storage. In cross pollinated species, if the genetic purity of a variety is to be maintained, flowers for seed collection should be bagged with a paper envelope before anthesis or opening to prevent pollen contamination from other nearby accessions/varieties of the same species. The bagged flower buds should be pollinated artificially at anthesis with pollen collected from the same accession or variety. If the accessions are grown at a sufficient isolation distance from each other, then bagging is not required. The physiologically mature fruits are harvested and air-dried under shade. Hanging the wet fruits can help hasten drying and prevent insect pest infestation.

Conclusion

Genetic diversity continues to meet farmers' needs and plays an important role in traditional agroecosystems. New materials were also incorporated into existing landraces, permitting the agricultural system to evolve without total replacement. The local landraces are also being replaced with high yielding landraces from neighbouring areas. It has now been well realized that in many areas farmers have developed distinct systems of classification and description of local landraces. The agriculture related indigenous knowledge possessed by the elderly people, mostly womenfolk, is facing series of challenges due to variety of factors and out-migration of younger generation to urban areas. The Uttarakhand Himalaya is one where people are still practicing traditional landrace-based cultivation. Farmers' dependence on varietal mixtures, multiple crops, intercropping, growing genetically diverse varieties of individual crops fits with high variability in their edaphic and biological environments and their limited access or inability to acquire purchased inputs. Gradual reduction in area of several traditional crops and farmers preferences for certain other traditional and introduced crops is induced by the economic and socio-cultural factors. The market forces are creating new preferences. It suggests that the hill agro-ecosystems with traditional crops are ecologically and economically viable and still have the potential to support the food requirements in the Himalayan region. *In situ* (on-farm) conservation will be most effective when targeted to specific areas with significant plant genetic resources and with communities who are willing to participate in conservation programmes. *In situ* and *ex situ* conservation of biodiversity for agriculture and forestry are complementary approaches. We can conserve our extinct perennial cucurbits genetic resources either by *in situ* or *ex situ* conservation mainly landraces / folk varieties / primitive form of crop plants.

References

Almekinders C.J.M. and Elings A. 2001. Collaboration of farmers and breeders: participatory crop improvement in perspective. Euphytica **122**: 425–438.

Alvarado, J.L., Lira, R. and Caballero, J. 1992. Palynological evidence for the generic delimitation of *Sechium sensu lato* (Cucurbitaceae) and its allies. Bull. Br. Mus. Nat. Hist. **22**:109-121.

Anonymous, 1996. Report on the state of the world's plant genetic resources for food and agriculture, prepared for the International Technical Conference on Plant Genetic Resources, Leipzig, Germany, 17–23 June 1996. Food and Agriculture Organization of the United Nations, Rome, Italy.

Arora, R.K. 1985. Genetic Resources of Lesser Known Cultivated Fruits Plants. NBPGR, Sci. Monogr. No. **9.**

Arora, R.K. 1995. Genetic resources of vegetable crops in India: Their diversity and conservation, In Genetic Resources of Vegetable Crops (Eds., R.S. Paroda, P.N. Gupta, Mathura Rai and S. Kochhar), NBPGR, New Delhi. pp. 29-39.

Arya, R. R. and Pandey, G.P.2000. "Diversity in horticultural wealth of UP Hills "in "Agri- biodiversity. International Conference on Managing Natural Resources for Sustainable Agricultural Production in the 21st Century": *Extended summaries: Volume: Natural Resource:* pp 827 - 830.

Aung, L.H., Ball, A. and Kushad, M. 1990. Developmental and nutritional aspects of chayote *(Sechium edule,* Cucurbitaceae). *Econ. Bot.* **44:** 157-164.

Bennet, E. 1970. Tactics of plant exploration. In: Genetic Resources in Plants. (Eds. O.H. Frankel and E. Bennet) Oxford, Blackwell p 157-179.

Browne, P. 1756. Civil and Natural History of Jamaica. London, England.

Brush, S.B. 1999. Genes in the Field: On-farm Conservation of Crop Diversity. Lewis Publishers, Boca Raton, Florida, USA.

Chakravarty, H.L. 1990. Cucurbits of India and Their Role in the Development of Vegetable Crops. *In* : Biology and Utilization of the Cucurbitaceae (eds.) D.M. Bates, R.W. Robinson, and C. Jeffrey, Cornell University Press, Ithaca, NY. pp. 325-334

Choudhury, B. 1967. Vegetables: National book trust, India.

Engels, J.M.M. and Jeffrey, C. 1993. *Sechium edule Qacq.* Swartz. *In:* Plant Resources of South-East Asia. No. 8. Vegetables (eds.) J.S. Siemonsma and K. Piluek Pudoc Scientific Publishers, Wageningen. pp. 246-248

Giusti, L., M. Resnik, T. del V. Ruiz and A. Grau. 1978. Notas acerca de la biologfa de *Sechium edule* 0acq.) Swartz (Cucurbitaceae). Lilloa **35:** 5-13.

Goldblatt, P. (ed.). 1981. Index to plant chromosome numbers (1975-1978). Monogr. Syst. Bot. Missouri Bot. Gard., Missouri. 195 pp.

Goldblatt, P. (ed.). 1984. Index to plant chromosome numbers (1979-1981). Monogr. Syst. Bot. Missouri Bot. Gard., Missouri. 152 pp.

Goldblatt, P. (ed.). 1990. Index to plant chromosome numbers (1988-1989). Monogr. Syst. Bot. Missouri Bot. Gard., Missouri, pp. 90.

Gupta, P. N. and Rai, M. 1995. Genetic resources of cucurbitaceous vegetables in India. In :Genetic Resources in Vegetable Crops Management Conservation and Utilization (Eds. R. S. Rana, P. N. Gupta, Mathura Rai, and S. Kochhar), NBPGR New Delhi pp 29-39.

Gupta, P. N., Rai, M. and. Rana, R. S. 1995. Centres of origin and genetic variability of vegetable crops. In: Genetic Resources in Vegetable Crops Management Conservation and Utilization (Eds.) R. S. Rana, P. N. Gupta, Mathura Rai, and S. Kochhar), NBPGR New Delhi pp 52-62.

Harlan J.R. 1992. Crops and Man. 2nd ed. American Society of Agronomy, Madison,WI, pp. 284.

Harlan, J.R. 1975. Practical problems in exploration of seed crops. In: Crop Genetic Resources for Today and Tomorrow. (Eds. O.H. Frankel and J G. Hawks), Cambridge University press, London , pp 111-115.

Jain, S.K. 1975. Population structure and the effect of breeding system. In: Crop Genetic Resources for Today and Tomorrow. (Eds. O.H. Frankel and J G. Hawks), Cambridge University press, London , p 15-36.

Jeffrey, C. 1990. Appendix: An outline classification of the Cucurbitaceae. *in* Biology and Utilization of the Cucurbitaceae. (D.M. Bates, R.W. Robinson and C. Jeffrey, eds.). Cornell University Press, Ithaca, NY. pp. 449-463

Kabitarani, A. And Bhagirath, T. 1991. Powdery mildew of cucurbits in Manipur. *Indian Phytopathol.* **44:**137-139.

Lira, R. and Bye, R. 1992. Las Cucurbitaceas en la Alimentacion de los Dos Mundos. *In* Simposio 1492. El Encuentro de Dos Comidas. Puebla, Mexico. 38 pp.

Lira, R. and Chiang, F. 1992. Two new combinations in *Sechium* (Cucurbitaceae) from Central America and a new species from Oaxaca, Mexico. Novon **2:**227-231.

Lira, R. and Soto, J.C. 1991. *Sechium hintonii* (P.G. Wilson) C. Jeffrey (Cucurbitaceae). Rediscovery and observations. *FAO/IBPGR Plant Genet. Res. Newsletter* **87:** 5-10.

McNeely J.A. 1988. Economic and biological diversity: developing and using economic incentives to conserve biological resources. International Union for Conservation of Nature and Natural Resources, Gland.

Newstrom, L.E. 1989. Reproductive Biology and Evolution of the Cultivated chayote, *Sechium edule* (Cucurbitaceae). In The Evolutionary Ecology of Plants. (eds.) J. H. Bock and Y. B. Linhart, Westview Press, Boulder, Colorado. pp. 491-509

Palacios, R. 1987. Estudio Exploratorio del Numero Cromosomico del chayote *Sechium edule* Sw. Tesis Lie. Ciencias Agrfcolas. Universidad Veracruzana. 59 pp.

Rai, M., Dwivedi, R. and Gupta, P.N. 1991. Variability and potentials of identified germplasm of Bael (*Aegle mormelos Corr.*) *Indian J. Pl. Gen. Resources.* **4(2):** 86-92.

Rai, M., Gupta, P. N. and Singh, B. 1994. Genetic Variability and sampling procedure in fruit crops *Indian J. Pl. Resources* **4(2):** 86-92.

Rai, M., Pandey, S.; Ram, D.; Kumarand, S. and Singh, M.2005. Cucurbits Research in India. An Overview. In Souvenir National Seminar on Cucurbits Department of Vegetable Science, College of Agriculture, GBPUA&T, Pantnagar, Uttaranchal pp 41-54.

Ram, Harihar. 1997. Vegetable genetic resources In: Vegetable Breeding Principals and Practices, Kalyani Publishers New Delhi. pp-32-35.

Ramirez, M.A., Valverde, E. and Saenz, M.V. 1990. Estudio de algunos factors que afectan la fertilization de flores y la abssicion de frutos de chayote *(Sechium edule* Sw.) en cultivo en Costa Rica. Turrialba **40:**340-345.

Sharma, M.D., Newstrom-Lloyd, L.and Neupane, K.R. 1995. Nepal's new chayote genebank offers great potential for food production in marginal lands. Diversity **11:** 7-8.

Sheshadri, V.S. 1987. Genetic resources and their utilization in vegetable crops, In Plant Genetic Resources: Indian Perspective (Eds., R.S. Paroda, R.K. Arora and K.P.S. Chandel), NBPGR, New Delhi. pp. 335-343.

Singh, A.K. 1990. Cytogenetics and Evolution in the Cucurbitaceae. In Biology and Utilization of the Cucurbitaceae. (eds.) D.M. Bates, R.W.Robinson and C. Jeffrey, Cornell University Press, Ithaca, NY pp. 10-28

Singh, B. P. 1993. Principal and procedure for exchange of plant genetic resources. In: Conservation and Management of Plant Genetic resources (eds. R S Rana, R. K. Saxena, Sanjeev Saxena and Vivek Mitter). NBPGR, New Delhi pp 31-38.

Singh,R.K.;.Verma,S.K.;Arya, R.R. and Muneem, K.C. 2002. Genetic variability in Chow-Chow (*Sechium edule*). *Prog. Hort.* **34(1):** 92-94.

Smale, M. and Bellon, M.R. 1999. A conceptual framework for valuing on-farm genetic resources. In: Wood

D. and Lenne J. (eds), Agrobiodiversity: Characterisation. Utilization and Management, CAB International, Wallingford, pp. 387–408.

Sobti, S.N. and Singh, S.D. 1961. A chromosome survey of Indian medicinal plants.Part I. Proc. Indian Acad. Sci. **54:** 138-144.

Sugiura, T. 1938. A list of chromosome numbers in Angiosperm plants. V. Proc. Impp. Acad. Japan **14:** 391-392.

Sugiura, T. 1940. Studies on the chromosome numbers in higher vascular plants. Citologia **10:** 363-370.

Swartz, O.1800. Flora Indiae Occidentalis. Leiden. Terraciano, N. 1905. Il *Sechium edule* Sw. e sua coltivazione in Napoli e dintorni. Atti. Real Inst. d'Inc. Napoli. Ser. VI(1)

Verma, S. K. (2005). Plant Genetic Resource Management in Cucurbitaceous Vegetables. In Souvenir National Seminar on Cucurbits Department of Vegetable Science, College of Agriculture, GBPUA&T, Pantnagar, Uttaranchal pp77-86.

Verma, S.K., Muneem K.C., Verma V.D.,.Negi K.S., Gautam P.L., Mishra K.K., Parakh D.B.,Gupta S.,Singh R.V., Arya R.R and Pandey G.P. 2000. Diversity in horticultural wealth of UP Hills. In international Conference on "Managing Natural Resources for Sustainable Agricultural Production in the 21st Century" at IARI, New Delhi –12 w.e.f. 14 – 18 Feb. 2000, appeared in Agri-biodiversity Extended Summary, Vol. **2** : pp. 827-830.

Zageja, S.W. 1970. Temperate zone tree fruits. In: Genetic Resources in Plant. (Eds. O.H. Frankel and E. Bennet) Oxford, Blackwell p 327-333.

Zevan, A.C. and Zhukovasky, P.M. 1975. Dictionary of Cultivated Plants and Their Centre of Diversity . Pudoc. Wageninger, 219p.

Zuniga, L.E. 1986. Aspectos economicos del cultivo del chayote *(Sechium edule)* en Costa Rica. Facultad de Agronomia, Universidad de Costa Rica.

❑❑❑

Chapter – 14

Genetic Resources of Vegetable Crops of North Eastern Himalayan Region

R.K. Yadav, H. Choudhary, S.K. Sanwal and D.K. Singh

Introduction

The North-eastern region comprises of eight states viz. Assam, Arunachal Pradesh, Meghalaya, Manipur, Mizoram, Nagaland, Tripura and Sikkim lying between 21.5° N - 29.5° N latitudes and 85.5 ° E - 97.3 ° E longitudes. It has a total geographical area of 262180 Km^2 which is nearly 8 % of the total geographical area of the country. In the whole of NE region, about 35 % area is plain and the remaining 65 % area is under hills. Whereas in Assam plains account for 84.44 % of its total geographical area and the remaining 15.56 % area is under hills. Net sown area is highest in Assam (34.12 %) followed by Tripura (23.48 %), however, Arunachal Pradesh has lowest net sown area in the region. Cropping intensity is highest in Tripura (173 %) followed by Manipur (152.1 %), Mizoram (136.36 %) and Assam (123.59 %). About 0.5 million hectare area is under shifting cultivation in the NE region. Out of 4.4 million hectare net sown area, roughly 1.4 million hectare lies in hilly sub region and at least 1.3 million hectare suffer from serious soil erosion problem.

The diverse agro climatic conditions, varied soil type and abundance of rainfall offers immense scope for cultivation of different types of horticultural crops, including fruits, vegetables, flowers, plantation crops, tuber and rhizomatous crops and crops of medicinal and aromatic importance. The region has rich diversity of different vegetable crops and both indigenous tropical vegetables and temperate vegetables are grown to a considerable extent. The major vegetables grown in the region are brinjal, cabbage, cauliflower, okra, onion, pea, potato, tomato, knol-khol, radish, carrot, French bean and different cucurbitaceous crops. Tuber and rhizomatous crops like tapioca (cassava), sweet potato, *Dioscorea*, colocasia, ginger and turmeric grow abundantly in the region.

Table 1. Demography of North Eastern states-1991

State	Geographical Area (sq. m)	Forest (%)	Net sown area (%)	Population			(%)
				Total (1991)	% Rural	% Urban	Schedule tribes
Assam	78,439	25.67	32.4	2,24,14,322	88.9	11.1	12.8
Arunachal Pradesh	83,743	93.79	3.37	8,64,588	87.2	12.8	63.5
Manipur	22,327	27.23	6.33	18,37,149	72.5	27.5	34.4
Meghalaya	22,429	41.72	9.64	17,74,778	81.4	18.6	85.5
Mizoram	21,081	75.77	5.17	6,89,756	53.9	46.1	94.8
Nagaland	16,579	56.11	14.63	12,09,546	82.8	17.2	87.7
Sikkim	7,096	36.20	13.38	4,06,457	90.9	9.1	22.4
Tripura	10,486	57.77	26.41	27,57,205	84.7	15.3	30.6
Total NE states	2,62,180	51.78	13.92	3,19,53,821	84.35	15.65	34.45
India	32,87,300	(19.35)*	—	84,63,02,688	74.3	25.7	8.0

*Values are in percentage

Area, production and productivity of vegetable crops

No systematic and accurate estimate of area and production of different vegetable crops in the North Eastern region are available and the estimates made by various sources vary considerably. North Eastern Council compiled the data available from different sources. According to Agricultural Research Data Book ICAR 2002 the total area under horticultural crops is around 822.5 thousand hectare, which is around 3.14 % of the total geographical area of the region and it gives total production of 6818.4 thousand tonnes. However, the area under various vegetable crops in the NE region was 367.9 thousand hectares and production was 4051.8 thousand tonnes with the productivity of 11.01 tonnes per hectare against the area under vegetable crops in the country was 5993.0 thousand hectare and production 90830.7 thousand tonnes with productivity level of 15.16 tonnes per hectare (Table 2). This shows that the productivity level of horticultural crops in the NE region is quite below the national productivity.

Table 2. State wise area and production of fruits and vegetables in NE region

Area-000 ha, Production-000 t, Yield-t/ha

State	Area production & productivity (yield	Vegetables	
		1996-97	1999-2000
Arunachal Pradesh	A	16.7	16.9
	P	80.5	80.9
	Y	4.82	4.79
Assam	A	223.2	255.9
	P	2074.1	3089.4
	Y	9.29	12.07
Manipur	A	8.0	9.0
	P	53.2	60.8
	Y	6.65	6.76
Meghalaya	A	41.8	29.2
	P	412.2	252.9
	Y	9.86	8.66
Mizoram	A	6.8	8.3
	P	49.6	56.3
	Y	7.29	6.78
Nagaland	A	19.3	20.9
	P	188.4	235.7
	Y	9.76	11.13
Sikkim	A	12.0	9.6
	P	54.0	43.0
	Y	4.50	4.48
Tripura	A	32.0	18.4
	P	358.5	232.8
	Y	11.20	12.65
NEH region	A	359.80	367.9
	P	3270.50	4051.8
	Y	9.09	11.01
India	A	5515.2	5993.0
	P	75074.6	90830.7
	Y	13.61	15.16

Source: Agril. Research Data Book ICAR-2002

Out of total area under different vegetable crops, the maximum area of about 113.2 thousand hectare is under potato only. Potato is a very important cash crop of the entire region. Area wise second most important crop is cabbage, covering about 18.5 thousand hectares, while sweet potato occupies 17 thousand hectares. Other important vegetable crops from area point of view are brinjal (12.5 thousand ha), cauliflower (12.5 thousand ha), onion (7.9 thousand ha) etc. (Table 3)

Table 3. Crop wise area, production and productivity of vegetable crops in NE region (1997-98)

Crop	NE states			India		
	Area (,000 ha)	Production (,000 tonnes)	Productivity (t/ha)	Area (,000 ha)	Production (,000 tonnes)	Productivity (t/ha)
Potato	113.2	1048.3	9.26	1208.9	17652.3	14.6
Cabbage	18.5	227.5	12.3	218.4	3861.7	17.7
Sweet potato	17	70.4	4.1	128.8	1171.0	9.1
Tapioca	7.8	55.6	7.1	264.3	6681.9	25.3
Brinjal	12.5	187.7	15.0	434.2	6443.1	14.8
Onion	7.9	18.1	2.3	338.5	3142.8	9.3
Cauliflower	12.5	120	9.6	220.0	2474.0	11.3

Source: Basic statistics of North Eastern Region 2000, North Eastern Council, Shillong, Ministry of Home affairs, GOI.

Bio-diversity of vegetable crops

The North Eastern region is considered to be the richest reservoir of genetic variability of large number of horticultural and plantation crops. The enormous diversity makes the region a gene pool for the varietal improvement but in spite of potentiality no worth mentioning development in the field of horticulture has taken place. It may be mentioned that in hill area particularly horticultural crop cultivation as an alternative to *jhuming* (shifting cultivation) may prove to be a boon in the regional economy. In NEH region, farming being the main stay of the people and development of horticulture will markedly improve the economy of the people. Establishment of orchards and planting of plantation crops on hill slopes will prevent soil erosion of shifting cultivation areas and migration of people to towns. The diversity among different vegetable groups is given below.

Solanum group: There is wide range of *Solanum spp.* found in the various parts of the region (Table-4). The local tribes grow a vegetables having red tomato like fruits slightly bitter in taste but related to brinjal and belonging to the genus S*olanum.* In Manipur another kind of brinjal having roundish fruit and intermediate in appearance between tomato and brinjal is grown.

Table 4. Occurrence of vegetable *Solanum* diversities in North East India.

Cultivated Species	Remark
Solanum macrocarpon L. or *diprenum*	Introduced in NE region. Used in birth control, very less Solasodine
Solanum xanthocarpum Schard & Wendl	Used as vegetables and medicinal purpose.
Solanum indicum L	Domesticated, used as vegetables and medicinal purpose.
Solanum mammosum L.	Possibly introduced, ornamental with high solasodine percentage.
Solanum khasianum Clarke	Wild and cultivated for Solasodine alkaloid.
Solanum orvum Swartz.	Wild, sold in the market of Mizoram. It is small fruited in cluster of 18-20
Solanum berbisetum Nees.	Ripe fruits are eaten.
Solanum ferox L	Wild, leaves are used medicinally.
Solanum spirale Roxb.	Wild but domesticated for medicinal use in Arunachal Pradesh
Solanum sisymbrifolium Lam.	Native of Africa, Wildly grown in Meghalaya
Solanum kurzii. Br.	Endemic in Garo hills, Meghalaya. It is also highly grooved like gilo.
Solanum gilo Raddi.	Introduced in NE region as vegetable
Solanum cthiopecum	A green brinjal available in market.

Tomato

Tomato is an introduced crop roughly in 18th century and most of the introductions are bred varieties, which have adapted to this region. Germplasm of wild species of tomato *L. pimpinellifolium*, has been found in NE region

Bird eye chilli (BEC)

Small, very pungent fruit easily separated from the calyx and are dispersed by birds. It is native to Central and South America.

Chilli

Chillies usually grown in warm to hot and humid climate in Manipur, Mizoram, Meghalaya, Nagaland, Tripura and Arunachal Pradesh in that order with respect to area under the crop. Due to the long history of cultivation, out crossing nature and popularity of the crop, large genetic diversity including local landraces have evolved. In hot chilli great range of variability for several attributes (fruit shape, size, colour, and bearing habit and semi-perennial, perennial and pungency) occurs throughout India, particularly in North Eastern Region.

Table 5. Importance of chillies and their characteristics.

Scientific name	Common name	Comments
C. annum L.	Sweet pepper, chillies, Hot pepper	Principle source if commercial dry chillies, non-pungent type.
C. annum L. var. *avicular*	Bird pepper	Wild type, said to be progenitor of bell pepper.
C. annum L. var *grossum* Sendt	Sweet pepper	Fruit contains less capsaicin
C. annum L. var *longum* Sendt	Pepper	Used to produce condiment
C. chinense Jacq.	Pepper	Closely related to *C. frutescens*
C. eximium Hunziker	Pepper	Wild type pseudo self incompatible, related to *C. pubescens*
C. frutescens L.	Tobacco pepper, Bird chilli	Widely cultivated in dry region of India. Highly pungent fruits used for sauce preparation, more pungent type.
C. minimum Roxb. Syn. *C. fastigiatum*	Bird-eye chilli	Cultivated all NE region but at very limited scale, closely resembles *C. annum.*
Bhumme *C. pubescens* Ruiz. and Paron	Pepper	Introduced in India for breeding purposes.

Cucurbitaceous Vegetables

This is large group of vegetable crops, consisting of more than 15 kinds, grown and consumed within the region. In North East many species of cucurbits are found as vegetables and fruits. These include *Cucurbita, Momordica, Luffa* and several lesser known cucurbitaceous crops.

Exotic	Indigenous
Bottle gourd, pumpkin, snake gourd, ash gourd, chow chow etc.	Cucumber, *Luffa* gourd, *Momordica* gourd, *Trichosanthes* gourd and tinda etc.

Table 6. Diversity of cucurbits in North East India.

Cultivated species.	Area of concentration for diversity	Range of diversity
C.maxima	Throughout the country	Extensive
C. moschata	Hilly areas	Moderate
C. ficifolia	Meghalaya	Introduced, neutralized
C. pepo	Meghalaya, Mizoram	Limited
C. grandis	Assam, West Bengal	Limited
C. sativus	Throughout the country	Wide
C. callosus	Foothill areas of Assam	Confined to limited pockets
Luffa acutangula	Tropical areas of Assam	Wide

Contd...

L. cylindrica	Tropical and sub tropical areas of Assam, Meghalaya, Manipur and West Bengal	Moderate
Momordica charantia	Throughout the country	Moderate
M. cochinchinensis	Assam, Meghalaya, Manipur and West Bengal	Limited
M. dioca	Garo hills	Rare
Trichosanthus anguina	Meghalaya, Tripura, Assam, West Bengal	Limited
T. dioca	Tropical areas of Assam and Tripura	Limited
Cyclanthera pedata	Hills of Meghalaya, Manipur, Nagaland and Arunachal Pradesh	Moderate
Benincasa hispida	Assam, Nagaland, Meghalaya	Wide
Lagenaria siceraria	Throughout the country	Wide
Sechium edule	High hills of Meghalaya, Manipur, Mizoram, Nagaland, Sikkim and Darjeeling of West Bengal.	Moderate

Pumpkin varieties abound in number with variation in fruit size, fruit skin, flesh colour, thickness, sweetness etc. The wild species *Cucumis hardwickii*, the likely progenitor of cultivated cucumber, is found growing in natural habitants in the foothills of Himalayas and NE region partivularly Meghalaya. *C. sativus* var. *sativus* is cultivated all North Eastern region in tropical and sub tropical condition. However, *C sativus* var. *sikkimensis* has adapted in temperate and humid climate. Among gourds, in North East India maximum variability has been recorded for bottle gourd in fruit shape and size. The Indian gene center has rich diversity in genetic resources of ridge gourd (*L. acutangula)* and sponge gourd (*L. cylindrica*) especially in North Eastern Region. In bitter gourd small as well large sized forms are available.

Crucifers

These are essentially cruciferous vegetables namely Cauliflower, Cabbage, Knol-khol, etc. they have been introduced from the days of East India Company in 14th-15th century when European traders visited this region. The genetic resources in these crops are essentially bred varieties introduced during the last 3-4 centuries and now acclimatization and adaptation have taken. Variability in Indian cauliflower widely exists in Assam and Meghalaya of North Eastern region.

Leguminous Vegetable

There are wide variability of French bean, cowpea, and Indian bean found in the various parts of the region. In French bean, climbing or pole type is popular among the tribal since it is used for mix cropping with maize, the stem of which act as the support for the bean. One of the interesting species of Vigna namely *V. vexillata* is grown by the tribals of Tripura. It is legume cum tuber crop with much variation in edible tubers. Sword bean (*Canavalia ensiformis* L.) is bushy plant of papilionaceae family also cultivated on limited scale in the North Eastern region of the country (CSIR, 1950). Winged bean is confined in humid sub tropical parts of NE region (Sarma, 2001).

Table 7. Occurrence of vegetable legumes diversity in North East India.

Cultivated species	Diversity in cultivars	Wild related species
Sem-*Dolichos lablab*	12	*Dolichos falcatus, D. bifilorus*
French bean *Phaseolus vulgaris*	14	—
Sword bean- *Canavalia ensiformis*	02	*Canavalia gladiata*

Leafy Vegetables

The important leafy vegetables include lai (*Brassica juncea* var.*rugosa),* lafa (*Malva verticillata*) and spinach (*Spinacea oleracea*). In addition to these a wide variety of indigenous leafy vegetables are also available. These are amaranth (*Amaranthus spp.*) poi sag (*Basella rubra and B. alba*), sorrel (*Rumex rasicarius)* etc. Other indigenous leafy vegetables used occasionally are jilmil sag (*Chenopodium album*) and Kolmou sag (*Ipomea reptans*). *Amaranthus viridis, A. lividus, A. retroflexus* and *A. spinosus* are important leafy types grown in North East India (Sarma, 2001)

Tubers and Rhizomatous Crops

Based on the colour and skin to broad types of sweet potato are grown in the region. These are the white skinned and red skinned varieties. A numbers of *Dioscorea* species *alata, bulbifera, brevipetiolata, esculenta, hamiltonii, hispida, kamaonensis, nummularia, pentaphylla, puber* and *quinata* were recorded in the Region. *D. hamiltonii* occurs in humid forest of NE hills (Sarma, 2001). Varieties of Tapioca or cassava like M-4 and hybrid H-97, H-165, H-226 have also been cultivated to some extent. In Colocasia also there is a wide variability even in one species such as *Colocasia esculenta* (Sarma, 2001).

Lesser Known Vegetables

In addition to the above there are a large numbers of indigenous vegetables crops that are used particularly by the tribal population. Maximum of such varieties are available in the Arunachal Pradesh. Tree Bean (*Parkia roxburgii* G. Don) is one of the most common of multipurpose tree species in the Manipur and Mizoram. In the hilly areas tree tomato (*Cyphomandra betacea*) a perennial shrub producing red tomato like vegetables is also grown and used as such. It is grown as backyard venture crop in Meghalaya. Another vegetable tree growing in the lower altitude zones and popular among the people is drum stick or horse raddish locally called Sajina (*Moringa oleifera*)

Cho-cho (*Sechium edule*) a native of tropical America is a very popular vegetable in the region commonly called squash and grown abundantly without much care and attention. *Flemingia vestita* known as Sohphlong, are consumed raw. It is a weak climbing/trailing type, under ground tubers, distributed the humid to sub tropical regions of NE India upto 1500 m (Sarma, 2001). Kakrol (*Momordica cochinchinesis*) and Kartoli (*M. dioica*) are widely spread in Assam, the Garo hills of Meghalaya (Ram *et al.,* 2002)

Germplasm Evaluation

In order to facilitate effective utilization of plant genetic resources, it is important that these are evaluated for productivity and its components, crop duration, resistance to biotic and abiotic stress and quality of produce. The germplasm material available at different centers has been evaluated and utilized for crop improvement.

Some of the brinjal varieties have excellent quality in having large size, soft flesh, and less seeds. An important species of medicinal importance (solasodine content) is *S. khasianum.* Another species *S. torvum* is extensively used in Ayurvedic medicine system. Three tomato varieties namely Manileima, Manikhamnu and Manithoibi were released by the State Variety Release Committee, Manipur and found suitable for rice based cropping system. *L. pimpinellifolium* is also good source of resistance to late blight and tomato leaf curl virus. In chilli, a collections from Tezpur (Assam) has been found to have the highest capsaicin content recorded so far anywhere in the world.

Table 8. Resistance source of Brinjal for disease/pest.

Traits	Promising accession(s)
Fusarium wilt/rot	*S. indicum*
Bacterial wilt	*S. torvum, S. xanthocarpum, S. sisymbriifolium*
Phomopsis fruit rot	*S.gilo*
Shoot and Fruit borer	*S. khasianum,S. sisymbriifolium, S. indicum, S. gilo.*

Source: Seshadri and Srivastava, 2002

Cho-cho (*Sechium edule*) produces large starchy edible roots in addition to fruits. The National Bureau of Plant Genetic Resources Regional Station, Shillong, is maintaining more than 50 germplasm of Indian bean were evaluated along with new introduced lines. Winged bean having excellent nutritional qualities particularly being very rich in protein (Rao and Dora, 2002). In the trial conducted by the ICAR Research Complex for NEH Region, Umiam varieties of Sweet potato such as Sonipat-2 and Sree Bhadra have been found to be promising. Recent studies indicate the Dioscorea hybrid H-312 and H-1687 are more adapted to wide range of conditions ensuring higher tuber yield. Tree tomato is consumed as delicious chatney when raw or after roasting and peeling off the skin. The tubers of *Vigan vexillata* are rich in carbohydrates and minerals. The tubers of *Flemingia vestita (Sohflong)* are rich in iron (2.64 mg), phosphorus (64.06 mg) and contain fair amount of protein (2.99 g), calcium (19.77) and carbohydrate (27.02 g) (Sarma, 2001). Turmeric variety Megha turmeric-1 (earlier known as RCT-1) and ginger variety Nadia were found suitable for the region. Lakadong is a turmeric variety found in this region, which has high percentage of curcumin content (7.4%).

Problems relating to diversity conservation

- **Land tenure issues:** Land tenure systems vary widely among different North-Eastern states, which are quite different from the rest of India. The complexity in land ownership and tenurial rights makes it difficult for survey, demarcation and consolidation of land. Therefore, cadastral survey and land demarcation are completely absent in the hill areas of north-east.
- **Gender and Equity issues in natural resources and diversity management:** Unequal distribution of land resources is responsible for increasing dependence on forests by certain sections of the society leading to diversity degradation. Resolving the gender and equity issues concerning natural resource management is equally important in North-East as in the other parts of the country.
- **Over exploitation of genetic resources**
- **Deforestation and soil degradation**
- **The adverse impact of development and increase in the population**
- **Significant numbers of species have been declared as rare and endangered.**
- **Inter-departmental coordination:** Needs a close inter-departmental coordination to the sustainable management of horticultural resources in the region.
- **Smuggling of timber across the international border:** The illicit felling of trees and timber smuggling across the international borders has been the most important cause of horticulture areas/forest degradation in border.
- **Shifting cultivation:** Unregulated shifting cultivation by the local tribal populations has been a major threat to sustainable diversity management particularly in unclassed and community forests of the region.
- **Inter-state border dispute:** There exist a lot of inter-state border disputes among the north-eastern states. Most of these border areas are forest lands and because of boundary disputes, such lands are often declared as "no man's land" hence, does not come under any form of management. This leads to the degradation of diversity in such areas.
- **Insurgency:** The long insurgency problem in some states such as Assam and Tripura has considerable impact on diversity conservation.

Conservation of Diversity

Although there are not many agencies/organizations working exclusively for diversity conservation in north-east *per se*, the activities taken up by many organizations including non-governmental and traditional institutions, government departments and scientific institutions have direct or indirect implications for diversity conservation.

- **State Government Agencies:** Many state agencies are now involved in such diversity conservation activities as establishment of germplasm banks for horticultural crops.
- **Research Organizations:** Many state and central government research organizations including universities of the region are engaged in research, inventory and conservation of diversity in the region. Such organizations are Botanical Survey of India, Shillong, G.B. Pant Institute of Himalayan Environment and Development, North-East Unit, Itanagar, Indian Council of Agricultural Research for North-Eastern Hill Region, Barapani, Shillong with campuses through out the north-east, State Forest Research Institute, Ita nagar, NBPGR, Shillong, North-Eastern Hill University, Shillong, Nagaland University, Kohima, Mizoram University, Aizawl, Arunachal University, Itanagar, Tripura University, Agartala, Assam University, Silchar, Tezpur University, Tezpur, Gauhati University, Guwahati, Assam Agricultural University, Jorhat, Regional Research Laboratory, Jorhat, Dibrugarh University, Dibrugarh.
- **Non-Governmental Organizations:** Many non-governmental organizations are now working for the conservation of diversity in north-east. Although most of them are local and at grassroots level.
- **International Donor Agencies:** International donor agencies in Meghalaya, Manipur, Assam and Nagaland have been playing crucial role in conserving the diversity through their respective projects.
- **International and National Policies and Conventions:** All the international treaties and national policies have significant impact on the conservation of diversity in the north-east.
- **Academic Institutions Including Schools and Colleges:** The educational curriculums in the universities, colleges and schools have an important role to play in diversity conservation.
- **Shifting Cultivators:** The shifting cultivators and other traditional farming communities of north-east have played a key role in conserving the rich horticultural crops germplasm of the region. In spite of the availability of many hybrid and high yielding varieties these farmers have been cultivating the traditional varieties for generations.

Gaps in Diversity Conservation

The depletion of diversity and inadequacy in actions to conserve the diversity of the region may be attributed to several factors, which range from inadequate knowledge about diversity and its components to adoption of wrong and inappropriate policies by the concerned stakeholders.

1. Gaps in Knowledge and Information

- Information on urban diversity is scanty
- Species inventory in inaccessible areas of Arunachal Pradesh, Nagaland, Karbi Anglong and North Cachar hills of Assam and parts of Mizoram and Manipur is yet to be made.
- Information on genetic diversity is extremely poor

2. Gaps in Vision

Most of the programmes and activities being undertaken by the state governments are shortsighted. Long-term planning based on sustainable development strategies and integration of diversity conservation issues with development planning is the need of the hour.

Monoculture plantations: In order to increase the revenue generation, the State Horticulture Departments pursue the policy of raising plantations of commercially important species by clearing and burning the natural diversity areas.

Introduction of high yielding varieties/hybrids of Crops: The horticulture departments are introducing various high yielding varieties/hybrids of cucurbitaceous crops. This is associated with increasing use of inorganic fertilizers and chemicals for plant protection. Such policies not only ignore the indigenous species and varieties but also have adverse effects on existing flora and fauna.

3. Gaps in Policies and Legal Structure

The wrong conservation policies with focus on economically important species have been harmful to diversity. Such policies as adopted in Tripura, Mizoram, Nagaland and Meghalaya has not only decreased the species diversity in natural/ rehabilitated forests but have also resulted in accelerated soil erosion and loss of soil moisture.

- The policy of rehabilitation of *jhumias* through rubber plantation as has been done in Tripura may prove to be a disaster for other floral species in such areas.
- The policy of promoting high yielding varieties and assessment of progress and success on the basis of consumption of fertilizer and plant protection chemicals has led to ignoring the indigenous varieties. The government subsidy and credit policy is instrumental in adopting these schemes.
- Through the Public Distribution System, only HYV are distributed. There is a need to include distribution of indigenous varieties too.
- The planners have not considered the role and value of diversity in preparing developmental plans. Such ignorance has been responsible for taking no efforts to conserve and enhance diversity.

- Most of the problems are related to increase in population. The rate of population growth in the northeast is unusually high. This causes tremendous strain on the natural resources and adoption of certain policies that are not very friendly to conserve diversity. No population policy has been adopted for future planning.
- Education policy does not include teaching on diversity conservation. The school curriculum should be able to mould the young minds in favour of diversity conservation.
- No policy as such to create awareness among masses for diversity.

4. Gaps in Institutional and Human Capacity

- The number of trained taxonomists in the region is grossly inadequate. This is one of the most important bottlenecks for completing the inventorization of diversity.
- Not all persons concerned with management of genetic resources understand the concept of diversity in proper perspective. Many of them suffer from biased attitudes. So it is imperative that those who plan, decide and implement the developmental programmes are adequately trained and educated in favour of diversity conservation.
- There are a number of institutions, departments, colleges, universities, NGOs, local community groups that follow certain programmes having bearing on genetic resources. While framing their programmes, these agencies are motivated to pursue their own goals in watertight compartments without considering their impact on other programmes or existing resources. There is no institution, which can make them sit together and discuss the programmes in a holistic manner.

5. Gaps in Diversity Related Research and Development

- Regeneration and cultural practices for many species need to be researched and standardized for their cultivation. Threatened species need immediate action for ensuring their continued existence.
- Identification and classification of threatened species need to be done.
- Richness of diversity of horticultural crop species is yet to be fully inventorized and documented.
- There is a serious gap between research and field needs. The established formal institutions like university departments, departmental research stations and others rarely consult the farmers and local communities about their problems while pursuing research. Need-based research needs to be encouraged.

Conclusion

Considerable diversity exists among the regional vegetable crops including variation in plant type, morphological and physiological characteristics, reactions to diseases and pests, adaptability and distribution. Apart from the nutritional value, many regional vegetable crops are used for medicinal purposes, income generating and poverty alleviation programmes in the rural areas. Problems relating to diversity conservation and development of horticulture in north eastern region are land tenure issues, gender and equity issues, inter-departmental coordination, smuggling of timber across the international border, shifting cultivation, inter-state border, insurgency etc., which are responsible for horticulture diversity degradation. Adequate attention was given for systematic management of the rich diversity available in this region with the establishment of NBPGR, ICAR, BSI, and various universities in north east has made tremendous efforts in collection, evaluation, conservation and utilization of regional germplasm for development of horticultural varieties in this region. Keeping in view the regional demand for horticultural crops more germplasm needs to be identified for collection particularly for high yield, quality, resistance to diseases and pests, tolerance to frost and acidity.

References

Annonymous, 2000. Basic Statistics of North Eastern Region 2000.North Eastern Councel, Ministry of Home Affair, GOI, Shillong.

Annonymous, 2002. Agril Research Data Book, ICAR, 2002.

Annonymous, 2002. National Horticulture Board, Year Book-2002. , New Delhi, 230p.

CSIR, 1950. The Wealth of India, CSIR, New Delhi. Vol. II, pp56.

Gosh, S. P. 1984. Horticulture in North East Region. Associated Publishing Company, New Delhi, pp.38.

Hore, D.K. 2001. North East India-A hot-spot for agrodiversity. *Summer school on agriculture for hills and mountain ecosystem*, pp 361-362.

Ram, D., Kalloo, G. and Banerjee, M. K. 2002. Popularizing kakrol and kartoli: the indigenous nutritious vegetables. *Indian Horticulture*, pp 6-9.

Rao, Srinivasa M., and Dora, Dillip. K. 2002. Lesser known vegetables for nutritious, health and economic security- Indian context. International Conference on Vegetables, Bangalore, pp 610-617.

Sarma, B. K. 2001. Under utilized crops for hills and mountain ecosystems. *Summer school on agriculture for hills and mountain ecosystem*, pp 308-314.

❑❑❑

Chapter – 15

Leafy Vegetables: Genetic Resources and Improvement

K.K. Gangopadhyay, Gunjeet Kumar and S.K. Yadav

Introduction

Leafy vegetables belong to an important group of vegetables and are highly nutritious due to their richness in Vitamin A (Carotene), Vitamin C, Folic acid, Riboflavin, Thiamine and minerals like Iron, Calcium and Phosphorus etc. Appreciable quantity of proteins is also found in these crops. They also provide a variety of phyto-nutrients including beta-carotene, lutein and zeaxanthin, which protect cells from damage, age-related problems, among many other effects. Dark green leaves even contain small amounts of *Omega-3* fat. They are available at cheaper rate in the market as compared to other vegetables. They provide roughage and have an important place in the balanced diet. It is recommended that a daily intake of 100 g and 40 g of leafy vegetables in the diet of woman and man respectively. India is the 2nd largest producer of leafy vegetables in the world after China, accounted for about 10% of the world production (Singh *et al.*, 2006) but, the production levels of these leafy vegetables are very low because of lack of availability of high yielding varieties. Leafy vegetables are those crops from which leaves and associated parts are harvested for use as vegetable. The leaves of about 700 different kinds of plants belonging to125 families are consumed in the form of vegetables. The major leafy vegetables grown in the country are amaranth, spinach beet (desi palak), spinach and fenugreek. In addition to this a number of under-utilized annual crops are also grown as leafy vegetables in specific regions. Leafy vegetables are usually grown in kitchen and market gardens. Tender stems and leaves of a number of perennial crops are rich sources of vitamins and minerals and are used for cooking. Though leafy vegetables have high nitrate and oxalate levels, but adverse nutritional effects are not to be feared with a consumption level of 100-200 g per day. The commonly grown major leafy vegetables and their wild relatives are listed below:

S. N.	Crop	Common name	Scientific name	Chromosome No.	Type	Use(s)
1.	Amaranth	Chaulai	*Amaranthus tricolor* *A. dubius* *A. blitum* *A. tristis* *A. hypochondriacus* *A. cruentus* *A. caudatus* *A. viridis* *A. spinosus* *A. retroflexus* *A. graecizans*	2n=32 2n=64 2n=34 2n=32 2n=34 2n=34 2n=32 2n=34 2n=34 2n=32 2n=32	Cultivated Cultivated Cultivated Cultivated Cultivated Cultivated Cultivated Wild Wild Weed Weed	Leaf & stem Grain Leaf Leaf
2.	Fenugreek	Common methi Kasuri/ Champa methi	*Trigonella foenum graecum* *T. corniculata* *T. angeuna* *T. arabica* *T. bolansae* *T. calliceras* *T. coerulea* *T. cretica* *T. gladiota* *T. glomerata* *T. gordej* *T. hamosa* *T. lypokyi* *T. monspeliach* *T. radiata* *T. stellata* *T. striata* *T. ornithopodiodes* *T. gemniflora* *T. grandiflora* *T. polycerata*	2n=16 2n=16 2n=18 2n=44	Cultivated Wild	Leaf, grain Leaf
3.	Spinach beet	Desi palak	*Beta vulgaris* var *bengalensis* Hort.	2n=18	Cultivated	Leaf

Contd.....

4.	Spinach	Vilayati	*Spinacea oleracea* L. palak	2n=12	Cultivated	Leaf
5.	Basella	Poi	*Basella alba* *B. rubra*	2n=44 2n=48	Cultivated	Leaf
6.	Chenopods	Bathua	*Chenopodium album*	2n=36	Cultivated	Leaf & young shoot

Amaranths (Chaulai)

Amaranths (*Amaranthus spp.*) constitute a single major group in leafy vegetables. In India, it is grown during summer and rainy seasons. Being a C_4 plant, it is highly efficient in biomass production. The harvest index is almost one in this crop. Amaranth leaf protein contains more lysine than the best high lysine corn and more methionine than soybean. The red dye extracted from amaranth leaves is used to colour alcoholic beverages. Even the leaves of grain types are edible at young age. It fits well in a crop rotation because of its very short duration nature and high yield of edible matter. Amaranth is rich source of iron, calcium and phosphorus. The fresh tender leaves and stem of amaranth are used as vegetable. The origin of various species of cultivated amaranth is complicated, because the wild ancestors are pantropical cosmopolitan weeds. *Amaranthus spinosus, A. hybridus and A. dubius* are typical tropical type. *A. retroflexus, A. viridis, A. lividus* and *A. graecizans* are temperate hot season weeds. The morphological characters of 8 different species (Table 1) are described Mallika (1987).

Table 1. Morphological characters of *Amaranthus* spp.

Species	Habit	Plant height (cm)	Leaf colour	Stem & petiole colour	Leaf shape	Inflorescence position	Density of florets	Seed colour
A. tricolor	Erect	100	Green with deep purple centre	Green	Broad ovate	Axillary and small terminal panicle	Dense	Black
A. dubius	Erect, often branched	120	Green	Pale green	Rhombiod ovate	Huge, terminal and axillary panicle	Dense	Black
A. lividus	Prostrate, well branched	40	Pale green	Pale green	Rhomboid	Axillary and panicle small terminal	Dense	Black
A. hypochon-driacus	Erect, unbranched	65	Purplish green	Purplish green	Elliptic	Huge terminal panicle	Dense	Cream

Contd.....

A. cruentus	Erect, unbranched	150	Red/ Green	Red/ Green	Ovate lanceolate	Huge terminal panicle	Dense	Brow-nish
A. candatus	Erect, unbranched	120	Pale green	Pale green	Ovate lanceolate	Huge terminal panicle	Lax	Pinkish
A. viridis	Erect, branched	50	Green	Green	Deltoid	Slender, terminal and axillary panicle	Lax	Black
A. spinosus	Erect branched	80	Dull red	Dull red	Ovate lanceolate	Slender, terminal and axillary panicle	Lax	Black

The main vegetable amaranth *A. tricolor* originated somewhere in South or South East Asia. Since, then several secondary centers of diversity have developed in main production areas. In all probability, it might have originated in India and spread in neighboring countries. In amaranth, 62 accessions comprising 60 accessions (EC519509-63, EC524455-59) from Russia and one accession each from Slovakia (EC 588196) and UK (EC 582728) were introduced during the last five years. Among them, five trait specific introductions (EC524455-59) are resistant to root knot nematode, potato leaf hopper and leaf spot.

Crop Improvement

The evaluation work of amaranth germplasm is being carried out at NBPGR and a total of 2350 accessions of amaranth (*Amaranthus* spp.) at Shimla and 1146 accessions at Akola. Wide variability has been observed for many of the characters like plant height, leaf shape, leaf size, leaf pigmentation (both sides), stem thickness, stem pigmentation, days to 50 % flowering, disease resistance against white rust etc. The main crop improvement work is being done at IARI, New Delhi; IIHR, Bangalore and TNAU, Coimbatore. The vegetable amaranth is predominantly self pollinated crop due to presence of 10-25 % male flowers/glomerule and more number of axillary glomerule. Grain amaranthus is also self pollinated with an average of 15-30 % outcrossing mainly through wind. The main breeding objectives are one time harvest, multi-cut, pulling type etc. The traits of breeding importance are leaf colour, leaf yield, leaf-stem ratio and weight of stem, white seed (dual type). Heterosis and combining ability studies revealed that maximum heterosis for yield (13.4 %) was recorded by the cross IIHR-22 x IIHR-69 and the best general combiners for plant height and leaf weight were IIHR-6 and IIHR-22. Genotypes and environment interaction studies indicated the significant environmental influence on the expression of leaf weight, stem weight and plant weight in amaranth. Purple and green stem colours were found to be dominant over white stem colour. Leaf number was under the influence of additive gene action, whereas the other characters *viz.,* plant height, leaf weight, stalk weight and plant weight were controlled by non-additive/dominance gene action in amaranth. Genetical studies indicated that the nitrate content was under the influence of additive gene action, whereas anti-oxidant capacity and oxalate content were controlled by non-additive/dominance gene action in amaranth. Among the germplasm lines, IIHR-7, IIHR-65 and IIHR-124 were the late bolters, flowering 55 days after sowing, which is a desirable character in leafy vegetables to obtain higher leaf yields. Three germplasm lines, IIHR-4, IIHR-31 and IIHR-38 were found suitable for "stem amaranth" type with thick (>7 cm stem girth) and tender stems (each weighing >120 g), where as IIHR-47, IIHR-56, IIHR-66, IIHR-100, IIHR-107 and IIHR-114 were found suitable for "multi cut" type of amaranth with very good regeneration after cutting. Most of the interspecific hybrids between eight species were found unsuccessful. *A. lividus* x *A. tricolor* resembled the male parent in morphological characters but was highly sterile. The interspecific hybrids within the section Amaranthus included *A. spinosus* x *A. hypochondriacus*, *A. spinosus* x *A. caudatus*, *A. spinosus* x *A. cruentus*, *A. cruentus* x *A. caudatus*. *A. cruentus* x *A. caudatus,* produced only female flowers but the hybrid was stunted and deformed. Other hybrids showed preponderance of characters of *A. spinosus* indicated by presences of spines, reduced inflorescence size and distinct placement of male and female flowers in the inflorescence. *A. spinosus* x *A. viridis*, the first successful hybrid between the sections had short and sturdy plant type with dominating characters of *A. spinosus*. Cytological studies revealed that cryptic structural changes and genetic drift were involved in cytogenetic differentiation of the two species. The primitiveness of *A. spinosus* was evident from the study and this cosmopolitan weed played a major role in the evolution of other *Amaranthus* spp. It is reported that 40 % of oxalates exist in Free State in amaranth. The oxalate and nitrate contents of various amaranth genotypes ranged from 0.94 % to 1.29 % and 0.55 % to 1.00 % respectively. The genotype A 43 contained lesser content of oxalates and nitrates (Devadas and Mallika, 1991).

The intra-specific hybridization should be attempted as inter-specific hybridization leads to sterility. The promising accessions for different characters are as follows:

Character	Promising accession(s)
Herbage yield (>130 g/plant)	IC536732, IC536728, IC536731, IC550145, IC550142
Dwarf	IC550145, IC 38541, IC 38577, IC 38598, EC 170304, EC 169626
Leaf length (> 15cm)	IC467894, IC540860, IC467908, IC467886, IC415297, IC421885, IC547395, IC396961, IC386984, IC324011, EC169652
Resistant to white rust	IIHR-37, IIHR- 40, IIHR- 49, IIHR-50, IIHR- 70, IIHR- 111, IIHR- 112, IIHR-119, IIHR- 122, and IIHR-124
High antioxidant capacity (>255mg/100g AEAC).	IIHR-74, IIHR-70, IIHR-65
Low oxalate and nitrate	IIHR-7 (108.38 mg/g d.w.b and 24.47 mg/ 10 g d.w.b respectively)

Selection has been one of the predominant methods used for developing the varieties. The popular varieties developed are as follows:

1. Arka Suguna (*A. tricolor*): Pureline selection from Taiwanese introduction (IIHR-13560) Multicut variety with broad light green foliage, succulent stem, rich in calcium and iron, yielding 30 t/ha in 6-8 cuts, 1st picking from 24 days to last picking 90 days.
2. Co-1 (*A. dubius*): Tetraploid variety from local germplasm through selection, leaf and stem are dark green, resistance to *Rhizoctonia* leaf blight
3. Co-2 (*A. tricolor*) : Stem and leaves are green and large
4. Co-3 (*A. tristis*) : Suitable for tender green leaves, leaves are small and green, clipping starts from 20 DAS and up to 10 clippings in 90 days
5. Co-4 (*A. hypochondriacus)* : Green leaves, high yield (80 t/ha), duel purpose
6. Co-5 (*A. tricolor*): Leaf double coloured with green and pink, free from fiber, rosette growth
7. Pusa Kiran (*A. tricolor*): Developed by Sirohi and Sivakami (1995) from a natural crossing between *A. tricolor* and *A. tristis*, more resembling *A. tricolor*
8. Pusa Lal Chaulai (*A. tricolor*): Deep red upper surface and lower surface purplish red
9. Pusa Kirti (*A. blitum*): Developed by Sirohi and Sivakami (1995). Green leaf with green thick stem, suitable for summer
10. Pusa Badi Chaulai (*A. tricolor*): Tall plant with thick stem and green leaf, suitable for cutting
11. Pusa Chhoti Chaulai (*A. blitum*)
12. Arun (*A. tricolor*): This is a multicut variety from KAU. Maroon red leaves, leaf length 25 cm, leaf width 11 cm. High yield (20.1 t/ha).
13. Mohini (*A. tricolor*): This is a multicut, green leaved variety from KAU. During initial stage, the stem is light purple, 8-10 cuttings can be taken. Yield 13.2 t/ha (Gopalakrishnan, 2004).
14. Renushree (*A. tricolor*): This is a multicut with green leaves and red stem variety from KAU. Leaf: Stem ratio is 1.50. Yields 15.50 t/ha.
15. Krishna Shree (*A. tricolor*): This is a multicut variety from KAU. Reddish green leaves. Leaf stem ratio is 1.10. Yields 14.80 t/ha.
16. Kannara Local *(A. tricolor):* High yielding (25 t/ha), short day local cultivar having red leaves.

In amaranth (*A. tricolor*), two accessions IC395324 and IC395327 have been registered with NBPGR for Dantu type (pulling type with stem) and resistance to white rust, respectively.

Spinach Beet (Desi Palak)

Palak or beet leaf is a very important leafy vegetable grown throughout the year. It is related to beet root and Swiss chard. It is mainly grown in winter in plains and summer in hills. It is the most popular leafy vegetable of North and East India. Spinach beet is most probably originated in Indo-Chinese region (Nath, 1976). The herbaceous plant is annual for leaf production and biennial for seed production. Variability for plant height, leaf length, leaf width, leaf mid rib colour (red and green) and veins has been observed. The main crop improvement work is being done at IARI Regional Station, Katrain, H.P.; MPUA&T, Udaipur; PAU, Ludhiana and IIHR, Bangalore. The promising accessions for foliage type are NIC14420, EC31894 and BDJ-516. IIHR Sel. 6 and IIHR Sel.9 were found to be moderately resistant to cercospora leaf spot (*Cercospora baticola* Sacc.) which is a serious disease during rainy season. Being a cross pollinated crop, mass selection, hybridization followed by pedigree method or bulk selection will be effective (Swarup, 1977). Polyploidy breeding was also attempted .Crosses between palak and beet root were attempted to obtain late bolting, better rejuvenation and wider adaptability.

Crop Improvement

The varieties developed through different breeding procedure using diverse germplasm are as follows:

1. All Green (Sel. from local germplasm at IARI): Developed by Singh and Joshi (1960). Green tender leaves, 6-7 cuttings at 15-20 days interval, yield 125 q/ha of green leaves. Bolts after 75 days of sowing.
2. Pusa Palak (Bulk sel. from Swiss chard x Local Palak): Uniform green leaves without purple pigmentation, late bolting type.
3. Pusa Jyoti (Selection from All Green after 2% colochicine treatment): It can be grown throughout the year. , Large green, giant sized, thick, tender, crisp and succulent leaves. Multi-cut (6-8 cuttings), yield 49 t/ha.
4. Pusa Harit (Bulk sel. from Sugar beet x Local Palak): Developed by Gill *et al.*, (1980). Plants are erect; leaves upright, thick, dark green, slightly wavy margins, short petiole; quick rejuvenation after cuttings; very late bolting. Under hill conditions, if grown during summer it continues to give leaf cuttings throughout the year. Yield 260 q/ha, adaptability to varying climate.
5. Pusa Bharti: Tender green flavored leaves.
6. Arka Anupama (Pedigree sel. from IIHR-10 x IIHR-8): Thick, medium large succulent green leaves and field tolerant to Cercospora leaf spot, yield 41 t/ha in 4 cuttings, late bolters.
7. Pant Composite-1: Heavy yielder and resistant to Cercospora leaf spot
8. Jobner green (Spontaneous mutation from local collection No.5): Tender green leaves having strong flavor.

9. HS-23 (Developed through mass selection at CCSHAU, Hisar): Dark green juicy leaves.
10. Punjab Selection: Light green, long, thin, narrow and smooth leaf with slightly sour taste.
11. Punjab green (Selection from local material): Semi-erect plants, leaves are succulent, shining dark green, thick, long, sweet and free from sourness, mild purple pigmentation on stem, first cutting after 30 days of sowing, slow bolter, low oxalates (0.38 g/100g on fresh wt. basis), yield 3 t/ha.
12. Palak No. 15-16 (Developed by Deptt. of Agriculture, Maharashtra): Suitable for multi-cut, late bolting type.
13. Banerjee's Giant (Sel. from Local Palak x Beet root): Large thick leaves with very succulent stem and fleshy root.
14. Ooty-1(Selection from Himachal collection): First harvest at 45 days and subsequent three cuttings at 15 days interval.

Spinach (Vilayati Palak)

This crop is not grown commonly in India and mainly cultivated in USA and Europe. In India, its cultivation is limited to home garden of hilly region and used as a cooked vegetable. It is usually dioecious but, monoecious plants are rarely located. Variability is noted mainly in leaf shape, leaf size and seed prickliness. The crop improvement programme in this crop has been only done by IARI with the development of two varieties. The objectives of breeding are similar to spinach beet. However, spinach is only grown successfully in winter and cultivars should be developed which can withstand higher temperature. Moreover, spinach has high oxalic acid contents than spinach beet, hence, there is need to develop cultivars with low oxalic acids. Two types of spinach varieties round seeded and ii) prickly-seeded are grown in India. The first flourishes well in plains and the later in hills (Thakur *et al.*, 1989). Two popular varieties developed at IARI Regional Station, Katrain are described below:

Virginia Savoy: It is an introduction from USA. Plants upright, vigorous, round-seeded, late bolter; leaves blistered, crumpled, thick and dark green. Good yielder over an extended period; suitable for sowing in September and October in the northern plains. Yield 125 q/ha.

Early Smooth Leaf: This is a smooth seeded cultivar and produces small light green leaves with pointed apex.

Resistance breeding against mildews and viruses and development of late bolting types were also done. The first hybrid spinach resistant to downy mildew viz. EH 7 was developed. Another popular hybrid was Melody. This hybrid was with Savoy leaved, fast growing, late bolting suited to spring and fall crop. Yield varies from 10-15 t/ha.

Fenugreek (Methi)

Fenugreek, commonly known as '*Methi*' is mainly grown in North India. The leaves are used as vegetables and the seeds as condiments having several medicinal properties. It is also used for industrial purposes such as dye and for extraction of alkaloids and steroids. It is grown especially in areas having cool and dry climate. Fresh green leaves, tender pods and stems are used for cooking. The leaves are very rich in protein ranging from 18-41 % at different stages of growth on dry weight basis. Of late, *methi* is being exported in the form of dried leaves or powder to Middle East countries where it is added in various vegetarian and non-vegetarian curries to impart pleasant *methi* flavour to the dishes. Dry matter content of these high yielding lines ranged from 15-16%; hence they are suitable for the preparation of dry leaf powder also. Southeastern Europe and west Asia are considered to be the origin of this crop. In north western India, it is found growing wild. The variability is found for plant type, growth habit, foliage characteristics, duration of vegetative phase and seed characters. Although several species have been reported but, only two species i.e. *T. foenum-graecum* (common methi) and *T. corniculata* (Kasuri methi) are of economic importance. . The salient features of the two species are listed below:

Character	Common methi (T. foenum-graecum)	Kasuri methi (T. corniculata)
Growth Habit	Quick grown with erect shoot upto 1.0 meter height.	Initially slow growing and remain rosette during vegetative growth. Leaves are generally scented types.
Flower	Small white to creamy white coloured flower	Yellow to bright orange coloured flower
Pods	Pods are slender, long & boat shape	Pods are sickled shaped and small.
Seed	Seed size is medium to bold.	Seed size is very small.

Two accessions (EC520254-55) were introduced from Eritrea during the last five years. The promising accessions for high foliage and high yield are IC143855, IC143967, IC143890, and IC144225.

Crop Improvement

Some of the popular varieties developed through utilizing germplasm are as follows:

1. Pusa Early Bunching (Selection at IARI, New Delhi): It is quick growing variety, produces upright shoots, early and suitable for many cuttings, bold seeded and heavy yielder. It takes about 125 days for its seed maturity.
2. Kasuri selection (Selection at IARI, New Delhi): It is a late flowering variety, forms rosette. It gives several cuttings and is heavy yielder. Leaf has got special fragrance and is preferred for its flavour. It takes about 156 days for seed maturity.

3. Hisar Sonali (Developed at CCSHAU, Hisar): Dual purpose
4. Hisar Suvarna (Developed at CCSHAU, Hisar): Dual purpose
5. RMt-1 (Developed at SKM college of Agriculture, Jobner): Late type, dual purpose, Tolerant to root rot and moderately resistance to powdery mildew
6. Kasuri Methi (Selection from local material at Department of Agriculture, Punjab): Plants are dwarf and give rosette appearance.
7. Methi No. 47 (Developed by Department of Agriculture, Maharashtra): High yielding varieties with broad leaves, high content of vitamin-C
8. Methi No.14 (Developed by Department of Agriculture, Maharashtra): High yielding varieties with broad leaves, high content of vitamin-C
9. ML-150 (Sel. from T-8 x T-36) : Early and more pods
10. Lam selecton -1 (Selection at RARS, Lam): Dual purpose, suitable for irrigated condition, yield 740 kg of dry grains and 10-12 tonnes of greens/ha under irrigated condition.
11. Co-1 (Developed at TNAU, Coimbatore): High yield, dual purpose variety, tolerant to root rot
12. Rajender Kranti (Developed at RAU, Dholi): Medium tall, bushy plants, dual purpose
13. UM-50: Dual purpose
14. UM-112: Erect, tall and dual purpose, rich in diosgenin content (7.5mg/g dre seed weight)
15. Prabha (Pureline selection from local collection of Nagpur): Semi erect growth habit with pest and disease resistance
16. IC-74 (Landrace): Erect, green and more branching type, dual purpose
17. Barbara: Yellow seeds, erect plants, high trigonellin content in seeds

In fenugreek (*T. foenum-graecum*), two accessions IC296791 and IC296792 have been registered with NBPGR for downy mildew resistance, quick germination and tall, erect plant with multiple branch, powdery mildew resistance, and downy mildew tolerant respectively.

Basella (Poi)

Basella commonly known as Poi, Malabar Nightshade or Indian spinach is a popular summer leaf vegetable grown in North East and South India. It has been originated in Asia particularly in India. The plant has fleshy stem and leaves and is of trailing habit. The fresh tender stem and leaves are consumed after cooking. The colouring matter present in red cultivar has been used as dye in China. Deep violet colouring matter present in ripe fruits is also used for colouring food. Its young leaf and shoots are rich in terms of salts and vitamins. Besides this it also has medicinal properties and juice is used for constipation. There is only one species *Basella rubra* which is very variable. The species *B. alba,* is considered as a form of B. *rubra*. The variability is found for leaf shape, size, pigmentation and fleshiness; stem pigmentation and fleshiness. The major objective for basella improvement is large, thick, fleshy leaf and stem along with high yield. It is also desired to develop varieties having photo insensitive and late bolting type which can be grown through out the year. It is sensitive to extreme temperature and come up under partial shade.

Chenopod (Bathua)

Bathua (*Chenopodium album*), a nutritious leafy vegetable, is commonly grown in summer and winter. Tender shoots are cooked as vegetable. The stem is slender, angled and green, red or purple colored. Wax coated leaves are dull green in colour with pale green centre and waxy margin. Flowers are seen in cluster forming a compact or loosely panicled spike in axils. Two leafy varieties namely Pusa Bathua from IARI, New Delhi and Ooty-1 from TNAU, Coimbatore are released. Ooty-1 shows resistance to *Cercospora* leaf spot, *Colletotrichum* leaf spot and *Macrophomina* foot rot. In chenopod (*Chenopodium album*), one accessions IC258253 has been registered for brown seed.

Documentation

Two catalogues have been published on the basis of the characterization and evaluation of crops germplasm in leafy vegetables one each on Amaranth (400 accessions) and fenugreek (171 accessions) for 31 and 27 descriptors respectively.

Conservation

The details of the leafy vegetable crops germplasm conserved in long term storage at –20°C in National Gene Bank (NGB) are listed in Table 2.

Table 2. Leafy Vegetable Crops Germplasm Status in National Seed Gene Bank (as on 30-04-2009)

Crop	Accs.	Crop	Accs.
Amaranths (5131)		Spinach Beet (Desi Palak)	75
Amaranthus tricolor	74		
A. dubius	9	Spinach (Vilayati Palak)	130
A. blitum	3		
A.spinosus	8	Fenugreek (698)	
A. amora	3	Common methi	636
A. hypochondriacus	2160	Kasuri methi	62
A. cruentus	79		
A. caudatus	103	Poi (*Basella alba*)	1
A. viridis	21		

Contd..

A. gangeticus	14	Bathua (*Chenopodium album*)	180
A. retroflexus	3		
A. hybridus	24		
A. calliticus	1		
Amaranthus spp.	2521		
A.leucocarpus	2		
A.paniculata	13		
A.polygonoides	4		

Future Thrusts

- Augmentation of germplasm with special emphasis on trait specific germplasm from hot spot areas, fragile eco-system, unexplored areas and inaccessible areas for rare, endangered, threatened species.
- Multi-location evaluation with common checks and uniform data format for biotic and abiotic stresses in the identified hot spots, value added traits particularly nutritional value in leafy vegetable crops.
- Use of molecular markers for genotyping of selected accessions thorough diversity analysis and identification of genes through conventional and molecular techniques for unique traits in leafy vegetables
- Identification of the stable donors for desirable traits on the basis of multi-location evaluation.
- Development of core-sets in the crops having large genetic variability.
- Development of a national PGR database having linkages with all stakeholders.

References

Devadas, V. S. and Mallika, V. K. 1991. Review of research on vegetables and tuber crops - Amaranthus. Kerala Agricultural University, Vellanikkara.

Gill, H. S., Singh, J. P. and Lakhanpal, K. P. 1980. 'Pusa Harit' a variety of sugar beet. *Indian Horticulture*, **24:** 13

Gopalakrishnan, T. R. 2004. Three decades of vegetable research in Kerala Agricultural University, Vellanikkara, pp.105-111.

Mallika, V. K. 1987. Genom analysis in the genus Amaranthus. Ph.D (Hort.) thesis, Kerala Agricultural University, Vellanikkara.

Nath, P. 1976. Vegetabl for the Tropical Region. ICAR, New Delhi, Low Priced Book series. **No.2.**

Singh, H. B. and Joshi, S. A. 1960. Our Leafy Vegetables. *ICAR Farm Bull.*, **No 56.**

Singh, U., Sagar, V. R., Behera, T. K. and Kumar, P. Suresh. 2006. Effect of drying conditions on the quality of dehydrated selected leafy vegetables. *J. Food Sci. Technol.*, **43:** 579-582.

Sirohi, P. S. and Sivakami, N. 1995. Vegetable amaranth varieties from IARI. *Indian Horticulture*, **40:** 17-20.

Swarup, V. 1977. Breeding procedures for cross pollinated vegetable crops, ICAR, New Delhi, India, pp 47.

Thakur, P. C., Verma, T. S. and Joshi, S. 1989. Technology for producing more spinach seed. *Indian Horticulture*, **36:** 29-31.

❑❑❑

Chapter – 16

Genetic Resources for Cowpea Breeding

B. B. Singh

Introduction

Cowpea, {*Vigna unguiculata* (L.) Walp.} is an important food legume in the semi-arid tropics covering Asia, Africa, Southern Europe and Central and South America (Henriet *et al.*, 1997; Mortimer *et al.*, 1997; Van Ek *et al.*, 1997). It is a drought tolerant and warm weather crop. Therefore, well adapted to the drier regions of the tropics where other food legumes do not perform as well. It fixes atmospheric nitrogen and grows well even in the poor soils with more than 85% sand and with less than 0.2% organic matter and low levels of phosphorus (Kolawale *et al.*, 2000 and Sanginga *et al.*, 2000). Also, it is shade tolerant and therefore, compatible as an intercrop with maize, millet, sorghum, sugarcane and cotton as well as with several plantation crops (Singh and Emechebe, 1998). Coupled with these attributes, its quick growth and rapid ground cover checks soils erosion and in situ decay of its roots and nitrogen rich residues improves soil fertility and structure which together have made cowpea an important component of cropping system in the dry savannas of the sub-Saharan Africa (Carsky *et al.*, 2001 and Mortimer *et al.*, 1997).

Cowpea is cultivated in about 14 million ha world wide with about 5 million tons annual production. However, a substantial part of the cowpea production comes only from few countries. Nigeria is the largest producer and consumer of cowpea with about

5 million ha area and about 3 million tons production annually. Niger Republic is the next largest producer with 3 million ha and over 1 million ton production. Northeast Brazil grows about 1.5 million ha of cowpea with about 500.000 tons production. In the southern USA about 40,000 ha of cowpea is grown with an estimated 45,000 tons annual production of dry cowpea seed and a large amount of frozen green cowpeas. Even though exact statistics are not available, India is the largest cowpea producer in Asia and together with Sri Lanka, Pakistan, Nepal, Bangladesh, Thailand, Indonesia and other far eastern countries, there may be over 1.5 million hectares under cowpea in Asia.

Cowpea as Food and Fodder Crop

Cowpea is consumed as food in many forms: young leaves, green pods and green seeds are used as vegetables and dry seeds are used in various food preparations (Nout, 1996; Nielsen *et al.,* 1997). With 25% protein (on dry weight basis) in its seeds and tender leaves (Bressani, 1985, Nielsen *et al*., 1997), cowpea is a major source of protein, minerals and vitamins in the daily diets and thus it has positive impacts on the health of women and children. Bulk of the diet of rural and urban poor in Africa consists of starchy food made from cassava, yam, plantain and banana, millet, sorghum and maize etc. Addition of even a small amount of cowpea ensures nutritional balance of the diet and enhances the protein quality by the synergistic effect of high protein and high lysine from cowpea and high methionine and high energy from the starchy foods. Trading of fresh produce and processed cowpea foods and snacks provides rural and urban women opportunity for earning cash income.

Cowpea haulms also serve as nutritious fodder for the livestock particularly in the dry savannas of West Africa (Singh and Tarawali, 1997). In West Africa, the mature cowpea pods are harvested and the haulms are cut while still green and rolled into small bundles containing the leaves and vines. These bundles are stored on roof tops or on tree forks for use and for sale as 'Harawa' (feed supplement) in the dry season. On the dry weight basis the price of cowpea haulms ranges between 50 to 80% of the grain price and therefore haulms constitute an important source of income. Just like cowpea grains and leaves, the nutritive value of cowpea haulms is also very high. The crude protein content range from 13 to17% in cowpea haulms with high digestibility and low fibre and thus cowpea fodder is a good protein supplement with cereal stalks for the feeding livestock.

Variety Requirements

The plant type, seed type, maturity and use pattern of cowpea are extremely diverse in different countries and also vary from region to region. In West Africa, the preferred varieties have large white and brown seeds and a rough seed coat. In contrast, people in Central America and the Caribbean prefer varieties with red, black or white with a smooth seed coat or blackeye types with large seeds. The preference in Mexico, Guatemala, Nicaragua, Costa Rica and Cuba is for black seeds, while in Honduras, El

Salvador, Venezuela and Jamaica it is for red. In East Africa and Asia, any color other than black is acceptable, but tan and red seeds with a smooth seed coat are preferred in East Africa as a substitute for red kidney beans and white and cream are preferred in Asia.

Origin and Distribution

Initially, both Asia and Africa were thought to be independent centers of origin of cowpea. However, the Asian center of origin could not be supported due to the absence of wild cowpeas as possible progenitors. Systematic explorations conducted by the International Institute of Tropical Agriculture (IITA) strongly indicate that the region encompassing Namibia, Botswana, Zambia, Zimbabwe, Mozambique, Switzerland and South Africa have the highest genetic diversity in respect of primitive wild forms of cowpea (Padulosi *et al.*, 1990 & 1991; Padulosi, 1993). Based on this, Padulosi and Ng (1997) suggested that Southern Africa may be the origin of cowpea and from there primitive forms moved to other parts in Southern and Eastern Africa, and from there to Asia and West Africa. Since cowpea was known in India before Christ and it has Sanskit name in early treatise dating back to 150 BC (Steel and Mehra, 1980), cowpea moved from East Africa to Asia more than 2000 years ago where human selection have led to modified forms of cowpea different from Africa. Ng and Maréchal (1985) suggested that cowpea probably moved from Eastern Africa to India before 150 BC, to West Asia and Europe about 300 BC and to Americas in 1500 AD. Since Western Asia and Europe do not have desired climatic conditions for cowpea, not much variability and selection occurred as it happened in South Asia and South East Asia where small seeded and vegetable cowpeas were selected. Probably, the wild cowpeas were distributed by bird and humans in West Africa much before 2000 BC and therefore the presence great diversity and secondary wild forms are found. Selections for larger seeds and better growth habits from natural variants in wild cowpeas, must have led to diverse cultigroups and their domestication in Asia and in Africa (Ng, 1995). Using chloroplast DNA polymorphism, Vaillancourt and Weeden (1992) suggested Nigeria to be the center of domestication in West Africa.

Taxonomy and Gene Pool

Cowpea belongs to the order Fabaceae, subfamily Faboideae (Syn. Papillionoideae), tribe Phaseoleae, subtribe Phaseolinae and genus *Vigna* (Verdcourt 1970; Maréchal *et al.*, 1978). Vigna is a pantropical and highly variable genus with several species. The genus *Vigna* has been subdivided into seven subgenera: *Vigna*, *Sigmoidotropis, Plectotropis, Macrophynca, Ceratotropis, Haydonia* and *Lasiocarpa* (Marechal *et al.,* 1978). The subgenus *Vigna* has been further subdivided into six sections: Vigna, Comosae, Macrodontae, Liebrechtsia, and Catiang. Cowpea belongs to section Catiang which includes two species, *Vigna unguiculata* and *Vigna nervosa*.

Vigna unguiculata (L.). Walp. has been further subdivided differently by different researchers but they all agree that subspecies *unguiculata* contains the cultivated cowpeas. The cultivated cowpeas were earlier divided into four subspecies but these have now been accepted to be four cultigroups within the subspecies *unguiculata*. These are cultigroups *unguiculata*, *biflora* (or *cylindrica*), *sesquipedalis*, and *textilis* (Westphal, 1974; Maréchal *et al.,* 1978; Ng and Maréchal, 1985). A brief description of the four cultigroups under subspecies *unguiculata* is presented below:

1. cultigroup *unguiculata*: pods 10-30 cm long, pendant, seeds 5-12 mm long, originally from Africa, and now popular world over.
2. cultigroup *biflora or cylindrica*: pods 7.5-13 mm long, usually erect, seeds 5-6 mm long, from India mostly grown for fodder and seeds in Asia and used as pulse.
3. cultigroup *sequipedalis*: very long and succulent pods (30cm to 60cm), also called yard-long bean and asparagus bean, of India, south-east Asia and China with seeds usually 8-12 mm long mostly grown in Asia as vegetable.
4. cultigroup *textiles*: long fibrous peduncles (50-80 cm long), mostly grown in northern Nigeria for fiber.

The classification and nomenclature of the immediate wild progenitors of cultivated cowpeas was first proposed by Marechal *et al.,* 1978 and subsequently revised by Piennar and van Wyk (1992), and Pasquet (1993a). An extensive work on characterization of over 400 wild V. *unguiculata* accessions was conducted at IITA (Ng and Padulosi, 1991; Padulosi, 1993). This work, coupled with surveys of live materials in the field and specimens in major herbaria in Europe and Africa, as well as cytological studies, have led to the description of new species and a change of nomenclature of some species (Padulosi, 1993 and Ng, 1995). In the classification system proposed by Padulosi (1993), the three subspecies *dekindtiana, tenuis*, and *stenophylla* proposed by Maréchal *et al.* (1981) were retained, but var. *protracta* and var. *pubescens* were raised to the level of two distinct subspecies, because of their very distinctive hairy characteristics in pods and other plant parts, morphology of their flowers, pollen, grains, and leaves, as well as their root systems. Within the subspecies *protracta*, three varieties, namely var. *protracta*, var. *rhomboidea*, and var. *kgalagadiensis,* were distinguished. Similarly, three varieties *tenuis, oblonga,* and *parviflora* were recognized within the subspecies *tenuis*, while four new varieties namely var. *huillensis*, var. *congolensis*, var. *ciliolate*, and var. *grandiflora*, have also been proposed and added to the subspecies *dekindtiana*.

Ng (1995) proposed to reinstate var. *rhomboidea* to a species ranking in its own right, because of its strong incompatibility with other taxa within *V. unguiculata*. Pasquet (1993 a) proposed that the name subspecies *unguiculata* var. spontanea be used to describe all the weedy forms and the intermediates between truly wild var. *dekindtiana* and cultivated cowpea. Great variability in plant morphology has been observed in wild cowpea (Vaillancourt and Weeden 1992; Vaillancourt *et al*., 1993; Panella *et al*., 1992; Pasquet, 1993b). Subsp. *pubescens* and *protracta* are pubescent, with their stems leaves, and pods covered with hairs. The hairs are silky, straight, soft, and appressed to the surface of the stems and pods. On the other hand the hair type of subsp. *protracta* is bristly, erect, straight, and harshly stiff.

Cowpea Gene Pool

Knowledge of crop taxonomy contributes to an efficient crossing program for breeding and the results of hybridizations and cross compatibilities provide the basis for improving plant classification. This approach has led to the genepool concept of Harlan and De Wet (1971) which groups germplasm resources into primary, secondary or tertiary gene pools. The primary gene pool includes both the freely hybridizing cultivated and the wild forms. The secondary and the tertiary genepools consist of all the species among which gene flow is possible through interspecific hybridization but with increasing degrees of difficulty and sterility. In general, when the cross is successful but the F1 hybrid is less fertile or presents moderate chromosomal heterology, the donor species is considered to belong to the secondary gene pool. Thus, the secondary gene pool comprises other species that are relatives of the crop and are suitable for interspecific hybridization. When a cross between a cultivated and a donor species is possible with difficulty and the F1 hybrid is highly sterile or presents major meiotic irregularities because of marked structural differentiation between the two genomes, then the donor species is considered to belong to the tertiary gene pool of the cultivated species. Thus, the tertiary gene pool involves still greater barriers to hybridization compared to the secondary gene pool, embracing species that display either unviable or sterile hybrids with the cultivated plant and do not permit gene flow by conventional methods of introgression.

Using this criterion, the primary gene pool of cowpea comprises of 4 cultigroups of the subspecies *unguiculata* and most of the varieties of subspecies *dekindiana, stenophylla* and *tenuis*. However, some of the wild subspecies like *pubescence* are not freely crossable but need embryo rescue and show some level of pollen sterility (Fatokun and Singh, 1987). Therefore, some of the subspecies of V. *unguiculata* may constitute a secondary gene pool. Recent attempts to cross *Vigna vexillata* and *Vigna radiata* with *Vigna unguiculata* (Barone *et al.,* 1992; Gomathinayagam *et al*., 1998; Fatokun, 1991 & 2002) indicate the possibility of a tertiary gene pool. Barone *et al*., (1992) observed occasional fertilization between *V. vexillata* and *V. unguiculata* but the fertilized embryos degenerated within a few days. Gomathinayagam *et al.* (1998) reported successful cross between *V. vexillata* and *V. unguiculata*. However, the resulting F_1 hybrid and backcross plants looked more like *V. vexillata* plants but these were not followed systematically to ascertain the veracity of the cross. Fatokun (2002) reported various levels of pollen tube elongation in crosses between *V. vexillata* and *V. unguiculata* and tried to rescue embryos but no success was achieved. Tyagi and Chawla (1999) reported successful cross between *V. radiata* and *V. unguiculata* using in vitro culture technique. About 10% of the total embryos cultured resulted in plantlet formation but they did not report further growth and culture of these plantlets to confirm the true nature of the cross. *V. vexillata* belongs to the subgenus Plectotropis and *V. radiate* belongs to subgenus Ceretotropis within the genus *Vigna*. Using isozyme and RFLP analysis, Fatokun *et al.* (1993) and Sonnante *et al*., (1997) found *V. vexillata* to be closer to *Vigna unguiculata* than other species. There is a need to continue efforts to cross these and other species with *V. unguiculata* and determine if these constitute a tertiary gene pool for cowpea.

Germplasm Resources

The details of cowpea germplasm collection, evaluation, preservation and use for improving cowpea cultigens have been reviewed (Steele *et al.,* 1985 and Ng, 1987). Prior to 1967, a few national agricultural research programs such as Nigeria, Senegal, Tanzania and Uganda in Africa; India in Asia and USA in Americas, had some level of cowpea improvement program and they were maintaining some collection of cowpea germplasm. However, with the establishment of the International Institute of Tropical Agriculture (IITA) in 1967 and its global mandate for the improvement of cowpea, IITA has made a collection of cowpea germplasm exceeding 15,100 accessions of cultivated varieties drawn from over 100 countries and 560 accessions of wild cowpeas (Table 1).

Table. 1 .Cowpea germplasm accessions at IITA collected from different countries

Country	No. Ass.	Country	No. Ass.
Central and West Africa		Asia (contd.)	
Burkina Faso	222	Israel	8
Cameroon	600	Ian	29
Central Africa Republic	183	Japan	2
Chad	268	Laos	1
Gambia	4	Nepal	3
Ghana	282	Pakistan	7
Guinea	2	Papua New Guinea	12
Côte d'Ivoire	134	Philippines	114
Liberia	9	Sri Lanka	2
Mali	293	Syria	8
Niger	976	Thailand	1
Niger (+ IITA)	3,221	Turkey	47
Republic of Benin	331	USSR	42
Senegal	290	North and South America	
Serierra Leone	13	Argentina	1
Togo	103	Brazil	171
Zarïe	15	Canada	182
East and Southern Africa		Colombia	2
Angola	1	Cuba	1

Contd...

Botswana	457	El Salvador	1
Congo	47	Guatemala	11
Ethiopia	7	Honduras	1
Madagascar	34	Jamaica	1
Malawi	494	Mexico	23
Lesotho	42	Nicaragua	2
Swaziland	19	Paraguay	12
Somalia	3	Peru	3
Tanzania	495	Suriname	14
Uganda	71	USA	828
Zambia	587	Venezuela	3
Zimbabwe	158	Europe	
North Africa		UK	282
Algeria	1	Hungary	36
Egypt	347	Portugal	5
Asia		Italy	93
Afghanistan	65	Australia	23
Bangladesh	1	Unknown	610
China	35	Ex-mix	630
India	2,075	Wild Cowpeas	560
Indonesia	4	**Total**	**15,660**

The germplasm lines maintained at IITA have been numbered as TVv (Tropical *V. unguiculata i.e.* TVu-1, TVu-2 etc.). These have been characterized and evaluated for desirable traits and being preserved and used in the breeding program at IITA as well as in national breeding programs (Ng and Padulosi, 1991). In the collection extreme cowpea genotypes have been observed with respect to many traits such as plant pigmentation, plant type, plant height, leaf type, growth habit, photosensitivity and maturity, nitrogen fixation, fodder quality, heat and drought tolerances, root architecture, resistance to major bacterial, fungal and viral diseases, resistance to root-knot nematodes, resistance to aphid, bruchid and thrips, and resistance to parasitic weeds such as *Striga gesnerioides* and *Alectra vogelii*. Pod traits, seed traits and grain quality. These large and diverse cowpea germplasm collections are available to all the cowpea researchers around the world, especially in Africa, to exploit the valuable genes to improve cowpea cultivars.

Sources of Resistance to Insect Pests

Through intensive screening of the existing collections at IITA, many sources of resistance to insect pests like cowpea aphid (*Aphis craccivora*), leafnolpers (*Empoasca signata* and *E. dolichi*), legume bud thrips (*Megalurothrips sjostedti*), pod borers

(*Maruca vitrata*), pod-sucking bugs (*Clavigralla sp.*), and bruchids (*Callosobruchus maculates*) have been identified (Singh, 1977 & 1980; IITA, 1978 & 1983; Singh and Allen, 1979; Singh *et al.*, 1983, Singh *et al.*, 1885). These sources are summarized in Table 2.

Table 2. Cowpea germplasm lines resistant to various insect pests

Pest	Sources of resistance (IITA's TVu No.)
Aphid	TVu 35, 57, 134, 157, 191, 200, 310, 308, 410, 801, 3000, IT84S-2246, IT98K-205-8
Leaf hoppers	TVu 50, 59, 123, 305, 418, 662, 1037, 1045, 1190
Legume bud thrips	TVu 1509, 2870, TVx 3236, IT98K-205-8 (moderate resistance)
Legume pod borers	TVu 946, 13271, VITA-5 (low level of resistance)
Cydia ptychora	TVu 4579, 4328 (moderate resistance)
Pod-sucking bugs	TVu 1977, 7274
Bruchids	Tvu 2027, 11952, 11953, IT84S-2246, IT89KD-288, IT90K-277, IT98K-205-8

Sources of Resistance to Diseases and Parasitic Plants

The cowpea germplasm has also been screened for sources of resistance to major diseases (Williams, 1975, 1977a &1977b; Ladipo and Allen, 1979; IITA, 1978; Singh and Allen, 1979; Singh *et al.,* 1983) and these resistant sources are listed in Tables 3a and Table 3b.

Table 3a. Cowpea germplasm lines resistant to major diseases*

Resistant germplasm accessions Disease Locality of screening (IITA's TVu No)		
Brown blotch	Nigeria	TVu 201, 1977
Scab	Nigeria	TVu 843, 1404, 1433, 1977
Bacterial blight	Nigeria	TVu 347, 410, 456, 483-2, 726, 745, 1190, 1977
Septoria	Nigeria	TVu 456, 483, 486, 726, 853, 1433, 1583, 2455, 1977
Fusarium wilt	Nigeria	TVu 109, 347, 984, 1000, 1016
Fusarium root rot	Puerto Ricc	TVu 202, 231, 243, 266, 274, 320, 316, 393,
		408, 1563
Phakopsora rust	Nigeria and Uganda	TVu 612, 1258, 1962, 2455, 4540
Web blight	Nigeria	TVu 317, 2483, 4539 (all are moderately resistant)
Synchytrium false rust	Uganda	TVu 43, 222, 612, 4535, 4537, 4569, 6666
Target spot	Nigeria	TVu 1190
Root knot nematodes	Nigeria	TVu 264, 401, 857, 1560

Contd...

Bacterial blight + scab +		
Septoria + brown blotch	Nigeria	TVu 1977
Scab + Septoria	Nigeria	TVu 853, 1433
Bacterial blight + Septoria	Nigeria	TVu 456, 4832, 726
Anthracnose + bacterial pustule +	Nigeria	TVu 201, 347, 408, 410, 537, 697, 746
Cercospora leaf spot + rust +		TVu 1190, 1283, 2430, 3415, 310, 345,
Cowpea yellow mosaic virus		393, 645, 990, 1452, 1980, 2755, 3565
Major Viruses	Nigeria	201, 410, 1190
Striga and Alectra	Nigeria	TVu 9238, 11788, 12415, 12432, 12470, B301

*Ng and Padulosi (1991) and Singh and Emechebe (1997)

Table 3b. Improved cowpea breeding lines with resistance to major diseases.

Diseases Sources of Resistance
Anthracnose TVx 3236
Cercospora IT89KD-288, IT97K-1069-8, IT97K-556-4
Smut IT97K-556-4, IT95K-1090-12 IAR-48, IT97K-506-6
Rust(Uromyces) IT97K-1042-8, IT97K-569-9 IT97K-556-4, IT97K-1069-8 IT95K-238-3, IT97K-819-118 IT90K-277-2, IT96D-610, IT86D-719
Septoria IT90K284-2, IT95K-1090-12, IT97K-556-4 IT97K-1021-15, IT98K-205-8 IT98K-476-8, IT97K-819-118, IT95K-193-12
Scab. TVu 12349, IT95K-1090-12, IT98K-476-8, IT97K-1069-8 TVx 3236, IT95K-398-14 Ascochyta TVu 11761
Bacterial Blight IT81D-1228-14, IT95K-1133-6 IT97K-556-4, IT97K-1069-8, IT90K-284-2,

During the field evaluation of cowpea germplasm collection, four lines, TVu-201, TVu- 1190, TVu-1977 and TVu-4577 were found to be resistant to many diseases and had very high yield potential. These were named as VITA-1, VITA-3, VITA-4 and VITA-5 (Vigna IITA-1, 3, 4, 5) respectively and released in many countries (Singh and Ntare, 1985). These VITA lines have also been extensively used as parents in cowpea breeding programs.

Cowpea Genetics

Cowpea genetic research began in early 1900s (Harland, 1919a, 1919b & 1920) and after a long gap it picked up again in early 60s (Saunders, 1960a & 1960b) and continued thereafter. These studies have elucidated inheritance of many traits and given suitable gene symbols. A detailed review of cowpea genetics was first published in 1980

(Fery, 1980) and subsequent supplements were published in 1985 (Fery, 1985), in 1997 (Fery and Singh 1997) and in 2002 (Singh, 2002). Combined over all the studies, a total of 207 genes have been identified which control plant pigmentation, plant type, plant height, leaf type, growth habit, photosensitivity and maturity, nitrogen fixation, fodder quality, heat and drought tolerances, root architecture, resistance to major bacterial, fungal and viral diseases, resistance to root-knot nematodes, resistance to aphid, bruchid and thrips, and resistance to parasitic weeds such as *Striga gesnerioides* and *Alectra vogelii*, pod traits, seed traits and grain quality. Quantitative genetic studies on heritability of yield and yield components and protein content etc. and heterosis for important traits and improved breeding methods have facilitated selection of parents and increased efficiency in breeding programs. Some of the genes controlling important traits are listed in Table 4.

Table 4. List of gene symbols for important traits in cowpea.

Trait	Gene symbol	Reference
Vining growth	Vi-1, Vi-2, Vi-3	Brittingham (1950), Norton (1961) Singh and Jindla (1971)
Hastate leaf (narrow)	Ha (incomplete dominance)	Ojomo (1977)
Early flowering	Ef-1, Ef-2	Ojomo (1971)
Male sterility Adu-	ms1, ms2, pms	Sen and Bhowal (1962), Rachie *et al.,* 1975, Singh and Dapaah (1998)
Photogensitivity	Ps	Ishiyaku and Singh (2001)
Smooth seed texture	Rt1, Rt2	Singh and Ishiyaku (2000)
Resistance to:		
-Drought stress	Rds1, Rds2	Mai-Kodomi *et al.,* (1999)
-Bacterial pustule	Bptl-1, Bpl-2	Patel (1982)
-Bacterial canker	BC-1, bc-2	Singh and Patel (1977), Fery (1980)
-Verticillium wilt	Vw	Moore (1974)
-Cercospora leafspot	Cls-1, Cls-2	Fery and Duke, (1977)
-Striga gesnerioides	Rsg-1, Rsg-2, Rsg-3	Singh and Emechebe (1990), Atokple (1995)
-Alectra vogelii	Rav-1, Rav-2	Singh *et al.,* (1993)
-Septoria leafspot	Rsv-1, Rsv-2	Abadassi *et al.,* (1987)
-Uromyces rust	Uv-1, Uv-2	Chen and Heath (1993)
-Root-knot nematode	Rk, rk	Fery and Dukes (1980, 1982)
-Black eye mosaic	blc	Walker and Chambliss (1981)
-Southern bean mosaic	Sbm	Brantley and Kuhn (1970)

Contd..

-Tobacco ring spot	Tr	Dezeeuw and Ballard (1959)
-Cowpea severe mosaic	ims	Jimenez *et al.*, (1989)
- Aphid	Rac	Bata *et al.*, (1987)
- Bruchid	rcm-1, rcm-2	Adjadi *et al.*, (1985)

Cowpea Breeding

Cowpea breeding programs and their progress in different regions of the world have been reviewed by Singh *et al.*, 1997, Hall *et al.*, 1997, Singh, 2002, Singh and Ajegbe, 2002a, Singh and Matsui, 2002b and Singh 2005. International Institute of Tropical Agriculture (IITA) has been the center of excellence for cowpea research. However, significant cowpea improvement activities are carried out at the University of California, Riverside in the USA, EMBRAPA, Brazil and in Nigeria, Burkina Faso, Senegal, Mali, and India.

Production Constraints

The average cowpea yields are generally low due to spreading plant type, photosensistivity and late maturity of the local varieties and their susceptibility to several biotic and abiotic constraints as well as due to cultivation of cowpea as an intercrop with cereals in marginal environments where soils are infertile and rainfall is scanty.

The major diseases of cowpea are anthracnose, web blight, brown blotch, *Cercospora* leaf spot, *Septoria*, scab, *Macrophomina*, bacterial pustule, bacterial blight, cowpea yellow mosaic, cowpea aphid borne mosaic, blackeye cowpea mosaic, cowpea severe mosaic and southern bean mosaic. Nematodes are important in some areas and parasitic weeds such as *Striga gesnerioides* and *Alectra vogelii* are important in Africa.

The major insect pests of cowpea are aphid (*Aphis craccivora*), thrips (*Megalurothrips sjostedti*), Maruca pod borer (*Maruca vitrata*), a complex of pod sucking bugs (*Clavigralla* spp., *Acanthomia* spp., *Riptortus* spp.) and the storage weevil (*Callosobruchus maculates)*. Of these, thrips and *Maruca* cause major damage in Asia and sub-saharan Africa. There are some location specific insect pests such as *Lygus* in Americas, bean fly in Asia and East Africa and ootheca beetles in wetter regions of the tropics.

The major abiotic constraints for cowpea production are drought, low fertility especially low phosphorus, low pH and Aluminium toxicity in some areas. Early maturing cultivars escape terminal drought but if exposed to intermittent moisture stress during the vegetative or reproductive stages, they perform very poorly. Cowpea is inherently more droughts tolerant than other crops but it still suffers considerable damage due to frequent drought in the drought prone regions where rainfall is scanty and irregular.

Breeding Strategy

In view of the diverse variety requirements and numerous biotic and abiotic constraints, cowpea breeders need to design and develop specific plant types with desired maturity, pest resistance and seed types to suit different cropping systems and agro-ecological niches (Singh and Sharma 1996). Therefore, the general strategy has been to develop a range of cowpea varieties differing in growth habit and maturity, seed type and for sole crop and intercrop in different agro-ecologies with a focus on following types of varieties:

1. Extra-early maturing (60-70 days) grain type, for use as sole crop in multiple cropping systems and short rainy seasons.
2. Medium-maturing (75-90 days) grain type, for use as sole crop and intercrop.
3. Late-maturing (85-120 days) dual-purpose (grain + leaf) types, for use as sole crop and intercrop.
4. High-yielding, bush-type vegetable varieties.
5. Desirable seed types with high protein and micronutrient contents.
6. Resistance to major diseases, insect pests, and parasitic weeds.
7. Tolerance to heat, drought, and adaptation to sandy soils with low fertility.

Progress in Cowpea Breeding

Breeding for improved grain type and dual-purpose cowpea Varieties

Using the vast genetic pool and useful genes already identified, a great deal of progress has been made in breeding a range of high yielding cowpea varieties with combined resistance to major diseases, insect pests, Striga and Alectra and drought tolerance through conventional breeding program. Combining erect plant type with early maturity and resistance to major pests new extra-early cowpea varieties have been developed which yield up to 2.5 tons/ha within 60 days compared to less than 1ton/ha of the local varieties which mature in 100 to 140 days. Similarly, a number of medium maturing dual purpose cowpea varieties have been developed which yield over 2.5 t/ha grain and over 3t/ha fodder in 75-80 days. These varieties have been tested and over 40 improved cowpea varieties have been released in over 60 countries covering Africa, Asia and Central and South America.

Some of the important released varieties in different regions are listed below:

- **Asia:** VITA-3, VITA-4, VITA-5, IT82D-752, IT82D-789, IT82D-889, IT85F-2020, IT86D-1010, IT98K-205-8.
- **East and Southern Africa:** IT82E-16, IT82E-18, IT82D-889, IT85F-2020, IT86D-1010, IT87D-611-3, IT89KD-245, IT93K-2046-2, IT97K-568-18, IT97K-499

- **West and Central Africa:** TVx 3236, IT81D-985, IT81D-994, IT83S-818, IT83S-728-13, IT84S-2246-4, IT86D-719, IT86D-721, IT87D-453-2, IT89KD-245-1, IT89KD-288, IT88D-867-11, IT89KD-374-57, IT90K-76, IT90K-82-2, IT90K-277-2, IT90K-372-1-2, IT93K-452-1, IT97K-499-35, IT98K-205-8, IT98K-1111-1.
- **Central and South America:** VITA-3, VITA-7, IT82D-889, IT83D-442, IT84D-449, I84D-666, IT86D-1010, IT87D-697-2.

Bush type vegetable cowpea varieties

Several countries grow cowpea as a vegetable crop. The most preferred types are the yardlong cowpeas with fleshy tender pods, but these cultivars need staking to keep pods from touching the ground and rotting, which involves extra cost and thus restricts the area under cultivation. Bush-type vegetable cultivars with 30-cm long succulent pods have been developed, such as IT81D-1225-10, IT81D-1228-14, IT81D-1225-15, and IT86D-880, which yield up to 18 t ha^{-1} green pods with 3 to 4 pickings starting at 45 days after planting. These cultivars have semi-erect growth habit with extra-long peduncles (40-50 cm long), protruding well over the canopy and holding the pods above the ground. Picking green pods periodically reduces the weight on peduncles and they remain upright all the time. Frequent picking also stimulates further flowering and podding on the same peduncles, which ensures a continuous supply of green pods for a 6 to 7 weeks period after the start of picking, provided soil moisture is not limiting (Singh *et al.*, 2003).

Breeding for Resistance to Diseases and Nematodes

Using a combination of field and laboratory screening, a number of new sources of resistance to major diseases like Cercospora, smut, rust, Septoria, scab, Ascochyta blight and bacterial blight, Macrophomina, anthracnose, have been identified and used in the breeding programs (Abadassi *et al.*, 1987). Several cowpea breeding lines have been developed with combined resistance to cowpea yellow mosaic, blackeye cowpea mosaic and many strains of cowpea aphid borne mosaic. Among these, IT82D-889, IT86D-880, IT86D-1010, IT86F 2089-5, IT87D-611-3, IT90K-76, IT90K-277-2, IT90K-284-2 have been tested in many countries and found resistant. Good progress has also been made in breeding for resistance to nematodes. Some of the improved breeding lines with nematode resistance are IT849-2049, IT89KD-288, IT86D-634, IT87D-1463, IT95K-398-14, IT96D-772, IT96D-748, IT95K-222-5, IT96D-610, IT87K-818-18, IT97K-556-4. Among these, IT89KD-288 is a high yielding variety with high level of resistance to nematodes in Nigeria (Singh *et al.*, 1997, Mustapha *et al.*, 2000) and it is the most popular variety in the dry season in northern Nigeria. IT89KD-288 was also found to be resistant to 4 strains of *Meloidogyne* incognita in USA (Ehlers *et al,* 2000a).

Breeding for resistance to Striga/Alectra

Good progress has been made in breeding improved cowpea varieties with combined resistance to the two parasitic weeds, *Striga gesnerioides* and *Alectra vogeli* which have recently become important in Africa. A local landrace, B 301 from Botswana, confers complete resistance to *Striga* and *Alectra* in Burkina Faso, Mali, Cameroon, Niger and Nigeria but only moderate level of resistance to the strain from Benin Republic. Other lines such as IT81D-994, IT89KD-288, 58-57 and Gorom local confer complete resistance to strains from Benin Republic and Burkina Faso. Therefore, by using these complementary parents in crosses, a number of new varieties have been developed with combined resistance to *Alectra* as well as all the five strains of *Striga*. The most promising new cowpea varieties are IT93K-693-2, IT95K-1090-12, IT97K-499-35, IT97K-499-38, IT97K-499-39, IT97K-497-2, IT97K-819-118, IT97K-819-154and IT98K-205-8 with combined resistance to *Striga* and *Alectra*. Some of these lines are also resistant to bacterial blight, aphid, bruchid, thrips, and viruses with much higher yield potential than the local varieties (Singh, 2002; Carsky *et al*., 2003). Most of these lines also serve as a false host for *S. hermonthica* reducing its seed bank in the soil when grown as intercrop or in rotation with cereals.

Breeding for insect resistance

Insect-pests are the major constraint in cowpea production. Several improved cowpea varieties with combined resistance to aphid, thrips and bruchid have been developed (Adjadi *et al*., 1985; Bata *et al*., 1987; Singh, 2002, Singh, 2005). Of these, IT90K-76, IT90K-59, IT 89KD-288 and IT90K 277-2 are already popular varieties in several countries. Among the new varieties IT97K-207-15, IT95K-398-14 and 98K-506-1 have high level of bruchid resistance. Shade *et al*., (1999) reported a virulent strain of bruchid. (*Callosobruchus maculatus*) which was able to cause severe damage to Tvu 2027, which is otherwise resistant to normal bruchid strain. Yunes *et al.*, (1998) observed that the 7s-storage protein, 'vicillin' is responsible for bruchid resistance in cowpea lines related to Tvu 2027. Only low level of resistance has been bred for Maruca pod borer and pod bugs. This is because none of the cultivated cowpea germplasm lines and cross-compatible wild cowpeas are resistant to Maruca pod borer. A distant wild relative of cowpea *Vigna vexillata* has shown high level of resistance to Maruca pod borer and bruchid but all the efforts made at IITA to transfer Maruca resistance genes from *Vigna vexillata* to cowpea has not been successful (Fatokun, 1997). However, by evaluating new breeding lines without insecticide sprays and using the best lines in successive cycles of crossing and selection, some progress has been made for field resistance to Maruca and pod bugs. The most promising varieties with some level of field resistance to major insects are IT90K-277-2, IT93K-452-1, IT95K-321-1, IT95K-193-12, IT97K-837, and IT97K-499-38 which yield up to 500 kg/ha without any chemical protection compared to less than 50 kg/ha yield of the local varieties in the same trials. These lines are also resistant to major foliar diseases, aphid, thrips and bruchid with pods at wide angle and suffer less damage due to Maruca. IT94K-437-1 and IT97K-499-38 also have combined resistance to *Striga* and *Alectra*. Developed through conventional breeding approaches, the new field resistant lines require only 1 or 2 sprays of insecticide for normal yield of 1.5 to 2.5 tons compared to 5-6 sprays needed for the susceptible varieties.

Pyramiding available genes for resistance

Over the years, efforts have been made to systematically incorporate genes for resistance in improved breeding lines as well as some selected varieties as recurrent parents. Starting with a susceptible but high yielding variety, Ife Brown, as a base in 1973, successive improved but related varieties like TVx 3236, IT82D-716, IT84S-2246-4, IT90K-59, IT90K-76, IT97K-499-35 and IT00K-1251 have been developed with successive addition of new genes for resistance such that the newest varieties are resistant to all the major pests except for Maruca and pod bugs (Table 6).

Table 6. List of different countries which have released IITA developed improved cowpea varieties

Country Variety released

Angola TVx 3236

Argentina IT82D-716

Benin VITA-4, VITA-5, IT81D-1137,

Bolivia IT82D-889, IT83D-442

Republic IT84S-2246-4

Botswana ER-7, TVx 3236

Brazil VITA-3, VITA-6, VITA-7, TVx 1836-01J

Burkina TVx 3236, VITA-7 (KN-1)

Burma VITA-4 (Yezin-1)

Faso Cameroon IT81D-985 (BR1),

Central VITA-1, VITA-4, VITA-7, VITA-5, IT81D-994, (BR2),

African TVx 1948-01F, IT81D-1137, IT83S-818, TVx 3236,

Republic IT82E-18, IT81D-994, IT88D-363 (GLM-92), IT90K-277-2 (GLM-93)

Colombia IT83S-841

Costa Rica VITA-1, VITA-3, VITA-6

Cote IT88D-361, IT88D-363, VITA-7

D'IvoireCuba IT84D-449 (Titan)

Cyprus IT81D-1137, IT85D-3577, IT84D-666 (Cubinata-666) IT86D-314 (Mulatina-314) IT86D-368, (IITA-Precoz)

El Savador TVx 1836-013J (Castilla deseda),IT86D-782 (Tropico-782) VITA-3 (TECPAN V-3),IT86D-792 (Yarey-792) VITA-5 (Tecpan V-5)IT88S-574-3 (OR 574-3)

Equador VITA-3

Equatorial IT87D-885

Ethiopia TVx 1977-01D, IT82E-16,

Guinea IT82E-32

Contd..

Egypt TVu 21, IT82D-716

Fiji VITA-1, VITA-3 IT82D-709, IT82D-812, IT82E-16

Gambia IT84S-2049, (Sosokoyo) IT83S-728-13

Ghana IT82E-16 (Asontem) IT83S-728-13 (Ayiyi)

Guinea IT81D-879, IT83D-340-5, IT83S-818 (Bengpla)

Konakry IT82E-16, IT85F-867-5 (Pkoku Togboi) TVx 1843-1C (Boafa) IT85F-2805, IT83S-990, TVx 2724-01F (Soronko) IT87S-1463, IT84S-2246-4

Guinea IT82E-9, IT82D-889

Guyana ER-7, TVx 2907-02D, TVx 66-2H,

Bisau VITA-3

Guatemala VITA-3

Haiti VITA-4, IT87D-885

India VITA-4, TVx 1502,

Jamaica VITA-3, ER-7, IT84S-2246-4,IT 98K-205-8 (Pant Lobia-1) IT82E-124

Lesotho IT82E-889, IT87D-885

Liberia IT82D-889, TVx 3236, VITA-5,IT82E-16, IT82E-32 VITa-4, VITA-7

Malawi IT82D-889, IT82E-16

Mali Tvx 3236, IT89KD-374 (Korobalen)IT82E-25 IT89KD-245 (Sangaraka)

Mozambique IT82D-812, IT83S-18, IT85F-2020

Mauritius TVx 3236

Namibia IT81D-985, IT89KD-245-1,

Nicaragua VITA-3 IT87D-453-2

Nepal IT82D-752 (Aakash) IT82D-889 (Prakash)

Nigeria TVx 3236, IT81D-994, IT86D-719, IT88D-867-11, IT89KD-349 IT86D-721, IT88D-867-11,

Niger IT89KD-374, IT90K-372-1-2 IT82E-60, IT89KD-374, IT90K-277-2, IT90K-82-2, IT89KD-288

Pakistan VITA-4

Paraguay IT86D-1010, IT87D-378-4, IT87D-697-2, IT87D-2075

Panama VITA-3

Philippines IT82D-889

Peru VITA-7

Senegal TVx 3236

Sierra TVx 1990-01E, IT86D-721,

Somaila TVx 1502, IT82D-889

Leone IT86D-719, IT86D-1010, IT82E-32 IT82E-32, TVx 3236, TVu 1990, VITA-3

South VITA-5, VITA-7

South Korea VITA-5

Yemen Sudan IT84S-2163

South IT90K-59, (Daha ElGoz = Gold from sand)

Africa IT82E-16 (Pannar 311)

Swaziland IT82D-889 (Umtilane), IT82E-18,IT82E-27, IT82E-71

Sri Lanka IT82D-789 (Wijaya)

Thailand VITA-3, IT82D-889 IT82D-889 (Waruni) TVx 309-01EG, VITA-4 TVx 930-01B, (Lita)

Contd..

Uganda TVx 3236, IT82E-60
Suriname IT82D-889, IT82D-789
USA IT84S-2246-4, IT84S-2049, (for nematode resistance)
Tanzania TKx 9-11D (Tumaini) TVx 1948-01F (Fahari)
Yemen TVx 3236, IT82D-789, VITA-5 IT82D-889 (Vuli-1) IT85F-2020
Venezuela VITA-3, IT81D-795,
Togo VITA-5, TVx 3236, IT82D-504-4 TVx 1850-01E, IT81D-985, (VITOCO)
Zambia TVx 456-01F, TVx 309-01G, IT82E-16 (Bubebe)
Zaire VITA-6, VITA-7 IT89KD-349,
Zimbabwe IT82D-889 IT89KD-389, IT89KD-355

In addition to the lines mentioned in Table 6, there are lines with specific resistances. For example, cowpea breeding lines IT86D-880 and IT86D-1010 are resistant to all the three isolates of blackeye cowpea mosaic and 5 strains of cowpea aphid borne mosaic (Van Boxtel *et al*., 2000).

Breeding for drought, heat and cold tolerance

Since cowpea is grown in varied environments it encounters different types of stresses including drought, heat and cold temperatures. Early maturing varieties escape terminal drought (Singh, 1987), but if exposed to intermittent moisture stress during the vegetative or reproductive stages, they perform equally poorly. Also, cowpea suffers due to high temperatures in the Sahelian region. Using a simple screening method for heat and drought tolerance and root architecture, major varietal differences for all the three traits have been identified and incorporated into improved lines (Singh *et al*., 1999, Mai-Kodomi *et al*., 1999a, 1999b, Singh, 2001a, Singh and Matsui, 2002. Good progress has also been made on water use efficiency, heat tolerance and chilling tolerance (Hall *et al*, 1997 Ismail and Hall 1998,). The best drought tolerant varieties are IT89KD-374-57, IT88DM-867-11, IT98D-1399, IT98K-131-1, IT97K-568-19, IT98K-452-1, IT98K-241-2 and the best heat tolerant lines are IT93K-452-1, IT98K-1111-1, IT93K-693-2, IT97K-472-12, IT97K-472-25, IT97K-819-43, IT and IT97K-499-38.

Breeding for enhanced N-fixation and efficient use of phosphorus

Significant variation in cowpea rhizobium strains has been observed for nodulation in cowpea but the local rhizobia invariably outpopulates the introduced strains. Therefore, in recent years, major efforts are concentrated to exploit genetic variability in cowpea as a host for effective nodulation and nitrogen fixation. Sanginga *et al.,* (2000) screened 94 cowpea lines and observed major varietal differences in cowpea for growth, nodulation and arbuscular mycorrhizal fungi root infection as well as for performance under low and high phosphorus. The improved cowpea variety IT86D- 715 showed equally good growth under low as well as high phosphorus levels. It also showed better N-fixation than others. Based on its adaptability to grow in low P soils and overall positive N balance, they recommended cultivation of IT86D-715 cowpea variety in soils with low fertility. Kolawale *et al.*, (2000) screened 15 cowpea varieties for tolerance to aluminum and to determine the effect of phosphorus addition on the performance of Al-tolerant lines. The results indicated IT91K-93-10, IT93K-2046-1 and IT90K-277-2 cowpea varieties to be tolerant to aluminum and they gave higher response to phosphorus fertilization when grown in soils with aluminum toxicity problems. Recently 200 lines were screened and the best lines were IT90K-372-1-2, TN5-78, IT98D-1399, TN27-80, IT99K-1060, IT89KD-374-57, TN 256-80, IT03K-314-1, IT03K-351-2, IT97K-568-19, IT97K-568-11 and IT97K-1069-6.

Breeding for higher protein and micronutrient contents and cooking quality

There is a widespread malnutrition and iron deficiency in the sub-Saharan Africa and parts of South Asia. Increased consumption of cowpea with higher contents of minerals and vitamins would greatly minimize the current deficiencies and ensure general improvement in the health of children and nursing mothers. Considerable genetic variability exists for protein and other quality traits in cowpea (Nielsen *et al.*, 1993). Therefore, a bio-fortification project funded by Gates Foundation and World Bank is focusing on the development of high yielding cowpea varieties with higher levels of protein and micronutrients. The project started in 2003 and considerable progress has been made. A total of 50 selected cowpea varieties and breeding lines grown in 2003 were analyzed at Waite Analytical Services, Australia in 2004. The mean and range values in the 50 cowpea lines, were 24.4% and 21 to 31% for protein, 59.4 ppm and 46 to 79 ppm for iron, 829 ppm and 545 to 1330 ppm for calcium, 38 ppm and 23 to 48 ppm for zinc and 14721 ppm and 12750 to 16250 ppm for potassium. The best variety in respect of high protein and high iron and zinc was IT97K-1042-3 with 30% protein, 70 ppm Fe and 46 ppm zinc followed by IT98K-205-8 with 28% protein and high iron and zinc.

Considerable differences have also been observed for physical quality of cowpeas also. The relative density of cowpea seed ranged from 1.01 to 1.09, hardness (crushing weight) ranged from 3.96 kg for IT89KD-288 to 8.4 kg for Aloka local. The seed coat content ranged from 5.7% in IT95K-207-15 to 13.8% in TVu 12349 and cooking time ranged from 27.5 minutes for IT90K-277-2 to 57.5 minutes for Aloka local. The seed hardness was positively correlated with cooking time (Singh, 2001b). Varietal differences for taste and acceptability have also been observed. Akara prepared from high yielding new cowpea varieties Ayiyi (IT83S-728-13) and Bengpla (IT83S-818) were the best in Ghana. (Nout (1996). Similarly, the Women in Agriculture (WIA) section of the Kano Agricultural and Rural Development Authority KNARDA (Nigeria) observed the improved variety IT90K-277-2 as the best for four local dishes compared to other varieties. IT90K-277-2 has already become very popular in Nigeria and Cameroon as a high yielding variety.

Global Impact of Cowpea Research

Variety released by national programs

The collaborative interactions between the IITA, Bean/Cowpea CRSP and the national program scientists have been very effective and over 60 countries have identified and released improved cowpea varieties for general cultivation (Table 5). Some countries, have given specific names to the new varieties and, in some areas, farmers themselves have given names and facilitated farmer to farmer diffusion of seeds. A few examples are 'Big Buff' in Australia; 'BR-1' in Cameroon; 'Bira' in Angola, 'Titan and Cubinata' in Cuba; 'Asontem' and 'Bengpla' in Ghana 'Akash' (sky) and Prakash (light) in Nepal; 'Sosokoyo' in Gambia; 'Pkoko Togboi' in Guinea Conakry; 'Korobalen and 'Sangaraka' in Mali; Dan IITA (son of IITA) and 'Dan Bunkure' in Nigeria;; 'Pannar 31' in South Africa; ' Vuli-1' in Tanzania; 'Dahal Elgoz' (gold from the sand) in Sudan Umtilane in Swaziland; and 'Bubebe' in Zambia. Several other varieties have been released in different countries such as 'Charodi-1' and 'Vamban 1' (IT85F-2020) (Vishwanathan *et al.*, 1997), IT98K-205-8 (Pant Lobia-1) in India ; 'Big Buff' (IT82E-18) and 'Ebony PR' in Austrialia; IT83S-852 and IT82D-889 (Lee *et al.*, 1996) in South Korea; Melakh and Mouride (Cisse *et al.*, 1997) in Senegal; IT87D-611-3 in Guyana; Cream 7 in Egypt; IT90K-76, IT90K-277-2, IT90K-82-2 in Nigeria, Sangaraka (IT89KD-374-57) and Korobalen (IT89KD-245) in Mali (Toure and Singh, 2005); INIFAT 93 (Diaz *et al.*, 1997) in Cuba and GLM 93 (IT90K-277-2) in Cameroon. California Blackeye No. 27' (CB27) is a new blackeye cowpea cultivar released by the University of California-Riverside, in 1999. CB27 has high yield, heat tolerance, strong broad-based resistance to root-knot nematodes, resistance to two races of Fusarium wilt, excellent canning quality, and a brighter white seed (Ehlers *et al.*, 2000b). Brazil has released 18 varieties in the last twelve years for the Northern region.

Increased global cowpea production

The availability and release of the high yielding diseases and insect resistant varieties with desired seed and growth types is quietly catalyzing rapid increase in cowpea cultivation including its extension in non-traditional areas. Similarly, due to *Striga* resistant and other improved cowpea varieties, there has been rapid increase in cowpea cultivation in Brazil, Mali, Burkina Faso, Cameroon and Senegal. Also, the bruchid resistant varieties have significantly reduced storage losses of cowpea. As a result, the world cowpea production has increased from about 1.1 million tons in 1981 to over 5 million tons in 2007. There has been a revolutionary increase in cowpea production in many countries in West Africa where cowpea is an important crop. Nigeria produced over 3 million tons of cowpea in 2007 compared to only 580,000 tons in 1981. Similarly, Niger produced over 1 million tons cowpea in 2007 compared to only 300,000 tons in 1981.

Intensive Cropping in the Tropics using 60-day Cowpea

The new improved 60-day cowpea varieties have enabled farmers in Nigeria to produce two crops in the same season and use these varieties in improved intercropping systems for maximizing cereals, root crops and legume productivity (Singh and Ajeigbe, 2002a). Experiments conducted using 60-day cowpea varieties in northern Niogeria have clearly demonstrated that an intensive cropping system involving 'wheat-cowpea-rice' can be successfully grown each year using supplementary irrigation for wheat and cowpea with a total output of about 8 tons of food/ha comprising about 3 tons of wheat, 4 tons of rice and about 1-1.5 tons of cowpea (Singh and Ajeigbe, 2002a). Recent experiments at G. B. Pant University, Pantnagar, Uttarakhand, India have shown successful cultivation of 60-day cowpea varieties in wheat-rice system with 1.5 t/ha yield of cowpea within 60-70 days when planted after the harvest of wheat. Based on these results, the Uttarakhand Government has released IT98K-205-8 as 'Pant Lobia-1' for cultivation in its plain and hill regions. The additional cowpea crop in the summer season after wheat harvest (April-June) not only provides extra employment but it also provides nutritious grain and fodder which fetch higher prices at that time. Including cowpea in systems also makes the crop rotation more sustainable and productive because it improves the soil fertility by fixing atmospheric nitrogen and provides nutritious fodder for the livestock. Northern India has an area of over 10 million ha in wheat-rice system and thus a potential of producing additional 10-15 million tons of cowpea by utilizing this niche.

Future Priorities in Cowpea Breeding

Using biotechnological tools for insect resistance

The level of resistance to insects like thrips, Maruca pod borer, pod bugs and bruchid in the improved varieties is still low and therefore, all the improved varieties still require 2-3 sprays of insecticides to protect against the insects particularly, Maruca pod borer. Recent advances in biotechnology have provided new powerful tools to transfer desirable genes beyond the conventional species boundaries. Therefore, efforts are now underway to exploit biotechnological tools to complement conventional methods for improving insect resistance and quality traits in cowpea. The most successful examples are the transfer of insect resistance genes from the bacterium *Bacillus thuringiensis* (*Bt*) to cotton (Bt-cotton) and to maize (*Bt*-maize) which condition very high level of protection against cotton bollworm and maize stalk borer respectively. Laboratory studies have shown that products of the same *Bt* genes are highly effective against *Maruca* pod borer in cowpea. Incorporation of *Bt* genes for insect resistance in cowpea will reduce losses due to *Maruca* and revolutionize cowpea production providing tremendous benefit to resource poor farmers of West Africa and Asia who do not have access to farm chemicals. Therefore, concerted and joint efforts are being made by IITA, Bean/Cowpea CRSP, advanced laboratories in the USA and Australia, African Agricultural Technology Foundation (AATF), Network for Genetic Improvement of Cowpea for Africa (NGICA) and Monsanto Corporation to transform cowpea with *Bt* gene for *Maruca* resistance. It is expected that very soon transformed cowpea with Bt gene would be available and this trait would be transferred to a number of improved cowpea varieties. This combined with strategic research in cowpea IPM would greatly minimize yield losses in cowpea.

Marker Assisted Selection for Relevant Traits

Efforts are underway to develop markers and protocols for marker assisted selection (MAS) for *Striga* resistance in cowpea. There are five strains of cowpea *Striga* which are location specific. Therefore, the new *Striga* resistant breeding lines have to be tested at different locations. Marker assisted selection would eliminate the need for multilocation testing for *Striga*. Already two markers for aphid resistance (Myers *et al.*, 1996) and *Striga* resistance have been identified (Boukar *et al.*, 2004; Ouadraogo *et al.*, 2000 & 2002; Timko and Singh 2008) and these are now being used for developing appropriate methods for MAS. Similar efforts would also be made for resistance to bacterial blight and bruchid and other relevant traits.

Breeding for High Protein and High Micronutrients and Faster Cooking

Cowpea being the main source of protein and minerals for the urban and rural poor in West Africa, there is a need to initiate a systematic breeding program to develop cowpea varieties with higher protein and minerals as well as faster cooking property. Recent screening of selected germplasm lines has indicated considerable variability for theses traits with protein values up to 30% compared to 22-24 % in the cultivated local varieties. A beginning has been made under the Bio-fortification Challenge Project funded by Gates Foundation and World Bank and it should be strengthened.

Breeding for higher grain and biomass

The major success in the past has been the development of a range of cowpea cultivars with diverse maturity and plant type with combined resistance to major biotic and abiotic stresses to ensure yield stability in sole crop as well as intercrop. Little or no efforts were made in the past to breed for high yield per se. The current improved varieties have the maximum yield potential of about 2.5 to 3.0 t/ha within 60 to 75 days. This is quite acceptable yield level on per day productivity basis but it can be raised further to 4 t/ha by breeding a better plant type with higher canopy height, enhanced photosynthesis and translocation efficiency and suitable for higher density cultivation.

Breeding for Specialty Foods

Cowpea is already consumed in many forms: as green pods for vegetable, as green canned beans, as leaf spinach, dry boiled beans and split dahl. Recently there is a growing need for using cowpea for baby foods and specialty snacks. Cowpeas also have a potential for nutraceutical uses as it has many antioxidants. As cowpea production and utilization becomes widespread and more popular, there would be a need for breeding specific cowpea varieties for different uses.

References

Abadassi, J.A., Singh, B.B., Ladiende, T.A.O., Shoyinka, S.A. and Emechebe, A.M. 1987. Inheritance of resistance to brown blotch, septoria leaf spot and scab in cowpea (*Vigna vexillata*) *Indian Journal of Genetics,* **47:** 299-303.

Adjadi, O., Singh, B.B. and Singh, S.R.. 1985. Inheritance of bruchid resistance in cowpea. *Crop Science,* **25:** 740-742.

Atokple, I.D.K., Singh, B.B. and. Emechebe, A.M 1995. Genetics of resistance to *Striga* and *Alectra* in cowpea. *J. Heredity,* **86:**45-49.

Barone, A., Guidice, A. del and Ng., N.Q. 1992. Barriers to interspecific hybridization in V. unguiculata and V. vexillata. Sexual Plant reproduction, **5:**195-200

Bata, H.D., Singh, B.B., Singh, S.R.and Landeinde. T.A.O. 1987. Inheritance f resistance to aphid in cowpea. *Crop Scienc.,* **27:** 892-894.

Boukar, O., Kong, L., Singh, B.B., Murdock, L. and Ohm., H.W. 2004. AFLP and AFLP-derived SCAR markers associated with *Striga gesnerioides* resistance in cowpea (*Vigna unguiculat* (L.) Walp.). *Crop Science,* **44:** 1259-1264

Brantley, B.B., and Kuhn, C.W. 1970. Inheritance of resistance to southern bean mosaic virus in southern pea (*Vigna sinensis*). *Proceedings of the American Society of Horticultural Science,* 95, 155-158.

Bressani, R. 1995. Nutritive value of cowpea. In: Singh, S.R., Rachie, K.O. (Eds.) . Cowpea research, production and utilization. John Wiley and sons. Uhichester, U.K pp.353-359

Carsky, R.J., Singh, B.B. and Oyewole, B. 2001. Contribution of early season cowpea to late season maize in the savanna zone of West Africa. *Biological Agriculture & Horticulture,* **18:**303-316.

Carsky, R.J., Akakpo, C., Singh, B.B. and Detongnon, J. 2003. Cowpea yield gain from resistance to *Striga gesnerioides* parasitism in southern Benin. Experimental Agriculture, **39:** 327-333.

Chen, C.Y. and. Heath, M.C. 1993. Inheritance of resistance to *Uromuyces vignae* in cowpe and correlation between resistance and sensitivity to acultva-specific elicitor of necrosis. *Phytopathlogy,* **83:** 324-230.

Cissse, N., Ndiaye, M., Thiaw, S. and Hall, A.E. 1997. Registration of "Melakh" cowpea. *Crop Science.* **37:**6, 1978.

De Zeeuw, D.J., and Ballard, J.C. 1959. Inheritance in cowpea of resistance to tobacco ringspot virus. *Phytopathology,* **49:** 323-334.

Diaz, M., Shagarodsky, T., Lastres, N., and F., Puldon, G. 1997. INFAT 93, a new variety of cowpea (*Vigna unguiculata*). *Agrotecnia-de-Cuba.,* **27:**1, 148.

Ehlers, J. D., Matthews, W. C., Hall, A. E. and Roberts, P. A. 2000a. Inheritance of a broad-based form of nematode resistance in cowpea. *Crop Sci.* **40:**611-618.

Ehlers, J. D., Hall, A. E., Patel, P. N., Roberts, P. A. and Matthews, W. C.. 2000b. Registration of 'California Blackeye 27' Cowpea. *Crop Sci.,* **40:**854-855.

Fatokun, C.A. and Singh, B.B.1987. Interspecific hybridization between V. pubescence and V. unguiculata through embryo rescue. *Plant Cell, Tssue and Organ Culture,* **9:** 229-233.

Fatokun, C.A. 1991. Wide hybridization in cowpea: problems and prospects. *Euphytica* **54:** 137-140.

Fatokun, C.A., Danesh, D. I. Menancio-Hautea, and N.D. Young. 1993. A linkage map for cowpea (*Vigna unguiculata*) [L.] Walp.) based on DNA markers. Pages 256-258 in Genetic maps, edited by S.J. O'Brien. Locus Maps of Complex Genomes. VI Edition. Cold Spring Harbor Laboratory Press, USA.

Fatokun, C.A. 1997. Breeding cowpea for resistance to insect pests : attempted crosses between cowpea and V. vexillata. Pages 52-61 *In:* Advances in cowpea research. Singh B.B., D.R. Mohan Raj, K. Dashiell and L.E.N. Jackai (eds.) 1997. Copublication of International Institute of Tropical Agriculture (IITA) and Japan International Center for Agricultural Sciences (JIRCAS), IITA, Ibadan, Nigeria.

Fatokun, C.A. 2002. Breeding cowpea for resistance to insect pests: attempted crosses between cowpea and *Vigna vexillata*. pages-52-61. In: Fatokun, C.A., S.A. Tarawali, B.B. Singh, P.M. Kormawa and M. Tamo (editors).2002. Challenges and opportunities for enhancing sustainable cowpea production. IITA, Ibadan, Nigeria.

Fery, R.L., and Dukes, P.D. 1977. An assessment of two genes for Cercospora leaf spot resistance in the southern pea [*Vigna unguiculata* (L.) Walp.] *HortScience,* **12:** 454-456.

Fery, R.L. 1980. Genetics of *Vigna*. Pages 311-394. In Janick, J. (ed.) 1980 *Horticultural Reviews,* Westport, CT, USA, AVI Publishing.

Fery, R.L. and Dukes, P.D. 1980. Inheritance of root-knot resistance in the cowpea [*Vigna unguiculata* (L.) Walp.]. *Journal of the American Society of Horticultural Science,* **105:**671-674.

Fery, R.L. and Dukes, P.D. 1982. Inheritance and assessment of a second root-knot resistance factor in southernpea [*Vigna unguiculata* (L.) Walp.]. *HortScience,* **17:**152 (abstract).

Fery, R.L. 1985. The genetics of cowpea. A review of the World Literature Pages 25-62. In Singh, S.R. and K.O. Rachie (eds.) 1985 Cowpea Research, Production and Utilization. John Wiley and Sons. Chichester, U.K.

Gomathinayagam, P., Ram, S.G., Rathnaswanmy, R., and Ramaswamy, N.M. 1998. Interspecific hybridization between *Vigna unguiculata* (L.). Walp and *V. vexillata* (L.). A. Rich, through in vitro embryo culture. *Euphytica,* **102(2)**:203-209.

Hall, A. E., Singh, B. B. and Ehlers, J. D. 1997. Cowpea breeding. *Plant Breeding Reviews,* **15:** 215-274.

Harlan, J.R. and De Wet, J.M.J. 1971. Toward a rational classification of cultivated plants. *Taxon,* **20:** 509-517.

Harland, S.C. 1919a. Inheritance of certain characters in the cowpea (*Vigna* sinesis). *Journal of Genetics,* **8:**101-132.

Harland, S.C. 1919b. Notes on inheritance in cowpea. *Agricultural News,* **18:**20.

Harland, S.C. 1920. Inheritance of certain characters in the cowpea (*Vigna* sinensis). *Journal of Genetics,* **10:** 193-205

Henriet J., Van Ek, G.A., Blade, S.F. and Singh, B.B. 1997. Quantitative assessment of traditional cropping systems in the Sudan savanna of northern Nigeria: I. Rapid survey of prevalent cropping systems. *Samaru Journal of Agricultural Research,***14:**37-45.

IITA. 1978. Annual Report for 1977. IITA, Ibadan, Nigeria

IITA. 1983. Annual Report for 1982. IITA, Ibadan, Nigeria

Imrie, B.C. 1995. Register of Australian grain legume cultivars. Cowpea cv. Big Buff. *Australian J. of Experimental Agri.,* **35:** 678

Ismail, A. M. and Hall, A. E. 1998. Positive and potential negative effects of heat-tolerance genes in cowpea. *Crop Sci.,* **38:**381-390.

Jimenez, C.C.M., Borges, de F.O.L. and Debrot, C.E.A. 1989. Herencia de la resistancia del frijol (*Vigna unguicalata* [L.] Walp.) al virus del mosaico severo del caupi. *Fitopathologia Venezolana,* **2(1)**:5-9, (English summary).

Kolawale, G.O., Tian, G., and Singh, B.B., 2000. Differential response of cowpea varieties to Aluminum and phosphorus application. *Journal of Plant Nutrition, 23*:731-740

Ladipo, J.L. and Allen, D.J. 1979. Identification of resistance to southern bean mosaic virus in cowpea. *Trop. Agric. Trinidad,* **56:** 33-40.

Lee-SangMoo, Koo-Jae Yun, Jeon-ByonTae, Lee, S.M., Koo, J.Y. and Jeon, B.T. 1996. Studies on the growth characteristics and yield of cowpea cultivars for silage. *Journal of the Korean Society of Grassland Science*, **16(2):** 105-112.

Mai-Kodomi, Y., Singh, B.B. Myers, O. Jr., Yopp, J. H., Gibson, P.J. and Terao, T.. 1999a. Two mechanisms of drought tolerance in cowpea. *Indian J. Genet.*, **59:** 309-316.

Mai-Kodomi, Y., Singh, B.B. Terao, T. Myers, O. Jr., Yopp, J.H. and Gibson, P.J. 1999b. Inheritance of drought tolerance in cowpea. *Indian J. Genet.*, 59: 317-232.

Maréchal, R., Mascherpa, J.M. and Stainier, F. 1978. Etude taxonomique d'un groupe d' especes des geners Phaseolus et Vigna (Papilionaceae) sur la basedennees morphologiques, traitees, pour l'analyse informatique. *Boisseira*, **28:**1-273.

Maréchal, R., Mascherpa, J.M, and Stainier, F. 1981. Taxonometric study of the Phaseolus-Vigna complex and related genera. In Polhill, R.M., and Raven, P.H. Advances in Legume Sysematics. Royal Botanic Gardens, Kew, UK.

Moore, W.F. 1974. Studies on techniques of hybridization and the inheritance of resistance to *Verticillium* wilt in protepea (cowpea). State College, MI, Missdissipi State University, (Ph.D. thesis).

Mortimore, M.J., Singh, B.B., Harris, F. and Blade, S.B. 1997. Cowpea in traditional cropping systems. *In:* Advances in cowpea research. Singh B.B., D.R. Mohan Raj, K. Dashiell and L.E.N. Jackai (eds.) 1997. Copublication of International Institute of Tropical Agriculture (IITA) and Japan International Center for Agricultural Sciences (JIRCAS), IITA, Ibadan, Nigeria. 99-113

Mustapha, O.T., Mohammed, S.G. and Singh, B.B. 2000. Inheritance of resistance to root-knot nematode in cowpea (*Vigna unguiculata* (L.) Walp. *Biosience Research Communications* **12:**287-290

Myers, G.O., Fatokun, C.A. and Young, N.D. 1996. RELP mapping of an aphid resistance gene in cowpea. *Euphytica,* **91:** 181-187.

Nielsen, S.S., Brandt, W.E. and Singh, B.B., 1993. Genetic variability for nutritional composition and cooking time of improved cowpea lines. *Crop. Sci.* **33:** 469-472.

Nielsen, S.S., Ohler, T.A., and Mitchell, C.A. 1997. Cowpea leaves for human consumption: production, utilization and nutrient composition. *In:* Advances in Cowpea Research. Singh, B.B., Moham Raj D.R., Dashiell, K.E., and Jackai, L.E.N. (eds.).Copublication of International Institute of Tropical Agriculture (IITA) and Japan International Research Centre for Agricultural Science (JIRCAS). IITA, Ibadan, Nigeria. pp.326-332

Ng. N.Q. and Marechal, R. 1985. Cowpea taxonomy, origin and germplasm. In Singh, S.R., and Rachie, K.O. (eds.) pages 11-21. Cowpea Research, Production and Utilization. John Wiley, New York, USA.

Ng, N.Q. 1987. The gene bank at IITA and its significance for crop improvement in Africa, with special reference to cowpea germplasm. *In*: Proceedings of International Symposium on Conservation and Utilization of Ethiopian Germplasm, Engels. J.M.M. (ed.) Addis Ababa, October 1986. PGRC/E Addis Ababa, Ethiopia

Ng, N.Q. 1987. Genetic studies and interspecific crosses in cowpeas. In Genetic Resources Unit Annual Report 1987. IITA, Ibadan, Nigeria.

Ng, N.Q. and Padulosi, S. 1991. Cowpea genepool distribution and crop improvement. Pages 161-174 in Crop genetic resources of Africa Vol. II., edited by N.Q. Ng, P. Perrino, F. Attere, and H. Zedan. IITA, CNR, IBPGR, and UNEP. IITA, Ibadan, Nigeria.

Ng, N.Q. 1995. Cowpea *Vigna unguiculata* (*Leguminosae-Papilionoideae*). Pages 326-332 in Evolution of crop plants, 2nd edition, edited by J. Smart and N.W. Simmonds. Longman, Harlow, UK.

Nout, M.J.R.,1996. Suitability of high yielding cowpea cultivars for koose, a traditional fried paste of Ghana. Tropical Science (UK) v. **36(4):** 229-236.

Ouedraogo, J.T., Maheshwari, V. Berner, D.K. St-Pierre, C.A. Belize, F. and Timko, M. P. 2001. Identification of AFLP markers linked to resistance of cowpea to parasitism by Striga gesnerioides. *Theoretical and Applied Genetics,* **102:** 1029-1036.

Ouedraogo, J.T., Gowda, B.S. ,Jean, M., Close, T.J., Ehlers, J.D., Hall, A.E., Fillaspie, A.G., Roberts, P.A.Ismail, Bruining, A.M., Gepts, G., Timko, P. and Belzile, F.J.. 2002. An improved genetic linkage map for cowpea (*V.unguiculata*) combining AFLP, RFLP, RAPD, biochemical markers and biological resistance traits. *Genome* **45:** 175-188

Padulosi, S., G. Laghetti, N.Q, and P. Perrino. 1990. Collecting in Swaziland and Zimbabwe. *FAO/IBPGR Plant Genetic Resources Newsletter,* **78/79:** 38.

Padulosi, S., Laghetti, G., Pienaar, B., Ng, N.Q. and Perrino, P.. 1991. Survey of wild Vigna in southern Africa. *FAO/IBPGR Plant Genetic Resources Newsleeter* **83/84:** 4-8.

Padulosi, S. 1993. Genetic diversity, taxonomy and ecogeographic survey of the wild relatives of cowpea (*V. unguicullata*). Ph.D. Thesis. University Catholique Lovain-la-Neuve, Belgique.

Padulosi, S. and N.Q, Ng. 1997. Orgin, taxanomy and morphology of *Vigna unguiculata* (L.). Walp pages 1-2. In Singh, B.B., D.R. Mohan Raj, K.E. Dashiell, and L.E.N. Jackai (eds.) 1997. Advances in cowpea research. Copublication of International Institute of Tropical Agriculture (IITA) and Japan International Research Center for Agricultural sciences (JIRCAS). IITA, Ibadan, Nigeria.

Panella, L., Kemi, J. and Gepts, P. 1992 Vignin diversity in wild and cultivated taxa of V. unguiculata. *Economic Botany,* **47**:371-386

Pasquet, R.S. 1993a. Classification infraspecifique des formes spontanées de Vigna unguiculata (L.). Walp. (Fabacceae) a partir de données morphologiques. Bulletin du Jardin Botanique *National de Belgique,* **62:** 127-173.

Pasquet, R.S. 1993b. Variation at isozyme loci in wild *Vigna unguiculata* (Fabaceae Phaseoleae). *Plant Systematics and Evolution,* **186:** 157-173.

Piennar, B.J. and Van Wyk, A.E. 1992. The *Vigna unguiculata* complex (Fabaceae) in South Africa. *South Afric. J. Bot.,* **58**:414-429

Sanginga, N., Lyasse, O., and Singh, B.B. 2000. Phosphorus use efficiency and nitrogen balance of cowpea breeding lines in a low P soil of the derived savanna zone in West Africa. *Plant and Soil,* **220:** 119-128

Saunders, A.R. 1960a. Inheritance in the cowpea. 2: seed coat color pattern; flower, plant and pod color. *South African Journal of Agricultural Science,* **3:** 141-142.

Saunders, A.R. 1960b. Inheritance in the cowpea 3: mutations and linkages. *South African Journal of Agricultural Science,* **3:** 327-348.

Shade, R.E., Murdock, L.L. and Kitch, L.W. 1999. Interactions between cowpea weevil (Coleoptera: Bruchidae) populations and *Vigna* (Leguminosae) species. *Journal of Economic Entomology,* **92(3):**, 740-745.

Singh, D. and Patel, P.N. 1977. Studies on resistance incrops to bacterial diseases in India. 7: inheritance o resistance to bacterial blight in cowpea. *Indian Phytopahology,* **30:** 99-102.

Singh, B.B., Adjadi, O., and Singh, S.R., 1985. Bruchid resistance in cowpea. *Crop Science,* 25, 736-739.

Singh, B.B., and Ntare, B.R. 1985. Development of improve cowpea varieties in Africa. In Singh, S.R., and Rachie, K.O. (eds.) Cowpea Research, Production and Utilization. John Wiley, New York, USA.

Singh, B.B. 1987. Breeding cowpea varieties for drought escape. In (Menyonga *et al* (eds.) Food Grain Production in Semi-Arid Africa. OUA/STRC-SAFGRAD. p. 299-306.

Singh, B.B. and Emechebe, A.M.. 1990. Inheritance of *Striga* resistance in cowpea geotype B301. *Crop Science,* **30:** 870-881.

Singh, B.B., Emechebe, A.M. and Atokple, I.D.K. 1993. Inheritance of *alectra* resistance in cowpea genotype B301. *Crop Science,* **33:** 70-72.

Singh, B.B. and Sharma, B. 1996. Restructuring cowpea for higher yield. *Indian J. Genet.* **56:**389-405.

Singh, B.B. and Tarawali, S.A. 1997. Cowpea and its improvement: Key to sustainable mixed crop/livestock farming systems in West Africa. pages 79-100. in: Renard, C. (ed.) Crop Residues in Sustainable

Mixed Crop/livestock Farming Systems CAB International in association with ICRISAT and ILRI, Oxon, U.K.

Singh B.B. and Emechebe, A.M. 1997. Advances in research on cowpea *Striga* and *Alectra* pages 215-230. *In:* Singh B.B., D.R. Mohan Raj, K. Dashiell and L.E.N. Jackai (eds.) 1997. Advances in cowpea research. Copublication of International Institute of Tropical Agriculture (IITA) and Japan International Center for Agricultural Sciences (JIRCAS), IITA, Ibadan, Nigeria.

Singh, B.B., Chambliss, O. and Sharma, B. 1997a. Recent advances in cowpea breeding, pages 30-49. in: Singh B.B., D.R. Mohan Raj, K. Dashiell and L.E.N. Jackai (eds.) 1997. Advances in cowpea research. Copublication of International Institute of Tropical Agriculture (IITA) and Japan International Center for Agricultural Sciences (JIRCAS), IITA, Ibadan, Nigeria.

Singh, B.B. and Emechebe, A.M. 1998. Increasing productivity of millet cowpea intercropping systems. Pages 68-75 in: pearl millet in Nigeria agriculture: production, utilization and research priorities. Proceedings of the Pre-season Planning Meeting for the Nationally Coordinated Research Programme for pearl Millet, Maiduguri, 21-24 April 1997. (Emechebe, A.M., Ikwelle, M.C., Ajayi, O., Aminu Kano, M., and Anaso, A.B. eds.). Lake Chad Research institute, P.M.B. 1239, Maiduguri, Nigeria.

Singh, B.B., Mai-Kodomi, Y. and Terao, T. 1999. A simple screening method for drought tolerance in cowpea. *Indian J. Genet.*, **59:** 211-220.

Singh, B.B. 2001a. Genetic variability for drought tolerance, heat tolerance and root architecture in cowpea. *African Crop Science Proceedings,* **5:** 47-50.

Singh, B.B. 2001b. Genetic variability for physical properties of cowpea seeds and their effect on cooking quality. *African Crop Science Proceedings*, **5:** 43-46.

Singh, B.B. 2002. Breeding cowpea varieties for resistance to *Striga gesnerioides* and *Alectra vogelii*. Pages 154-166. In Fatokun, C.A., S.A. Tarawali, B.B. Singh, P.M. Kormawa and M. Tamo (editors).2002. Challenges and opportunities for enhancing sustainable cowpea production. IITA, Ibadan, Nigeria.

Singh, B.B. and Ajeigbe, H.A.. 2002. Improving cowpea-cereals-based cropping systems in the dry savannas of West Africa. Pages 278-286. In Fatokun, C.A., S.A. Tarawali, B.B. Singh, P.M. Kormawa and M. Tamo (editors).2002. Challenges and opportunities for enhancing sustainable cowpea production. IITA, Ibadan, Nigeria.

Singh, B.B. and Matsui, T. 2002. Breeding cowpea varieties for drought tolerance. Pges 287-300. In Fatokun, C.A., S.A. Tarawali, B.B. Singh, P.M. Kormawa and M. Tamo (editors).2002. Challenges and opportunities for enhancing sustainable cowpea production. IITA, Ibadan, Nigeria.

Singh, B.B., Hartmann, P., Fatokun, C., Tamo, M., Tarawali S.A. and Ortiz, R. 2003. Recent progress in cowpea improvement. *Chronica Horticulturae,* **43:**8-12

Singh, B.B. 2005. Cowpea [*Vigna unguiculata* (L.) Walp. Pages 117-162. In Singh, R.J. and Jauhar, P.P. (editors). Genetic Resources, Chromosome Engineering and Crop Improvement. Volume 1. 2005. CRC Press, Boca Raton, Florida, USA.

Singh, S.R. 1977. Cowpea cultivars resistant to insect pests in world germplsam collection. Tropical Grain Legume Bulletin **9:** 3-7.

Singh, S.R. 1980. Biology of cowpea pests and potential for host plant resistance. In Harris, M.K. (eds.) Biology and Breeding for Resistance to Atrhropods and Pathogens in Agricultural Plants. Texas A and M University Bulletin, USA.

Singh, S.R., and Allen, D.J. 1979. Cowpea Pests and Diseases. Manual Series No 2. IITA, Ibadan, Nigeria.

Singh, S.R., Singh, B.B, Jackai, L.E.N. and Ntare, B.R. 1983. Cowpea Research at IITA. Information Series No 14. IITA, Ibadan, Nigeria.

Sonnante, G., Piergiovanni, A.R. Ng, N.G. and Perrino, P. 1997. Isozyme markers and taxonomic relationships among *Vigna* species. Pages 58-65. In: Singh B.B, Mohan Raj., D.R., Dashiell, K., and Jackai, L.E.N. (Eds.) 1997. Advances in cowpea research. Copublication of International Institute of Tropical Agriculture (IITA) and Japan International Centre for Agricultural Sciences (JIRCSA), IITA, Ibadan,

Nigeria.

Steele, W.M. and Mehra, K.L. 1980. Structure, evolution and adaptation to farming system and environment in *Vigna*. Pages 393-404 in Advances in legume science, edited by R.J. Summerfield, and A.H. Bunting. Her Majesty's Stationery Office (HMSO), London, UK.

Steele, W.M., Allen, D.J. and Summerfield, R. 1985. Cowpea (*Vigna unguiculata* L. Walp.) In Summerfield, R.J. and Roberts, E.H. (eds.) Grain Legume Crops. Collins, London, UK.

Toure, M.A. and Singh, B.B. 2005. Registration of 'Sangaraka' cowpea. *Crop Science* **45 :** 2648

Boxtel,J., Singh, B.B. Thottappilly, G . and Maule, A.J. 2000. Resistance of (*Vigna unguiculata* (L.) Walp.) breeding lines to blackeye cowpea mosaic and cowpea aphid borne mosaic poty virus isolates under experimental conditions. *Journal of Plant Disease and Protection,* **107:** 197-204.

Tyagi, D.K. and Chawala, H.S.1999. Effect of season and hormones on in vitro hybrid development between *Vigna radiata* and *Vigna unguic. Hungarica,* **47:** 147-154.

Van Boxtel, J., Singh, B.B., Thottappilly, G. and Maule, A.J. 2000. Resistance of (*Vigna unguiculata* (L.) Walp.) breeding lines to blackeye cowpea mosaic and cowpea aphid borne mosaic poty virus isolates under experimental conditions. *Journal of Plant Disease and Protection,* **107:** 197-204.

Van EK, G.A., Henriet, J., Blade, S.F. and Singh, B.B. 1997. Quantitative assessment of traditional cropping systems in the Sudan savanna of northern Nigeria: II Management and productivity of major cropping systems. *Samaru Journal of Agricultural Research,* **14:** 47-60.

Verdcourt, B. 1970. Studies in the *Leguminosae-Papilionoideae* for the Flora of Tropical East Africa. IV. *Kew Bull.,* **24:** 507-69.

Vaillancourt, R.E. and N.F. Weeden. 1992. Chloroplast DNA suggests Nigerian center of domestication for the cowpea, *V. unguiculata*, Leguminlsae. *Am. J.Bot.*, **79:** 1194-1199

Vaillancourt, R.E., Weeden, N.F. and Barnard, J. 1993. Isozyme diversity in the cowpea species complex (*Vigna unguiculata*). *Crop Science,* **33:** 606-613.

Viswanathan, P.L., Murugesan, S., Ramamoorty, N., Veerabadran, P. Jehangir, K.S., Natarajan, N., Dhanakodi, C.V., Vamban, P.P.R.C. 1997. Vamban 1, a new cowpea variety for Tamil Nadu. *Madras Agricultural Jouranl.* **84(5):** 271-272.

Walker, C.A. Jr. and Chambliss, O.L. 1981. Inheritance of resistance to blackeye cowpea mosaic virus in cowpea [*Vigna unguiculata* (L.) Walp.]. *Journal of the American Society of Hortcultural Science* **106 (4)**, 410-412.

Westphal, E. 1974. Pulses in Ethiopia. Their taxonomy and agricultural Research Report, Center for Agricultural Publishing and Documentation, Wageningen, Netherlands.

Williams, R.J. 1975. International testing program: Aims, progress and problems seen from IITA. In Luse, R.A., and Rachie, K.O. (eds.) Proceedings of IITA Meeting on Grain Legume Improvement June 1975. IITA, Ibadan, Nigeria.

Williams, R.J. 1977a. Identification of multiple disease resistance in cowpea. *Trop. Agric Trinidad* 54: **53-60.**

Williams, R.J. 1977b. Identification of resistance to cowpea yellow mosaic virus. *Trop. Agric. Trinidad* **56;** 61-67.

Yunes, A.A., Andrade, M.T. – de, Sales, M.P., Morais, R.A., Fernandes, K.V.S., Gomes, V.M., Xavier-Filho, J., De Andrade, M.T. 1998. Legume seed cicilins (7S storage proteins) interfere with the development of the cowpea weevil (Callosobruchus maculates (F). *Journal of the Science of Food and Agriculture*, **76:** 1, 111-116.

□□□

Chapter – 17

Genetic Resources for Cole Crops Improvement

S.R. Sharma and Chander Parkash

Introduction

'Cole crops' is a general term used to describe several vegetables in the Brassicaceae family. It is an important and highly diversified group of vegetable crops grown worldwide that belongs to Brassica oleraçea (Monteiro and Lunn, 1998) and includes particularly six Brassica crops (cabbage, cauliflower, Brussels sprouts, broccoli, kale and knol kohl). All of these crops can trace their history to a common ancestry of wild cabbage originating in the Mediterranean and Asia Minor regions. Cole crops are one of the oldest cultivated plant groups in the world. They are cultivated in Europe since very ancient time from where they have spread to other parts of the world (Nieuwhof, 1969). All the cole crops (except few cauliflower, broccoli and cabbage types) require chilling temperatures for about two months to transform from vegetative phase into reproductive phase. However, the period of chilling requirement differs according to the kind of crop and varieties. Cole crops are also commonly called as European or biennial vegetables due to their cool temperatures requirements and two season requirement to complete their life cycle.

Major Resources of Cole Crop Germplasm

Germplasm is any form of the hereditary material from an organism. Genetic resources encompass all the various forms of germplasm that are available for collection, storage and use. The genetic diversity of crops, represented by traditional local cultivars and wild relatives has been disappearing rapidly during recent decades. Initial concern was noted by H.V. Harlan in the 1930s (Harlan and Martini, 1936) but by the 1960s, modern plant breeding and land-use changes accelerated loss and stimulated the realization that genetic resources in regions of diversity were being lost at an alarming rate. To conserve the genetic variation within the cole group, a comprehensive base collection of the cultivated forms of Brassica oleracea has been established at the vegetable gene bank of Horticulture Research International (HRI), Wellesbourne, (U.K.). Gene banks have also been established at Instiuut de veredeling van Tuinbouwgewssen, Wageningen, the Netherlands and the *Instituto del Germoplasm*, Bari, Italy (Van der Meer *et al.*, 1984) for conservation of genetic resources. The International Board for Plant Genetic Resources (IBPGR) has recognized HRI as an international centre for conservation of Brassica oleracea (Innes, 1975). Large number of collections of cole crops is also available with the United States Department of Agriculture, Plant Introduction Service. In India, National Bureau of Plant Genetic Resources (NBPGR), New Delhi has been assigned the duty of conserving the germplasm of all crops, including vegetables. Germplasm of all the cole crops is being maintained at the Indian Agricultural Research Institute, New Delhi (tropical types) and its regional station at Katrain (temperate types). Other major institutes maintaining germplasm of cole crops include Punjab Agriculture University, Ludhiana, G. B. Pant University of Agriculture & Technology, Pantnagar, Indian Institute of Vegetable Research, Varanasi, Dr. Y.S. Parmar University of Horticulture and Forestry, Solan and SKUAS&T, Srinagar.

Collections in gene banks are genetic resources, generally available for research and breeding purposes. The European database for Brassica, abbreviated 'Bras-EDB', was developed by the Centre for Genetic Resources the Netherlands (CGN). The objectives of this database are to support rationalization of Brassica germplasm conservation and to improve the access to the Brassica germplasm for users. The database focuses on passport data of the genus Brassica maintained in European collections. Also characterization data of a restricted number of descriptors are currently included. By September 1997 the database contained 13000 accessions, from 21 collections in 17 countries. The species B. oleracea is represented with the highest number of accessions (57 %), followed by *B. napus* (17 %) and *B. rapa* (16 %). The database includes cultivated as well as wild material.

Crisp *et al.* (1989) proposed schematic screening of genetic resources for pests and diseases for various Brassica oleracea crops in relation to club root resistance. Broccoli germplasm have been evaluated for resistance to downy mildew. It is possible that similar schemes could be developed for all brassica crops. Crisp and Astley (1985) reported that limited taxonomic work on cultivated vegetables is hindering the effective management of genetic resources.

A genetic material known as Kose cabbage is raised locally in Trachya region of Turkey. A breeding study was conducted on this local variety and some plants with superior characteristics were developed, but it couldn't be developed as a new variety. Study on another local head cabbage population in Van province in 'Ercis' Local population, the highest ratio of head formation, number of outer leaves covering head and leaf thickness were found as 86.6 %, 2-3 and 0.58-9.02 mm, respectively (Yasar *et al.*, 1995). In another study, the cabbage gene sources of Turkey were determined during 1998-2000 (Yanmaz *et al.*, 2000).

Kale is widely grown in the Black Sea Region of Turkey. In the region, kale is one of the most important crops, mainly for use as green vegetable as fresh or cooked and also farmers often use the most tender leaves for feeding of farm animals (Balkaya, 2002). A high degree of diversity is maintained within kale populations in the Black Sea Region and therefore, it is still possible to collect valuable germplasm. Kale genetic resources were collected in 2001 from the Black sea Region and evaluated according to their morphological characters (Balkaya *et al.*, 2003). A totally of 127 kale genetic materials were investigated in Samsun province in the Black-Sea Region. In the following observations 22 genotypes were selected as superior types using weight based ranking method. The other genetic resources were taken preservation as breeding materials at Turkey seed gene banks in Izmir-Menemen. Now, selfing is continued for developing new kale varieties in kale breeding program.

Concerted international activities to collect and preserve crop genetic resources were initiated by the Food and Agriculture Organization (FAO) of the United Nations. Technical conferences organized by FAO in 1961, 1967 and 1973 stimulated the global awareness with regard to conservation of genetic resources. In 1971, the Consultative Group on International Agricultural Research (CGIAR) was formed with co-sponsorship from FAO, the United Nations Development Program (UNDP) and the World Bank. The 1973 FAO Technical Conference and a United Nations Environment Conference in Stockholm in 1972, led to recommendations for a global program. The International Board for Plant Genetic Resources (IBPGR) was subsequently established in 1974 by the CGIAR as one of 13 International Agricultural Research Centres (IARC). As a secretariat organization, IBPGR was charged with salvaging threatened germplasm and ensuring that it was described, documented and conserved (Williams, 1989).

Germplasm Introductions into India

The historical facts reveal that most of the European (temperate) vegetables were introduced in India by the Portuguese during the 18th Century and little later by the English during 18th and 19th Centuries. Dr. Jenson, a botanist from 'KEW' introduced cauliflower varieties in 1822 and grew it in Botanical Garden, Saharnpur. That remains the basis for bringing out Indian cauliflower varieties. Though the credit for developing heat tolerant tropical cauliflower commonly known as Indian or Asiatic cauliflower goes to the farmers of eastern Uttar Pradesh and Bihar. The first four Indian varieties listed by Sutton and Sons in 1929 were Early and Main Crop, Patna and Early and Main Crop, Banaras. The Royal Agri-Horticultural Society, Calcutta introduced seeds of English vegetables from South Africa during 1824. English rulers also brought seeds of European vegetables to India during their regimes.

European vegetable breeding in India has a short history of about 5 decades plus. Earlier to Second World War (1939-45), the seeds of all temperate vegetable crops were being imported from abroad mainly by England based seed company M/S Sutton's and Sons. Later they established their office in India in 1916. During the war, there were difficulties in the import of seeds and therefore, the then Imperial Government in India established a research station at Quetta (Baluchistan) in 1942-43. After the partition of India in 1947, the Quetta centre went to Pakistan and the Government of India on the recommendations of an Expert's committee set up a Central Vegetable Breeding Sub-Station at Katrain in Kullu valley (H.P.) in May 1949. The main objectives were production and distribution of quality seeds of temperate vegetables including cole crops, and to give advice to the growers on these aspects.

In view to intensity the improvement work on temperate vegetables, the Katrain Station was transferred to the Indian Agricultural Research Institute, New Delhi in 1955 with its changed name as Vegetable Research Station and now as IARI, Regional Station, Katrain (Kullu Valley) H.P.-175129. Soon after the transfer of the station to IARI, breeding work was taken up under the leadership of Late Dr. Har Bhajan Singh (1955-1957) who also contributed greatly to the standardization of seed production technology. In 1958, the late cauliflower seed was produced for the first time in India. Indigenous production of temperate vegetable seeds is a unique example of import substitution leading to the development of viable vegetable seed industry in Himachal Pradesh. Government of India launched National Seed Project in 1976-77 with the financial aid from the World Bank under which the Katrain and Solan centres were strengthened for temperate vegetable seed production and processing. Besides, Kullu Valley and Kashmir Valley, seed production was also done in Kalpa Valley (H.P.), Kalimpong (Darjeeling), Saproon Valley (H.P.) and Nilgiri hills (T.N.) during 1950 and 1960 but this was not in organized manner.

Vegetable research further received a boost with the establishment of the Division of Horticulture at IARI, New Delhi in 1956-57 and subsequently by the Division of Vegetable Crops and Floriculture which came into existence in 1970 and Division of Vegetable Crops in 1982, and Katrain Centre was strengthened. The National Seed Corporation was established in 1963 to produce foundation and certified seeds and took seed production of temperate vegetables in Kashmir and Kullu valleys. The organizations like National Horticultural Research and Development Foundation (NHRDF) founded in 1977 by NAFED has been doing seed Production in Kullu valley. Under ICAR, All India Coordinated Vegetable Improvement Project came in to being in 1970-71 and gave further fillip to temperate vegetable research having main centres at Katrain, Srinagar and Solan and voluntary centres at Almora (VPKAS), Pithoragarh (DARL) and Ranichauri (GBPUA&T). The Vivekanand Parvatiya Krishi Anusandhan Research Sansthan, Almora after transfer to ICAR in 1974, ICAR Complex for NEH region, Shillong; National

Bureau of Plant Genetic Resources, New Delhi established in 1976 at its Shimla centre and others; Central Institute of Temperate Horticulture, Srinagar created in 1996 have been doing researches on temperate vegetables.

The private seed companies have been developing their own R & D programmes. Many of them are getting temperate vegetable seeds in OGL from abroad and are selling. Similarly NGOs have their involvement. National Horticulture Board, Gurgaon has been financing to the different government and non-government organizations to create facilities for R&D. The visionary National Agricultural Technology Project (NATP), the largest project ever funded by the World Bank also included some research programmes on temperate vegetable research particularly a project on improvement of cole crops entitled "Development of insect pests and disease resistant superior varieties of cabbage and cauliflower".

India has collaborated programme with AVRDC, China; and USSR. There is a South Asian Vegetable Research Network (SAVERNET) of the countries like India, Pakistan, Bangladesh, Srilanka, Nepal and Bhutan which deals with the testing of elite vegetable entries.

A good number of collections have been made on cole crops at the IARI, RS, Katrain since its inception till date (Verma and Chander Parkash, 2006). Forty nine germplasm of cabbage collected from HRI, Wellesbourne (U.K.) were evaluated for important horticultural traits and resistance to black rot (Chander Parkash, 2005). Some of the promising collections/introductions are listed below:

S.No.	Crop	Total no. of collections made	Promising ones
1.	Cabbage	1050	Golden acre, Pride of India, Early Drum Head, Copenhagan Market, Late Drum Head, Spitzkool, MR-1, AC-204, AC-208, AC-238, EC-240613, EC-173419 (CMS), EC-490162, EC-490165, EC-490176, EC-490185, EC-490191, EC-490200, EC 187228-187230 (club root resistant), EC 287707-287708 (resistant to yellows and tolerant to black rot)
2.	Cauliflower	724	Snowball No.16, Janavon, EC-103576, EC-162587, EC-12012, Puakea, EC 205372, 205373, EC 175800-175806 (multi-disease resistant)
3.	Brussels sprout	35	Hild's Ideal, Rubine
4.	Knol-kohl	39	White Vienna, Purple Vienna
5.	Broccoli	97	EC-005559
6.	Kale	22	K-64, K-5, K-1, K-Green

The main countries from where the cole crops have been introduced consist of Russia, Canada, USA, Australia, Brazil, France, U.K, Netherlands, Germany, Italy, Poland, Hungary, Denmark, Turkey, Yugoslavia, Egypt, Zambia, Israel, Korea, Japan, Taiwan etc.

Characterization & Evaluation of Germplasm for Biotic Stresses

Breeding for Disease Resistance

Black Rot

In cauliflower, Sn 445, MGS 2-3, Puakea, EC 162587, Lawyana and RBS-1 (Sharma *et al.,* 1972 & 1995), Sel-12 (Gill,1993), Sel-6-1-2-1 and Sel-1-6-1-4 (Chatterjee, 1993), Sakata 6, Takki's February, Nazarki Early, Henderson's Y-76 and Y-77 (Moffett *et al.,* 1976), Avans and Igloory (Dua *et al.,* 1978) have been identified as resistant sources against black rot. Pusa hybrid-2 (Singh *et al.,* 1994) and Pusa Snowball K-25 have been released as resistant cultivars.Green Hammer, EC-10109, Spitzkool, MR-1, AC-204 and AC-208 are some of the sources for black rot resistance in cabbage. Pusa Mukta (EC 24855 x EC 10109) has been released as resistant cultivar.

Downy Mildew

Cauliflower genotypes, Igloo, Snowball Y, Dok Elgon and RS 355 (Kontaxix *et al.,* 1979), BR-2, CC and 3-5-1-1; EC 177283, 191150, 191157, 191140, 191190, 191179, Kibigiant, Merogiant and Noveimbrina (Singh *et al.*, 1987; Mahajan *et al.,* 1991); MGS 2-3, 1-6-1-4, 1-6-1-2 and 12 C (Chatterjee, 1993), KT-9 (Sharma *et al.,* 1991), Early Winter Adam's White Head (Sharma *et al.,* 1995): CC-13, KT-8, xx, 3-5-1-1, CC (Trivedi *et al.,* 2000), Perfection, K-1079, K-102, 9311 F_1 and 9306 F_1 (Jensen *et al.,* 1999), Kunwari-7, Kunwari-8, Kunwari-4 and First Early Luxmi (Pandey *et al.*, 2001) were reported resistant to moderately resistasnt. Pusa Hybrid-2 of Indian Cauliflower (Singh *et al.,* 1994) and Pusa Snowball K-25 of Snowball type having resistance to downy mildew were released for cultivation in India. Cabbge genotype MR-1 and spitzkool have been found resistant to downy mildew.

Sclerotinia Rot

Cauliflower genotypes, EC-103576, EC- 162587, EC-131592, Janavon, Early Winter Adam's White Head, EC-177283 and Kn-81 possesses moderate resistance to sclerotinia rot (Baswana *et al.*, 1991; Sharma *et al.,* 1995 and 1997). Pusa Snowball K-25 developed by using EC 103576 as resistant source with Pusa Snowball-1 has been released for commercial cultivation which possesses field resistance against this disease. In cabbage, MR-1 has shown field resistance to sclerotinia rot.

Other Diseases

MR-1, cabbage carries multiple resistances to soft rot, wire stem, black rot and sclerotinia rot. AC 238, Spitzkool and EC 93559 have been found resistant against cabbage yellows. MGS 2-3, Puakea, 1-6-1-4, 6-1-2-1 and Pusa Subhra in cauliflower have resistance against alternaria black spot disease. Cauliflower selection-1301 has high degree of tolerance to diseases like sclerotinia rot, downy mildew and black rot coupled with desirable horticultural traits.

Breeding for Pest Resistance

In cabbage, cultivars Large Blood Red and Red Pickling (all red types) were identified to possess preferential resistance against cabbage butterfly caterpillars (Pieris brassicae) and susceptible to aphid (Brevicoryne brassicae). The white cabbage varieties All Season and Round Sure Head showed resistance to these pests and vice versa. Cabbage IRCH-4 to 6 and KK Cross have shown tolerance to aphids.

Under the National Agricultural Technology Project entitled" Development of insect pests and disease resistant superior varieties of cabbage and cauliflower" a total of 118 genotypes of cauliflower and 21 of cabbage have been deposited with the NBPGR, New Delhi. This included sources of resistance and lines having resistance/tolerance to diseases viz., black rot, sclerotinia rot, downy mildew and atlernaria leaf spot and important pests like diamondback moth and cabbage butterfly. Morphological characterization of all the notified varieties of cabbage and cauliflower has been undertaken under the DUS testing project.

Utilization of Genetic Resources for Cole Crop Improvement

1. Cauliflower (*Brassica oleracea* var. *botrytis*)

Among the cole crops, cauliflower follows cabbage in importance with regards to area and production in the world. However, in India cauliflower is more widely grown both in the hills and plains from 10^0 N to 35^0 N. Cauliflowers cultivated in the Indian plains can be broadly classed into four maturity groups (Group I, II, III and IV) depending upon the time of curd availability of these groups, the first three are typically Indian cauliflowers while the fourth is of Snowball, Erfurt or Alpha types. Seeds of Indian cauliflowers can be produced in the northern plains but snowball types do not set seed there and require mild temperatures at the time of flowering and seed setting.

Crop improvement work on Indian cauliflower is going on in most of the Agricultural universities and ICAR institutes. However, the major centres are IARI, New Delhi, PAU, Ludhiana, and GBPUA&T, Pantnagar. The work for improvement of snowball type is mainly done at IARI, Regional Station, Katrain and Dr. Y. S. Parmar UHF, Solan. Important varieties of cauliflower belonging to different maturity groups and recommended by the AICRP (VC) are listed below:

Important varieties of cauliflower of different Maturity group

Variety	Maturity Group	Developing Centre	Year of Identification
Early Kunwari	Early	PAU, Ludhiana	1975
327-14-8-3	Early	IARI, New Delhi	1975
Pusa Deepti (351-4-1	Early	IARI, New Delhi	1975
Improved Japanese	Mid early	IARI, New Delhi	1975

Contd....

Pusa Synthetic (Synthetic-1)	Mid late	IARI, New Delhi	1975
Pusa Snowball-2 (EC 12012	Late	IARI, RS, Katrain	1975
Pusa Snowball-1	Late	IARI, RS, Katrain	1975
Pusa Snowball K-1	Late	IARI, RS, Katrain	1979
Pant Gobhi (114-5-1)	Early	GBPUA&T, Pantnagar	1981
Pusa Subhra (Line 6-1-2-1)	Mid late	IARI, New Delhi	1985
Early Synthetic	Early	IARI, New Delhi	1990
Pant Subbra (235-S)	Mid late	GBPUA& T, Pantnagar	1990
Pusa Hybrid-2	Mid early	IARI, New Delhi	1992
Pusa Snowball K-25	Late	IARI, RS, Katrain	2002
Pusa Kartik Sankar (DCH-541)	Early	IARI, New Delhi	2003
SYCFH 202	Early	Syngenta	2004
Summer King	Early	Sungrow	2004
SYCFH 203	Early	Syngenta	2005
DC-76	Mid late	IARI, New Delhi	2008

Varieties

Cauliflower is a thermo sensitive crop. Varieties differ in their temperature requirement for curd formation and development. They have been classified into different maturity groups according to their temperature requirement. The sowing and transplanting time have to be adjusted so that the varieties are ready for harvest at specified period in the north Indian plains (Table 1).

Table 1. Maturity groups in Cauliflower

Maturity		Sowing time*	Trans-planting time*	Temperature for curd development	Varieties
Early	September maturity (mid Sep.- mid Oct.) October maturity (mid Oct.- mid Nov.)	Mid-May May end-mid June	July beginning Mid July	22°C - 27°C 20°C - 25°C	Early kunwari*, Pusa Early Synthetic, Pusa Meghna , Pusa Kartik Shankar, Pusa Deepali, Pant Gobhi 2, Pant Gobhi-3, Pusa Katki
Mid early	November maturity (mid Nov.- mid Dec.)	July end	September beginning	16°C - 20°C	Pusa Sharad, Improved Japanese, Pusa Hybrid-2, Pant Gobhi-4,
Mid late	December maturity (mid Dec.- mid Jan.)	August end	September end	12°C - 16°C	Pusa Synthetic, Pusa Himjyoti, Pusa Shubhra, Pusa Paushija (DC-76), Pusa Shukti (DC-5)
Late	Snowball	Sept. end- mid	Oct. end-mid Nov.	10°C - 16°C	Pusa Snowball-1, Pusa Snowball-2, Pusa Snowball K-1, Pusa Snowball K-25

*Under North Indian plains conditions

Early Maturity Group

Early Kunwari: Early variety suitable for growing in Punjab, Haryana, Himachal Pradesh and Delhi. Ready to harvest from mid September to mid October. Curds are semi-spherical, loose and yellowish in colour.

Pusa Early Synthetic: Plants erect with bluish green leaves, curd small to medium in size, flat, creamy white and compact. Average yield is 11.7 t/ha.

Pant Gobhi 3: A synthetic variety. Plants with long stem, semi-erect leaves and hemispherical creamy white, medium compact non ricey curds. Yield 12 t/ha. Curds are ready for harvest in September.

Pusa Meghna: Belongs to extra early maturity group, suitable for growing under hot and humid climate. Curds compact, creamish white and medium in size, weighing about 350-400g.

Pusa Kartik Sankar: A hybrid variety resistant to downy mildew and can tolerate high temperature and high rainfall during its vegetative growth. Curds are medium sized, semi dome shaped, compact, retentive white with fine texture, weighing about 475g. It is free from bracts and riceyness.

Pusa Deepali: Plants medium tall, erect, bluish green and waxi leaves, curds compact, retentive white and medium in size with an average yield 10-12 t/ha.

Pant Gobhi 2: Recommended for cultivation in northern plains of the country. Curds are medium compact and yellowish. Yield potential is 10t/ha. Available in October in the plains.

Mid Early Maturity Group

Improved Japanese: An introduction from Israel. Plants erect, leaves bluish green, curds compact and creamish-white. Average yield is 16-18 t/ha.

Pusa Hybrid 2: First F_1 hybrid released by a public sector organization. Plants semi-erect with bluish-green upright leaves, resistant to downy mildew. Curds are creamy white, very compact, with an average yield of about 23 t/ha.

Pusa Sharad: Foliage bluish-green, leaf with narrow apex and prominent mid rib. Semi-dome shaped white and very compact curds. Average yield is 24 t/ha.

Pant Gobhi 4: A variety released for November maturity. It has medium long stem, semi-erect leaves, hemispherical creamy white, medium compact, non ricey curds. Average yield is 14 t/ha.

Mid Late Maturity Group

Pusa Synthetic: A synthetic variety, plants erect, frame narrow to medium, curds creamy white to white and compact. The yield potential is 27 t/ha.

Pant Shubhra: Released for cultivation in Bengal Assam basin and Sutlej Ganga Alluvial plains. Curds compact, slightly conical, retentive, creamish-white in colour, non ricey and non leafy. The yield potential is 25 t/ha.

Pusa Himjyoti: Erect bluish-green leaves and waxy coating; curds retentive white, self blanched, solid and 500-600g in weight. This is the only variety which can be grown from April-July in the hills. Also suitable for growing in December maturity group in north Indian plains.

Pusa Shukti (DC-5): High yielding variety and is recommended for growing in north Indian plains and hills. This variety produces attractive compact white curds. Tolerant to major pest and diseases and low frosting temperature. Average curd yield is about 30-33 t/ha.

Pusa Paushija (DC-76): High yielding variety maturing during 2nd fortnight of December to first fortnight of January. It has distinguished bluish green, narrow conical leaf top. Curds are compact, retentive white in colour weighing about 900 g with average curd yield of 31-41 t/ha. This variety is highly fertilizer responsive and gives stable performance under wider environmental conditions.

Late Maturity Group

Pusa Snowball-1: A late variety suitable for cool season. Leaves are straight and inner leaves cover the curd. Curds are compact, medium and snow white in colour. Ready for harvest in January- February in north Indian plains and during March- April in the hills. Yield potential is about 22-25 t/ha.

Pusa Snowball-2: Outer leaves upright, inner leaves cover the curd, milky white curd, very compact, availability from January to March, matures in 90-95 days. Yield 27 t/ha. The variety could not become popular because of its shy seed setting.

Pusa Snowball K-1: Among the snowball types grown in the country, it has best quality curds which are snow white and retain it even if harvesting is delayed. The leaves are puckered, serrated and light green in colour. It is late in maturity by about a week than Pusa Snowball-1. It is tolerant to black rot disease. Average yield is 25-30 t/ha.

Pusa Snowball K-25: Leaves waxy, upright, slightly bending towards inner side with puckered margins inner leaves cover the curds initially. Curds very solid, medium sized, white with good staying power. Very late and the curds available from end of January to end of March when the temperature ranges from 10-15^0 C. Yield 175-300 q/ha. Suitable for transplanting from October to early November in North India. Resistant to black rot and tolerant to sclerotinia rot diseases.

Varieties on the Basis of Colour

Orange cauliflower: Orange cauliflower (*B. oleracea* L. var. *botrytis*) contains 25 times the level of Vitamin A of white varieties. This trait came from a natural mutant found in a cauliflower field in Canada. Cultivars include 'Cheddar' and 'Orange Bouquet'.

Green cauliflower: Green cauliflower is sometimes called broccoflower. It is available both with the normal curd shape and a variant spiky curd called "Romanesco broccoli" Both types have been commercially available in the US and Europe since the early 1990s. Romanesco's head is an example of a fractal image in nature, repeating itself in self-similarity at varying scales. Green curded varieties include 'Alverda', 'Green Goddess' and 'Vorda'. Romanesco varieties include 'Minaret', and 'Veronica'.

Purple cauliflower: Purple cauliflower also exists. The purple color is caused by the presence of the antioxidant group anthocyanin, which can also be found in red cabbage and red wine. Varieties include 'Graffiti' and 'Purple Cape'. In Great Britain and southern Italy, a broccoli with tiny flower buds is sold as a vegetable under the name "purple cauliflower." It is not the same as standard cauliflower with a purple curd.

Though number of hybrids are marketed by Private seed companies it is heartening to note that no hybrid has been released by the public sector in the snowball group but efforts are being made at Katrain centre to develop S.I and CMS lines and also at HPKV, Palampur to develop S.I. lines. Some of the hybrids are Himani, Guardian (Indo-American), Candid Charm, White Flesh, Cashmere(Sakata), Serrano (Sandoz), Early Himlata, Early Himanagiri (Century), Nath Ujwala, Nath Shweta (Nath), Namdhari 84 (Namdhari), Pusa Hybrid-2 (IARI), Pusa Kartik Sankar (IARI) ofcourse belong to variable groups.

S.I. lines are available in early and mid groups of cauliflower but the success is meagre in snowball types. Pusa Hybrid-2 and Pusa Kartik Sankar have been developed by using S.I. lines. The Ogura CMS system has been successfully transferred into various lines of different groups of cauliflower. Sera K-1, Sera K-25 and Call B-1 having Ogu CMS system have been developed in snowball types at the IARI, RS, Katrain. These lines are being used in development of hybrids and two hybrids KTH-1 and KTH-2 are already under testing in AICRP (VC) trials. In early maturity group 3 CMS lines (MS 01, MS 04 and MS 05) and in mid group 2 CMS lines (MS 09 and MS 10) have been developed and are being used in development of hybrids in respective groups.

2. Cabbage (*Brassica oleracea* var. *capitata L*)

Cabbage is an important vegetable crop grown in many countries of the world. Historical evidence indicates that modern hard-head cabbage cultivars are descended from wild non-heading brassicas originating in the eastern Mediterranean. It is commonly accepted that the origin of cabbage are North European Countries, the Baltic Sea coast (Baldwin, 1995; Monteiro and Lunn, 1998) and the Mediterranean region. The cultivated types found in cabbage group have been spread out to other regions of the World from this region. The Latin name Brassica is derived from the Celtic word bresic, meaning cabbage (Dickson and Wallace, 1986).

Cabbage is the one among cole crops where F_1 hybrids have become very popular. Public sector has also come forward and the first indigenous hybrid 'KGMR-1' developed by utilizing SI line has been identified for release during 2005 under the AICRP (VC) from the IARI, RS, Katrain. The requirement of F_1 hybrid seed is very high (> 150 q) and being taken care of through OGL by the private seed merchants and National and State Seed Corporations. Some of hybrids have been identified through AICVIP i.e. Pusa Synthetic (from IARI Katrain in 1992), Sri Ganesh Gol (from MAHYCO in 1992), Nath-401 (from Nath Seeds in 1993), BSS-32 in the name of Swarna (from Bejo Sheetal in 1995), Nath 501 (from Nath Seeds in 1997), and Quisto (from Syngenta in 1998). The other hybrids (Bahar, Pragati and Unnati from Pro Agro; Kalyani, Kranti and Hari Rani Gol from Mahyco; Hero, Mitra and Aditya from Sungrow; Yamuma, Ganga and Kaveri from IAHS; Masrgan, Meenaxi and Kuwaxi from Century; Vishesh and Uttam from Hindustan Lever; H-30, 50, 10 and 20 from NSC; Gloria, Runa and Ratan from Dechan-Feldt; Rare Ball from Kaneko; Green Challenger from Hungnong are brought under cultivation without testing under AICVIP.

Similarly, many heat tolerant cabbage hybrids were introduced e.g. Golden Cross, KK Cross, OS Cross, Resistlaka, Green Cornet, Autumn Queen, Green Ball (Takii); Green Boy, Green Express, Herculis, Stone Head, Regalia (Sakata). Japanese seed companies have also undertaken seed multiplication of their S.I. lines in South Asian Countries. Another significant achievement in cabbage breeding is the development of tropical lines e.g. DTC-507-4, 513, 528 (released as Pusa Ageti) etc. at the IARI, New Delhi. Cytoplasmic male sterility system (CMS) was transferred successfully from EC 173419 to all the cole crops at IARI Katrain but this system suffered from many defects (Chander Parkash, 2008) and therefore discarded. New improved ogura CMS system obtained from IARI, New Delhi has now been introduced into 5 promising lines/varieties (MS-01, MS-02, MS-03, MS-04 and MS-05) of cabbage and is ready to be utilized for developing hybrids. Five stable S.I. lines (CS-831, CS-835, CS-208, CS-5 and CS-6) of cabbage have been isolated and being utilized for the development of F_1 hybrids at the IARI, RS, Katrain.

List of popular cabbage hybrids in India

	Hybrid	Developed by
Through AICRP (VC)	Green Emperor	Tokita Seeds
	KGMR-1	IARI, RS, Katrain
	Quisto	Syngenta
	BSS-32	Bejo Sheetal
	Nath 501	Nath Seeds
	Nath-401	Nath Seeds
	Sri Ganesh Gol	MAHYCO
Others		
	Green Express, Regalia, Stone	Sakatia
	OS Cross, KK Cross, Green Cornet	Takii
	GS-403, Goldy Ball	Golden Seeds
	Bahar, Unnati, Pragati	Pro Agro
	Kranti, Sumit, Hari Rani Gol	MAHYCO
	Varun	Choice

Varieties

Many open pollinated varieties of cabbage differing in maturity, head shape and size, colour and shape of leaves are available. The popular varieties grown are listed below:

Golden acre: This is an introduction. The plants have shorter frame, few outer leaves, short stem and small cup shaped leaves. Heads is round compact with interior white of excellent quality. It is an early variety and yields 25 t/ha.

Pusa Drum Head: It is a late maturing variety taking 80-90 days for head formation. Plants have wider frame, light green leaves, few outer leaves and short stem. Head is partially covered, less compact and flat weighing 3-4 kg. It possesses field resistance to black leg (Phoma lingam).

Pusa Mukta: This variety was developed at IARI, RS, Katrain by hybridization of EC-24855 x EC-10109. Plants have short stalk, medium plant spread, few outer leaves having wavy margin and puckered leaf blade. Heads are not perfectly round but slightly flat, moderately resistant to black rot. The variety gives higher yield than Golden Acre but bursts early in the field.

September: An introduction from Germany is most popular in the Nilgiri hills. Recommended by Tamil Nadu State Department of Agriculture. It has solid, round to slightly oblong heads weighing 3-5 kg. Leaves are dark green with wavy margin. The stalk is long and heads usually tilt on one side. It has a very good staying power in the field. It takes around 100 days to maturity.

Pusa Ageti: A tropical variety produces seeds under sub-tropical conditions, forms marketable heads at a temperature range of 15-30^0 C but day temperature should not be above 35^{0}C. Heads weigh 600-1200 g, gets ready for harvest in 75-90 days and yields 11-33 t/ha depending upon the time of planting.

Pride of India: The plant type of this variety in Similar to Golden Acre but it is about a week later in maturity. Head weighs between 1.0 kg-1.5 kg. This is an introduction recommended by Dr. Y. S. Parmar University of Horticulture and Forestry, Solan.

Kinner Red: This is a red cabbage variety recommended by the Dr. Y.S. Parmar UHF, Solan. Head is oval shaped and weighs 500-600 g and takes 70-80 days for head formation

Hybrid Varieties

KGMR-1: First public sector cabbage hybrid developed at Katrain was identified for release for Zone I and IV. Proposed name for release is Pusa Cabbage Hybrid-1. Its leaves are slightly cut at the margin, round, medium waxy and green. Head is very compact, attractive, fully covered with good staying capacity in the field and takes 60 days after transplanting to first harvest. It carries field resistance to black rot disease with the average yield of 40 t/ha.

BSS-32: This is from Bejo Sheetal Seeds Pvt. Ltd. Heads are round with average weight of 3 kg. It matures in 100 days.

Sri Ganesh Gol: This is from Maharastra Hybrid Seeds Co. Ltd. (MAHYCO). It has round, compact, attractive and bluish green heads with better staying power. It takes 90-95 days after sowing.

Kranti: Plants have shorter frame, heads are flattish, compact and weigh around 1 kg. It has also been developed by MAHAYCO.

Nath-401: This is from Nath seeds Ltd. It has uniform compact heads with better shelf life. Its yield is 50-70 t/ha.

BSS-50: This is from Bejo sheetal Seeds Pvt. Ltd. It is early in maturity (65 days) with dark green smooth and highly compact heads having average weight of 1 kg. It can tolerate high temperature of up to 36^0 C and is resistant to fusarium wilt. It can stay in the field for 45 days after maturity without splitting.

Quisto: It has been developed by Syngenta. It is a late variety but has tolerance to high temperatures. Heads are compact and weigh 1.5 – 2.0 kg.

Red Queen: This is a red cabbage hybrid produced by Crystal Crop Sciences Pvt. Ltd. by using CMS system. Heads are very compact, dark purple and weigh around 1.5 kg.

3. Kale (*Brassica oleracea* var. *acephala* L)

Kale is one of the oldest forms of cabbage, originating in the eastern Mediterranean. Kale is thought to have been used as a food crop as early as 2000 B.C. Theophrastus described a savoyed form of kale in 350 B.C. Travellers and immigrants through the ages have introduced this green vegetable to many parts of the World (Hodges, 2004).

Kale is commercially cultivated in Kashmir and to a limited extent in Jammu, Assam and Himachal Pradesh. Khanyari, Kawdari and Jamadary are some of the local cultivars grown in different parts of Kashmir. Other genotypes which showed promise for different characters across environments on the basis of stability analysis performed by Khan *et al.* (2008) for various economic traits were SH-K-28, SH-K-33 and SH-K-21. The variety Westo has been recommended from IARI Katrain. Recently some promising genotypes viz., K-64, K-5, K-1, K-Green have been selected from a collection of 22 germplasm of kale made under the ICAR network project "Improvement of under utilized vegetable crops".

4. Knol kohl (*Brassica oleracea* var. *gongylodes* L)

Knol kohl, a member of Brassica group, is a vegetable species with an edible knob on its stem. West Europe countries are considered as the homeland of this vegetable. White Vienna, Purple Vienna, Kyote No. 3 and King of North are the introduced promising varieties. Recently a new variety Pusa Virat (KKS-1) having larger knob size and little or no fibre at maturity has been developed at Katrain.

KKS-1 (Pusa Virat): This variety has shorter frame, dark green leaves, globular round knobs weighing 700-800 g and having little or no fibre at maturity. It is 7-10 days late in maturity than White Vienna but gives higher yield of 20 t/ha.

White Vienna: Its plants are dwarf, leaves medium green, knobs globulor and light green, flesh creamy white, tender with delicate flavours. It is an early variety maturing in 55-60 days.

Purple Vienna: An early variety which takes 55-60 days for knob formation. Foliage purple, knobs globular round, large with purple skin and light green flesh.

King of North: This variety 60-65 days to harvest after transplanting. It has dark green, flattish-round knobs. Dark green leaves are well spread over the knobs.

Palam Tender: It has been developed at CSKHPKV, Palampur through selection and released for cultivation in Himachal Pradesh.

5. Broccoli (*Brassica oleracea* var. *etalica* L)

The recommended varieties are Pusa Broccoli KTS-1(selection from EC-005559) from IARI Katrain, Palam Samridhi from CSKHPKV, Palampur, Punjab Broccoli-1 from PAU, Ludhiana, Green Head from UHF, Solan and Italian Green. All are green sprouting broccoli and have been developed from the introduced exotic material. A cauliflower x broccoli hybrid (EC 243384 from Taiwan) produced greenish cauliflower heads with taste and flavour of broccoli (Goutam *et al.*, 2000). Many hybrids like Southern Comet, Premium Crop, Clipper, Laser (early), Corsair, Excalibur Cruiser, Emerald Carona (mid season), Late Carona, Stiff, Kayak, Green Surf (late) are being imported for cultivation in India. Aishwarya, Fiesta, Green are some of the popular F_1 hybrids of sprouting broccoli from the private sector in India. In coloured broccoli types, Palam Vichitra (purple) and Palan Kanchan (yellowish) have been recommended by the CSKHPKV, Palampur. Under the ICAR network project "Improvement of under utilized vegetable crops" 25 germplasm of broccoli were collected and evaluated at the IARI, RS, Katrain. Besides improved CMS has also been transferred to 4 promising varieties.

Pusa Broccoli KTS-1: This is the first variety of sprouting broccoli developed in India by selection from a segregating exotic germplasm. Plants are medium tall having dark green and waxy foliage with slightly wavy margins. Heads are solid, green with small buds and slightly raised at the centre and weigh around 350-450 g. Matures in 90-105 days. The maturity period is reduced by 5-10 days in plains. Average yield is 16 t/ha.

Punjab Broccoli-1: Developed through selection by Punjab Agricultural University, Ludhiana and recommended for growing in Punjab. Foliage is dark green with smooth leaf surface and wavy margin with bluish tinge. Plant branched and bears large number of sprouts in the axil of leaves. It takes about 65-70 days from transplanting to harvest. Head and sprouts are dark green with bluish tinge. Average yield is 7 t/ha.

Palam Samridhi: A selection from exotic material, developed by CSKHPKV, Palampur. Compact, green heads free from yellow eyes and bracts, average head weight is 300-400 g. Plants branched, also bears sprouts in axils of leaves which develop in to small heads. Average yield varies from 15-20 t/ha. Ready for harvest in 85-90 days after transplanting. Mainly recommended for subtropical conditions.

Green Head: This variety has been recommended by the Dr. Y. S. Parmar University of Horticulture and Forestry, Solan. Heads are compact and dark green and weigh around 300 g.

6. Brussels sprouts (*Brassica oleracea* var. *gemmifera* L.)

The recommended varieties are Hild's Ideal and Rubine. The imported hybrids are Jade Cross, Doreman, Alkazar, Meron, Poster, Fortress, Predora, Rovoka, Somora, Ladosa etc.

Hild's Ideal: It has a plant height of 55-60 cm with 45-55 sprouts and 45-55 leaves/plant. The diameter of the sprouts is 7-8 cm each weighing about 6-7 g. The sprouts are compact and have good flavour. Average yield is 16 t/ha. First picking may be taken in 115 days and hereafter 3-4 pickings may be taken at 10 days intervals. It is the only recommended variety for cultivation in northern plains and hills of India.

Jade Cross (F_1 hybrid): Sprouts are firm, dark green, closely packed on long stems, can be grown under wide range of growing conditions, takes 90 days for first picking.

Future Prospects

Some of the unique genetic resources may not be favourable to standard varieties in terms of productivity and plant characteristics, but should be rescued from extinction. We must consider that conservation and maintenance of the valuable genetic materials is necessary, because these populations are an important source of diversity, which could be used in breeding programmes.

References

Baldwin, B., 1995. The history of cabbage. 2p. http://gardline.usask.ca/veg/cabbage.html

Balkaya, A., 2002. Kale growing. (In Turkish) Türk- Koop Ekin Dergisi. Yýl:**5 (18)**, 19s.

Balkaya, A., Yanmaz, R., Demir, E., Ergün, A., 2003. A research on collection of genetic resources, characterization of kale (*Brassica oleracea* var. *acephala*) and selection of suitable types for fresh consumption in The Black Sea Region. The Scientific and Technical Research (TUBÝTAK- Project No: TOGTAG 2826). 4th Research Development Results. (in Turkish)

Baswana, K.S., Rastogi, K.B. and Sharma, P.P. 1991. Inheritance of stalk rot resistance in cauliflower(Brassica oleracia var. botrytis). *Euphytica*, **57:** 93-96.

Chatterjee, S. S. 1993. Cole crops. *In*: Vegetable crops. Bose, T.K., Som, M.G., and Abir, J.(eds). Naya Prakash. India pp 125-223.

Crisp, P. and Astley, D. 1985. Genetic resources in vegetables. p. 281-310. In: G.E. Russel (ed.). Progress in Plant Breeding – 1. Butterworths, London.

Crisp, P. Crute, I.R. Sootherland, R.A., Angell, M. Bloor, K., Burgess, H. and Gourdan, P.L. 1989. The exploitation of genetic resources of Brassica oleracia in breeding for resistance to club root.(Plasmodiophora brassicae). *Euphytica,* **42:** 215.

Dickson, M. H. and Wallace, D. H., 1986. Cabbage Breeding. Breeding Vegetable Crops (Edited by M.J. Bassett). 395-432.

Dua , I. S., Suman, B. C. and Rao , A. V. (1978). Resistance of cauliflower (Brassica oleracia var. botrytis) to Xanthomonas compestris influenced by endogenous growth substances and relative growth rates. *Indian J. Exptl. Bio.* **16:** 488-491.

Gill, H. S. (1993). Improvement of cole crops. *In:* Advances in Horticulture vol. 5-Vegetable crops: Part 1. Eds.: K.L. Chadha and G. Kalloo. Malhotra Publishing House, New Delhi, India.

Harlan, J. R. and M. L. Martini. 1936. Problems and results in barley breeding. USDA Year book 1936. p. 303-306

Hodges, L., 2004. Kale: The "New" Old Vegetable. Commercial Vegetable Production. http://ianrpubs.unl.edu/horticulture/nf51.htm

Innes, N.L. 1975. Genetic conservation and the breeding of field vegetables for the United Kingdom. *Outlook Agric.* **8:** 301-305.

Jensen, B. D., Hockenhull, J. and Munk, L. (1999). Seedling and adult plant resistance to downy mildew in cauliflower (Brassica oleracea var. botrytis). *Pl. Pathology,* **48:** 604-612.

Kalloo, G. (1988). Vegetable breeding. Volume III. CRC Press, Inc. Boca Raton, Florida.

Khan, S. H., Ahmed, N., Jaheen, N., wani, K. P. and Hussain, K. (2008). Stability analysis for economic traits in kale (*Brassica oleracea* var. *acephala* L.). *Indian J. Genet.* **68(2):** 177-182.

Kontaxis, D. G., Mayberry, K. S. Rubatzky, V. S.(1979). Reaction of cauliflower cultivars to Downey mildew in imperial valley. *California Agriculture* **33:** 19

Mahajan, V., Gill, H. S. and Singh, R. (1991). Screening of cauliflower germ plasm against Downey mildew. *Cruciferae Newl.* **14/15**, 148-149.

Moffett. M. L., Trimboli, D. and Bonner, I. A. (1976), A bacterial leaf spot disease of several Brassica varieties. *Apps Newsletter*, **593:** 30-32.

Monteiro, A. A. and Lunn, T., 1998. Trends and Perspectives of Vegetable Brassica Breeding World-Wide. World Conference on Horticultural Research. 17-20 June 1998, Rome, Italy.

Nieuwhof, M. (1969). Cole crops. Leonard Hill, London.

Pandey, K. K., Pandey, P. K., Singh, B., Kalloo, G. and Kapoor, K. S. (2001). Sources of resistance to Downey mildew disease in Asiatic group of cauliflower. *Veg.Sci.* **28:** 55-57.

Parkash Chander, 2005. Performance of exotic collections of cabbage in the northwestern mid hill conditions of India. *Indian J. Pl. Genet. Resour.* **18(1):** 88-90.

Parkash Chander. 2008. Effectiveness of self-incompatibility and cytoplasmic male sterility systems in developing F_1 hybrids of cabbage. *Int. J. Plant Sci.* **3(2):** 468-470.

Sharma B. R., Swarup, V. and Chatterjee, S. S. (1972). Inheritance of resistance to black rot in cauliflower. *Can. J. Gen. Cytol.* **14:** 363-370.

Sharma, B.R. Dhiman, J. S.,Thakur, J. C., Singh, A. and Bajaj, K. L. (1991). Multiple disease resistance in cauliflower. *Adv. Hort. Sci.* **5:** 30-34.

Sharma, S. R., Gill, H. S. and Kapoor K. S. (1997). Inheritance of resistance to white rot in cauliflower. *Indian J. Hort.* **54:** 86-90.

Sharma, S. R.., Kapoor, K. S. and Gill, H. S. (1995). Screening against sclerotenia rot (Sclerotenia sclerotiarum), downey mildew (Peronospora parasitica) and black rot (Xanthomonas compestris) in cauliflower (Brassica oleracia var. botrytis sub variety cauliflora DC). *Indian J. Agric. Sci.* **65:** 916-918.

Singh, R, Trivedi, B.M., Gill, H.S. and Sen, B. (1987). Breeding for resistance to black rot, downy mildew and curd blight in Indian cauliflower. Eucarpia Cruciferae Newsl. **12:** 96-97.

Singh, R., Gill, H.S. and Sharma S.R. (1994). Breeding of Pusa Hybrid-2 cauliflower. Veg. Sci. **21:** 129-131.

Trivedi, B. M., Sen, B., Singh, R., Sharma, S. R., and Verma, J. P. (2000). Breeding multiple disease resistance in mid season cauliflower. Proc. Indian Phytopath. Soc. Golden Jubilee International Conference on Integrated Plant Disease Management for Sustainable Agriculture **2:** 699-700.

Vander Meer, O. P., Toxopeus, H., Crisp, P. H. and Astley, D. (1984). The collection of land races of cruciferous crops in E C countries .Final report of EC research programme 0890.Instituut Voor de veredeling van tuinvouwgewassen ,wageningen, Netherlands. Rapport 198-210.

Verma T. S. and Chander Parkash (2006). Temperate vegetables: Germplasm resources, their utilization and achievements in India. *Indian J. Pl. Genet. Resour.* **19(3):** 397-403.

Williams, J. T. 1989. Plant germplasm preservation: A global perspective *In:* Beltsville Symposium in Agricultural Research. XIII. Biotic Diversity and Germplasm Preservation-Global Imperatives. May 9-11, 1988. (In review).

Yanmaz, R., Kaplan, N., Balkaya, A., Apaydýn, A., Kar, H., 2000. Investigation on the identification of cabbage (Brassica oleraceae var. capitata sub.var. alba) gene sources of Turkey. III. Sebze Tarýmý Sempozyumu . 11-13 Eylül Isparta. 160-166.

Yaþar, F., Türkmen, Ö., Akýncý, I. E., Karata°, A., 1995. Ercis local cabbage selection studies. Türkiye II. Ulusal Bahçe Bitkileri Kongresi, pp265-268.

❑❑❑

Chapter – 18

Management of Parthenocarpic Genotypes of Cucumber

D.K. Singh and H. Choudary

Introduction

Cucumber is one of the most popular vegetable of cucurbitacae family and grown worldwide for its freshly harvested fruits. Fruit yield in cucumber is suppressed due to its fruiting habit. Fruits developing from the first pollinated flower inhibit the development of subsequent fruits. This phenomenon is broadly known as crown fruit dominance or first fruit inhibition. Therefore, in mechanical once over harvest of seeded cucumber, only one to two fruits per plant are developed. Fruit set inhibition is less in seedless cucumber when compared to its seeded counter part. Parthenocarpy is the development of fruit without fertilization. This phenomenon was first studied in cucumber by Noll (1902) who introduced the term parthenocarpy. In parthenocarpic lines, a larger proportion of photosynthates could be diverted to fruit tissue instead of producing seeds. Partheonocarpic habit has become indispensable to greenhouse cucumber production because it alleviates the need for bees. Individual plants in the greenhouse produce many fruits simultaneously, with many fold higher yield in comparison to field grown crop.

Greenhouse or parthenocarpic cucumber production is very popular in many areas of the world. The fruits of parthenocarpic cucumber are mild in flavour, seedless and have thin edible skin that requires no peeling. Fruits are generally 10-14 inches in length and 200 g in weight. The parthenocarpic varieties require no pollination for fruit setting.

If pollination occurs, the fruit will form seeds, shape of fruit will be destroyed and develop bitter taste. It is therefore essential to prevent bees and other pollinators entering the green houses. Cucumber cultivars are selected for commercial use based on quality and yield characteristics. An important quality factor is the shape of the cucumber referred to as the length to diameter (L/D) ratio. This becomes important to meet specifications of the number of whole cucumber units per container. The cool marine environment causes the L/D ratio of many cultivars to be undesirably long and therefore many of the useful cultivars for other production regions have been eliminate. The presently available parthenocarpic cucumber cultivars have been bred and developed in Europe under similar climatic conditions and produce fruit with desirable L/D ratios. The parthenocarpic cultivars have potential for improvement of both the L/D ratios and fruit interior quality.

Almost all slicing cucumber varieties cultivated in greenhouse in Western Europe are parthenocarpic, and parthenocarpic pickling cucumber varieties are also available in the market. In Germany the use of parthenocarpic varieties in production for the processing industry is a major part of the total cultivation area of pickling cucumber. Parthenocarpic fruit setting results in an earlier and much more regular production rhythm (Zwinkels, 1987). Parthenocarpy circumvents the inhibitory effect of seed formation on subsequent fruit development. Parthenocarpy must be combined with constant femaleness, because the fruits formed after fertilization in genetic parthenocarpic varieties become misshapen, are of no economical value and lead to loss of production. It is possible to induce artificially parthenocarpic fruit set by treatment with morphactin (Chlorfluorenol) as reported by Robinson and Cantliffe (1971).

Genetics of Parthenocarpy

Parthenocarpy can be regarded as a qualitative character i.e. the ability to develop fruits without pollination (apomixes excluded). For breeding purposes parthenocarpy should preferably be regarded as a quantitative character i.e. the ability to produce a certain yield of parthenocarpic fruits. There has been conflicting reports regarding the inheritance of parthenocarpy in cucumber. Wellington & Hawthorn (1928) reported that there was incomplete dominance of parthenocarpy which they later concluded as recessiveness (Hawthorne & Wellington, 1930). Pike and Peterson (1969) studied parthenocarpy as a qualitative character and claimed to be governed by one incompletely dominant gene in the homozygous condition PP produces parthenocarpy fruit early, with the first fruit developing generally by 5th node. Heterozygous Pp plant produces parthenocarpic fruit later than homozygous plant and fewer in number. The homozygous recessive pp produces no parthenocarpic fruits. Trapping of pistillate flowers was effective in identifying homozygous pp plant but failed to identify heterozygous plants. Screen cages and field isolation of gynoecious lines are effective means for accurately identifying all parthenocarpic plants.

Juldasheva (1973) communicated that one recessive gene was responsible for parthenocarpy. De Ponti and Garretsen (1976) explained the inheritance of parthenocarpy by three independent, isomeric major genes with additive action, together with a non-allelic interaction of the homozygote- heterozygote type. This interaction must be explained by the phenomenon that in genotypes with one + allele for parthenocarpy, this allele dose not causes any phenotypical expression. In view of the rather larger variance in parthenocarpy expression of segregating population, there is a possibility that minor genes also contribute to the degree of parthenocarpy. Indications have been found for linkages between genes that govern parthenocarpy, femaleness and the spined /hairy fruit-character. They reported the probability that one of the genes for parthenocarpy and one of those for femaleness (Kubicki, 1969) were located on the same chromosome and that a second gene for parthenocarpy and the gene for the spine/hairy fruit character are both located on another chromosome.

Breeding of Parthenocarpic Pickling Cucumbers

A programme for development of parthenocarpic hybrid pickling cucumber for field production was started at Department of Horticulture, University of Wisconsin. The parthenocarpic germplasm was derived from a gynoecious population referred previously as gynoecious synthetic (GS). This population was developed in 1977-78 from a series of gynoecious lines collected from Europe, US and America through numerous breeding programmes. In early 1980, a half sib recurrent selection programme was combined with a pedigree selection programme to develop parthenocarpic population from gynoecious synthetic material. The parthenocarpic population was screened for parthenocarpy under three different photoperiods. Initial selection was made for Parthenocarpy on the basis of yield, fruit quality with little emphasis on disease resistance and other characters. After several generations of breeding, methodology was redirected to inbred line, F_1 hybrid development. The inbreds were selected for parthenocarpy, gynoecious stability, fruit quality, yield, disease resistance and other horticultural traits.

Development of parthenocarpic cucumber genotypes at Pantnagar

The parthenocarpic cucumber lines PPC 1, PPC 2 and PPC 3 have been isolated from more than 300 accessions collected from India and abroad. These lines were maintained by the use of plant growth regulator like GA_3, $GA_3 + GA_7$ and silver nitrate ($AgNo_3$) at 2-3 true leaf stage. Four sprays of these PGRs at 4 days interval was found effective for the induction of male flowers on the parthenocarpic cucumber lines. Induction of male flower depended upon the plant growth regulator concentration and number of the sprays. All the treatments having only one spray were ineffective to produce the male flowers. The treatment with 100 ppm failed to induce the male flowers. All three parthenocarpic cucumber lines bear dark green fruits in clusters. The fruit length was 18 cm (PPC 1), 23 cm (PPC 2) and 20 cm (PPC 3). The estimated yield was 1100 to 1200 q/ha. The four parthenocarpic cucumber hybrids were developed namely PPC 1 x PPC 2, PPC 1 x PPC 3, PPC 2 x PPC 3 and PPC 1 X Poinsette. The performance of these hybrids was better than their respective parents. It was also observed that no male flower appeared on plants of all parthenocarpic pure lines and hybrids up to 30-35° C temperature during the month of May and plants also showed resistance to CMV, MMV and powdery mildew under polyhouse condition .

Maintenance of parthenocarpic cucumber lines

The parthenocarpic lines are usually maintained by use of plant growth regulators and chemicals, e.g. $AgNO_3$ and GA_3. Kasrawi (1988) found that greatest number of staminate flowers are produced on plants when $AgNO_3$ was sprayed twice with 300 ppm concentration at one week interval with initial spray at first true leaf stage. He also found that 2-3 sprays of 200-500 ppm $AgNO_3$ starting at first true leaf stage produce the maximum number of staminate nodes. When the plants were sprayed at first true leaf stage, production of staminate flowers began with the I^{st} node and at the time of onset of flowering. The effect of $AgNO_3$ for induction of staminate flowers of gynoecious parthenocarpic lines had been reported and it was observed that $AgNO_3$ concentration and number of sprays affected induction of staminate flowers. All the treatments using only one spray were ineffective for producing staminate flowers. The treatment with 100 ppm $AgNO_3$ failed to induce staminate flowers. The greatest numbers of staminate nodes were produced on plant by 2-3 sprays of $AgNO_3$. The initial spray was applied at first true leaf stage and subsequent treatment was applied at one week interval. He also reported that plants showed injury a few days after spray $AgNO_3$ and recovered within 7-10 days. Some plant growth regulators may be used for induction of male flower in gynoecious parthenocarpic line e.g. GA_3. Staminate flowers were induced in parthenocarpic lines through use of plant growth regulator (GA_3) and by silver nitrate ($AgNO_3$) by four sprays at 4 days interval at Pantnagar. (Singh and Ram, 2004).

Production Technology for Greenhouse Cucumber Cultivation

Cucumber is a warm season crop and requires high temperature of 20°C - 30°C and plenty of sun light to produce a good crop. The optimum soil temperature should be 15 -18° C, lower temperature will delay plant growth and development of fruit especially in greenhouse cucumber. Prolonged temperature above 35°C should also be avoided as fruit quality and production are reduced. Greenhouse cucumber is established in greenhouse as transplants. The cost of seed is high while germination percentage is nearly to 100 %. So, one seed/hill or bag is sufficient. The seeds of cucumber germinate rapidly within 3 - 4 days when the optimum temperature is available in greenhouse. During seedling or transplanting, the plant should never be stressed for nutrient and water. Plants are ready for transplanting at 3 - 4 true leaf stage. Greenhouse cucumber plants have very large leaves, grow vigorously and require large amount of sun light. In good sunlight conditions or in warm weather 150-200 cm^2 space should be provided to plants in greenhouses but where light conditions are poor, more spacing is required for good growth of plants. Exact spacing between row to row and plant to plant depends upon variety, soil fertility, light requirement, training and production system of crop.

Training and pruning are also needed for greenhouse cucumber. The basic principle of training system is to uniformly maximize the leaf interception of sunlight throughout the greenhouse. Generally 2 types of training systems are used. First one is vertical cordon system. In this system, plants are trained vertically to an overhead wire. Once the plants reach the wire, they are topped and then pruned using an umbrella system. The second one is V cordon system. In this case, the plants are trained up the strings and grow in an incline way from low centre. Cucumbers are trained on plastic wires running horizontally. The base of string can be anchored loosely to base of stem. As the stem develops it can be fastened to the string with clips. Only one stem is allowed to develop, removing all laterals and tendrils as they develop. Fruit buds should be removed from first five leaf nodes. Thereafter, fruits are allowed to develop but all laterals and tendrils are removed. Old leaves should be removed from older part of stem.

Varieties

Cultivar selection is one of the most important decisions made during crop production process. Several factors should be considered while selecting cultivars, including disease resistance, plant vigour, earliness yield, fruit size, quality and colour of fruit. The available parthenocarpic cultivars are generally produced by European Seed Companies, however, a few of companies have sales and technical support representative in other parts of world and Asia. Some of the popular parthenocarpic varieties are:

Kalunga, Bellissma, Millagon, Discover, Marianna, Fitness, Aramon, Fidelio, 90-0048, E 1828, B 1157 , Sweet Success (AAS), County Fair, Bush Crop, Space Master, Patio Pick, Bush Whopper, Bush Champion, Bush Pickler, Euro-American, Adrian, Al Rashid F_1, Mustang, Bronco, Sandra, Boneva, Daleva, Padex, Ferti Ja, Factum, Fanspot, Femfrance, Toska 70, Farbio, Corona, Sweet Slice, Radja, Bella, La Reine.

Nutrient

Greenhouse seedless cucumbers have a high nutrient requirement and grow very rapidly when supplied with sufficient nutrients. As a result, growers must plan for an optimum nutrient programme making adjustments in the programme as the crop demands change. The greatest demand for nutrients is during the peak fruit production period. Nitrogen and potassium are required in the greatest amounts, however, a complete nutrition programme including essential minor elements also. Designing a fertility programme varies depending on the production system desired and extreme caution must be used when interpreting or comparing research or articles from one production system to another. Florida greenhouse producers are using soilless production systems. In these systems, a complete nutrient solution is used to supply the needed nutrients to the crop. Soilless culture increases the grower's ability to control the growth of the plant, but it also requires management to achieve success. Soilless grower must supply their crop with 6 macronutrients (nitrogen, phosphorus, potassium, calcium, magnesium and sulphur) and 7 micronutrients (iron, manganese, copper, zinc, molybdenum, boron and chlorine).

It is especially important in a soilless system to test the source of water prior to developing a nutrients programme. The target pH of the nutrient solution supplied to the plants should be between 5.5 and 6.0. Either nitric, sulphuric or phosphoric acid is recommended for pH control. If it is necessary to raise the pH, potassium hydroxide is usually used. If the source water is alkaline due to high bicarbonate concentrations, the pH should be adjusted before the fertilizer salts are added to prevent precipitation. Plant tissue testing can be a useful tool in addition to a good soil testing or nutrient solution programme to monitor the fertility. Several research articles report a sudden temporary wilt of greenhouse cucumber plants when using NFT. Cucumber roots have a greater oxygen requirement when compared with roots of tomatoes. Cucumbers develop a large root system in the growing tubes and as a result may become stressed for oxygen. Because of this concern, measures may need to be taken to provide improved oxygen supplies to the root zone. High solution temperatures also reduce the amount of dissolved oxygen in the solution. Growers must avoid high temperatures in the solution.

Fruit development

A fruit may develop at each node and more than one fruit may begin to develop at some nodes. It is usually best to thin these multiple-fruit clusters to a single fruit, however, vigorous plants can sometimes have more than one cucumber at a node. Any distorted fruit should be removed immediately. The greatest growth of the fruit occurs between day 6 and 14 after the bloom opens (anthesis). Maximum fruit length occurs at day 14 followed by diameter increase. The shape of the fruit is somewhat tapered being largest at the stem end prior to day 10 after bloom, however, the fruit becomes uniformly cylindrical by day 14. During the spring season, commercially acceptable fruit size is usually reached by the 11th day after the bloom opens.

Harvesting

With good management each plant may produce about 6 - 7 kg fruit over a 3 - 4 month period. Cucumber fruit is harvested when they attain a size of 10-14 inches long. After harvesting fruits are stored at 10° C with 70 - 80 % relative humidity. Incorporating intensive culture of trickle irrigation and fertigation with plastic mulch has increased the hand harvest yield of parthenocarpic cucumber.

References

De Ponti, O. M. B.. 1976. Breeding parthenocarpic pickling cucumber *(Cucumis sativus* L.), necessity, genetical possibilities, environmental influences and selection criteria. *Euphytica,* **25:** 25-29.

De Ponti, O. M .B. and Garretsen, F. 1976. Inheritance of parthenocarpy in pickling cucumbers *(Cucumis sativs* L.) and linkage with other characters. *Euphytica,* **25:** 633-639.

Hawthorn, L. R. and Wellington R. 1930. Geneva: a greenhouse cucumber that develops fruit without pollination. *Bull. N. Y. St. Agric. Exp. Stn.*, **580:** 11.

Juldasheva, L. M. 1973. Inheritance of the tendency towards parthenocarpy in cucumbers (Russian). *Byull.*

Vsesoyuznogo ordena Lenina Inst. Rastenievodstva Imeni N.I. Vavilova, **32:** 58-59.

Kasrawi, M. A. 1988. Effect of silver nitrate on sex expression and pollen viability in parthenocarpic cucumber *(Cucumis sativas* L.). *Dirast Jorolan.,* **15 (11):** 69-78.

Kubicki, B. 1969. Investigations on sex expression in cucumber (Cucumis sativus L.). *Genet. Polo.,* **10:** 5-143.

Noll, F. 1902. Fruchtbilding ohne vorausgegangene Bestaubung (Parthenocakarpie) bei der Gurke. Sitzungsber. Niederrhein. Ges. Nat. Heilk. Bonn: 149-162.

Pike, L.M. and Peterson, C.E. 1969. Inheritance otparthenocarpy in the cucumber *(Cucumis sativus* L.). *Euphytica,* **18:** 101-105.

Robinson, R.W. and Catliffe, D.J. 1971. Morphactin-induced parthenocarpy in cucumber. *Science,* **171:** 1251-1255.

Singh, D.K. and Ram, H.H. 2004. Isolation of parthenocarpic cucumber line. First Indian Horticulture Congress on Improving Productivity, Quality, Post Harvest Management and Trade in Horticultural Crops Organized by Indian Society of Horticulture, held at IARI New Delhi, 6th -9th November 2004. Book of Abstracts pp 87.

Strong, W. J. 1921. Greenhouse cucumber breeding. *Proc. Am. Soc. Hort. Sci.,* **18:** 271-273.

Strong, W.J. 1932. Parthenocarpiy in cucumber. *Scient. Agric.,* **12:** 665-669.

Sturtevant, E.L. 1890. Seedless fruits. *Mem. Torrey bot. club.,* **1:** 141-185.

Wellington, R. and Hawthorn, L. R. 1928. A parthenocarpic hybrid derived from a cross between an English forcing cucumber and the Arlington White Spine. *Proc. Am. Soc. Hort. Sci.,* **25:** 97-1000.

Zwinkels, L.A.M. 1987. The present situation and fruit development in breeding of parthenocarpic pickling cucumber. *Acta Horticulture,* **220:** 129-130.

❑❑❑

Chapter – 19

Breeding Potential of Indigenous Germplasm of Cucurbits

D.K. Singh

Introduction

Cucurbits, belonging to the family Cucurbitaceae are among the most ancient cultivated plants. The term cucurbit was coined by Liberty Hyde Bailey for cultivated species of Cucurbitaceae. Other vernaculars applied to the family and several of its members are 'gourd', 'melon', 'cucumber', 'squash' and 'pumpkin'. The Cucurbitaceae consists of two well defined sub families, eight tribes representing varying degrees of circumscriptive cohesiveness and about 118 genera and 825 species (Jeffrey, 1990). It is mostly tropical and sub tropical in distribution and only a few species occur in temperate regions. Some are semi desert or saprophytic in nature. The Indian sub continent is considered to be the centre of origin for a number of wild and cultivated crops. Chakravorty (1982) reported that 36 genera and 100 species of cucurbits are found in India. Cucurbits are cultivated in 8.5 m ha area in world giving a production of 17.9 mt. While in India a total of 4.5 mt cucurbits are produced from an area of 0.42 m ha area (FAO, 2004).

Cucurbitaceae is one of the most genetically diverse groups of plants in the plant kingdom. As a family and as individual crop, cucurbits epitomize adaptive differentiation and evolutionary divergence. Not only may cultivars within a crop vary significantly in their characteristics, but the same cultivar grown in distinct areas can have different needs in response to disparate local growing conditions. Cultures and ethnic groups may have different cultivar preference and horticultural practices, which also increase morphological diversity within a crop.

Cucurbits are of tremendous economic importance as food plants. They are extensively grown in mixed cropping of more than one kind in long and meandering river beds. Of various improvements approaches possible, breeding of superior pure lines and hybrid cultivars of cucurbits utilizing locally adopted land races are the most important methods which will have far reaching impact on increasing the production and productivity of cucurbits.

Cucurbits in India: Collection and Evaluation

Cucurbitaceae is one of the most homogenous families. The plants are generally herbaceous, flowers are constantly unisexual, ovary is inferior with one or two exceptions, seeds have scanty or no endosperm. This group consists of a wide range of vegetables either used as salad (cucumber) or for cooking (all gourds) or for pickling (cucumber, bitter gourd) or as dessert fruits (water melon and musk melon) or candied or preserved (ash gourd). These crops have the potential to be grown on lands which are otherwise unsuitable for raising other crops. This feature has made them popular among small farmers. The river beds of Yamuna, Ganga, Gomti, Saryu and other distributaries in Haryana, Uttar Pradesh and Bihar; Banas river bed in Tonk district of Rajasthan; Narmada, Tawa and Tapti river beds of Madhya Pradesh and Maharashtra; Sabarmati, Panam, Vartak and Orsung of Gujarat and Tungabhadra, Krishna, Hundri Pennar river beds of Andhra Pradesh are some of the important areas where cucurbits are extensively grown. These river beds and the adjoining areas are rich sources of promising land races of cucurbits which can be improved upon through breeding programmes. List of cucurbits found in India is given in Table 1

Table 1. Cucurbits found in India

Common name	Botanical name	Frequency of cultivation	Usage
Pumpkin	*Cucurbita moschata*	Common	Food
Winter squash	*Cucurbita maxima*	Common	Food
Summer squash	*Cucurbita pepo*	Common	Food, medicinal
Buffalo gourd	*Cucurbita ficifolia*	Localized	Food
Cucumber	*Cucumis sativus*	Common	Food
Muskmelon	*Cucumis melo*	Common	Food
Snap melon	*Cucumis melo* var *momordica*	Sporadic	Food
Long melon	*Cucumis melo* var *utilissimus*	Localized	Food
Netted melon	*Cucumis melo* var *reticulatus*	Sporadic	Food
Pickling melon	*Cucumis melo* var *conomon*	Rare	Food
Cantaloupe	*Cucumis melo* var *cantalupensis*	Localized	Food
Mango melon	*Cucumis melo* var *chicko*	Rare	Food
Gherkin	*Cucumis anguria*	Localized	Food
Wild cucumber	*Cucumis sativus* var *hardwickii*	Rare	Food

Contd....

Serpent melon	*Cucumis melo* var *flexuosus*	Sporadic	Food
Pomegranate melon	*Cucumis melo* var *dudain*	Sporadic	Food
Winter melon	*Cucumis melo* var *inodorus*	Sporadic	Food
Teasel gourd	*Cucumis dipsaceus*	Sporadic	Ornamental
Sponge gourd	*Luffa cylindrica*	Common	Food, utilitarian
Ridge gourd	*Luffa acutangula*	Common	Food
Satputia	*Luffa hermaphrodita*	Sporadic	Food
Bottle gourd	*Lagenaria siceraria*	Common	Food, utilitarian, ornamental
Water melon	*Citrullus lanatus*	Common	
Tinda	*Praecitrullus fistulosus*	Localised	Food
Colocynth	*Citrullus colocynthoides*	Sporadic	Food
Snake gourd	*Trichosanthes anguina*	Frequent	Medicinal
Pointed gourd	*Trichosanthes dioica*	Localized	Food
Bitter gourd	*Momordica charantia*	Common	Food, medicinal
Balsam apple	*Momordica balsamina*	Frequent	Food
Kakrol	*Momordica dioica*	Localized	Food
Sweet gourd of Assam	*Momordica cochinchinensis*	Sporadic	Medicinal
Ash gourd	*Benincasa hispida*	Frequent	Food
Ivy gourd	*Coccinia grandis*	Sporadic	Food
Chow-Chow	*Sechium edule*	Common	Food
Mitha Karela	*Cyclanthera pedata*	Sporadic	Food

India holds a prominent position as primary/secondary centre of origin for several cucurbits. Therefore, there is rich diversity available in all the cucurbit crops including ash gourd, bottle gourd, bitter gourd, cucumber, muskmelon, watermelon, ridge gourd, sponge gourd, pumpkin and long melon in India (Dhillon *et al*., 2001). Due to several factors such as monoculture, adoption of improved open pollinated varieties and hybrids, urbanization, fragmentation of habitat and deforestation, several cucurbits in India are at the verge of extinction and considered endangered (Table 2).

Table 2. Rare and endangered cucurbits of India

Species	**Biogeographic Zones**
Corallocarpus gracillipes	Western Ghats
Gomphogyne macrocarpa	East Himalaya
Indogevillea khasiana	North East India
Luffa umbellata	Western Ghats
Melothria amplexicaulis	Deccan Peninsula
Momordica sub sp. *angulata*	Deccan Peninsula, Western Ghats
Trichosanthes lepiniana	Deccan Peninsula, Western Ghats
Trichosanthes perrotteliana	Western Ghats

Source : Rai, et al., 2006

Plant genetic resources constitute a reservoir of gene and gene complex and are the raw materials for improvement of crop plants. It is well known that transition from primitive to advanced cultivars has the effect of narrowing the genetic base. Diverse genetic materials are, therefore, required to meet the ever changing demands of plant improvement. Collection and evaluation of germplasm is a pre-requisite for their utilization.

In India, Indian National Plant Genetic Research System (INPGRS) under the aegis of the Indian Council of Agricultural Research (ICAR) with National Bureau of Plant Genetic Resource (NBPGR) as the apex body plans, conducts, promotes, coordinates and takes lead in the activities related to plant germplasm collection introduction, exchange, evaluation, documentation, conservation and sustainable management of germplasm of crop plants and their wild relatives. NBPGR has collected the germplasm from all over India and has around 7,000 accessories of various cucurbits (Table 3).

Table 3. Accessions of various cucurbits at NBPGR

S.N.	Crop	Collection No.
1.	Ash gourd	397
2.	Bitter gourd	728
3.	Bottle gourd	968
4.	*C. sativus* var *hardwickii*	58
5.	Cucumber	545
6.	Ivy gourd	149
7.	*Kachari*	342
8.	Muskmelon	383
9.	Pointed gourd	304
10.	Pumpkin	1085
11.	Ridge gourd	511
12.	Round melon	64
13.	Snap melon	656
14.	Sponge gourd	722
15.	Watermelon	113
	Total	**6717**

Source : Malik *et al.*, 2001

The NBPGR has also registered several germplasm which have certain unique traits (Table 4). These germplasm lines have been collected by different institutes and SAUs, from the farmers' field at different places all over India. These germplasm could be utilized in the crop improvement programme by various centres.

Table 4. Registered germplasm of cucurbits

Cucurbits	Name of line	Natl. identity No.	Registered Unique Traits
Pointed gourd	IIVR PG-105	INGR-03035	Seedless fruits (parthenocarpic)
Bitter gourd	GY-63	INGR-03037	Gynoecious sex with high yield
Water melon	RW-187-2	INGR-01037	High yield and yellow coloured flesh
	RW-177-2	INGR-01038	Leaf mutant with simple unlobed leaves
Bottle gourd	Androman-6	INGR-99009	Andromonoecious sex
	PBOG-54	INGR-99022	Segmented leaves
Cucumber	AHC-2	INGR-98017	High yield and long fruit
	AHC-13	INGR-98018	High yield, small fruit, drought and high temperature tolerance
Kachari	AHK-119	INGR-98013	High yield and drought tolerance
Round melon	HT-10	INGR-99038	Intermediate, semi-spreading vine, tolerant to downy mildew and root rot wilt
Snap melon	AHS-10	INGR-98015	High yield and drought tolerance
	AHS-82	INGR-98016	High yield and drought tolerance

To strengthen the germplasm collection, evaluation and maintenance programme at G. B. Pant University of Agriculture & Technology, Pantnagar, Pantnagar Centre for Plant Genetic Resources (PCPGR) was established in 1999 with the financial assistance of World Bank through Diversified Agricultural Support Project. Over 1,600 accessions of different cucurbits have been collected by PCPGR from various places like Bareilly, Lucknow, Allahabad, Varanasi, Gorakhapur, Faizabad, Azamgarh, Meerut, Moradabad district of Uttar Pradesh, Kumaon and Garhwal Division of Uttaranchal, West Bengal, Madhya Pradesh, Rajasthan, Karnataka and some parts of Bihar.

Table 5. Cucurbits germplasm collection at PCPGR

S.N.	Crop	Collection No.
1.	Ash gourd	62
2.	Bitter gourd	154
3.	Bottle gourd	219
4.	Chow chow	01
5.	Cucumber	190
6.	*Cucumis sativus* var *hardwickii*	08
7.	Long melon	36
8.	Muskmelon	485
9.	Pointed gourd	02
10.	Pumpkin	171
11.	Ridge gourd	141
12.	Round gourd	11
13.	Snake gourd	24
14.	Snapmelon	23
15.	Sponge gourd	133
16.	Water melon	237
17.	Wild Ivy gourd	02
	Total	**1899**

After collection of germplasm, its evaluation using morpho-agronomic traits is the next most important step. This evaluation of germplasm results in establishment of core set. The available information may further be used for efficient management and effective utilization of assembled germplasm. At the PCPGR cucurbits germplasm has been evaluated for the agronomic traits such as yield and yield contributing traits like number of fruits per plant, fruit length, fruit girth and fruit weight. The promising genotypes which could further be utilized in crop improvement programme are presented in Table 6.

Table 6. Promising genotypes of cucurbits evaluated at PCPGR

Bitter gourd	PBIG-1, 2,11,17,42,44,45,46,56,68,71,77,78,79,91,123,
Bottle gourd	PBOG-13, 22, 40, 61, 62, 70, 73, 76, 86, 88, 98, 106, 103, 114, 115, 117, 118, 119, 122, 125, 127, 132, 159, BGLC2-1
Cucumber	PCUC-8, 11, 16, 21, 2 3, 25, 26, 28, 29, 31, 38, 46, 58, 61, 64, 65, 68, 73, 81, 82, 94
Muskmelon	PMM-12, 33, 42, 43, 55, 81, 101, 113, 133, 135, 150, 166, 171, 197, 202, 205, 211, 220, 241, 251, 226, PMM-96-22, 97-22, 97-41, 97-43, 97-50, 2000-11, 2000-19
Watermelon	PWM-61, 72, 75, 81, 88, 91, 94, 121, 153, 154, 2000-1, 2000-2, 2000-3, 2000-7, 2000-12
Pumpkin	PPU-39, 46, 56, 103, 104
Long melon	PLM-1, 2, 44, 52
Ash gourd	PAG-3, 5, 8, 15, 18, 21, 38, 43, 44, 72, 96-1, 96-2
Smooth gourd	PSM-35, 39, 44, 113, 115
Ridge gourd	PRG-6, 7, 48, 54, 88, 92

Source : Singh et al., 2003

Indigenous Germplasm and its Breeding Potential

The Indian sub-continent is considered to be the centre of origin for number of wild and cultivated cucurbits. There are certain known geographical areas all over India having a lot of useful genetic variability in form of land races, traditional cultivars wild edible forms and related non-edible wild and weedy species, which can be tapped for breeding purposes. Accordingly, local collections of important cucurbits exiting in the form of land races were subjected to inbreeding, selection and hybridization to develop promising pure lines/ hybrid cultivars. These local collections through times would have been distributed geographically by migration trade and subjected to local selection pressures in their area of new environments. These localised selection termed landraces are genetically diverse populations selected under low input agriculture for local preferences and dependability rather than productivity. Further, these land races/local collection can be used as sources to donors for various horticultural traits as high yield, resistant to diseases and insect pests and certain quality parameters (Table 7).

Table 7. Sources/donors identified for various traits in different species of cucurbits

Crops/trait (S)	Donor (S)
***Cucumis sativus* and related species**	
Extended shelf-life	IC203838, IC203839
High yield	IC203838, VJ/98-176, VJ/98-151 PI 197087
Downy mildew resistance	PI197087
Powdery mildew resistance	
Cucumber scab resistance	PI79376
Angular leaf spot resistance	
Bacterial wilt resistance	PI169400, PI200815, PI200818
Cucumber green mottle mosaic resistance	*Cucumis anguria*, *C. africanus*, *C. ficifolius*, *C. asper*, *C. dinteri*, *C. sagittatus*
White fly resistance	
***Cucumis melo* and related species**	
Downy mildew resistance	*Cucumis callosus*
Powdery mildew resistance	PMR 45,
Fusarium wilt resistance	*Cucumis melo* var. *chito*, *C. indorus*, *C. flexuosus*, *C. reticulatus*
Watermelon mosaic virus resistance	PI 414723, *Cucumis metuliferus*
Gummy stem blight resistance	
Citrullus lanatus	
High yield	K3228, OC92910
Powdery mildew resistance	Arka Manik
Lagenaria siceraria	
Summer type	BDJ628, BDJ864, ICI612, U8-217, UIO-315, U16-102
Rainy season type	BDJ536, BDJ1733, IC92904, IC144389
Early type	IC92426, IC94404, NIC13188, NICI3182
High yield	IC29404, IC92429, ICIIO389
Heavy fruit	IC92429, IC92490, IC201150, IC201159
Cucurbita moschata	
High yield	IC90512, IC92499, IC92533, IC92571, NICIO170, NICIO174
Luffa cylindrica	
Early type	IC92604, IC92779, IC92797, IC93445, NIC597, NICIO23, NICI3236, NICIIO235
High yield	NIC23288, NIC23291, NIC23292
Long and heavy fruit	IC92475, IC92721, IC92727, IC92751, IC201217, IC201229
***Cucurbita* spp.**	
Watermelon virus resistance	*C. ecuadorensis*, *C. joetidissima*
Bacterial wilt resistance	*C. ficifolia*, *C. lundelliana*
Drought resistance	*C. foetidissima*
Powdery mildew resistance	*C. lundelliana*, *C. okeechobeensis*
Powdery mildew, squash and watermelon virus resistance	*C. martinezii*

Several indigenous germplasm accessions collected by NBPGR, IIVR and State Agricultural Universities have been directly adopted by the farmers for large scale cultivation. On the other hand, a number of varieties have been developed through selfing and selection from indigenous germplasm. These varieties have been release by Central Varietal Release Committee (CVRC) or State Varietal Release Committee (SVRC) and are becoming popular among cucurbits growers (Table -8).

Table 8. List of open-pollinated varieties identified/released in India.

Cucurbits	National level	State level
Muskmelon	Pusa Sharbati, Hara Madhu, Pusa Madhuras, Arka Rajhans, Arka Jeet, Durgapura Madhu, NDM-15	Punjab Rasila, Arka Rajhans, Hisar Madhur, RM-43, RM-50, Kashi Madhu
Watermelon	Durgapura Meetha, Arka Manik	Durgapura Kesar, Durgapura La
Bitter gourd	Priya, RHRBG-4-1, KBG-16, Pant Karela-2	Coimbatore Long, Pusa Do Mausmi, Pusa Vishesh, Punjab-14, Kalyanpur Baramasi, CO-1, CO-2, Balam Pear, Coimbatore Green, Arka Harit, Kalyanpur Sona, Hirkani, Prithi Priyanka, Punjab-14, Pant Karela-1
Pumpkin	CM-14, Pusa Vishwas, Arka Chandan, Arka Suryamukhi, CM-350, NDPK-24	CO-1, CO-2, Narendra Amrit, Kashi Harit, Azad Kaddoo-1, Pusa Vikas, Suvarna
Cucumber	Swarna Ageti, Swarna Sheetal,	Japanese Long Green, Pusa Uday, Himangi, Swarna Poorna, Sheetal, CO-1, Pant Khira-1
Ridge gourd	Swarna Manjari, Arka Summet	Swarna Uphar, CO-1 PKM-1, Arka Sujat, Pusa Nasdar, Punjab Sadabahar, Haritham, Hisar Kalitori
Bottle gourd	Pusa Naveen, Narendra Jyoti, NDBG-132, Pant Lauki-3	Arka Bahar, Pusa Sandesh, Pusa Summer Prolific Round, Pusa Summer Prolific Long, Punjab Round, Punjab Long, Punjab Komal, Co-1, Narendra Dharidar, Narendra Shishir, Kashi Ganga, Kalyanpur Long Green, Azad Harit, Azad Nutan
Sponge gourd	Pusa Chikni, CHSG-1, JSGL,	Pusa Supriya Pusa Sneha, Rajendra Nenua-1 Kalyanpur Torai Chikni, Azad Torai-1, pant chikni Torai-1
Ash gourd	Pusa Ujjwal, Pant Petha-1	CO-1, CO-2, Mudliar, Indu, KAU Local, Kashi Dhawal, PAG-3
Long melon	—	Arka Sheetal, Punjab Long Melon-1, Pant Kakari-1
Round melon		Arka Tinda, Punjab Tinda
Snake melon	—	CO-1, CO-2, PKM-1, Kaumudi, Baby, H-8, H-371, H-372, IIHR-16A
Summer melon	—	Punjab Chappan Kaddu-1, Patty Pan, Australian Green
Snap melon	—	Pusa Shandesh, Grism Bahar, Kwari Bahar

The Department of Vegetable Science of G. B. Pant University of Agriculture & Technology, Pantnagar has also released seven open-pollinated varieties of different cucurbits through continued selfing and selection of superior genotypes. These varieties have been tested at multilocation traits and released by CVRC/SVRC. The varieties released are as listed below.

Table 9. Open Pollinated Varieties Released from Pantnagar

Crop	Variety	National ID	Parentage	Characteristic features	Releasing agency
Ash gourd	Pant Petha 1	—	PAG 72	Fruits cylindrical with white wax coating, first harvest in 115 days. Yield 600 q/ha.	CVRC
Bitter gourd	Pant Karela 1	541199	PBIG 3	Fruit thick, 15 cm long with tapering ends, suitable for planting in hills. Total yield 150 q/ha	SVRC
	Pant Karela 2	541200	PBIG 1	Fruits relatively thin, 10 cm long tapering at both ends, dark green. First harvest in 60 days. Yield 150-200 q/ha	CVRC
Bottle gourd	Pant Lauki 3	541198	PBOG 61	Fruit straight, 45 cm long, slightly club shaped. First harvest in 60 days. Yield 400-500 q/ha.	CVRC
Cuc-umber	Pant Khira 1	541203	PCUC 28	Fruits 20 cm long, cylindrical with light white stripes. First picking in 50-60 days. Yield 150 q/ha.	SVRC
Long melon	Pant Kakri 1	541205	PLM 1	Fruits long light green, straight. First picking 50 days after sowing, total yield 300q/ha. Free from common diseases and insect pest.	SVRC
Ridge gourd	Pant Torai 1	541204	PRG 7	Fruits 15-20 cm long, club shaped. Suited to rainy season. Yield 100 q/ha. Vine 5m long.	SVRC

Hybrid Breeding

The cucurbits though cross pollinated in nature, do not show any loss of vigour due to inbreeding. Therefore, in these crops, inbreeding along with individual plant selection is routinely practiced as a method of breeding as effectively as in any self-pollinated crop (Swarup, 1991). However, considerable heterosis has been noted by several workers and F_1 hybrids are commercially important in most of the cucurbits. First F_1 hybrid of water melon (1930) followed by cucumber (1933) was developed during 1930s. Hybrid cultivars are commercialized in selected cucurbits, which express desirable amount of heterosis for yield (Table 10).

Table 10. Manifestation of desirable heterosis yield in important cucurbits

Cucurbits	Heterosis (%) over better Parent
Muskmelon	313.8 (Mishra and Sheshadri, 1985), 48.86 (Munshi and Verma, 1997)
Watermelon	87.00 (Nath and Dutta, 1970)
Bitter gourd	98.17 (Ram *et al.*, 1997), 64.47 (Singh *et al.*, 1992)
Cucumber	187.80 (Singh *et al.*, 1999)
Bottle gourd	62.34 (Padma *et al.*, 2003), 64.45 (Sirohi *et al.*, 2005), 91.02 (Dubey and Maurya, 2003)
Ridge gourd	47.92 (Acharya *et al.*, 2006), 86.15 (Purohit *et al.*, 2006)
Pumpkin	24.6 (Sirohi, 1994), 126.75 (Pandit *et al.*, 2006)
Ash gourd	90.65 (Pandey *et al.*, 2005), 96.6 (Sureja *et al.*, 2006)

At Pantnagar, several indigenous germplasm lines of different cucurbits have been crossed in various combinations to test the extent of heterosis. In bitter gourd, better parent heterosis was reported for fruit characters (65-80%) and yield and yield related attributes (60-70%). Vr, Wr graphical analysis indicated the involvement of recessive genes for earliness. Based on these observations PBIG 2 was adjudged as the potential donor for genes for earliness (Chaubey *et al.,* 2004). In bottle gourd, Dubey (2004) reported better parent heterosis for a number of traits such as fruit length (74.5%), fruit weight (48.0%), days to first harvest (35.4%), number of fruit per plant (150.0%) and fruit yield (90.3%). Vr, Wr graphical analysis indicated involvement of dominant genes for earliness and recessive gene for fruit diameter. PBOG 13 was identified as potential donor for greater fruit diameter. PBOG 13 X PBOG 61, PBOG 13 X PBOG 76 and PBOG 61X PBOG 76 were found to be best heterotic combination alongwith good SCA effects for important economic traits and are worth exploiting on commercial scale. Similar results were obtained by Maurya *et al.* (2003) PBOG 61 X PBOG 88 and PBOG 81 X PBOG 88 were found to be best heterotic combination along with good SCA effects. In cucumber also heterosis has been reported for a number of traits. For earliness, PCUC 25 X DC 1, PCUC 8 X DC 1and PCUC 83 X PCUC 25 were found to be best. PCUC 83 X PCUC 25, PCUC 83 X PCUC 15 and PCUC 25 X PCUC 15 were reported to be best combinations for yield and yield related traits. PCUC 25 and DC 1 were adjudged to be the potential donor for earliness and PCUC 25, PCUC 15 and PCUC 83 were adjudged to be the potential donor for yield (Bairagi *et al.,* 2005). Based on these observations, three F_1 hybrids have been related from Pantnagar for commercial cultivation at national level. The details are as below:

Table 11. F_1 hybrids released from Pantnagar

Crop	Hybrid	National ID	Parentage	Characteristic features
Bottle gourd	Pant Sankar Lauki 1	541196	PBOG-22 × PBOG-40	Fruits are 35 cm long and green in colour vine length is 5.5 m. First picking in 60 days. Yield potential is 400 q/ha.

Contd....

	Pant Sankar Lauki 2	541197	PBOG-22 x PBOG-61	Fruits about 40 cm long, club shaped and smooth green in colour. First harvest in 65 days. Yield potential is 400 q/ha. Suitable for both plains & hills.
Cucumber	Pant Sankar Khira 1	541201	PCUC-28 x PCUC-8	Fruit are long, cylindrical and green with light stripes. First picking in 50 days. Yield potential is 200 q/ha.

Apart from these hybrids, one more hybrid, Pant Sankar Khira 2 (PCUC 15 × PCUC 25) has been identified at All India Coordinated Vegetable Improvement Project meeting for release.

In addition to these, several hybrids have been developed and released both at National and State level (Table 12).

Table 12. List of cucurbits hybrids identified at National (through AICRP) and state levels

Cucurbits crops	Name of identified hybrids	
	At national level	**At state level**
Muskmelon	MHY-5, Pusa Rasraj	Punjab Hybrid-1, MHY-3, MHL-10, DMH-4
Watermelon	Arka Jyoti	RHRWH-12
Cucumber	Hybrid No.1	Pusa Sanyog, AAUC-1, AAUC-2
Bottle gourd	NDBH-4	Pusa Manjari, Pusa Hybrid-2, NDBGH 7, Kashi Bahar, Azad Sankar-1
Bitter gourd	Pusa Hybrid, NBGH-167	RHRBGH-1
Summer squash	—	Pusa Alankar
Pumpkin	—	Pusa Hybrid-1, NDPKH-1

Resistance Breeding

Cucurbits are highly susceptible to several biotic and abiotic stresses. The first requirement of resistant breeding is to find out the resistance sources. Resistance sources are generally present in landraces and wild relatives. Resistance to downy mildew (*Pseudopernospora cubensis*) is reported in snap melon (*Cucumis melo* var. *momordica*), resistance to fruit fly is reported in *Cucumis callosus* etc. Most of the resistant varieties in cucurbits have been developed by simple selection. The crossability relationships between the cultivated and potential wild species are known, which can be utilized for interspecific hybridization followed by backcross to transfer resistant genes.

Table 13. Major biotic stresses and their sources of resistance

Crop/Biotic Stress	Resistance Sources
Muskmelon	
Powdery mildew	PMR-45, PMR-450, PMR-5, PMR-6, PI-124111
Downy mildew	MR-1, PI414723, DMDR-1, DMDR-2
CGMMV	DVRM-1,2, *Cucumis africanus, C. ficifolium, C. anguria*
CMV	PI 161375
WMV	PI 414723
Fruit fly	*Cucumis callosus*
IPM, DM, MMV	MR-12
Nematode	*Cucumis metuliferus*
White fly	*Cucumis asper, C. denteri, C. dipsaceus, C. sagittatus*
Water melon	
Anthracnose	PI 189225
Bottle gourd	
CMV, SMV, WMV	PI 27 1353
Cucumber	
Anthracnose,	PI 175111, PI 175120, PI 179676, PI 182445,
Downy mildew	B-184, B-159
Powdery mildew	PI 200815, PI 200818, *Cucumis hardwickii*
CMV	MR-12, SMR-15, SMR-18
Pumpkin	
PM and Viruses	*C. lundelliana, C. martenzii*
ZYMV, WMV	*C. ecuadorensis, C. faetidistima, C. martenezii*

Characterization of Indigenous Germplasm

1. Through morphological markers

In hybrid seed production of all the crops including cucurbits, the possibility of inclusion of accidental selfed seeds in the hybrid seed can not be ruled out. But if the male parent has a novel seedling morphological trait showing dominance over the counterpart normal trait of female parent, the F_1 seedlings will show the novel phenotype and the selfs will show normal phenotype. One such trait, segmented leaf has been detected for the first time in bottle gourd at Pantnagar and the variant genotype has been registered as PBOG-54 with National Bureau of Plant Genetic Resources, New Delhi.

Generally genotypes of bottle gourd have normal leaves with entire margin. But PBOG-54 which has been developed following pureline selection in the indigenous germplasm collection from eastern Uttar Pradesh has segmented leaves and fruits with mottled skin and somewhat crooked neck. Incidentally, this phenotype has been shown to be governed by one dominant gene (S) and thus holds promise to be used as dominant seedling marker trait in distinguishing hybrid *vs.* selfed seedlings in bottle gourd provided it is used as the male parent (Tiwari *et al.*, 2006).

2. Seed protein electrophoresis

The seed protein obtained by polyacrylamide gel electrophoresis is important in biosystematics for species and cultivar identification, studying the origin of polyploidy plants and evolution of crop plants. The polyacrylamide gel electrophoresis has been utilized in different cucurbits to characterize the germplasm lines. The techniques used for cultivar identification have been described by Cooke (1984).

In bottle gourd, thirty one genotypes could be resolved into a total of nineteen bands distributed in three zones i.e. A, B and C. PBOG 54 showed similar band pattern as normal leaf type genotype PBOG 52, PBOG 74, a round fruited genotypes, had profile similar to long fruited genotype PBOGI 61 and PBOG 62. PBOG 81, PBOG 117 and PBOG 128 which were all long fruited showed distinct banding pattern than other long fruited genotypes (Upadhyay and Ram, 2006). In cucumber, 19 genotypes could be resolved into 17 bands distributed into 3 zones i.e. A, B and C Zone A had 6 bands, zone B had 7 bands and zone C had 4 bands. These genotypes could be grouped into 8 dissimilar groups based on presence or absence of bands (Singh and Ram, 2001). In the study involving pureline and hybrid of cucumber (Pant Khira 1 and Pant Sankar Khira 1), both of them were found distinguishable from each other as Pant Khira-1 had 6 bands and Pant Sankar Khira 1 had 3 bands (Sharma and Ram, 2003). The three open-pollinated varieties, one each of ridge gourd (Pant Torai-1), bitter gourd (Pant Karela-1) and long melon (Pant Kakri-1) which have been developed from indigenous stock had distinct and distinguishable banding pattern. Pant Torai-1 had A_1, A_2, B_1, B_2, B_3, C_1, C_4 and D bands; Pant Karela-1 had A_1, B_1, B_2, B_3, C_3, C_4 and D while Pant Kakri-1 had A_1, B_2, B_3, C_2, C_4 and D bands (Sharma and Ram, 2003).

In similar study involving 33 germplasm lines of pumpkin, the seed proteins could be resolved into 11 bands distributed into four zones i.e. A, B, C and D zone. A comprised of 5 bands, zone B had 2 bands, zone C had 3 bands and zone D included single band (Kumar *et al.,* 2006).

In view of above, it could be concluded that electrophoresis of seed protein was successful in characterizing different genotypes in crops as cucumber, bitter gourd, ridge gourd, bottle gourd, pumpkin etc. while in certain cases as in bottle gourd it could not differentiate between two morphologically distinguishable genotypes.

3. DNA profiling

The use of DNA profiling technique and different approaches of its application for cultivar identification are increasingly being used. RAPD is widely used in labs and is likely to become important in the area of cultivar identification and verification.

In a study involving 38 genotypes of bitter gourd including few commercial cultivars collected from different parts of India were analysed for diversity study at molecular level using RAPD markers. It was found that among the 116 random decamer primers screened, 29 were polymorphic and informative enough to analyze these genotypes. A total of 208 markers generated, 76 were polymorphic and the number of bands per primer was 7.17 out of them 2.62 were polymorphic. Pair wise genetic distance (GD) based on molecular analysis ranged from 0.07-0.50 suggesting a wide genetic base for the genotypes (Dey *et al.,* 2006). Nine inbred lines of ash gourd were analyzed using 130 RAPD primers. It was found that out of 130 primers, 26 amplified 163 DNA marker bands. A total of 47 polymorphic bands were obtained with a mean of 1.8 per primer which, in combination, discriminated all the inbreds from each other. Pair-wise genetic distance measurements ranged from 0.056- 0.179, suggesting a narrow genetic base for the inbreds.

At Pantnagar, five parental lines viz., PBOG-13 (round fruit), PBOG-22, PBOG-61 and Pusa Naveen (long fruits) and PBOG-54 (almost long fruits with segmented leaves) were subjected to RAPD analysis using 10 decamer primers. Seven out of ten RAPD primers could amplify genomic DNA in the genotypes. Polymorphism was detected by scoring the presence and absence of bands. Out of the seven primers which could amplify the genomic DNA, primer 19940 gave maximum amplification meaning there by that all five parental lines were amplified. Thus, it could be concluded that primer 19940 could be successfully used to have different RAPD profiles in PBOG-54 *vs.* other normal leaf genotypes and this opens scope of searching RAPD markers having linage with segmented leaf morphology in bottle gourd (Ram *et al.*, 2006).

Therefore, it can be said that cucurbits make a big group and has large genetic diversity across the country which need to be collected, characterized, evaluated, conserved and utilized. If utilize its carefully these can play an important role in agriculture diversification. A good number of high yielding varieties and F_1 hybrids could be developed from this genetic stock. Molecular techniques may be utilized for identification of germplasm and study of inter-specific differentiation in cultivated species and wild relatives. These tools also need to be deployed to overcome crossability barrier and to enhance utilization of wild species.

References

Acharya, R. R.; Patel, A. D.; Patel, J. A.; Patel S. B. and Kathiria, K. B. 2006. Heterobeltiosis and standard heterosis in ridge gourd. *In:*. Cucurbits Breeding and Production Technology. Hari Har Ram and H. P. Singh (eds.). Proceedings of National Seminar on Cucurbits. Sep 22-23, 2005, Pantnagar. pp 301-306.

Bairagi, S. K.; Ram, Hari Har; Singh, D. K. and Maurya S. K. 2005. Exploitation of hybrid vigour for yield and attributing traits in cucumber. *Indian J. Hort.*, **62(1):** 41-45.

Chakravorty, H. L. 1982. Fascicles of flora India, Fascicle II, Cucurbitaceae, BSI, Howarh.

Chaubey, Arun Kumar and Ram, Hari Har 2004. Heterosis for fruit yield and its components in bitter gourd (*Momordica charantia* L.). *Veg. Sci.,* **31:** 51-53.

Cooke, R. J. 1984. The characterization and identification of crop cultivars by electrophoresis.

Electrophoresis, **5:** 59-72.

Dey, S. S.; Singh, A. K.; Chandel, D. and Behera, T. K. 2006. Genetic diversity of bitter gourd (*Momordica charantia* L.) genotypes revealed by RAPD markers and agronomic traits. *Scin. Hort.*, **109:** 21-28.

Dhillon, B. S.; Varaprasad, K. S.; Srinivasan, K.; Singh, M.; Archak, S.; Srivastava, U. and Sharma, G. D. 2001. National Bureau of Plant Genetic Resources: A Compendium of Achievements. National Bureau of Plant Genetic Resources, New Delhi. pp 329.

Dubey, R. K., 2004. Development of hybrids and seed protein electrophoresis in bottle gourd [*Lagenaria siceraria* (Mol.) Standl]. Ph. D. Thesis, G. B. Pant University of Agriculture and Technology, Pantnagar.

Dubey, S. K. and Maurya, S. K. 2003. Studies on heterosis and combining ability in bottle gourd [*Lagenaria siceraria* (Mol.) Standl.]. *Indian J. Genet.*, **63:** 148-152

FAO, 2004. www.fao.org.

Kumar, Jagesh; Singh D. K. and Ram, Hari Har 2006. Protein profiling study of indigenous germplasm lines of pumpkin (*Cucurbita moschata* Duch. ex. Poir.). *Veg. Sci.*, **33(1):** 10-12.

Malik, S. S.; Srivastava, Umesh; Tomar, J. B.; Bhandari, D. C.; Pandey, Anjula; Hore, D. K. and Dikshit, N. 2001. Plant exploration and germplasm collection. *In*: National Bureau of Plant Genetic Resources: A Compendium of Achievements. National Bureau of Plant Genetic Resources, New Delhi. pp 17.

Maurya, S. K.; Ram, H. H. and Singh, J. P. 2003. Hybrid breeding in bottle gourd. *Prog. Hort.*, **35:** 46-50.

Mishra, J. P. and Sheshadri, V. S. 1985. Studies on heterosis in muskmelon. *Genet. Agr.*, **39:** 367-376.

Munshi, A. D. and Verma, V. K. 1997. Studies on heterosis in muskmelon (*Cucumis melo* L.). *Veg Sci.*, **24(2):** 103-106.

Nath, P. and Dutta, O. P. 1970. Heterosis in water melon. *Indian J. Hort.*, **27:** 176

Padma, M.; Neerajam, G.; Padmavatamma, A. S.; Reddy, I. P. and Gautam, B. 2003. Evaluation of F_1 hybrids in bottle gourd. *Veg. Sci.*, **30:** 54-56.

Pandey, Sudhakar; Rai, Mathura; Singh, B. and Pandey, A. K. 2005. Heterosis and combining ability in ash gourd [*Benincasa hispida* (Thunb.) Cogn.]. *Veg. Sci.*, **32(1):** 33-36.

Pandit, M. K.; Hazra, P. and Dutta, A. K. 2006. Study on heterosis for yield and quality characters of pumpkin (*Cucurbita moschata* Duch. Ex. Poir.). Cucurbits Breeding and Production Technology. *In:* Proceedings of National Seminar on Cucurbits. Hari Har Ram and H. P. Singh (eds.). Sep 22-23, 2005, Pantnagar. pp 319-327.

Purohit, V. L.; Dhaduk, L. K.; Mehta, D. R. and Gajipara, N. N. 2006. Heterosis studies in ridge gourd [*Luffa acutangula* (Roxb.)L.]. Cucurbits Breeding and Production Technology. *In:* Proceedings of National Seminar on Cucurbits. Hari Har Ram and H. P. Singh (eds.). Sep 22-23, 2005, Pantnagar. pp 289-296.

Rai, Mathura; Pandey, Sudhakar; Ram, D.; Kumar, Sanjeet and Singh, Major 2006. Cucurbits Research in India. Cucurbits Breeding and Production Technology. *In:* Proceedings of National Seminar on Cucurbits. Hari Har Ram and H. P. Singh (eds.). 22-23 Sep. 2005, Pantnagar. pp 1-20.

Ram D.; Kalloo, G. and Singh, Major 1997. Heterosis in bitter gourd (*Momordica charantia* L.). *Veg Sci.*, **24(2):** 99-102.

Ram, Hari Har; Sharma, Kavita and Jaiswal, H. R. 2006. Molecular characterization of promising genotypes in bottle gourd including a novel segmented leaf type through RAPD. *Veg. Sci.*, **33(1):** 1-4.

Sharma, K. and Ram, H. H. 2003. Varietal variation in released varieties of vegetable crops from Pantnagar by SDS-PAGE of seed protein. Role of Indigenous Germplasm in Improvement of Horticultural Crops. *In:* Proceedings of National Seminar. Ranvir Singh, Hari Har Ram and J. P. Tiwari (eds.).June 24-25, 2003, Pantnagar. pp 157-163.

Sirohi, Sanjai; Rana, S.C. and Rajkumar 2005. Heterosis study in bottle gourd [*Lagenaria siceraria* (Mol.). Standl.]. *In:* Abstracts. National Seminar on Cucurbits. September 22-23, 2005, Pantnagar. pp. 39.

Singh, D. K.; Singh, R. D. and Singh, A. 1992. Heterosis in bitter gourd (*Momordica charantia*). *Narendra Dev J. of Agric. Res.*, **7(1):** 164-168.

Singh, A. K.; Gautam, N. C. and Singh, R. D. 1999. Heterosis in cucumber (*Cucumis sativus* L.). *Veg Sci.*, **26(2):** 126-128.

Singh, D. K. and Ram, Hari Har 2001. Characterization of indigenous germplasm lines of cucumber (*Cucumis sativus* L.) through SDS-PAGE. *Veg. Sci.* **28:** 22-23.

Singh, M; Kumar, Gunjeet; Abraham, Z; Phogat, B. S; Sharma, B. D. and Patel, D. P. 2001. Germplasm Evaluation. *In*: National Bureau of Plant Genetic Resources: A Compendium of Achievements. National Bureau of Plant Genetic Resources, New Delhi. pp 138-139.

Singh, D. K.; Ram, Hari Har and Jaiswal, H. R. 2003. Indigenous cucurbits and their breeding potential. Role of Indigenous Germplasm in Improvement of Horticultural Crops. *In:*. Proceeding of National Seminar. Ranvir Singh, Hari Har Ram and J. P. Tiwari (eds.) June 24-25, 2003, Pantnagar. pp 94.

Sirohi P. S. 1994. Heterosis in pumpkin (*Cucurbita moschata* Duch. Ex. Poir.). *Veg Sci.*, **21(2)**: 163-165.

Srivastava, Umesh 2006. Genetic Resources Management in Cucurbits. Cucurbits Breeding and Production Technology. *In:* Proceedings of National Seminar on Cucurbits. Hari Har Ram and H. P. Singh (eds.). Sep 22-23, 2005, Pantnagar. pp 55-56.

Swarup, V. 1991. Breeding Procedures for Cross Pollinated Vegetable Crops. ICAR Publication, New Delhi. pp 14-15.

Tiwari, Akhilesh; Ram, Hari Har and Jaiswal, H. R. 2006. Inheritance of segmented leaf in bottle gourd. *In:* Cucurbits Breeding and Production Technology. Proceedings of National Seminar on Cucurbits. Hari Har Ram and H. P. Singh (eds.). 22-23 Sep. 2005, Pantnagar. pp 219-226.

Upadhyay, Megha and Ram, Hari Har 2006. Characterization of bottle gourd germplasm based on SDS-PAGE of seed protein. *Veg. Sci.,* **33(1):** 18-20.

❑❑❑

Chapter - 20

Plant Genetic Resource Management of Tomato

D.K. Singh and H. Choudhary

Introduction

Tomato is one of the most important vegetable crops which is grown throughout the world. It is a rich source of minerals, vitamins and antioxidants and considered as a protective food. It is used in several forms ranging from raw to processed one and is widely used as one of the food ingredient. Its popularity is due to its versatility and the variety it lends to the human diet. Tomato leads all other vegetables in terms of total global production except potato which is often classified as starchy staple. India stands fourth in terms of global tomato production after China, USA and Turkey. The total production of tomato in India during 2007-08 was 10.26 million tones which accounted for 8.9 % of total vegetable production. Though concrete historical records of its first introduction into the country from its primary centre of diversity in South America do not exist, tomato is presumed to have been brought here during the second half of the 18th century through Far Eastern countries. It is believed that tomato was introduced in India during British period in the year 1828 by Royal Agri -Horticultural Society, Calcutta. The story of tomato transformation from an exotic fruit to a popular dietary item and a major item of commerce all over the world is one of the most fascinating events in the saga of crop domestication. As recently as 1900, tomato was avoided in the belief that it was poisonous because of its known relation to the nightshades and other toxic members of the nightshade family. The tomatine is a predominant alkaloid mainly present in foliage and green fruits. However, at the stage of ripening, tomatine is degraded into an inert compound which is not toxic.

Centre of Origin and Domestication

Tomato belongs to the genus *Lycopersicon* of the family Solanaceae. It is generally accepted that the cultivated tomato originated from the new world (the Americas) since all the wild relatives of the cultivated tomato are native of western South America along the coast and high Andes from central Ecuador, through Peru to northern Chile and in the Galapagos islands. From these native habitats, tomato was taken to Europe soon after the discovery of the New World. It was possibly in already a fairly advanced stage of domestication. The most likely ancestor of cultivated tomato is the wild cherry tomato (*L. esculentum* var. *cerasiforme*) which is spontaneous throughout the tropical and subtropical America and has later spread throughout the tropics of the old world. The primitive genetic materials of (*L.esculentum* var.*cerasiforme*)was further domesticated in Mexico. From here, tomato was introduced into the Old World (Europe) from where it was distributed worldwide in different times. Selection over many generations presumably led to increase in fruit size and higher fruit ratio of fruit weight to seed content that are typical of the modern day tomato. The name tomato comes from Nahutal language of Mexico and variants of this name have followed the tomato on its spread throughout the world.

Taxonomic and Biosystematic study

Tomato is a member of Solanaceae or the nightshade family, a collection of some 1500 tropical and subtropical species probably originating in Central and South America. Since the tomato was introduced to Europe in the sixteenth century, early botanists recognized the close relationships of tomato with the genus *Solanum*, and commonly identified them as *S. pomiferum*. Tournefort (1694) considered the multilocular character of the fruit as a criterion to differentiate it from the genus *Solanum* and put in the other genus *Lycopersicon.* Linnaeus (1753) again classified tomatoes in the genus *Solanum* and under the name *Solanum lycopersicum.* On the other hand Miller (1754) reconsidered Tournefort's claasification and formally described the genus *Lycopersicon.* This classification continued prevailing treatment by several classical and modern authors. In the recent time, the phylogenetic relationships within the Solanaceae have been examined with molecular data and these molecular studies unequivocally supported tomato to be firmly interested in the genus *Solanum.* Based on these results a new phylogenetic classification has assigned tomato to the genus *Solanum* (Spooner, 2005). Although most taxonomists today place tomato in *Solanum*, most agronomists and horticulturists are reluctant to use the *Solanum* names on tradition or practical goal of maintaining familier names.

The genus *Lycopersicon* is a relatively small collection of species that are normally divided into two groups, the *Eulycopersicon* and the *Eriopersicon*. Fruits of the *Eulycopersicon* are usually red or yellow in colour when ripe. The cultivated tomato, *Lycopersicon esculentum*, belongs to this group. In contrast, fruits of the *Eriopersicon* remain green or purple green throughout their development. The first generally accepted description of several species of tomato including common one was made by Miller in 1768, hence various texts refer to the Latin name of tomato as *L. esculentum* Mill. in honour of Miller. Rick (1976) proposed another sub generic classification based on the utilization of these species into two group (i) those species which can be easily crossed with the cultivated tomato- *esculentum* complex and those not crossable, *peruvianum* complex. The *esculentum* complex consisted of six species of *Lycopersicon*, out of which three have coloured fruits (*L. esculentum; L.pimpinellifolium; L. cheesmani*) and three with green fruits (*L. parviflorum, L. Chmielewskii, L.hirsutum*). The two species forming the peruvianum-complex are *L. chilense* and *L.peruvianum* and the latter is a highly polymorphic species.

Types of Tomato

Two main types of tomatoes may be distinguished based on growth habit viz. determinate type (or simply "bush" type) and indeterminate type. The bush type cultivars are widely grown for either fresh consumption or for processing into canned tomatoes, puree, soup, juice or ketchup. The indeterminate cultivars are grown mainly for fresh consumption. Very rarely are they grown for processing as the processing industry has relied increasingly on high volume production over a short time span, a trait that was impossible to select and breed from indeterminately growing genotypes. Modern processing cultivars are suitable for mechanical harvest (once-over harvest), have the firmness of fruit necessary to withstand rough handling by machine, and have the required processing quality, viz. high soluble solids and good color.

Genetic Diversity

Several research study on the nature of genetic diversity of *L. esculentum* has facilitated the understanding of the evolution of the cultivated forms and their further genetic improvement. The general agreement from such studies is that the cultivated and wild forms of *L. esculentum* are remarkably homogenous outside their native region of western south America. In contrast, more variation is present in the native region both within and between accessions. Several factors have been postulated to account for these differences. Interbreeding with wild species in the native region, greater activity of insect pollinators promoting out crossing, effects of artificial selection, longer period of existence in the native region allowing the accumulation of more variants, and genetic drift in the long migration from South America through Central America and Mexico coupled with natural selection for limited number of successful genotypes. All members of the genus *Lycopersicon* are annual or short-lived perennial herbaceous diploids with a somatic chromosome number of 24. Tomato is the most intensively studied *Solanaceae* genome due to its simple diploid genetics, short generation times, routine transformation technology, and the advantage of readily available rich genetic and genomic resources. The tomato genome size (1C amount) is generally considered as approx.95 pg of DNA.

The genus as a whole tolerates a wide range of climatic conditions. For instance, *L. pennellii* survives in very dry, rolling hills of western Peru whereas *L. cheesmanii* has been observed to grow in slat-laden atmospheres, just a few meters above the high water mark of the northwestern shores of the Galapagos islands. In general, the relatives of the cultivated tomato have proven to be invaluable sources of desirable genes for further genetic improvement. Some of the examples of useful characters derived from wild relatives are as follows:

L. pimpinellifolium : *Fusarium* & *Verticillium* wilt, septoria leaf spot, leaf mould, root rot, late blight, bacterial wilt, bacterial canker and tobacco mosaic virus

L. peruvianum: septoria leaf spot, leaf mould, root rot, tobacco mosaic virus, leaf curl virus, tomato etch virus, spotted wilt virus, and root knot nematode

L. hirsutum : *Fusarium* wilt, septoria leaf spot, leaf mould, early blight, bacterial canker, tomato etch virus, pin worm, olorado potato beetle, spider mite, aphids and leaf miner

L. hirsutum var. glabratum: root rot, tobacco mosaic virus, leaf curl virus, fruit worm, pin worm, horn worm, spider mite, leaf miner, cold and frost

L. glandulosum: septoria leaf spot, and root rot

L. cheesmanii: alkalinity and salinity

L. pennellii ; spider mite, aphids and drought conditions

L. chilense: curly top virus

Genetic Resources

A large number of germplasm of tomato is being maintained at AVRDC, Taiwan, USDA (USA), VIR (Russia), IVT(Netherlands), DHUNA (Peru) and NIAS (Japan). The tomato Genetic Stock Centre (TGSC) at he University of California, Davis(USA) maintains a large collection of *Lycopersicon* and *Solanum* species, genetic stocks. Some of the germplasm lines useful for various stresses and maintained at TGSC are given below. For detailed list of tomato germplasm for special trait and botanical identification see the available information on the web site of the Tomato Genetics Resource Center (TGRC, University of California, Davis) at http://tgrc.ucdavis.edu/key.html).

Drought tolerance

L. pennellii (general feature): LA0716, and others

L. chilense (esp. coastal sites): LA1958, LA1959, LA1972, and others

S. sitiens (general feature): LA1974, LA2876, LA4105, and others

Flooding tolerance

L. esculentum var. *cerasiforme* (wet tropics): LA1421, and others

S. juglandifolium, *S. ochranthum* (probably a general feature): LA2120, LA2682

High temperature tolerance

L. esculentum cv.s Nagcarlang (LA2661), Saladette (LA2662), Malintka-101 (LA3120), Hotset (LA3320)

Chilling tolerance

L. hirsutum (from high altitudes): LA1363, LA1393, LA1777, LA1778

L. chilense (from high altitudes): LA1969, LA1971, LA4117A

S. lycopersicoides (from high altitudes): LA1964, LA2408, LA2781

The National Bureau of Plant Genetic Resources (NBPGR), New Delhi has a large collection of tomato germplasm with an approximate number of 3000. Few important accession for specific traits are as below.

Early	EC99935, EC362940, EC362947
High TSS (>6^{0}B)	EC162519, EC113820, EC103460
Pear shaped	EC35332, EC367857-1
Processing type	Bulgarian, EC246028
Tolerance to high temperature	EC251674, HS-101, EC 321425-26, EC 347359-68, EC 399828-38
High yield (> 5 Kg)	EC35346, EC129154, EC13903
Stamen less	EC 346011
Corolla and stamen less	EC 346013
Gren pistillate	EC 346014
Long shelf life	EC 391024
Slow ripening	EC 310299
Paste type	EC 321425-26

Utilization of Genetic Resources

Introduction

Major boost to tomato cultivation in the country was provided by the introduction of high-yielding exotic cultivars like Sioux, Roma, Marglobe, Red Cloud, Best of All, La Bonita, Fireball, Moneymaker, Chico Grande, Balkan and few others from 1950 onwards. Over the years, indigenous high-yielding cultivars have been bred from the old local cultivars, the early introductions and more significantly, the newly introduced cultivars and breeding lines.

Selection

Individual plant selection in heterogenous population of local cultivar Meeruti grown around Meerut in U.P resulted in new cultivar Improved Meeruti. Similarly, Punjab Kesari and Angurlata were selected from local cultivars. Several improved cultivars were developed by selection in heterogenous populations of exotic cultivars, such as Arka Vikas from Tip Top, Arka Saurabh from V-685 of Canada, Arka Ahuti from UC 83B(USA), Co-1 from pearl Harbour and Co-2 from a Russian introduction.

Inter specific hybridization

Wild species have also been a source of nutritional traits which can be exploited for development of varieties having better quality traits. It has been utilized for the development of tomato genotypes having high vitamin C content and elevated level of TSS. The importance of the allied taxa of vegetable crops to combine resistance genes in commercial cultivars was realized long ago and practically interspecific gene transfer has been demonstrated in a few vegetable crops but its utilization is highly restricted. Some of the achievements in India has been given in following table.

Crop	Variety	Trait	Remarks
Tomato	Pusa Red Plum	Vit.C	*L. esculentum X L.pimpinnelifolim*
	Hisar Anmol	Leaf curl virus	Hisar Arun × *L. hirsutum* f *glabratum*
	H-86 (Kashi Vishesh)	TLCV	*L. hirsutum* f *glabratum* (B6013) X Sel 7 (*L. esculentum)*
	DVRT-1 (Kashi Amrit)	High Yield	*L. esculentum* (Sel 7) X *L. hirsutum* f. *glabratum* (B6013)
	DVRT-2 (Kashi Anupam)	High Yield	*L. esculentum* (Sel 7) X *L. hirsutum* f. *glabratum* (B6013)

Resistance Breeding

Virus resistance

Severe damage is caused by a group of geminiviruses transmitted by *B. tabaci* in tomato plantings throughout the Mediterranean region, the Middle East, north Africa, central Africa and southeast Asia. These related although distinct geminivirus species, are collectively referred to as *Tomato yellow leaf curl virus* (TYLCV). This geminivirus was accidentally introduced in the last decade into the Americas in the early 1990s where it has already caused millions of dollars worth of industrial and fresh tomato production losses. Early efforts to identify sources of resistance to TYLCV within *L. esculentum,* only revealed the existence of some moderately resistant or tolerant genotypes. Some wild relatives of tomato, namely *L. pimpinellifolium* and *L. peruvianum,* possessed a higher level of resistance to TYLCV, although they were not immune. Crosses between *L. esculentum* and *L. pimpinellifolium* (currant tomato/accession LA 121) and genetic analyses of F1–3 and backcross generations, indicated the existence of incomplete dominance of resistance over susceptibility, suggesting a monogenic control of resistance. A dominant gene (*Tylc*) was later proposed for the resistance gene in *L. pimpinellifolium* (Kasrawi, 1989). The progenies derived from this cross showed only moderate symptoms, but their yield was markedly reduced. Nevertheless, among the *Lycopersicon* species, *L. pimpinellifolium* is one of the most compatible for crossing with *L. esculentum* (Picó *et al.*, 1996). In contrast, the inheritance of tolerance to TYLCV in *L. peruvianum* (PI 126935) is controlled by five recessive factors. This breeding program initiated in 1977, resulted in the release of the commercial hybrid TY-20, in 1988. This hybrid delays symptom expression and viral DNA accumulation in infected plants, resulting in acceptable yields (Pilowski and Cohen, 1990). Other tolerant/resistant TY-lines generated by this breeding program are: TY172, TY197, TY198, and TY536 (Friedmann *et al.*, 1998). In 1991, other wild tomato species: *L. chilense* and *L. hirsutum*, besides *L. peruvianum* and *L. pimpinellifolium*, were examined for the presence of viral DNA and symptom expression following their inoculation with whiteflies removed from TYLCV-infected tomato plants. Approximately 85 days after inoculation, all of the above species had infected plants with detectable levels of viral DNA, but *L. chilense* and *L. hirsutum* were the most resistant species, with the majority of the inoculated plants remaining symptomless, and only few containing viral DNA (Zakay *et al.*, 1991). The TYLCV resistance gene in *L. chilense* was identified as *Ty-1* (Michelson *et al.*, 1994). The resistance to this virus in *L. hirsutum*, on the other hand, seems to be dominant and controlled by more than one gene (Mazyad *et al.*, 1982). *L. hirsutum* has been crossed with *L. esculentum*, yielding tolerant and immune lines. One of the immune lines was crossed with *L. esculentum*, to produce the hybrid FAVI-9 or Line F1-901. The immune reaction was associated with 2-3 additive genes (Vidavski and Czosnek, 1998). Another promising species evaluated for TYLCV resistance, *L. cheesmanii*, possesses recessive resistance to TYLCV. Breeding projects in the Mediterranean region have also used *L. cheesmani, L. peruvianum* and *L. pimpinellifolium* to control TYLCV in this region (Laterrot and Moretti, 1996). Some of the TYLCV-resistant lines obtained from this project are: Pimpertylc -J-13 and Chepertylc -92. Interespecific hybrids obtained from crosses between *L. pimpinellifolium, L. peruvianum,* and *L. hirsutum*, show transgressive segregation for their reaction to TYLCV, suggesting the existence of different, complementary genes (Kasrawi and Mansur, 1994).

Leaf curl begomoviruses cause serious yield losses to Indian tomato crops. The most definitive method utilises nucleotide sequences and genome organisation , which allows the following four Indian TLCB species to be demarcated: *Tomato leaf curl Bangalore virus* (ToLCBV), which has strains [Ban1] , [Ban4] , [Ban5] and [Kolar] ; *Tomato leaf curl Gujarat virus* (ToLCGV) and various strains thereof ; *Tomato leaf curl Karnataka virus* (ToLCKV) ; and *Tomato leaf curl New Delhi virus* (ToLCNDV) and strains thereof (Reddy, *et al.*,2005) In 1991, Muniyapa and coworkers reported that lines of *L. hirsutum* and *L. peruvianum* were resistant to another tomato geminivirus: *Tomato leaf curl virus* (ToLCV). The resistance mechanism in these wild species was subsequently associated with the presence of exudates from trichome glands on the leaf surface, in which whiteflies became entrapped (Channarayappa and Shivashankar, 1992). This is one of the few cases where genetic resistance to a viral disease has been achieved indirectly by incorporating genetic traits against *B. tabaci*. Nevertheless, there is sufficient evidence showing that different cultivars of plant species such as common bean and tomato, interact differentially with *B. tabaci*. For instance, In Sinaloa, northwestern Mexico, the common bean cultivar 'Azufrado Peruano-87', had 16% more nymphs/leaf than the geminivirus (BCaMV) resistant common bean cultivar 'Azufrado Higuera' (Lopez, 1996). Similar data has been obtained for tomato, although the preference shown by *B. tabaci* for some tomato cultivars, was not related to virus resistance/ susceptibility traits in the tomato cultivars evaluated (Avilés, 1996).

At Hisar L. *hirsutum f. glabratum* was found completely resistant whereas L. *peruvianum, L. pimpineliifolium* and *L. hirsutum* were resistant. Among cultivated varieties minimum incidence was found in HS 101 and maximum in Punjab Chhuhara. Wild spp. *L. hirsutum* (LA 386, LA 1777, PI 390513, *L. glandulosum* (EC. 68003) and *L. peruvianum* PI 127830 and PI 127831 were resistant. Resistance in L. *pimpinellifolium* was controlled by single incomplete dominance whereas, in *L. hirsutum* f. *glabratum* it was controlled by two complementary factors. Resistance in *L. pimpineltifolium* and *L. hirsutum* f. *glabratum* is linked with small fruit size and late maturity. Resistance to tomato leaf curl virus is associated with the phenolic compounds: resistant genotypes had more amounts of total phenols than susceptible genotypes. Furthermore, all the susceptible genotypes had lower content of non-reducing sugar than resistant cultivars. After disease incidence phenols were increased in susceptible lines whereas in resistant wild species *L. hirsutum* f. *glabratum* it remained almost static. TLCV resistance from *L hirsutum* f *glabratum* and *L. pimpinellifolium* to *L. esculentum* was transferred by back cross method. Resistant genes from *L. hirsutum f. glabratum have* been incorporated in the cultivated varieties and resistant lines 'H 2', 'H 11', 'H 17', 'H 23', and 'H 24' have been developed. Furthermore, tolerance to tomato leaf curl virus was transferred from *L. pimpinellifolium* to cultivated varieties and thus tolerant lines LCP 22, LCP2, LCP 3 and LCP 9 have been developed.

Tospoviruses similar or identical to tomato spotted wilt virus (TSWV) are recognized as infecting more than 1000 monocot and dicot species worldwide (Peters, 1998). Sporadic tospovirus epidemics have plagued tomato production around the globe. Epidemics in Hawaii (Cho et al., 1989), South America (Maluf et al., 1991), and South Africa (van Zijl et al., 1986) have made tospoviruses one the most important limiting factors in tomato production in subtropical and tropical areas. Plant resistance and various strategies that reduce viral transmission from plant to plant are the only effective methods of controlling tospoviruses (Mumford et al., 1996). Sources of tospovirus resistance have been identified within *Lycopersicon* species, and resistant tomato cultivars have been developed. As reviewed by Stevens et al. (1992, 1994) resistance has been identified in *L. pimpinellifolium* (Jusl.) Mill., *L. hirsutum* Humb. & Bonpl., *L. peruvianum* (L.) Mill., and *L. esculentum* cultivars 'Rey de los Tempranos' and 'Manzana.' The TSWV resistant cultivar 'Pearl Harbor' derived its resistance from *L. pimpinellifolium* and the TSWV resistant cultivars 'Anahu' and 'Stevens' have resistance derived from *L. peruvianum* (Stevens et al., 1992). Lines derived from 'Stevens,' with the single dominant gene (*Sw-5*), have demonstrated broad tospovirus resistance. Boiteux & Giordano (1993) found that *Sw-5* provided resistance to isolates of three species of tospovirus: TSWV, TCSV (tomato chlorotic spot virus), and GRSV (groundnut ring spot virus).

References

Channarayappa, A., Shivashankar, G., 1992. Resistance of *Lycopersicon* species to *Bemisia tabaci* , a tomato leaf curl virus vector. *Can. J. Bot.* **70**, 2184-2192.

Channarayapa, A., Shivashankar, G., Muniyappa, V., Frist, R.H., 1992. Resistance of *Lycopersicon* species to *Bemisia tabaci*, a tomato leaf curl virus vèctor. *Can. J. Bot.* **70**, 2184- 2192.

Czosnek, H. Laterrot, H, 1997. A worldwide survey of tomato yellow leaf curl viruses. *Arch. Virol.* **142:** 1391-1406.

Friedmann, M., Lapidot, M., Cohen, S., Pilowski, M., 1998. A novel source of resistance to tomato yellow leaf curl virus exhibiting a symptomless reaction to virus infection. *J. Amer. Soc. Hort. Sci.* **123:** 1004-1006.

Kasrawi, M.A., 1989. Inheritance of resistance to tomato yellow leaf curl virus (TYLCV) in *Lycopersicon pimpinellifolium*. Plant Dis. **73:** 435-437.

Kasrawi, M.A., Mansour, A., 1994. Genetics of resistance to Tomato yellow leaf curl virus in tomato. J. *Hort. Sci.*, **69:** 1095-1100.

Kalloo,G. 1991. Genetic improvement of tomato. Theor. Appl. Genet. Monogr.14, Springer Verlag, Berlin.

Kalloo, G. Srivastava, U, Singh, M and Kumar, S. 2005. Solanaceous Vegetables. pp 19-33. *In*: B.S. Dhillon, R.K.Tyagi, S.Saxena and G.J.Randhawa (eds.), Plant Genetic Resources: Horticultural Crops, Narosa Publishing House, New Delhi.

Laterrot, H., Moretti, A., 1996. Chepertylc lines. Tomato Yellow Leaf Curl Newsletter No. **8:** 4.

Mazyad, H.M., Hassan, A.A., Nakhla, M.K., Moustagfa, S.E., 1982. Evaluation of some wild *Lycopersicon* species as sources of resistance to Tomato yellow leaf curl. *Egypt. J. Hort.* **9:** 241- 246.

Michelson, I., Zamir, D., Czosnek, H., 1994. Accumulation and translocation of tomato yellow leaf curl virus (TYLCV) in a *Lycopersicon esculentum* breeding line containing the *L. chilense* TYLCV tolerance gene *Ty-1*. *Phytopathology* **84:** 928-933.

Ramappa, H.K., 1991. Reaction of *Lycopersicon* cultivars and wild accessions to tomato leaf curl virus. *Euphytica* **56,** 37-41.

Nakhla, M.K., Maxwell, D.P., Martinez, R.T., Carvalho, M.G., Gilbertson, R.L., 1994. Occurrence of the Eastern Mediterranean starin of tomato yellow leaf curl geminivirus in the Dominican Republic. *Phytopathology* **84:** 1072.

Picó, B., Díez, M.J., Nuez, F., 1996. Viral diseases causing the greatest economic losses to the tomato crop. II. The Tomato yellow leaf curl virus-a review. *Scientia Horticulturae*, **67:** 151-196.

Pilowski, M., and Cohen, S., 1990. Tolerance to tomato yellow leaf curl virus derived from *Lycopersicon peruvianum*. *Plant Dis.* **74:** 248-250.

Rick, C.M .1976. Tomato (family Solanaceae) pp. 268-273. *In*: N.W.Simmonds(eds.), Evolution of Crop Plants. Longmen Publications, U.K.

Rick, C.M; Fobes, J. F. and Holle, M. 1977. Genetic variation in *Lycopersicon pimpinnelifolium*: evidence of ebolutionary change in mating systems. *Pl Syst Evol* 127: 139-170.

Spooner, D. 2005. New species of wild tomatoes(*Solanum* section *Lycopersicon*: Solanaceae). *Systematic Botany* **30(2):** 424-434.

Swarup, V. 2006. Vegetable Science and Technology in India. Kalyani Publishers, India.

Vidavski, F., Czosnek, H., 1998. Tomato breeding lines immune and tolerant to tomato yellow leaf curl virus (TYLCV) issued from *Lycopersicon hirsutum*. *Phytopathology* **88:** 910-914.

Zakay, Y., Navot, N., Zeidan, M., Kedar, N., Rabinowitch, H., Czosnek, H., Zamir, D., 1991. Screening *Lycopersicon* accessions for resistance to tomato yellow leaf curl virus: presence of viral DNA and symptom development. *Plant Dis.* **75:** 279-281.

Chapter – 21

Breeding Approaches for Utilization of Wild Species in the Improvement of Vegetable Crops

H. Choudhary and D.K. Singh

Introduction

Global food productivity will have to rise significantly to satisfy an expanding world population. Increased vegetable production is one option towards the efficient conversion of natural resources into food stuffs. As a group, vegetables are a diverse range of herbaceous plants that do not belong to any particular botanical species, genus or family. Individual members may be at various stages of domestication. Vegetables are typically annual herbaceous species, whose plant parts (leaves, stems, flowers, fruits, roots or tubers) are consumed as fresh or as cooked product, and as an additional source of vitamins, minerals, fibre, phytochemicals, protein and/or calories. Health benefits are also claimed for some vegetable species, mainly due to antioxidant activity and capacity to scavenge free radicals.

Domesticated plants have been fundamentally altered from their wild relatives; these species have been moved into and adapted to new environments. They have become dependent on the farmer's hand and they have been reshaped to meet human needs and wants. Modern crops are the results of thousands of years of these evolutionary processes. Like all biological evolution, crop evolution involves two fundamental process; the creation of diversity and selection (Harris and Hillman, 1989). Crop evolution is distinguished by two types of selection; one natural and another artificial or conscious.

These evolutionary processes must continue in order for agriculture, a living and evolving system to remain viable.

In the hands of plant breeder's sexual hybridization is a powerful tool for producing superior plants by combining characters distributed in different members of a species or different species of a genus. Genetic variability within the species has been efficiently utilized by breeders in their efforts to improve crops. However, the existing variability in a breeding population may not be sufficient for modern plant breeding purposes, and thus great efforts have been made to broaden the existing gene pool of crops through utilization of wild germplasm.

Evolution of crop species

All crop plants grown today have their origin in wild taxa. The initial domestication of crop plants took place in pre historic times as a consequence of the collection of wild species for food. The ranges of wild species used by different pre-historic groups are a reflection of the ecotypes in which they existed. The hunter-gatherer-fisher communities occupied generalized ecosystems or the ecotones between major ecosystems. These environments would have provided the widest possible utilizable range of flora and fauna throughout the year, which promoted a settled existence. The move from semi-settled gatherer to agriculturist must have been a slow process of trial and error based on conscious and unconscious selections. The classic example of evolutionary change is the simple mutation to non-shattering heads in cereal grasses.

Frankel and Soule (1981) distinguished two different levels of adaptations, first as adaptations in wild species mediated by natural selection under natural conditions and secondly as adaptations under domestication resulting from natural selection and conscious selection by man. Once the second phase of adaptation was established then crops became transportable entities and would have been distributed via migration or trade into diverse geographical habitats. These movements will have the subjected domesticates to a range of environments and to contacts with related wilds species. The former provided opportunities for recombination and mutations to be selected within a gene pool and latter offered the chance of introgressive hybridization to enhance the overall gene pool. This continual selection of crops in a step-wise in process has produced an evolutionary continuum from pre-domesticates to modern cultivars. In many crops the wild/ weedy species or primitive cultivars representing these steps can be identified today thereby forming an evolutionary continuum in both time and space. Grubben (1977) estimated that about 1500 wild species may have been used by primitive man to supplement his diet and this number has been gradually reduced through increasing horticultural specialization to leave only 20 species in today's intensified cultivation. The categorization of vegetables based on the stages of crop evolution extant in present day horticulture has been defined by Crisp and Astley, 1985 which has been more clarified with the examples by Ram (1997). There are 4 categories of vegetables which are described below.

1. Vegetables which have changed little from their wild state. (Chives, Loofah).
2. Landraces (Coriander, Yard-long bean, wax gourd).
3. Advanced cultivars with narrow genetic base (Garlic, Globe Artichoke, Parsnip, Tomato and Cucumber).
4. \Advanced diverse cultivars (Onion, Lettuce, Radish, Carrot, Eggplant, Squash, French-bean).

Importance of wild germplasm

Wild relatives of crop species are a rich source of valuable traits, from which we have realized only a small fraction of the potential gains to crop improvement. The range of genetic variation for a trait is often much greater in wild germplasm than among cultivated types, as cultivated types are usually derived from a small number of ancestors and have been selected intensely over many centuries. Consequently, the current breeding pool for many crops, especially self pollinated species contains only a small fraction of the existing genetic variation. The vegetable crop species are rather limited with genetic diversity for some important traits, such as resistance to biotic and abiotic stresses. With the rapid industrial development and increasing population, the vegetable cultivation is shifting from peri-urban areas towards marginal land and new genes should be searched for their better adaptation and the improvement in productivity and quality as well. The diversity of the wild relatives has enabled them to survive longer than the oldest cultivated variety without any human assistance. Thus wild germplasm act as sources of resistance to stresses and they offer a treasure of genes for crop improvement programmes. Wild species have also been a source of nutritional traits which can be exploited for development of varieties having better quality traits. It has been utilized for the development of tomato genotypes having high vitamin C content and elevated level of TSS. The importance of the allied taxa of vegetable crops to combine resistance genes in commercial cultivars was realized long ago and practically interspecific gene transfer has been demonstrated in a few vegetable crops but its utilization is highly restricted. Biotechnology has extended possibilities to carry the desired gene across the sexual boundaries for sustainable development of agriculture. The different wild relatives of tomato having desirable traits are given below.

L. pimpinellifolium : *Fusarium* & *Verticillium* wilt, septoria leaf spot, leaf mould, root rot, late blight, bacterial wilt, bacterial canker and tobacco mosaic virus

L. peruvianum: septoria leaf spot, leaf mould, root rot, tobacco mosaic virus, leaf curl virus, tomato etch virus, spotted wilt virus, and root knot nematode

L. hirsutum : *Fusarium* wilt, septoria leaf spot, leaf mould, early blight, bacterial canker, tomato etch virus, pin worm, colorado potato beetle, spider mite, aphids and leaf miner

L. hirsutum var. glabratum: root rot, tobacco mosaic virus, leaf curl virus, fruit worm, pin worm, horn worm, spider mite, leaf miner, cold and frost

***L. glandulosum*:** septoria leaf spot, and root rot

***L. cheesmanii*:** alkalinity and salinity

S. pennellii ; spider mite, aphids and drought conditions

***L. chilense*:** curly top virus

The different species of vegetable crops having resistance to diseases and insect-pests are presented below in Table1.

Table 1. Wild species of vegetable crops as a source of resistance

Genus	Priority species	Source of Resistance
Solanum	*S. khasianum*	Shoot and fruit borer
	S. gilo	Phomopsis rot
	S. integrifolium	Phomopsis rot
	S. sisymbrifolium	Shoot and fruit borer, aphids, root knot nematodes
	S. macrosperma	Drought
Capsicum	*C. baccatum*	*Phytophthora* rot, cucumber mosaic virus
	C. chinense	Leaf curl virus, tobacco mosaic virus, *Verticillium* wilt
	C. frutescense	Leaf curl virus
Abelmoschus	*A. manihot sp. manihot*	YVMV
	A. tetraphyllus	YVMV
	A. caillei	YVMV, fruit borer, Macrophomina, Cercospora and jassid
Cucurbita	*C. lundelliana*	Powdery mildew
Cucumis	*C. callosus*	Downey mildew
	C. metuliferus	Root knot nematode, fruit fly and water melon mosaic virus
	C. melo var. momordica	Downey mildew
	C. anguria	
C. africanus	Fruit fly	
Vigna	*V. unguiculata sp. biflorus*	Golden mosaic virus, cowpea mosaic virus, Cercospora

Distant Hybridization

Hybridization or crossing in a broad sense can be defined as any natural or artificial fusion of two genetically different cells leading to hybrid progeny. Sexual crossability is the probability of obtaining hybrid progeny form the fusion in pairs of different specialized haploid gametes, and somatic crossability is the possibility of obtaining hybrid progeny from the fusion in pairs of different naked unspecialized somatic cells or protoplasts. When the probability is zero, the parents concerned display non-crossability or incongruity. Introduction of new traits has been based mainly on sexual crosses between different genotypes within or between closely related species. Crossing between different species of the same genus or different genera of the same family is called distant hybridization. Crosses between different species of the same or different genera are commonly known as wide crosses. Distant hybridization is of two types, viz. (1) Interspecific hybridization, and (2) Intergeneric hybridization

Interspecific Hybridization

Crossing or mating between two different species of the same genus is referred as interspecific hybridization. It has been used to improve crops by transferring specific traits, such as pest and stress resistance to crops from their wild relatives. In nature, about 30 to 35% of flowering plant species were created by interspecific hybridization, followed by chromosome doubling. Successful construction of an allopolyploid results in the creation of a new combination of genomes or the production of a species that did not exist previously. However, great efforts may be required to hybridize cultivated and wild species. The first man made interspecific hybrid was synthesized by Thomas Fairchild in 1717 between carnation *(Dianthus caryophyllus* L.) and sweet william (*Dianthus barbatus* L.). Since then several interspecific crosses have been attempted, but success has been rather limited.

Intergeneric Hybridization

Intergeneric hybridization refers to crossing between two different genera of the same family. Such crosses are rarely used in crop improvement because of various problems associated with them. It has been generally used in asexually propagated species. Intergeneric hybridization was used by some workers to develop new crop species. Perhaps the first intergeneric cross was made between radish and cabbage by Karpechenko in 1928 in Russia. The main objective was to combine root of radish with leaves of cabbage. The fertile amphidiploid was named as *Raphanobrassica* but the new species developed had roots like cabbage and leaves like radish which was a useless combination. It took plant breeders about 100 years to produce Triticale, a new crop species created from the cross of wheat and rye by Rimpau.

Barriers to Interspecific Hybridization

The interspecific transfer of resistance largely depends upon success of interspecific hybridization. Such crosses have immensely contributed to plant improvement. In recent years, wide crosses between crop plant and their wild relatives have attracted significant attention as sources of desirable characters for genetic improvement of crops. Sometimes it is imperative that genetic material from wide hybrids be developed and exploited in breeding programs. However, due to the presence of various reproductive barriers, gene transfer has been restricted to sexually compatible species, thus limiting the possibilities of modifying and improving crop plants. When two species have a normal pollen- pistil relationship, then after crossing ,an unimpeded chain of processes is initiated, based on a perfect interaction between pollen genes (or gene complexes) and matching pistil genes (or gene complexes). Incomplete matching means one or more missing links in the chain and thus partial or complete failure of hybridization & introgression. Interspecific crosses results ranging from no fruit set to poor fruit set, fruit set without seeds or parthenocarpic fruits, and shrunken and unviable seed formation. Some of the probable barriers are as follows

- Geographic isolation of species
- Genetic barriers
  - Ploidy level variation
  - Variation in chromosome structure
  - Cytoplasmic barriers
  - Involvement of lethal genes
- Flowering behaviour
- Presyngamic barriers
- Failure of pollen germination
- Slow, short and poor pollen tube growth
- Release of gametes for fertilization
- Postsyngamic barriers
  - Embryonic breakdown
  - Failure of zygote development
  - Abnormal fertilization
  - No endosperm development
  - Inhibition of embryo development

Unilateral Incompatibility

The success is attained only in one direction and reciprocal differences are observed between species with regard to crossability *Viz., Solanum, Lycopersicon, Capsicum, Abelmoschus, Cucumis*

Measures for Successful Hybridization

- Early pollination
- Repeated pollination
- Sufficiently large number of flowers are pollinated
- Application of growth regulators
- Determining the barrier
  - Species with shorter style should be used as female parent
  - A part of style may be cutoff to make it shorter: *S. pinnatisectum* × *S. bulbocastanum*

- Hybridization with more than one strain of each of the parental species (*L. peruvianum* × *L.esculentum,)*
- Autopolyploidy
- Hybridization under different environmental conditions

- When species with different ploidy levels are crossed
  - Crossed directly, species with higher ploidy level as female
  - Chromosome number of wild species or interspecific hybrid may be doubled
  - Chromosome number of polyploid species may be doubled to obtain semisterile interspecific hybrid
  - Chromosome number of higher ploidy level species may be reduced

Bridge cross

When two species are not crssable but a third species is crossable to both the species so third species can be utilized as a bridging species for transfer of desirable traits between earlier two species. The bridge cross can be used to overcome the barriers and it has been very much utilized in case of potato. *S. bulbocastenum* has some valuable traits but it is not crossable with *S. tuberosum. S.pinnatisectum* was found crossable with both *S. tuberosum* and *S. bulbocastenum* exhibiting the possibility to be used as bridging species. *S. Acaule* (2n=4x) and *S. phureja* (2n=2x) have been used as bridging species for transfer of resistant trait from *S. bulbocastanum* to *S. tuberosum* and the procedure has been shown below.

S. Acaule (2n = 4x) *X* *S. bulbocastanum* (2n = 2x)

↓

F1 (colchicine treatment)

↓

Fertile hexaploid hybrid *X* *S. phureja* (2n = 2x)

↓

Triple hybrid *X S. tuberosum*

↓

Quadriplex hybrid (Resistant to *P. infestans)*

Consequences of interspecific hybridization on the segregation pattern in F2 and later generations

- Enhances the possibility of selecting better recombinants
- Segregation ratio is highly disturbed and cannot be accounted for by classical Mendelian inheritance
- Frequency of recombination is greatly restricted mainly due to gametic and zygotic elimination and pleiotropy and linkage

- Selected recombinants may or may not be superior to the parents

Successful Use of Wild Species

The different species of wild relatives of vegetable crops have been utilized for incorporation of disease resistance and other traits which are presented below.

Crop	Species	Trait
Tomato	*L. pimpinellifolium*	Fungi, vitamin C
	L. cheesmanii	Low input, drought, color
	L. chmielewskii	Quality, soluble solid, color
	L. parviflorum	Soluble solid, color
	L. hirsutum	Fungi, insects, Carotene
	L. pennellii	Drought
	L. peruvianum	Fungi, virus, nematodes, pest, vitamin C
	L. chilense	Virus, drought
	L. esculentum cerasiforme	Foliar fungi
Potato	*S. demissum*	Late blight, virus, beetle
	S. tuberosum andigena	Late blight
	S. acule	Virus
	S. spegazinii	Virus, Globodera rostochiensis
	S. stoloniferum	Virus
	S. vallis-mexici	
	S. phureja	Late blight Globodera rostochiensis
	S. vernei	Virus
Brinjal	*S. sisymbriifolium*	Root knot nematode
Okra	*Abelmoschus manihot*	Diseases & pest
	A. tetraphyllus; crinitus	Powdery mildew; jassids

Achievements

Some varieties of vegetable crops have been developed in India with the utilization of wild species which are given below.

Crop	Variety	Trait	Remarks
Tomato	Hisar Anmol	Leaf curl virus	Hisar Arun × *L. hirsutum* f *glabratum*
Okra	Parbhani Kranti	YVMV	Pusa Sawani × *A. manihot*
	Punjab Padmini	YVMV	*A. esculentus* × *A. manihot ssp manihot*
	P7	YVMV	Pusa Sawani × *A. manihot*
	Sel 10	YVMV	*A. esculentus* × *A. manihot ssp tetraphyllus*
	Sel4(Arka Abhay)	YVMV	*A. esculentus* × *A. manihot ssp tetraphyllus*

Biotechnological Approaches

Exchange of genes between cultivated and wild species and transfer of resistant gene (s) from wild species to cultivated varieties are restricted due to certain barriers (presyngamic and postsyngamic) in the sexual hybridization. Abortion of embryos at one or the other stage of development is a characteristic feature of distant hybridization.

Therefore, the *in vitro* method has been successfully used to overcome such barriers in the distant hybridization. Generally, embryo culture & ovule culture have been employed to overcome the abortion of young hybrid embryos. The hybrid embryos are rescued from getting aborted by excising them from the ovaries or ovules. In some cases, however, it is not technically possible to take embryos out of the ovules. To rescue these embryos whole ovules or even ovaries containing them are cultured.

Embryo culture

Embryo culture involves isolating & growing an immature or mature zygotic embryo under sterile conditions on an aseptic nutrient medium with the goal of obtaining a viable plant. The basic premise for this technique is that the integrity of the hybrid genome is retained in a developmentally arrested or an abortive embryo and that its potential to resume normal growth may be realized if supplied with the proper growth substances. The technique depends on-

(a) Isolating the embryo without injury

(b) Formulating a suitable nutrient medium &

(c) Inducing continued embryogenic growth & seedling formation.

It would be pertinent to make a distinction, albeit arbitrary, between embryo culture & embryo rescue. The former is a broader term and covers the basic research as well as applied aspects, such as breaking seed dormancy, testing seed viability, shortening breeding cycle etc. On the other hand, the term embryo rescue is restricted to only those cases where the embryos, if not rescued are endangered and would not form seedlings. The situation is very frequently encountered when the embryos are the result of distant crosses. The application of this technique in vegetable crops has been presented in Table 2.

Table 2. Application of embryo rescue in vegetable crops

Interspecific crosses	Character transferred
Abelmoschus esculentus X *A. moschatus*	*Resist to pathogen*
A. esculentus X *A. ficulneus*	*Do*
A. ficulneus X *A. moschatus*	*Do*
A. tuberculatus X *A. moschatus*	*Do*
Allium cepa X Allium sativum	*Do*
Allium species	*Do*
Brassica oleracea X B. Campestris	*Do*
B. oleracea var. *Capitata* X *B. Parachinensis*	*Do*
Capsicum annuum L. X *capsicum pubescens*	*Do*
C. frutescens X C. Pubescens	*Do*
C. pendulum X C. annuum	*Do*
Cucumis sativus L. X *C. hystrix* Chakr.	*Resistance to pathogen*
Cucumis metuliferus X C. anguria	*do*
Cucumis metuliferus X C. séller	*do*
Cucúrbita pepo L. X *C. martinezii*	*Resistant to PM, CMV*
Lycoperlicon esculentum X L. Peruuanum	

The importance of wild cucumber species has long been recognized as they possess resistance to several pathogens and serious efforts have been made to create inter-specific hybrid. In 1859, Naudin first tried to cross melon with cucumber and other species. Historically various approaches (traditional and biotechnological)for inter-specific hybridization have been used in *cucumis* to overcome the fertilization barriers between cucumber and wild species but with only limited success. Few successful inter-specific hybridization studies between cultivated *cucumis* species and the wild relatives are presented in following table.

Table 3. Wide cross between cultivated and wild *Cucumis* species

Cross	Result	Reference
C.sativus x C. melo	Globular stage embryo	Niemirowicz-Szczytt and only Kubicki, 1979
C. zeyheri x C.sativus	Fruits with inviable seeds	Custers and Den Nijs, 1986
C.sativus x C. metuliferous	Embryos only	Franken and Soule 1988
C.sativus x C. hystrix	Sterile plants	Chen *et al.*, 1997
C.hystrix x C. sativus	Fertile plants	Chen *et al.*, 1998
C.sativus x C. metuliferous	Seeds obtained but not	Walters and Wehner,2002 viable

The wild relative of cucumber, *C. hystrix* is a potentially valuable source of many economic traits, including resistance to root knot nematode (Chen and Kirkbride 2000,) tolerance to low irradiance and temperature (Qian *et al.*, 2002; Zhuang *et al* 2002) and resistance to downy mildew. The cross between cucumber and *C. hystrix* Chakr (2n=24) was the first repeatable cross between a cultivated *Cucumis* species and a wild relative and represented a breakthrough in inter-specific hybridization in *Cucumis* (Chen *et al.,* 1997). The success of this cross is of much importance because parental species have different chromosome numbers. The original F_1 hybrid (2n=19) obtained by embryo rescue following pollination of *C. sativus* by *C. hystrix* has 7 chromosomes from *C. sativus* and 12 chromosome from *C. hystrix* and was both male and female sterile caused by lack of homology and improper pairing during meiosis. While the multiple branching habit, densely brown hairs on corolla and pistil, orange- yellow corolla and ovate fruit of F1 hybrid plants were similar to that of *C. hystrix* parent and the appearance of first pistillate flower was more similar to that of *C. sativus* parent. The chromosome number in the hybrid was doubled through somaclonal variation during embryo culture and regeneration process to restore the fertility. Pollen grains were produced by these progeny when *C.hystrix* was used as seed parent and plants produced fertile flowers and set fruits with viable seeds indicating that fertility was restored. This restoration of fertility marked the creation of a new synthetic species which has close phylogenetic relationships with its parental species but it is distinctively different from each. It has the genome HHCC and chromosome number 2n=4x=38. Nutritional analysis indicated that the synthetic species had higher protein (0.78%) and mineral (0.35%)content compared to normal pickling cucumber (0.62% and 0.27% respectively) (Chen and Qian, *et al.,* 2002) This synthetic species might be useful as a new *Cucumis* crop or may be useful as a bridging species for transfer of useful traits to cucumber. The progress has been made towards achieving gene introgression from *C. hystrix*, following synthesis of a new fully fertile amphidiploid species *C.hytivus* (HHCC, 2n=38) (Chen *et al.,* 2003a) and the production of partially fertile allotriploids hybrids from a mating between *C. hytivus* and *C. sativus* (HCC, 2n =26) (Chen *et al.,* 2003b). When this allotriploid genotype was treated with colchicines to induce polyploidy, two monosomic alien addition lines (MAALs) were recovered. Each of these plants was morphologically distinct from allotriploids and cultivated cucumbers. Alien chromosome addition lines harboring one single chromosome from the wild species *C. hystrix* might be used as a bridge to transfer genes of interest originating in *C. hystrix* to individual chromosomes of *C. sativus* via recombination or translocation events. Successful examples of gene introgression through MAALs have been obtained in several important crops (Chen *et al.,* 2004).

Somatic hybridization

Many desirable and agronomically interesting trails may only found in distantly related species or even in unrelated organisms. Since they constitute a genetic resource potential, considerable effort has been allocated to identify and isolate these genes and transfer them into crops. Where interesting genes have been identified and isolated, they can be transferred by transformation, but, for most traits the genes have not been identified, and somatic hybridization might then be the method of choice. Besides being of value for the transfer of unidentified genes, somatic hybridization is a tool for the modification and improvement of polygenic traits. Furthermore, the modification of organelles genetic material is possible via somatic hybridization since a mixture of the two fusion partners is obtained in the hybrid cell.

Somatic hybridization may be an important first step to introgression breeding. The pre & post zygotic crossability barriers to sexual hybridization treated so far do not apply to somatic hybridization. However, finding the right procedure for protoplast culture, fusion & selection of hybrids and for regeneration to plant is crucial for success. The rate of success is also very much dependent on plant family, genus, species and even cultivar. Because there is virtually no barrier, to the fusion of protoplasts the scope of hybridization and introgression is very much widened by somatic hybridization. The results of the fusion of two somatic cells of different species, genera, or family are considered in somatic hybridization. This is also known as protoplast fusion or parasexual hybridization. Somatic hybridization has a great advantage over sexual method for interspecific and intergeneric hybridization. A number of species and genera are incompatible to each other due to barriers in sexual hybridization. The common presyngamic & postsyngamic sexual hybridization barriers are overcome by this technique. This technique has been widely used for the improvement of vegetable crops. Its application in different crops has been given in a Table 4.

Table 4. Somatic hybrids in vegetable crops

Protoplast fused	Character
Brassica Oleracea + sinapis alba	Resist to Altenaria
B. oleracea + B. napus	Resist. gene to black rot
B. oleracea + B. rapa	'Anand' cytoplasm for CMS
B. oleracea ssp. *Capitata* + Ogura CMS broccoli line	cold-tolerant ogura CMS into cabbage
B. oleracea + B. carinata	Alternaria & phomalingum
B. oleracea Var. *Botrytis + Raphanus sativus*	Clubroot
Citrullus lanatus + Cucumis melo	—
Cucumis melo + C. sativus	—
L. esculentum + S. melongena	—
L. esculentum + S. ochranthum	Diseases, Insect pest
Solanum tuberosum + S. Pinnatisectum	Late blight
S. tuberosum + S. bulbocastanum	Root knot nematode.
S. commersonii + S. tuberosum	Frost tolerant
S. tuberosum + S. circaeifolium	Late blight
S. tuberosum + S. torvum	Verticilium
S. tuberosum + S. brevidens	Late blight

New Vegetable Developed by Somatic Hybridization

A new vegetable called "Senposai" (1000 treasure vegetable) which is being test marketed near Tokyo, is expected to hit the world market. It has been developed by Tokita Seed Company and the kirin Brewery company in Japan. 'Senposai" is a cross between an ordinary cabbage and 'komatsuna', a type of Chinese cabbage. The breeders have used somatic cell hybridization to produce this new vegetable, which incorporates the cabbage resistance to high temperature.

Hybrid

Hybrids can be produced, where nucleus is obtained from one parent and cytoplasm is derived from both, thus producing cytoplasmic hybrids which are also called cybrids. These cybrids can be produced using any one of the following methods

1. Fusion of normal protoplasts from one parent with enucleated protoplast from the other parent
2. Fusion of normal protoplasts from one parent and protoplasts containing non-viable nuclei from other.
3. Selective elimination of one of the nuclei from heterokaryon.

The technique of cybrid production can be utilized for transfer of cytoplasmic male sterility. Melcher et al (1994) has shown the transfer of CMS in tomato from the cytoplasm of *solanum acaule* by cybrid production.

Genetic Engineering

Genetic engineering refers to a set of technologies that are being used to change the genetic makeup of cells and move genes across species boundaries to produce novel organisms. The advent of recombinant DNA technology has opened up tremendous possibilities for transforming almost any plant by transferring any gene from any organism, across taxonomic barriers. Transgenic plants refer to plants carrying the stably integrated foreign genes into it own genome. A transgenic plant is one in which a transgene, that has been manipulated by recombinant DNA techniques has been integrated into the plant genome. The recombinant DNA technique involves the introduction of only one or a few genes, either genes derived from non related organism, which code for new characters, or gene constructs based on the plant's own genes but modified in such a way as to interfere with existing gene expression. Whereas in breeding programme with wild species, a series of backcross is needed to eliminate the bulk of wild characters introduced with the desired one, genetic transformation allows the direct introduction of genetic material into elite breeding lines.

Steps involved in Genetic Engineering

- Isolate the gene with the desired genetic characteristics.
- Make several copies of the isolated gene.
- Transfer the desired genes to the plant's own genome.
- Create a new plant from the genetically modified tissue.
- Check that the inserted genes function as expected.
- Check that the inserted gene appears in the plant's seed

Application

The first edible transgenic plant FLAVR SAVR tomato with delayed ripening reached the market in 1994 by Calgene in USA which was developed using antisense technology but it was not commercially viable. Transgenic potato expressing Bt toxin gene against Colorado potato beetle developed by Monsanto has been commercialized as New Leaf. The Bt gene has been engineered into many vegetable crops including tomato, brinjal, cabbage, broccoli, cauliflower. Bt brinjal developed by MHYCO seed company in India is under field evaluation trial. Genes coding for antifreeze proteins isolated from the arctic flounder (an edible fish) have been introduced into tomato and plants transformed with glycerol-3-P acyltransferase were found to be cold tolerant. AmA1 gene from Amaranthus has been incorporated into potato for high protein content. Tobacco osmotin gene for resistance to late blight has been transferred into potato.

Use of Molecular Markers

In case of interspecific breeding, when the chromosomes of different species are sufficiently alike to recombine more or less freely their recombinant products are likely to be greatly inferior to the parental combinations. In the classical approach of plant breeding, there is no way of controlling this total disruption of the parental genotypes which results from segregation. Molecular markers provide a means of controlling the degree of disruption of the parental genotype. One can use molecular markers to determine which chromosomal regions carry agriculturally desirable genes, and also to reveal which individuals carry those chromosomal regions. One can then restore most of the domestic parent genotype and consequently phenotype by backcrossing, while retaining the desired genes and traits from the wild parent by selecting for nearby genetic markers. Molecular markers may even be used to accelerate the restoration of the domestic parent's attributes, by selecting against markers from the wild parent outside the regions carrying target genes. RFLP marker has been utilized in tomato to eliminate undesirable regions of donor genome. QTL mapping showed that a segment of chromosome 1 from *Lycopersicon chmielwskii* was valuable in elevating the total soluble solids concentration of *L. esculentum* . This marker was used to identify chromosome segments from *L. chmielwskii* and one plant in the first backcross was found which carried only 5 undesirable segments of *L. chmielwskii* genome in addition to the desired region of chromosome 1. After two backcross to *L. esculentum* , using RFLP marker to select these undesirable segments, a single plant was found which ws carrying only the segment of interest. (Paterson, *et.al,* 1990)

Future Perspective

Many simply inherited traits such as disease and insect resistances and even quantitative traits such as high TSS in tomato have been extracted from wild germplasm. In addition to agriculturally valuable traits, wild species exhibit many properties which are important to survival in their natural environment, but are ill suited to agricultural productivity. Some examples of such properties are seed dormancy, shattering, tall stature, excessive vegetative growth, unpleasant flavour or odour, toxins, thorns, small seedy fruit, and nonuniform maturity. Breeders are often reluctant to make crosses with wild germplasm for fear that such undesirable traits will be difficult to separate from valuable attributes. With the advancement in biotechnological tools especially rapid development of molecular markers may help in utilization of wild relatives for the improvement of vegetable crops. The potential of these new tools for utilizing wild germplasm can only be realized if the germplasm is readily available. Although extensive collections of certain species of wild plants have been made, many species are less extensively collected and in most species there are surely novel variants not yet found. The potential value of wild germplasm, and the substantial risk of extinction of as yet uncollected types in some areas, create an urgent need for accelerating the collection and characterization of wild plants.

References

Astley, D. (1987). Genetic resource conservation. *Expl. Agric.*, **23**: 245-257.

Chen, J.F. and Kirkbride, J.H. Jr. (2000). A new synthetic species *Cucumis* (*Cucurbitaceae*) from interspecific hybridization and chromosome doubling. *Brittonia,* **52**: 315-319.

Chen, J.F.; Luo, X.D.; Staub, E.; Jahn, M.M.; Qian, C. h. T.; Zhuang, F.Y. and Ren, G. (2003b). An allotripoid derived from a amphidiploidxdiploid mating in *cucumis* I production, micropropagation and verification. *Euphytica* **131**: 235-241.

Chen, J.F. and Qian, Ch. T.2002). Studies on Chromosome preparation using plant tendril as source tissue. *Acta Hortic Sin,* **29**: 378-380.

Chen, J.F., Staub, J.E.; Qian, Cht.; Jiang, J.M., Luo, X.D. and Zhuang, F.Y. (2003a). Reproduction and cytogenetic characterization of interspecific hybrid derived from cucumis hystrix *Chakr. X C. sativus* L. *Theor Appl genet.*, **106**: 688-695.

Chen, J.F.; Staub, J.E.; Tashiro, Y. and Miyazaki, S. (1997). Successful interspecific hybridization between *Cucumis sativus* L. and C. hystrix Chakr. *Euphytica*, **96**: 413-419.

Chen, J. F. and Adelberg, J. (2000). Interspecific hybridization in *Cucumis*- progress, problems and perspectives. *Hort Sci.,* **35(1):** 11-15.

Chen, J. F. and Staub, J. E. (1997). Attempts at colchicines doubling of an interspecific hybrid of *Cucumis sativus* L. *X C. hystrix Chakr. Cucurbit Genet. Coop. Rpt.*, **20:** 24-26.

Chen, J. F.; Adelberg, J. W.; Staub, J.E.; Skorupska, H.T. and Rhodes, B.B. (1998). A New synthetic amphidiploid in *cucumis* from a *C. Sativus X C. hystrix* F_1 interspecific hybrid, p. 336-339. In: J. McCreight (ed.), cucurbitaceae '98-Evaluation and enhancement of cucurbits germplasm. ASHS Press, Alexandria, Va.

Chen, J.F.; Luo, X.D.; Qian, C.T.; Jahn, M.M.; Staub, J.E.; Zhuang; F.Y.; Lou, Q.F. and Ren, G. (2004). *Cucumis monosomic* alien addition lines: morphological, cytological, and genotypic analyses. *Theor Appl Genet.,* **108:** 1343-1348.

Chen, J.F.; Isshiki, S.; Tashiro Y.; and Miyazaki, S. (1997a). Biochemical affinities between *C. hystrix* and the two cultivated *Cucumis* species. *Euphytica,* **97:** 139-141.

Crisp, P.C & Astley, D. (1985). Genetic resources in vegetables. In progress in plant breeding **1:** 281-310 (Ed. G.E. Russell). London pp. 281-310.

Cross, R.J. (1998). Global genetic resources of vegetables. *Plant Varieties and Seeds.*, **11:** 39-60.

Custers, J.B.M. and Den Nijs, A. P. M. (1986). Effects of aminoethoxyvinylglycine (AVG), environment and genotype in overcoming hybridization barriers between *Cucumis* species. *Euphytica,* **35:** 639-647.

Frankel, O.H. & Soule, M. E. (1981). Conservation and Evolution Cambridge: Cambridge University press.

Grubben, G. J. H. (1977). Tropical Vegetables and their genetic resources. Rome: International board for plant genetic resources.

Harris, D.R. and Hillman, G.C. (1989). Foraging and farming. The evolution of plant exploration London.

Kalloo, G. (1998). Vegetable Breeding. Panima Educational Book Agency, New Delhi. Pp 135-170.

Naudin, C. (1859). Revue des Cucurbitaceae cultivates au museum en 1859. *Ann. Sci. Nat. Ser. 4 Bot.*, **12:** 79-164.

Niemirowicz-Szczytt. K. And Kubicki, B. (1979). Cross fertilization between cultivated species of genera Cucumis L. and Cucurbita L. *Genetica Polonica,* **20:** 117-125.

Paterson, A. H.; Lancder, E.S.; Hewitt, J. D.; Peterson, S.; Lincoln, S. E and Tanksley, S. D. (1988). Revolution of quantative traits into Mendelian factors, using a complete linkage map of restriction fragment length polymorphisms. *Nature,* **335:** 721-726.

Paterson, A.N.; Deverna, J.W.; Lanini, B. and Tanskey, S. (1990). Fine mapping of quantitative trait loci using selected overlapping recombinant chromosomes in an inter species cross of tomato. *Genetics,* **124:** 735-742.

Paterson, A. H.; Tanksley, S.D. and Sorrells, M.E. (1991). DNA markers in plant improvement. *Advances in Agronomy,* **46:** 39-89.

Qian, CT.; Chen, J.F.; Zhuang, F.Y.; Xu, Y.B. and Li, S.J. (1985). Electrophotosynthetic characters of the synthetic species *Ciucumis hytivus* Chen and Kirkbride under weak light condition. *Plant physiol Commum*, **38:** 336-338

Ram, H. H. (1997). Vegetable breeding: principles and Practices. Kalyani Publishers, Ludhiana.

Stalker, H.T. (1980). Utilization of wild species for crop improvement. N.C. Brady(Ed.). *Advances in Agronomy*, **33:** 111-147.

Zhuang, F.Y.; Chen, J.F.; Qian, Ch.t.; Li, Sh. J.; Ren, G. and Wang, Zh. J. (2002). Responses of seedlings of *cucumis xhytivus* and progenies to low temperature. *J. Nanjing Agric Univ.*, **25:** 27-30.

❑❑❑

Chapter – 22

Biochemical Characterization of Indigenous Genetic Resources of Vegetable Crops

A.K. Gaur and D. Khokhar

Introduction

Germplasm identification as a part of a philosophy of consumer protection extends through the entire seed trade and allied industries, from plant breeders to food consumers. Increasingly, countries are introducing schemes for Plant Breeders (or germplasm) Rights. These reward breeders financially for their efforts and offer protection for their germplasm, but in turn require that new germplasm are distinct from others and also uniform and stable in the expression of their characteristics. Since, the genetic variations in plants and plant populations are of considerable interest to have desirable traits to develop a commercial variety. Biochemical and molecular biology based data and its analysis for information regarding selection have been realized of utmost importance.

Identification is interpreted broadly to include the following:

1. Identification in the strict sense.
2. Germplasm discrimination.
3. Germplasm characterization.

Germplasm identification has been traditionally carried by what might be called a classical taxonomic approach. This requires a detailed study of seeds and /or growing plants and the observation, recording and analysis of a number of morphological characters or descriptors. In practice, such an approach is extremely successful and largely forms the basis, for instance, of current D, U and S testing procedures. However, it can be time consuming and expensive. Large areas of land are needed in addition to highly skilled and trained personnel, making what are often subjective decisions.

Many of the morphological descriptors used are multigenic, quantitative or continuous characters, the expression of which is altered by environmental factors. Again in some species, the number of descriptors is limited or no longer sufficient for discrimination between all varieties. There are thus good grounds for finding alternative procedures to augment the morphologically based approach. The appeared physical appearance of a plant arises from an interaction between its genotype and environment. In general, it is increasingly desirable objective to reduce or eliminate the environmental influence, so that the genotype of a germplasm can be observed and utilized more directly.

Modern diagnostic techniques which can be applied to the various facts of germplasm identification are of following types:

1. The use of computerized systems to capture and process morphological information (Image analysis).
2. The use of biochemical methods to analyze various components of plants (chemo taxonomy).

Image analysis (IA) is also known as machine vision or computer vision or robot vision. It comprises of image capture, image processing and extraction of information from the processed image, pattern recognition and decision making. The technique has a number of attractions such as speed, objectivity, possibility of measuring new characters or features, electronic transmission of data and it is nondestructive and can be used remotely. Although, the use of IA has been limited to only a few crops like wheat, onion and beans. A great deal of interest has been expressed in the use of this technique for germplasm identification. However, as mentioned above that IA measures only morphological characteristics of the grain, bulb or pod and the morphological characteristics are governed by the environmental factors. Thus augmentation of morphologically based approach with biochemical approaches that are least affected by environmental factors is needed in germplasm identification.

Chemotaxonomists have identified two groups of compounds for the classification of plants and other organisms. These episemantic (secondary) compounds, such as, fatty acids, anthocyanins and other pigments and semantic (sense carrying) molecules, i.e. proteins and nucleic acids, A number of rapid chemical tests have been developed for germplasm identification (ISTA Handbook of Variety Testing, 1992a). Rapid tests rely on the presence of certain, not always known chemicals which elicit color changes in the part of seed or seedling in some varieties and not in others or some are based on particular enzyme activities e.g. seed coat peroxidase test in soybean (AOSA, 1988). But these rapid techniques have very limited discriminating power hence use of more sophisticated techniques in germplasm identifications is indispensable.

Chromatographic techniques like GLC and HPLC have been successfully used to separate and quantify episemantic compounds especially in Brassica varieties, Gerbera and Japanese Garden Iris (Yabuya, 1991). Use of these techniques in India has started only recently. HPLC is also being used to separate proteins for germplasm identification in a range of crops e.g. wheat, barley, maize, sorghum, oats, rice, soybeans and legumes (Morgan, 1989). HPLC based comparison of protein profiles allows varieties to be distinguished from each other and to be identified and the technique can be extremely discriminating. Further the qualitative nature of protein chromatogram is not affected by environment, although occasionally quantitative variations have been noted. Nonetheless the HPLC separation of proteins has an attraction of speed, automation, discrimination possibilities and above all the reduction of environmental effects. A typical analytical separation takes nearly an hour, but this is for only one sample and analysis of a large number of samples will inevitably require a very long time. Moreover its use outside the field is not possible.

Electrophoretic techniques have been most extensively employed in germplasm identification, mainly because of their easy operation and involvement of less expensive equipment. Electrophoretic study of proteins and enzymes of *Sclerotinia* species was employed for obtaining informations for the relationships of different species of this fungus (Wong and Willetts 1973). Basically two approaches have been used:

(1) A direct comparison of varieties for their protein profiles i.e. analysis of polymorphic proteins which are encoded by multiple loci, for example prolamines in cereal seeds and

(2) An indirect comparison of frequency of occurrence of protein band phenotypes within different varieties, i.e. isozymic patterns.

Enzymes are protein molecules that have catalytic functions. To date more than 3000 enzymes are known. A bacterium cell contains nearly 3000 proteins while a eukaryotic cell may synthesize 50,000 proteins; majority of them are enzymes.

Most enzymes are multimeric

Enzymes are proteins. Few enzymes consist of single polypeptide chains. However, a large number of enzymes consist of more than one polypeptide chains, hence called oligomeric or multi subunit enzymes. The subunits in oligomeric enzymes may be identical or non-identical on separation of subunits; the catalytic activity of the enzyme is lost.

Isozymes may be primary and secondary

Nearly half of the enzymes investigated occur in multiple molecular forms i.e. isozymes. These usually differ in electrophoretic mobility due to variation in charge and/or size and may have slightly different catalytic activities. The multiplicity can depend on genetic factors either (a) directly, or (b) indirectly. The isozymes that depend directly on genetic factors are divided into (1) isoenzymes and (2) alloenzymes; isoenzymes are also called as primary isozymes. Multiplicity arising due to chemical modification is referred to as secondary isozymes and as per Enzyme Commission recommendations these are not considered as isozymes.

Isoenzymes help determine multiple loci

Isoenzymes may be distributed in different cell compartments e.g. cytosol, chloroplasts, mitochondria etc. Such enzymes are encoded in different genes and gene loci of these enzymes are often located on different chromosomes (soluble form of malate dehydrogenase is coded by a gene in chromosome 2 while mitochondrial form in chromosome 7 in humans). If the isoenzymes from different cell compartments are mulitmeric their submit do combine to give hybrid forms. Enzymes of identical substrate specificity coded by multiple loci have occurred due to gene duplication during the course of evolution. As a result of point mutation there have been alter atrons in amino acid sequences and composition leading to different enzyme forms separable by electrophoresis.

Multiple Allelism results in differences between individual members of a species

At any given locus, a number of different alleles may occur in a population of individuals. If such alleles code for a structurally distinct enzyme version, these will differ from one individual to another. Such enzyme forms are called alloenzymes or allozymes. The numbers of elecrtrophoretically detectable allozymes depend on:

(a) The number of differently charged polypeptides evolved, and

(b) The number of subunits which form a catalytically active enzyme unit.

Individuals with homologous alleles, with respect to a certain enzyme, produce a single species of enzymes, but this enzyme differs from one individual to another, according to the particular alleles they happen to carry at the locus in question. Allozyme inheritance has been used as subunit structure of enzymes. If each of two homozygous individuals synthesize an enzyme form of different negative charge the number of isozymes in heterozygous individuals depend on the subunit composition of enzyme. With monomeric enzymes two enzyme forms will occur in heterozygotes (A and B), with dimeric enzymes there are three forms (AA, BB, and AB), with trimeric enzymes there are four forms (AAA, BBB, AAB and AMM) and with tetrameric enzymes there are five different forms (A_4, A_3B, A_2B_2, AB_3 and B_4), and so on so forth. Since every gene can mutate it is assumed that stable isozymes represent the most profitable forms proved to the progress of evolution.

Secondary Isozymes are generated by post evolutional modification

There are nine different mechanisms that lead to formation of secondary isozymes. These are:

1. Aggregation and polymerization,
2. Oxidation or reduction,
3. Limited proteolysis,
4. Magnitude of glucosylation,

5. Deamidiation,
6. Aggregation of substrate or co-substrates,
7. Temperature effects,
8. pH effects
9. Conformational Isomerism

Isozymes have been useful tool in plant breeding and genetics

Isozymes have been widely used to screen the variability present in plant populations and to select desirable genotypes for plant improvement. Linkages between specific isozyme alkles and agronomically important genes have contributed to simply inherit agronomic traits. Where genetic linkage exists, an isozyme is much easier to be assayed than testing susceptibility to pest susceptibility can be conducted at seedling stage and is unambiguous.

Isozymic studies in plants have proved that isozyme techniques for identifying and mapping desired loci for quantitative and qualitative traits are very effective. The alleles at most isozyme loci are co-dominent, permitting distinction between heterozygotes and homozygotes. Isozyme markers rarely exhibit epistatic interactions; hence any number of such markers segregating simultaneously.

Although the isozyme markers are limited compared with DNA based markers. They have contributed to identify appropriate genotype for disease resistance and otherwise. The use of assay of isozyme provides an attractive means for transferring a linkage block involving a resistance gene to produce a new variety. In the case of accumulating multiple resistance genes in to one germplasm, the tight linkage of each gene to a suitable isozyme allele provides important diagnositic markers for achieving the desired new germplasm.

Isozymes are named according to their electrophoretic patterns

The extent of migration in an electrophoretic support medium has been used to name the different isozymes. Naming on the basis of time distribution (eg. brand type etc) has a disadvantage that the distribution varies with species or within the species with the stage of development. Numbering starts from the anode and may be grouped into several classes with a number of subtypes or the patterns are devided into regions with letters designating regions and number designating the bands within a region.

Although a large number of elctrophoretic methods have been used by different workers for different crops (Cooke, 1988 and Lokhart, 1991) and sometimes it poses a problem of comparison, the actual results given by the various methods should be similar if unknown sample is the same as the reference sample of a given germplasm. Hence various electrophoretic procedures like, starch gel electrophoresis, native PAGE (Polyacrylamide Gel Electrophoretic), SDS-PAGE, IEF (Isoelectric Focussing) and 2-dimensional PAGE (i.e. a combination of SDS-PAGE and IEF) have been employed successfully in germplasm identification of self-pollinated as well as cross pollinated crops (ISTA Handbook of Variety Testing, 1992b; White and Cooke, 1992, Greneche *et al.,* 1991). Attempts have also been made to develop immunodiagnostic procedures by raising polyclonal or monoclonal antibodies against the distinct polypeptides among varieties for identification of a particular specific character.

Despite of the fact that HPLC and elctrophoretic methods reveal variability at a number of loci, these techniques may not discriminate the complete variation in underlying a genetic material. Since not all variations occur in expressed regions of DNA, it is probable that there will be more DNA variability. Such variations can be distinguished by techniques that can reveal and utilize DNA polymorphism. Two approaches have been applied to reveal DNA polymorphism and to exploit it for germplasm identification:

1. Use of probe-based technology. Such as, RFLP has also been applied for genetic variations within the genus with a remarkably higher amount of variations which could be of interest for aiming desired hybridization for generating well saturated genomic maps and inter species transfer of desirable genes (Fatokun *et. al.* 1993)
2. The unusual origin of the polymerase chain reaction by Muilis (1990) gave the way for making unlimited copy of DNA fragments Use of this amplification technology made it possible the popular technique like RAPD which involves PCR amplification of simple sequence repeats. Welsh and McClelland, (1990) used fingerprint genomes using PCR with arbitrary primer and reported simple and reproducible finger prints of complex genome with no primer sequence information. It involves two cycles of low stringency amplification followed by PCR at higher stringency. Such amplified DNA shows polymorphism and could be used to construct genetic map, such polymorphisms are called RAPD markers after random amplified polymorphic DNA (Williams *et. al.* 1990). Qualitative and quantitative characterization of RAPD variations among different genotypes have been used widely for genetic diversity analysis in different crops including vegetable crops which further determine genetic relationships among populations which can allow breeders to select the species for inter and intra specific hybridization (Bai *et. al.* 1998). Beside RAPD analysis is also be utilized for somaclonal variations during micropropagations in somaclonal progenies (Hashmi *et.al.* 1997)

References

Association of Official Seed Analysis (1988). Rules for testing seed. *J. Seed Technol.*, **12:** 1-126.

Bai Y., Michaels T.E. and Pauls K.P. 1998. Determination of genetic relationships among *Phasiolus vulgaris* populations in a conical cross from RAPD marker analyses. *Molecular breeding,* **4:** 395-406

Cooke, R.J. (1988). Electrophoresis in plant testing and breeding. *Advances in Electrophoresis,* **2:** 171-261.

Fatokun C.A., Danesh D., Young N.D., and Stewart E.L. 1993. Molecular taxonomic relationship in the genus Vigna based on RFLP analysis. *Theortical and applied genetics*, **86:** 97-104

Greneche. M., Lallemand, J. and Migaud, O. (1991). Comparision of different enzyme loci as a means of distinguishjing rye grass varieties by electrophoresis. *Seed Sci. Technol*, **19:** 147-158.

Hashmi G., Huettel R., Meyer R., Krusberg L. and Hammerschlag F. 1997. RAPD analysis of somaclonal variants derived from embryo callus cultures of peach. *Plant cell reports* **16:** 624-627.

International Seed Testing Association (ISTA) (1992a). Handbook of Variety Testing-Rapid Chemical Identification Techniques (Payne, R.C., ed) Zurich, pp 1-52.

International Seed Testing Association (ISTA) (1992b). Handbook of Variety Testing-Electrophoresis Testing (Cooke, R. J. ed) Zurich, pp 1-42.

Lookhart, G.L. (1991). In Hand book of Cereal Sci. Technol. Marcel dekker, New York, pp 441-668.

Morgan, A.G. (1989). Cultivar identification in *Brassica napus* L. by reversed-phase high-performance liquid chromatography of ethanol extracts Plant Varieties and Seeds, **2:** 34-45

Muilis K. V. 1990. The unusual origin of the polymerase chain reacthion. *Scientific American,* April 56-65

Welsh J. and McClelland M. 1990. Fingerprinting genome PCR with Arbitrary primer. *Nuclic Acid Research,* **18:** 7213-7218.

White, J, and Cooke, R.J. (1992). A standard classification system for the identification of barley varieties electrophoresis. *Seed Sci. Technol.,* **20:** 663-676.

Williams J. G.K., Kubelik A.R., Livak KJ, Rafalski J.A. and Tingey S.V. 1990. DNA Polymorphisms amplified by arbtary primer are used by genetic markers. *Nuclic Acid Research,* **18:** 6531-6535.

Wong A.L. and Willetts H.J. 1973. Electrophoretic study of soluble proteins and enzymes of Sclerotinia species. *Trans. Br. Mycol. Soc.,* **61:** 167- 178

Yabuya, T. (1991). High-performance liquid chromatographic analysis of anthocyanins in Japanese garden iris and its wild forms. *Euphytica,* **52:** 215-219.

❑❑❑

Chapter – 23

Role of Gene Bank in Plant Genetic Resource (PGR) Management of Vegetable Crops

K.S. Negi, K.C. Muneem, S.K. Verma, A.S. Rana and P.S. Mehta

Introduction

The Indian gene centre has rich diversity in vegetable crops. This diversity includes crops native to India like eggplant, parval, ridge gourd, sponge gourd, *Colocasia* sp., elephant yam and for centre of diversity for other vegetable crops i.e., Okra, Cucumber, Chayote, Chilli, Cucurbita etc. Rich diversity also occurs in introduced crops such as tomato, french bean, cowpea, leafy brassica, amaranth, yam and in cucurbits, bottle gourd, bitter gourd, coccinea, ash gourd, snake gourd and in bulbous crop like onion and garlic.

Genetic Resources of Vegetable Crops in India

Their diversity and conservation about 400 species constitute the global diversity in vegetable crops. This diversity is mainly distributed in seven-eight geographical regions, which represent the centres of origin and diversity as well. Among these, the regions possessing maximum diversity are the tropical American, Tropical Asian and the Mediterranean (q.v.) regions. In the tropical Asian region, both India and China hold maximum diversity.

The Indian sub-continent is one of the centre of origin and diversity in vegetable crops. Around 80 species of major and minor vegetables, apart from several wild and other kinds occur (Choudhury, 1967; Seshadri, 1987). India is a primary centre of diversity for crops such as eggplant, smooth gourd, ridge gourd and cucumber, and a secondary centre for cowpea, okra, chillies, pumpkin and several Brassicae. Overall around 20-25 vegetable crops are commercially important and these include both the indigenous and exotic species variability; the introductions being well adapted to local needs and agro-ecology. Vegetable crops of Indian onion are presented in table 1.

Table 1. Vegetable crop species originated in Indian Sub continent

Crop group	Crops
Vegetable crops	Egg plant/ brinjal (*Solanum melongena*), tori/ridge gourd and smooth gourd (*Luffa spp.*),, tinda / round gourd (*Citrullus lanatus*), parval/ pointed gourd (*Trichosanthes dioica*), arvi/taros (*Colocasia esculenta*), yams (*Dioscorea spp.*), zmikand (*Amorphophallus campanulatus*), kundru (*Coccinia indica*), cucumber (*Cucumis sativus*), radish (*Raphanus caudtaus*)-mungra type

Three / four categories of genetic resources predominate and the promising diversity in vegetable crops represented in each category is given below:

1. Introduced germplasm: Tomato, peas, french bean, scarlet bean, lima bean, winged bean, faba bean, cauliflower, cabbage, knol-kohl, turnip, radish, carrot, bell pepper, chayote potato, meetha karela, *Allium* spp., onion, leek, shallot, garlic, asparagus, artichoke and parsley.

2. Indigenous germplasm: Egg plant, smooth gourd, ridge gourd, snake gourd, round gourd, musk melon, phunt, kakri, cucumber, leafy vegetables-*Basella, Chenopodium, Amaranthus, Brassica,* spinach, *Rumex* / sorrel, taro, yam, elephant foot yam, ginger, *Allium* spp., *Caralluma, Anethum.*

3. Germplasm of exotic origin for which India is a centre of diversity: Okra, pumpkin, chilli, cowpea, chayote and coriander.

4. Genetic resources for which precise centre of origin is still disputable but rich diversity occurs: Bitter gourd.

Underutilized /Underexploited Vegetables

The distribution and extent of diversity discussed above mainly deals with major vegetable crops. In tribal areas of India, under traditional / subsistence agriculture and in kitchen gardens or home stead gardening, many minor vegetables are grown. Some of these are now better known such as amaranth

and chenopods, and several others now in process of domestication such as *Phytolacca acinosa* and still others that in semi-wild/protected state and /or are gathered locally. The genetic resources of such minor vegetables include over 50 species and prominent ones are listed below:

Allium clarkei, A. griffthianum, A. hookeri, A. wallichii, Amaranthus spp., Apium graveolens, Asparagus officinalis, Basella alba/ rubra, Bamboo spp.,

Caralluma spp., Centella asiatica, Emilia sonchifolia, Fagopyrum cymosum, Hibiscus sabdariffa, Ipomoea aquatica, Lactuca sativa, L.indica, Monochoria vaginalis, Malva spp., Moringa oleifera, Neptunia prostrata, Nelumbo nucifera, Ocimum spp., Oenanthe javanifa, Parkia roxburghii, Polygonum spp., Pilea spp., Portulaca oleracea, Rotippa indica, Sesbania grandiflora, Solanum torvum, Sonchus spp., Spilanthus acmella, Trichosanthes spp., Trigonella spp., Vigna spp., Zizania latifolia, avoids and yams.

Principal Wild food Plants including Vegetables

Wild edible plants have been documented recently with data focusing on over 1,000 species used for various purposes:

1. Roots/tubers 145
2. Leafy vegetables/green 521
3. For Buds/flowers 101
4. For Fruits 647
5. For seeds/nuts 118

Wild relatives and related species

The wild relatives and related species of vegetable crops belong to the categories, legumes–31 spp., vegetables–54 spp. and spices and condiments–27 spp. (Arora and Nayar, 1984) and these are distributed in the Western and Eastern Himalayas, North-eastern region, Gangetic plains, Indus plains and in the Western and Eastern Peninsular regions.

Table 2. Diversity of wild relatives of vegetable crops in different phyto-geographical regions of India

	Vegetables
Western Himalayas	*Abelmoschus manihot (tetraphyllus forms), Allium hookerii, Cucumis hardwickii, C, trigonus, Luffa echinata, L. graveolens, Solanum incanum, Trichosanthes multiloba, T. himalensis*
Eastern Himalayas	*Abelmoschus manihot, Cucumis trigonus, Luffa graveolens, Neoluffa sikkimensis*
North – eastern region	*Abelmoschus manihot (pungens forms), Alocasia macrorrhiza, Amorphophallus bulbifer, Colocasia esculenta, Cucumis hystrix, C. trigonus, Dioscorea alata, Luffa graveolens, Moghania vestita, Momordica cochinchinensis, M. macrophylla, M. subangulata, Trichosanthes cucumerina, T. dioica, T. dicaelosperma, T. khasiana, T. ovata, T. truncata, Solanum incanum, S. indicum*
Indus plains	*Momordica balsamina, Citrullus colocynthis, Cucumis melo var. momordica, Cucumis prophetarum*

Contd..

Western peninsular tract	*Abelmoschus angulosus, A. moschatus, A. manihot (pungens form), A. ficulneus, Amorphophallus campanulatus, Cucumis setosus, C. trigonus, Luffa graveolens, Momordica cochinchinensis, M. subangulata, Solanum indicum, Trichosanthes anamalaiensis, T. bracteata, T. cispidata, T. horsfieldii, T. perottitiana, T. neriifolia, T. villosa*
Eastern peninsular tract	*Amorphophallus campanulatus, Abelmoschus manihot, A. moschatus, Colocasia antiquorum, Cucumis hystrix, C. setosus, Luffa acutangula var. amara, L. graveolens, L. umbellate, Momordica cymbalaria, M. denticulate, M. dioica, M. cochinchinensis, M. subangulata, Solanum indicum, S. melongena* (insanum type), *Tricsosanthes bracteata, T. cordata, T. lepiniana, T. himalensis, T. multiloba*

Table 3. Wild relatives and related endemic / rare species and other types

Vegetables	*Abelmoschus tuberculatus, Allium rubellum, Cucumis hardwickii, Curcuma amarissima, Luffa hermaphrodita, Luffa umbellate, Moghania vestita, Neoluffa sikkimensis, Solanum melongena var. insanum, S. melongenaa var. potangi, Trichosanthes khasiana, majuscula, ovata, tomentasa, Zingiber intermedium*

Table 4. Promising wild vegetable diversity in India

Use category		Genera	Species
Roots/tubers		98	145
	Legumes: *Moghania vestita, Vigna capensis* Wild yam: *Dioscorea hamiltonii*		
Leafy vegetables / greens		377	52
	Allium stracheyi, A. rubellum, Bamboo (Bambusa tulda), Caralluma edulis, Polygonum spp., Clerodendrum spp., Moringa oleifera, Nasturtium officinale, Phytolacca acinosa, Portulaca oleracea, *Rumex dentatus*		
Buds / flowers		88	101
	Bauhinia, Capparis		

Botanically these can be classified as follows:

Family		**Species**
Brassicaceae	:	*Brassica, Lepidium*
Malvaceae	:	*Abelmoschus, Hibiscus, Malva*
Fabaceae	:	*Canavalia, Cicer, Dolichos, Lablab, Mognahia, Trigonella, Vigna*
Cucurbitaceae	:	*Citrullus, Coccinea, Luffa, Momordica, Neoluffa, Trichosanthes*
Apiaceae	:	*Carum*
Asteraceae	:	*Cichorium*
Solanaceae	:	*Solanum*
Amaranthaceae	:	*Amaranthus*
Chenopodiaceae	:	*Cehnopodium*
Polygonaceae	:	*Fagophyrum, Polygonum, Rumex*
Zingiberaceae	:	*Alpinia, Curcuma, Zingiber*
Dioscoreaceae	:	*Dioscorea*
Alliaceae	:	*Allium*
Araceae	:	*Alocasia, Amorphophallus, Colocasia*

Important genera with number of species occurring in India (given in parenthesis) are as follows:

Category		Genera/Species
Legumes	:	*Canavalia* (4), *Cicer* (1), *Dolichos* (6), *Lablab* (1), *Trigonella* (12), *Vigna* (10)
Vegetables	:	*Abelmoschus* (6), *Alocasia* (1), *Amaranthus* (40), *Chenopodium* (8), *Citrullus* (2), *Coccinia* (1), *Colocasia* (1), *Cucumis* (5), *Dioscorea* (>40), *Fagopyrum* (4), *Luffa* (4), *Malva* (5), *Moghania* (20), *Momordica* (4), *Neoluffa* (1), *Polygonum* (80), *Rumex* (13), *Solanum* (40), *Trichosanthes* (21)
Spices & Condiments	:	*Allium* (30), *Alpinia* (15), *Amomum* (10), *Carum* (3), *Cichorium* (3), *Curcuma* (18), *Zingiber* (4)

Germplasm Maintenance and Conservation

Over 21,000 germplasm accessions are held by NBPGR of which about 50% are exotic introductions. Thus there is sizable diversity of both indigenous and introduced material. These collections are rich in Solanaceous crops (egg plant, chillies, tomato), cucurbits (cucumber, *Cucubita spp.*, *Luffa spp.*, bottle gourd, pointed gourd, bitter gourd, legumes (dolichos/lablab bean, french bean, cowpea) and okra, apart from several roots and bulbous crops. The diversity held includes both temperate and tropical crops and accordingly is maintained at different locations viz., NBPGR Regional Stations such as: a) for French bean, faba bean and peas at Shimla and Bhowali; b) winged bean, okra and lablab bean, at Akola; c) okra, cluster bean, and cowpea at Jodhpur; d) winged bean, cowpea and okra at Trichur and sizable collections at Issapur Farm, New Delhi such as of tomato, onion and cowpea. The major collections (accessions) are maintained at the above different locations as per crop agro-ecology and are of French bean (>4500), cowpea (>3400), peas (>4000), *Dolichos* / Lablab bean (>1100), okra (>4000), egg plant (>2800), onion (>1300), garlic (>1000), cucurbits (>1100), Brassicae (>800), faba bean (>600) and winged beans (>400) (Singh, 1993).

In 142 exploration undertaken by NBPGR, R/S Bhowali and its collaborative members i.e., VPKAS, Almora, DARL, Pithoragarh, GBPUA&T, Pantnagar and GBPIHED, Kosi Katarmal from 1985 to December 2007, a total of 15792 accessions in which 2600 acc. of various vegetables germplasm have been collected for different districts of Uttarakhand (See table 5 to 6).

Table 5. Status of germplasm holding in the NGB at -18^{0}C (up to December 2006)

Crop group	Total
Cereals	1,34,421
Millets and forage	48,657
Pseudo cereals	5492
Grain legumes	52,844
Oil seeds	47,515
Fiber crop	9546
Vegetables	
Brinjal (3720)	
Chillies (1997)	
Others (15480)	21201
Fruits	265
Medicinal & Aromatic plants	5333
Spices and condiments	1632
Agro forestry	2053
Duplicate samples (pigeon pea & lentil)	10235
Others	1,44,421
	3,39,194 Includes Released Varieties (2535) & Genetic Stocks (1312)

Table 6. Details of Germplasm of Different Agricultural- Horticultural Crops and Economic Species Maintained in MTS of NBPGR, R/S Bhowali (Up to December, 2007)

S. N.	Name of Crop Group	Name of Crops	No. of Accessions
1.	Cereals	—	3644
		Barley	642
		Maize	296
		Paddy	670
		Wheat	2036
2.	Pseudo-cereals	—	585
		Amaranthus	554
		Buckwheat	31
3.	Millets and Minor millets	—	539
		Barnyard millet	165
		Finger millet	320
		Foxtail millet	32
		Proso millet	20
		Sorghum sp.	02
4.	Pulses	—	3239
		Adzuki bean	08
		Black gram	96
		Cowpea	80
		Faba bean	12
		French bean	951
		Gram	02

Contd...

		Green gram	03
		Horse gram	193
		Khesari	08
		Lentil	1231
		Moth bean	08
		Mucuna sp.	04
		Pea	209
		Pigeon pea	28
		Rice bean	47
		Soybean	359
5.	Oil seeds	—	533
		Brassica spp.	400
		Groundnut	82
		Linseed	01
		Sesame	48
		Sun flower	02
6.	Vegetables	—	266
		Ash gourd	04
		Bottle gourd	30
		Brinjal	01
		Carrot	01
		Chenopod	06
		Okra	36
		Radish	118
		Ridge gourd	15
		Sem bean	27
		Spinach	27
		Turnip	01
7.	Spices and Condiments	—	1484
		Chamsur	12
		Chillies	1132
		Coriander	181
		Fenugreek	91
		Hemp	22
		Onion	09
		Perilla	37
8.	Medicinal and Aromatic Plants	—	173
9.	Wild Relatives of Crops	—	301
		Aegilops sp.	97
		Allium sp.	48
		Avena sp.	32
		Cucumis hardwickii	26
		Hordeum spp.	04
		Abelmoschus sp.	06
		Triticum sp.	87
		Tree tomato	01
10.	Ornamental Crops	—	22
		Tagetes spp.	22
		Grand Total	10,786

Collection, Characterization and Evaluation of Vegetable crops Germplasm

A good number of vegetables are of Indian/ Asian origin and a good amount of diversity exists in various parts of the country. These includes eggplant, lablab bean, cluster bean, radish, few cucurbits (cucumber, round melon, sponge gourd, ridge gourd, snake gourd, pointed gourd, musk melon, snap melon, long melon, Indian squash), leafy vegetables (Indian spinach, climbing spinach) and tuber vegetables (*Colocasia* and Elephant foot yam). However, some of the vegetables are introduced and have acclimatized/adapted throughout the country leading to good amount of variability viz., cauliflower, cabbage, french bean, tomato, pumpkin, ash gourd, bottle gourd, bitter gourd, garlic and potato. However, these are being constantly threatened due to high yielding varieties / hybrids and also due to shrinkage of natural yielding. The biologists/ scientists throughout the World have shown great concern over the conservation and preservation of environment, biodiversity and ecosystem.

Germplasm is the basic tool and had played most vital role in the improvement of cultivated plants including vegetables. However, augmentation of germplasm through exploration and collection particularly in developing countries is still limited despite the wide recognition (Gill 1984). Characterization, preliminary and further evaluation of germplasm is the basic need to get acquainted with the germplasm for further utilization. Characterization and evaluation provide scope for the utilization of germplasm (Arora & Nayar, 1984; Rana *et,.al*. 1994a, b, 1995; Singh & Srivastava 2004).

Seed Increase

The first stage in PGR activity is seed increase and it is essential to increase sufficient seed in one cycle after keeping part of it as reference sample. The site of initial seed increase should be as close as to the original site to avoid the risk of loosing a particular accession due to poor adaptation, disease and pest damage. In vegetable crops this aspect is most essential because it consist of large group of crops including annual, biennial as well as requiring cold and dry climate for seed production. One to three rows depending on the quality of seeds and nature of the species should be grown for seed increase. However, for cross pollinated crops, subbing is required to avoid genetic drift.

For proper characterization and evaluation, adoption of descriptor list is important. It is better to use internationally approved list for uniformity in data processing and storage. IPGRI has developed over 62 descriptor for different agri-horticultural crops including 15 for vegetable crops

a. **Passport data:** Data collected by collector which is very important for identification and help in designating core collection, identifying duplicates and in planning future explorations.

b. **Characterization:** These consist of recording those characters which are highly heritable, can be easily seen by eye and are expressed in all conditions/ environments. These are usually qualitative, environmentally stable, mono/ oliogenic and easily manipulated in breeding.

c. **Preliminary evaluation:** Consist of recording a limited number of additional traits, thought desirable by the consensus of users of the particular crops/ species which help in identifying useful germplasm. The important ones include, site data, data on plant, leaf, flower, fruits, seeds and reaction to pest and diseases.

d. **Further or detailed evaluation:** Consists of recording number of additional descriptors thought to be useful in crop improvement. These include important agronomic traits, stress tolerance, disease and pest resistance and quality characters etc.

Observations are recorded on qualitative or observable characters and quantitative or non-observable traits. The qualitative characters include morphological, physiological and bio-chemical traits and scored once using a number of checks to determine variation within and among the traits. Quantitative characters are subject to environmental factors and are responsible for adoption and productivity. For preliminary evaluation locally adapted cultivar for detailed evaluation, known resistant stocks are used as check.

Augmented Block Design is used when size of germplasm is large for evaluation. Care should be taken to minimize contamination due to natural cross pollination and erroneous labeling. While regenerating, care should be taken to preserve original structure and integrity of accession due to inappropriate handling of materials during sowing/ planting, harvesting, threshing, cleaning, sub-sampling, packing and labeling. Frequent regenerations should also be avoided, for which sufficient seeds to be produced in the initial seed increase and stored in gene bank.

For fruit quality comparison, sample having similar stage of maturity should be taken to avoid environmental and stage of maturity differences. This may be difficult to achieve in vegetable germplasm having wide difference in maturity period. The simple procedure involves harvesting all ripe fruits to encourage fruiting and fruits having similar maturity stage should be taken to ensure more number of uniform fruits for quality determination.

Promising selections from exotic introduction

A good number of exotic germplasm have been utilized directly as promising varieties under Indian conditions. In this process, the augmented germplasm are evaluated to identify the promising types having higher quality yield in comparison to available check varieties. Some of the promising direct introductions used for cultivation are listed below.

(i) **Tomato:** Sioux, Fireball, Marglobe, Best of all, Roma, Dwarf, Money Maker (Ex Isreal), La Bonita (Ex USA), etc. Among these some have pear/ oval shape fruits, long keeping quality, profilic in bearing with less seed, suited for long distance transportation.

(ii) **Cauliflower:** Snow ball-16 (Holland), Improved Japanese (Israel), 96D (Israel) and Early Snow ball (Japan) have established well.

(iii) **Cabbage:** Golden Acre, Drum Head, Danish Ball Head and Late Flat Dutch are the promising exotic introductions under cultivation.

(iv) **Radish:** China Red, Rapid Red, Japanese White, Scralet Globe, French Breakfast and Rapid Red White Tipped are the exotic promising introductions.

(v) **Onion:** Early Grano from USA; Yellow Skinned, Bermuda Yellow from Philippines are famous for yield and quality.

(vi) **Cowpea:** Two promising exotic introductions of cowpea released under the name of Pusa Barsati (E. Philippines) and Pusa Phalguni (ex Canada) are very popular.

(vii) **Okra:** In okra, Perkins Long Green and Velvet Green were recommended for direct cultivation. Ghana Red (ex Ghana) is another important cultivar.

(viii) **Summer Squash:** Australian Green, Zucuni Long and Petty Pan are promising varieties suited to Indian condition.

(ix) **Pumpkin:** Large Cheese, Kentuky Field, Butternut, African Bell are promising exotic introductions.

Promising indigenous collections

A large number of vegetable crops have originated in Indian Centre of origin of cultivated plants and enormous variability has been accumulated as a result of selection and also by natural cross pollination. A number of landraces / primitive cultivars have been utilized directly as variety and also a donor parent after proper characterization and evaluation. Some of these are listed below:

Brinjal: Pusa Purple Long, Pusa Purple Round, Pusa Purple Cluster and Pusa Bhairav (IARI); Arka Sheel, Arka Kusumkar and Arka Navneet (IIHR, Banglore); CO-1, MDU-1 (TNAU); Junagarh Long (GAU), Pant Samrat (GBPUA&T), T-3 (CSAUA&T), H-4 (HAU) and Krishnagar Purple and Green Long (West Bengal) are important varieties released by different organizations.

Tomato: HS 101, HS-102, HS-110 from HAU, Hissar, S-12; Punjab Kesari from PAU, Ludhiana; Kuber, Kalyanpur T-1, KS-1, KS-2, Angoorlata from CSAUA&T, Kanpur, CO-1, CO-2, CO-3 from TNAU, Coimbatore have been found promising selections.

Cauliflower: A number of varieties such as Pusa Kekti, Sel-1, Sel-5, Sel-9, Line-294, Line-256, Line-357 and Line 328 by IARI, New Delhi; Hissar-1 by HAU, Hisar; Kanwari by PAU, Ludhiana; Kalyanpur Medium by CSA University Kanpur; Late Dania by Kalimpong, West Bengal have been released for cultivation, based on selection from indigenous and exotic germplasm.

Cabbage: Pride of India, a mid season variety for hills as well as for plains, and September an early season variety suitable for Nilgiri Hills has been developed. Besides

Sel-8 from Katrain and ARU-Glory from Almora are promising selections under cultivation.

Radish: Besides Pusa Desi from IARI, New Delhi and Arka Nishant by IIHR, Banglore several other landraces viz., Newari, Jaunpuri Giant, Kannauju, Long White, Bombay Red have been found promising and are under cultivation.

Onion: Pusa Red is a selection from local collection developed by Plant Introduction Division, IARI, New Delhi. Line 106, Line 133 and Line 127 from Delhi, Hissar-11 from HAU, Hissar, Udaipur-102 from Rajasthan, CO-1, CO-2, CO-3 and CO-4 from T.N.A.U., Coimbatore and N-53 holds good for cultivation.

Garden Pea: Selection from IARI viz., NP-26, NP-29 and Sel.-18, 30 and 35 are the promising selections out of indigenous collections. VL-2, VL-3 from Almora, suitable for hilly condition has been released. Jawahar Matar 10 a mid season variety has been developed by JNKVV, Jabalpur.

Dolichos bean: Pusa Early Prolific are selection by PID, IARI, New Delhi; Kalyanpur T-1 a pole type developed by CSAUA&T, Kanpur and TDL-37 a high yielding cultivar developed by JNKVV, Jabalpur have been released out of indigenous germplasm for cultivation.

Okra: Pusa Makhmali evolved by PID, IARI, New Delhi, Sel-2 by NBPGR, New Delhi; and Punjab Selection No. 13 by PAU, Ludhiana, Kanpur are selection from the indigenous germplasm.

Pumpkin: Arka Chandan is selection from indigenous germplasm from Rajasthan by IIHR, Banglore; CO-1 and CO-2 by TNAU, Coimbatore and Pusa Vishwas by IARI, New Delhi.

Bottle gourd: Pusa Summer Profilic Long and Pusa Summer Profilic Round are high yielding varieties released by IARI.

Sponge gourd: Pusa Chikni is a selection from indigenous germplasm collected from Patna by IARI, New Delhi.

Ridge gourd: Pusa Nasdar is a selection from the germplasm collected from Neemuch, Madhya Pradesh by PID, IARI, New Delhi.

Bitter gourd: Pusa Do Mausami, an early variety having dark green long fruit with medium thick skin.

Palak: In Palak, All Green and Pusa Jyoti varieties evolved by IARI, New Delhi; Jobner Green by University of Udaipur; H-23 by HAU, Hisar, Palak No. 51 by Dept. of Agri. Maharastra State and from indigenous germplasm.

Crop Responsibilities and Germplasm Activities at NBPGR

A. Solanaceous vegetables: Brinjal (*Solanum melongena*), Tomato (*Lycopersicon esculentum*), Chillies (*Capsicum annuum*).

B. Cucurbitaceous vegetables: Pumpkin (*Cucurbita sp.*), Melons (*Citrullus spp. and Cucumis spp.*), Gourds (*Luffa* spp., *Lageneria* spp., *Trichosanthes* spp., *Momordica* spp., *Benincasa* spp.), Cucumber (*Cucumis* spp.)

C. Leguminous vegetables: Cowpea (*Vigna unguiculata*), Pea (*Pisum sativum*), Lablab bean (*Lablab perpureus*), Winged bean (*Psophocarpus tetragonolobus*), Faba bean (*Vicia faba*), French bean (*Phaseolus vulgaris*), Mucuna (*Mucuna spp.*), Canavalia (*Canavalia ensiformis*)

D. Bulb crops: Garlic (*Allium sativum*), Onion (*Allium cepa*).

E. Root vegetables: Radish (*Raphanus sativus*), Carrot (*Daucus carota*), Turnip (*Brassica rapa*)

F. Okra: Lady Finger (*Abelmoschus esculentus*)

G. Miscellaneous vegetables: Cole crops (*Brassica spp.*), Chinese cabbage (*Brassica spp.*), Spinach beet (*Beta vulgaris*), Spinach (*Spinacia oleracea*)

Vegetables Crops in different agro-ecological zones of India

Western Himalaya: *Abelmoschus manihot* (tetraphyllus forms), *Cucumis hardwickii, C. trigonus, Luffa graveolens, Solanum incanum, Trichosanthes multiloba, T. himalensis.*

Eastern Himalaya: *Abelmoschus manihot, Cucumis trigonus, Luffa graveolens, Neoluffa sikkimensis*

North–eastern region: *Abelmoschus manihot* (pungens forms), *Alocasia macrorhiza, Amorphophallus, Colcasia esculenta, Cucumis hystrix, C. trigonus, Dioscorea alata, Luffa graveolens, Moghania vestita, Momordica cochin-chinensis, M. macrophylla, M. subangulata, Trichosanthes cucumerina, T. dioica, T. dicaelosperma, T. khasiana, T. ovata, T. truncata, Solanum indicum, S. torvum.*

Gangetic plains: *Abelmoschus tuberculatus, A. manihot (tetraphyllus forms), Luffa echinata, Momordica cymbalaria, M. diocia, M. cochinchinensis, Solanum incanum, S. indicum, S. toravum.*

Indus plains: *Momordica balsaminia, Citrullus colocynthis, Cucumis prophetarum.*

Western peninsular tract: *Abelmoschus angulosus, A. moschatus, A. manihot pungens forms, A. ficulneus, Amorphophallus campanulatus, Cucumis setosus, C. trigonus, Luffa graveolens, Momordica cochinchinensis, M. subangulata, Solanum indicum, S. torvum, Trichosanthes anamalaiensis, T. bracteata, T. cuspidate, T. horsfieldii, T. perottitiana, T. nerifolia, T. villosa.*

Eastern peninsular tract: *Amorphophallus campanalatus, Abelmoschus manihot, A. moschatus, Colocasia antiquorum, Cucumis hystrix, C. setosus, Luffa acutangula var. amara, L. graveolus, L. umbellate, Momordica cymbalaria, M. denticulate, M. dioicam, M. cochichinensis, Solanum indicum, S. melongena* *(insanum types), Trichosanthes bracteata, T. cordata, T. lepiniana, T. himalensis, T. multoloba.*

Seed Technology Research in Relation of Seed Storage

Seed technology refers to all those activities which are aimed at production, testing and maintenance of genetically pure, disease-free, high quality seed of plant species. Plant breeders develop varieties with desirable qualities / attributes. To make the seed of these varieties to the farmer / grower for cultivation, is the responsibility of Seed Technologists.

Purpose of Seed Storage

The first and foremost purpose to store seed is to make it available, with high level of germination and vigour, at the time of sowing in the forthcoming or subsequent season(s). For this, agricultural seeds need to be stored for one or two cropping season or years.

The second and equally important purpose of seed storage is to maintain plant genetic resources for conservation and to make them available for crop improvement programmes. For this, seeds of a large number of species and cultivars are stored for a prolonged period of time, normally extending to hundreds of years. Two kinds of collections of genetic resources can be recognized:

(a) Base collection for long-term storage, a reserve against genetic losses and

(b) Active collection for medium-term storage, seed supply, evaluation documentation and regeneration.

Besides, some quantity of seed is also stored for 2-3 years (or more) which is utilized for the production of seed to be used commercially, e.g. nucleus or breeder seed, seeds of the parental lines (of hybrids) etc.

Indian National Gene Bank

(i) Seed repository: Samples of orthodox type seeds are dehydrated at 15°C and 15% RH to around 5% moisture content, sealed in laminated aluminum foils under vaccum and deposited in cold storage at –20°C.

(ii) Tissue culture repository: Disease free tissue cultures of recalcitrant seeds (those not capable of withstanding dehydration and very low temperatures) and vegetatively propagated plant species, developed from meristem explants, are maintained using appropriate growth-limiting media and by sub culturing at suitable intervals.

(iii) Cry preservation facility: Seed samples of selected species, 'synthetic seeds', embryos and gametes are being preserved using cryopreservation technique, either by immersion in liquid nitrogen (–19°C) or by keeping in its vapour phase (around –150°C) following appropriate protocols and using computerized freezing-thawing rates and chemical cryoprotectants.

(iv) **Clonal repository:** Germplasm collections of vegetative propagated crops like turmeric, ginger, sweet potato and banana are being maintained under field conditions following suitable crop rotations and backed up adequately by tissue-culture safety duplication.

Gene Bank Standards

(i) **Seed sample size:** About 3,000 viable seeds in case of self-pollinated crops and twice of that number for cross-pollinated species are required to be deposited in the gene bank. One third of this sample is sealed in a separate packet and stored alongside the original sample under identical conditions (at –20°C) for monitoring germinability at regular intervals.

(ii) **Viability testing:** Initial germination tests are carried out on a minimum of 200 seeds, drawn randomly from the original samples. Acceptable initial germination percentage should be higher than 90% for most species but lower levels are acceptable in case of vegetables and wild forms, 50 to 100 seeds, drawn randomly from samples, are used for subsequently viability monitoring tests.

(iii) **Seed drying:** Seeds are dried in seed-drying cabinets, operated at 15°C and 15% relative humidity, using silica gel as desiccant and based on adsorption system and appropriate airflow.

(iv) **Seeds storage conditions:** Seeds are dried to 5% moisture content and sealed in laminated aluminum foils. Base collections are stored in modules operating at –20°C. Active collections are stored at 4°C and 35% relative humidity.

(v) **Regeneration:** Regeneration is required when seed viability declines to 85% of the initial value. Preferable 100 plants or more are grown for regeneration.

Table 7. Some methods of regeneration in vegetable crops

S. N.	Crop/Species (Out breeding rates)	Prop. Size	Isolation	Inter pollination	Harvesting	Location or author
1.	*Brassica* species [Self –incompatible: up to 100%]	507	Insect cage	Honey bess	Bulk	Ames, lowa, NCRPIS
2.	*Curcubia* [Monoecious: high]	24	Male & female flowers 'tied off'	Hand pollination	Bulk	Ames, lowa, NCRPIS
3.	*Cucumis* [Diclinous: high]	24	Insect cages	Honey bess (nucleus hive)	Bulk	Ames, lowa, NCRPIS
4.	*Vicia faba* (4-80%)	20-80	Spatial, barrier crops or artificial barrier	Insect pollination (honey bees)	Equal maternal contribution	De Pace *et al.*, 1983 (Witcombe, 1983., CARDA)

Contd....

5 (a)	Tomato spp. (Autogamous) [nil or low]	Enough for desired seed quantity	Nil	Automatic selfing	Bulk	US-CAC Report (NPGS)
(b)	Tomato spp. (Facultative inbreeders and self-incompatible) [Moderate to high]	As many as possible	Isolate in green house	Hand cross with pollen pool	Bulk	US-CAC Report (NPGS)

Source : From Breese, E.L. 1989

Guidelines for Sending Seeds to Gene Bank and Gene Bank Management

Before sending the seed for storage, following conditions should be observed.

1. Seeds should be well developed and properly (physiologically) matured and free from insect. Undersized, shriveled and immature seeds should be discarded.
2. As far as possible, harvesting in rain should be a avoided because seeds will absorb more moisture from the atmosphere and during storage, such seed will develop disease and become infested with insect pests. Before packing, visual inspection is advisable. Off types and discoloured seeds should be discarded.
3. Seed should be dried soon after the harvest (under shade only). If possible, dry the seeds in an air conditioned or dehumidified room; it will increase the life span of seed.
4. Sample should contain at least 4,000 seeds for self-pollinated crops and twice that number for cross-pollinated crops so as to fully represent variability of the original samples and also allow sufficient seeds for monitoring of viability during storage and subsequent regeneration. Sample size could be reduced to a greater extent in case of wild relatives where seed are available in limited quantity.
5. Seed should be packed in good quality paper pouches and wrapped in polythene bags. Entire seed lot after packing should be kept either in tin (metallic) boxes or cardboard boxes to minimize damage during transit. While packing, place label inside and also write the accession number outside the pouches for doubly ensuring the authenticity of the accession number. Packing in gunny bags should be avoided.
6. Only untreated seeds are accepted in the gene bank.
7. Seed lots must be accompanied by minimum passport data (e.g. name of the crop, location of collection, accession number etc. and evaluation data). Other specific attributes like stress tolerance, resistance to diseases and pets may be recorded against each accession.

8. After harvest, seed should be stored under ideal conditions like cooler place with less humidity to avoid spoilage due to pathogens and pests. Minimum time should be taken for dispatch to the gene bank.
9. Typed list of germplasm should be sent along with the material.
10. Prior information regarding the supply of seed material for storage should be sent to NBPGR (Germplasm Conservation Division) so that all arrangements can be made to accept the material.

Ensure the following points before dispatch

1. Month and year of harvest
2. Seeds are healthy and free from treatment
3. Seeds are sufficient in number
4. Seeds are properly dried and packed and accompanied with minimum passport and evaluation data

Short, medium and long term storage of PGR

The genetic variability is accumulated in the form of land races, advanced cultivars, wild relatives of domesticated plants and wild (i.e., non domesticated) species used by man. The conservation of plant species, population genotypes which are actually or potentially useful to man is immensely important as they form the basic genetic wealth on which plant breeders could operate for reconstructing the existing genotypes or synthesizing new once in accordance with requirements of time and space. Since it can not be determined what characteristics will be required in the future or even which of the current economic species will continue to be used, a reasonable strategy of conservation would include the types likely to be useful in the near future and representative samples of the genetic resources for use and study by the future generations. It follows from the qualification of usefulness that a tome scale of concern is always associated with the conservation process.

The conservation of plant genetic resources is both safest and cheapest through seeds. It is important first to understand the storage physiology of the seeds to be conserved. The seeds can be classified into two distinct groups according to their storage physiology. The first group is described as 'orthodox'. It includes the majority of arable and horticultural crop species. In these crops the seed ageing and loss of viability occurs as a function, not only of time, but also of temperature and moisture. Thus the storage life of these seeds can be increased if moisture content and temperature and reduced.

Seed storage

The genetic resources center undertakes the responsibility for one or more of the following collections.

(a) **The active collections:** These collections are for multiplication, regeneration, evaluation, documentation, distribution and use by plant breeders. These collections have a higher rate of usage and thus more frequent need for multiplication, the storage period for such collections is therefore, short (1-2 years) or at best medium (3-4 years).

(b) **The base collections:** The primary objective of a base collection is to preserve indefinitely specific germplasm accessions in order to maintain as broad a germplasm base as possible for the future. These collections are stored for long term periods and the conditions of storage are more exacting so as to reduce the risk of genetic changes introduced through frequent regeneration and prevent genetic erosion of the stored germplasm accessions resulting from excessive deterioration.

Short term storage

The storage period for such seeds is between I year and 18 months. In the arid and cool agro climatic zones, seeds of most cultivated species will maintain a high germination capacity for as long as 18 months with only the base cell of storage. However, in warm and humid environments the viability loss is very rapid in the simple seed warehouse. Also seeds of soybean, cotton, onion and several flower and tree species lose germination capacity rapidly under warm humid storage and may deteriorate in periods as short as 2 or 3 months in common storage under these climatic conditions.

Medium term storage

The size of the accessions kept in the medium term is generally larger than those meant for long term storage. These active collections have a higher rate of usage as they are distributed regularly for evaluation and breeding purposes. The accessions have to be regenerated periodically due to depletion of stocks, rather than from loss of viability. Therefore, the time period for storage is not more than 5 years and the pressure of extending the longevity as for the base collections does not exist. The active collections are stored at temperatures ranging from 0°C to 10°C. The seed samples in these collections are stored in various types of containers as cloth bags, metal cans or glass jars. If the containers are hermetically sealed after drying the samples then the relative humidity of the medium term storage is not important. If the unsealed containers are used than the relative humidity should be brought down. This shall, however, increase the running costs. The method adopted, thus shall be decided by the frequency of withdrawal of the samples. A temperature of 5°C and 35-40% RH is adequate for maintaining the viability of most orthodox seeds for 5-25 years. If unsealed containers are used then regular monitoring of seed moisture content is also very important.

Long term storage

The base collections and duplicate base collections are generally kept in stores where the strict storage environment is maintained. Although the working collections can also be kept in these stores but the cost considerations become prohibitive.

The conditions recommended by IBPGR for storage of base collections are probably most suitable and economical to maintain by conventional mechanical refrigeration methods. The preferred standard for base collections is storage at 5%± 1% seed moisture content in hermetically sealed, moisture proof containers at –18°C. In fact –18°C (0°F) is an arbitrary temperature. Most equipment will operate economically at -20°C. The 1976 Working Group of IBPGR suggested that the temperature standard could be relaxed to –10°C if the seed bank was restricted to a few species with good storage characteristics such as the common cereals. Thus any store which operates at –10°C is also referred as a long term store. Present indications suggest that accessions stored at –10°C will require monitoring and regeneration twice as frequently as those at –18°C. Consequently, it is recommended that new facilities should adopt –18°C or less, because any saving on the capital land running costs of the store will be outweighed in the long term by the increased staffing and costs required for the increased monitoring and regeneration.

General Considerations

NBPGR as a nodal organization is coordinating PGR activities in vegetables. It has identified areas and crop priority for further collecting, and germplasm being conserved at National Gene Bank. Also, working collections are maintained at Regional Stations. To enrich germplasm of vegetable crops for which rich diversity occurs in India and neighboring regions such as eggplant and okra, NBPGR has collaborated with International Plant Genetic Resources Institute (IPGRI). Through an IPGRI supported project, both cultivated and wild okra and eggplant has been collected from within India and from Bangladesh, Nepal and Sri Lanka. IPGRI has also been identified as a centre for holding global/ regional collections of okra and eggplant.

Conclusion

Genetic resources characterization and evaluation have played important in vegetable crops improvement leading to evolving high yielding varieties with wide adaptability. Better quality, primitive types of several vegetable crops are of immense importance for crop improvement, which is getting eroded due to release of new varieties and needs to be collected and conserved for further use in improvement. To meet the ever increasing demands of urban areas, leguminous and cucurbitaceous crops, should be tested. Activities of germplasm need to be strengthened. The assembled germplasm at different centres may be pooled, properly evaluated, documented and conserved under long term storage for the present and future use.

References

Arora, R.K. and Nayar, E.R. 1984. Wild relatives of crop plants in India, NBPGR Science Monograph No. 7. National Bureau of Plant Genetic Resources, New Delhi-110 012

Breese, E.L. 1989. Regeneration and multiplication of germplasm resources in seed gene banks. The scientific background. IBPGR, Rome.

Choudhury, B. 1967. Vegetables: National Book Trust, India.

Rana, R.S, Saxena, R.K., Tyage, R.K., Saxena Sanjeev and Mitter, Vivek 1994a. Ex-situ conservation of plant genetic resources. NBPGR, New Delhi-12.

Rana, R.S, Singh, B., Koppar, M. N., Rai, Mathura, Kochhar S. and Duhoon, S.S. 1994b. Plant genetic resources. Exploration, evaluation and maintenance. NBPGR, New Delhi-12.

Rana, R.S, Gupta, P.N., Rai, Mathura and Kochhar, S.1995. Genetic Resources of vegetable crops . Management, conservation and utilization. NBPGR, New Delhi-12.

Seshadri, V.S. 1987. Genetic resources and their utilization in vegetable crops, in Plant Genetic Resources: Indian Perspective. (Eds. R.S. Paroda, R.K. Arora and K.P.S. Chandel), National Bureau of Plant Genetic Resources, New Delhi-12. pp.335-343.

Singh, B.P. and Srivastava, U. 2004. Plant genetic resources in Indian perspective. Theory and practicals. ICAR, New Delhi-12.

Singh, K. 1993. Improvement of vegetable crops in India. Symposium on Horticultural Research-Changing Scenario. Horticultural Society of India, May 24-28, 1993, Banglore, India.

❑❑❑

Chapter – 24

Biometrical Approaches for Diversity in Vegetable Crops

D. Roy

Phenotypic Diversity

Phenotypic diversity measures the portion of genetic diversity which is expressed phenotypically. So it varies with the character under study and the genetic background and the environment in which it is measured. The measures of genetic diversity based on variance of quantitative traits may be unreliable indicators of diversity in a population at the level of individual gene. But diversity at the molecular level (isozymes, RFLPs) may be a reasonable estimate of phenotypic diversity. The degree of differentiation is related to several biological attributes such as breeding system, mode of pollination and seed dispersal, life history, geographical range, successional stages, etc. The outbreeding populations are much less differentiated whereas the selfers or apomictic are strongly differentiated. Populations of outbreeders contain different alleles especially the rare alleles which contribute little to the diversity. Inbreeders should have greater interpopulations diversity and less intrapopulation diversity than the comparable outbreeders. Further, inbreeders are more prone to show an even distribution of their genetic diversity among populations and this makes their optimum sampling more difficult and any evidence of the geographic pattern of this diversity levels would greatly improve sampling efficiency but geographical and environmental factors rarely explain more than 50% variation.

Evaluation of Germplasm

The germplasm collected should be properly evaluated. The characters to be observed can be anatomical, morphological, chemical, biochemical or physiological characters. One should also study the sources of cytoplasm and other traits such as mitochondria, chloroplast and polypeptides, resistance to biotic stresses (diseases and insects) and abiotic stresses (drought, flooding, heat, alkalinity, salinity, toxic elements). One of the objectives would be to measure the amount of variability. The variation at the level of genotype i.e. the variation in nucleotide sequence is called Genetic variation. The amount of variation at DNA level is very much greater than at the polyploid level. The variation at the level of phenotype is called Phenotypic variation. We can then measure the diversity as genetic diversity and phenotypic diversity.

Estimation of Gene Diversity

The measures of Genetic diversity are:

1. Percentage of polymorphic loci
2. Average heterozygosity per locus
3. Average number of alleles per locus
4. Nei's index of gene diversity
5. Shannon -Weaver information index
6. Wright's fixation index

of these 1, 3, 4 and 5 provide the statistics which are indices of genetic diversity and 2 and 4 provide measure of the distribution of genotypic frequencies in the population. All these statistics can be estimated from the isozyme data.

Allelic diversity or the allelic heterogeneity is measured using allele frequency. The larger the number of alleles and the closer they are to equal frequency, the greater is the allelic diversity. The observed frequency at each locus is used to calculate the expected frequency of heterozygotes using H. W. equilibrium. The problem of using such estimate as a measure of genetic diversity is that heterozygosity and polymorphism are quite inadequate descriptors of variation because they are heterogeneous. The observation of sequential gel electrophoresis, calibration experiments and of DNA sequencing has revealed two types of heterozygosity. At amino acid level, there is a class of major polymorphism of two or more alleles in roughly equally high frequency and there exists a vast array of minor polymorphism producing a long list of rare alleles which together constitute a significant heterozygote. Besides these two classes there is presence of some loci which are totally monomorphic lacking both kinds of heterozygosity. Thus any calculation of heterozygosity which confounds the two quite different phenomena confounds the different causes of variation. The amount of heterozygosity per nucleotide is greater for DNA as a whole than for protein coding region. Intervening sequences, exons and pseudogenes have the largest amount of heterozygosity. The less significant a nucleotide difference is the more likely it is to be heterozygous. Thus, neither the average heterozygosity nor the polymorphic loci will by itself yield complete picture of the variation in population. The number of alleles in the total population is a direct measure of the genetic polymorphism whereas the amount of differentiation among sub-populations is measured by the variance of allele frequency at a diallelic locus.

The genetic diversity measured by calculating the amount of actual or potential heterozygosity is as follows:

The total genetic diversity of the polymorphic loci (H_T) can be written as

$$H_T = Hs + D_{ST}$$

Where, H_S is the mean genetic diversity within population at polymorphic loci. A locus is polymorphic if the frequency of the most frequent allele is less than 99%, sometimes > 95%. D_{ST} is the genetic diversity among populations. The proportion of total genetic diversity found among populations is estimated as

$$G_{ST} = \frac{D_{ST}}{H_T} = \frac{H_S - H_T}{H_T}$$

The G_{ST} is thus a measure of the extent of the differentiation (Nei, 1973). Thus the total genetic diversity within a species can be partitioned into between races and within races. H_S measures the mean percentage of loci heterozygous per individual. It is a function of the number and evenness of allele frequency within populations. It represents the polymorphic index or expected heterozygosity. The difference in H_T primarily results from variation in the proportion of polymorphic loci. The actual or potential heterozygosity is calculated as

$$H = 1 - \Sigma p_i^2$$

It measures the genetic diversity whether or not the population is random mating. Here H is the probability that the two randomly chosen alleles are different and thus is a measure of actual heterozygosity. It is a function of gene frequency and the number of alleles in the population. The H_S and H_T are now estimated as

$$H_S = (1 - \sum_i p_i s^2)/n$$

and

$$H_T = 1 - \sum_i \bar{p}_i^2$$

where n is the number of sub-populations, *pi* is the average frequency of allele *Ai* in the entire population and *pi* is the frequency of allele *Ai* in the subpopulation.

Shannon-Weaver information index

Diversity can be expressed in terms of not only the number of species but also the relative abundance of each. S. W. measure of diversity is thus a function of gene-frequencies (allelic evenness) and number of alleles in the population (allele richness).

The diversity index is expressed as

$$H_S = -\sum_{i=1}^{s} p_i \log 2\, p_i$$

where S is the number of species, Hs is the amount of diversity in a group of S species, pi is the relative abundance of the ith species measured from 0 to 1.0 and log pi is the logarithm of p_i, it can be to the base 2 or 10. The value of Hs will be greatest if the species are all equally abundant. This measure of diversity has no upper limit i.e. if more and more species are included in the study, Hs will tend to increase. Also, if different independent classifications are used, the sum total of Hs obtained using each classification provides a measure of the total amount of diversity.

The diversity index can be estimated using the qualitative as well as the quantitative traits. The quantitative trait is analysed as qualitative trait and it is classified into 11 or more point scale. The diversity index then takes the form as

$$H^1 = -\sum_{i=1}^{n} p_i \log 2\, p_i$$

where n is the number of phenotypic classes for a trait and pi is the proportion of entries in the ith class. The variance of diversity index, VH^1, is calculated as

$$VH^1 = \frac{\Sigma p_i \log 2^2 p_i - (\Sigma p_i \log 2\, p_i)^2}{N} + \frac{n-1}{2N^2}$$

where N is the sample size. The test of significance is done using t test.

Determining representative variation within and among collection of genotype is key to the success of world gene bank. The Shannon-Weaver information index (H^1) provides a measure of phenotypic diversity based on frequency data. It can be used to evaluate geographical patterns of diversity and to determine specifically the relative contributions of various countries to the germplasm sources of an existing world collection. The estimate of diversity (H^1) ranges from 0 to 1.0. 0 indicates no variation whereas 1 shows maximum variation.

Estimates of the degree of similarity or differences between two populations

Nei's (1973) index provides the estimate of degree of similarity or difference between two populations. The index is expressed as

$$I_L = \frac{J_{XY}}{\sqrt{J_{xy} \cdot J_Y}}$$

where Jxy = ?$xiyi$, where xi is the frequency of ith allele in population X and yi is its frequency in population Y. Jx =?x_i^2 which is the sum of squared allele frequency in population X and Jy = ? y_i^2 is the sum of squared allele frequency in population Y.

The genetic distance D based on the identity of genes between the two populations is expressed as $D = -I_n I$

where I is the normalized identity of genes between two populations. It measures the accumulated allele differences per locus. If the rate of gene substitution per year is constant, it is linearly related to the divergence time between populations under sexual isolation. It is related to the geographical distance or area in some migration models (Nei, 1972). Nei's genetic distance (D) is said to measure a biological prosperity-the mean number of electrophoretically detectable substitution per locus that have accumulated since the two populations diverged from the common ancestors. The coefficient can be corrected for error due to sampling sizes but it is non-metric.

Considering 3 species A, B and C, the distance between A and C (A -C) must be

$$A - C = A - B + B - C$$

However, it does not satisfy the triangular inequality. The branch length may be negative, which is an undesirable and biologically unpredictable result for a coefficient used in restructuring phylogenes.

Nei (1972) noted that 'I' could also be computed as the arithmetic mean of Ij (I considering jth locus) over loci but the genetic interpretation of this statistic was less straight forward. Hillis (1984) has argued that the mean locus identity defined as

$$I^* = ?\ Ij\ |L$$

Where, L is the number of loci and its corresponding genetic distance $D^*= -\log eI^*$ may be more appropriate measures of diversity and distance than I and D, as is usually observed.

Wright's Fixation index

Wright (1943, 1965) showed that the variation in gene frequency among subpopulations may be analysed by fixation Indias and F stastics as follows

$$1\text{-}F_{IT}=(1\text{-}F_{IS})\ (1\text{-}F_{ST})$$

where F_{IT} and F_{IS} are the correlahons between two unity gamets relative to the total population and sub population respectively and F_{ST} is correlated between two gamets taken at random from each sub population F_{IT} and F_{IS} can take negative values whereas F_{ST} can not the F_{ST} thus, measures the gene differentiation among sub populations.

Genetic distance/Divergence

There are different indices of genetic distance (i) Morphological quantitative distance. (ii) Biochemical distance. The genetic distance from morphological data is particularly useful for inter group classification below the species level. As natural selection affects morphological traits linked to adaptive characters, genetic distances allow inferences about adaptation and co-adaptation pattern of populations. The divergence of quantitative traits on the basis of heterosis can be related to differences in allele frequency between populations and finally the genetic distances are relevant to evaluation of germplasm resources and in the context of breeding programmes. Genetic distances can also be used to derive a grouping of populations with regard to their genetic similarities or dis-similarities which is achieved through dendograms or dendrites (minimum spanning trees).

Biochemical and molecular traits (i.e. isozymatic variability) not affected by environmental variations are generally used to estimate genetic distances to establish phylogenetic pattern and to describe evolutionary process. Variation at enzyme level does not influence the quantitative traits to a major extent. That is why there is frequent discrepancies between morphological and biochemical distances which suggests that inferences concerning the rate of evaluation and variation apply only to a particular kind of character.

Estimation of genetic distance

Genetic distance can be estimated using qualitative traits or quantitative traits. The type of genetic distance statistics in allozyme studies cannot be used in the vast majority of quantitative traits because one can rarely determine the number of loci controlling a particular trait, much less the allele frequency. Using quantitative traits the genetic distance among the populations is calculated using D^2 statistics on the basis of the multiple characters. The phenotypic mean is taken as the best available estimate of genetic effects.

$$D_G^2 = X^T S^{-1} X$$

where X is a vector of phenotypic mean difference between sample i and j for n quantitative traits and S^{-1} is the inverse of the sample dispersion matrix. This method is based on the assumption that populations follow a multivariate normal distribution and that the dispersion matrices are relatively homogeneous. The statistical distance between cultivars/populations i and j, D_{ij}^2 is then simply the sum of squares of the difference for quantitative traits under consideration.

Simple phenotypic data can lead to biased estimates of genetic distances as morphological traits are more or less correlated and widely influenced by the environments. The original or correlated character means can be transformed to an un-correlated set of variables using the pivotal condensation method as described by Rao (1952). Canonical variate transformation permits an optimal visualization of the differences between cultivars/populations through a reduction of dimension that preserves most of the biological information.

Cluster formation

The clustering of varieties is done by a method suggested by Tocher (Rao, 1952). A pair of cultivars says i and j, with smallest D^2 value is selected first and its D^2 value is compared with an arbitrarily selected D^2 value to be decided by the worker. If this D^2 value is less than the selected D^2 value the two cultivars are clustered together. To this group of 2 another cultivar say k, is added which is next closest as far as the D^2 value is concerned. Now with 3 cultivars there would be combinations of cultivars such as i, j; i,

k and j, k with distances between them as respectively. After inclusion of the cultivar k, the average distance rises to $(D_i^2,k + D_{j,k}^2)/2$ and if this value is less than the selected D^2 value i, j and k cultivars form one cluster. With the inclusion of cultivar l there would be (4 x 3)/2 = 6 combinations of cultivars with D^2 values ($D_{i,j}^2$, $D_{i,k}^2$, $D_{j,k}^2$, $D_{i,l}^2$, $D_{j,l}^2$ and $D_{k,l}^2$). The denominator for estimating the average distance would be 3 with average D^2 value being $(D_{i,l}^2 + D_{j,l}^2 + D_{k,l}^2)/3$. This process of inclusion of a cultivar continues until the increase in average D^2 value is less than the arbitrarily fixed D^2 value.

Once a cluster is formed another pair of cultivars with least D^2 value is picked up and its D^2 value is compared with the same arbitrarily fixed D^2 value and if less than this value then the selected pair forms another cluster. Again another cultivar is added to this group of 2 and following the above procedure another cluster is formed. The criterion used in clustering of cultivars is that any two cultivars belonging to the same cluster must at least on the average show a smaller intracluster distance than intercluster distance.

Intra- and inter-cluster distances

The intra-cluster distance for a cluster of say n, cultivars is calculated as the mean of all D^2 values for all possible combinations of n cultivars. With n cultivars there will be $n(n-1)/2$ such combinations. The intercluster distance between two clusters say cluster I having n_1 cultivars (i,j, k, l) and cluster II having n_2 cultivars (m, n) is calculated as

$$[D_{im}^2 + D_{jm}^2 + D_{km}^2 + D_{lm}^2 + D_{in}^2 + D_{jn}^2$$
$$+ D_{kn}^2 + D_{ln}^2]/n_1 n_2$$

Contribution of individual character to genetic divergence:

One of the aims in D^2 analysis is to find a character which shows major contribution to overall genetic divergence or in other words to overall discrimination in the material under study and also a trait showing high correlation between genetic diversity and geographical diversity as the aim of D^2 analysis is to have a quantitative measure of the presumed association between geographical and genetic diversity.

A particular character's contribution can be said to be highest to genetic divergence if for that character maximum number of cross combinations yield highest values of D^2. For p characters each cross combination will have pD^2 values and rank 1 is associated with the highest D^2 value and the lowest rank is attached with the lowest D^2 value. The contribution of a character is calculated as the number of cross combinations yielding D^2 values of rank 1 to the total number of cross combinations.

Although experimental design can control most of the environmental variation and other non-genetic effects such as age or the developmental stage of a plant but the ontogenic factors, which bring about changes in variance and covariance and which is a general phenomenon in animal at least, can be difficult to control by means of experimental design. Thus individuals to be measured in the experiment should be at the same stage of development in order to avoid measuring differences (distances) arising from differential gene activity at different stages of development.

Use of genetic parameters in the estimation of genetic distance

As the morphological distances based on phenotypic data may lead to biased estimates of genetic distances, the interpretation of morphological distances in genetic terms is not easy so it would be better to measure genetic distance based on genetic parameters. Cultivars may differ in genetic effects such as additive, dominance, etc. or heterotic effect and these genetic parameters can be estimated from a suitable breeding design. The multivariate technique then allows the measurement of genetic distance (also called Mahalanobis distance) as a linear function of the estimated parameters (Camussi *et al.* 1985). The appropriate genetic distance matrix can be defined in terms of the estimated parameters. The use of T^2 statistics defined in terms of vector of contrasts specifying the distance permits the testing of the significance among distances between any pair of populations or lines. The linear transformation of the original data into canonical variates can be done for better representation of the results.

Literature Cited

Rao, C. R. (1952). Advanced Statistical Methods in Biometrical Research. John Wiley & Sons, Inc. New York.

Nei, M. (1973). Analys is of gene diversity in Subdivided Populations. Prod. Natt. Acad. Sci. USA. **70(12/1)** : 3321 - 3323

Brown, A.D.H. (1978). Isozymes, Plant Population Genetic Structure and Genetic Conservation. *Theor. Appl. Genet.* **52:** 145-157.

Roy, D. (2000). Plant Breeding Analysis and Exploirtation of Variation. Narosa, New Delhi and Alpha Science, UK.

Camussi, A., Ottaviano, E., Calinski, T. and Kaczamaren, Z. 1985. Genetic distance based on quantitative traits. *Gemets*. **111** : 9465-962.

Wright, S. (1943). *Genets,* **28** : 114-138.

Wright,s (1965). *Evoluation,* **19** : 395-420.

Nei, M. and Roy Choudhary, A. K. 1972. *Science.* **117** : 434-436.

Nei. M. (1972). *Amer. Naturalist.* **106** : 283-292.

□□□

Chapter – 25

Management of Genetic Resources of Vegetable Crops using Geographical Information System

Rajeew Kumar and H.Choudhary

Plant genetic resources are the most valuable and essential basic raw material to meet the current and future need of crop improvement programme. A wider genetic base assume the priority in plant breeding research aimed as developing new varieties for increased crop production. PGR for food and agriculture are a reservoir of genetic adoptability, which act as buffer against environmental changes and economical challenges. The erosion of these resources results in a severe threat to the world long term food securities. The fundamental objective of the genetic resource conservation is the maintenance of genetic diversity within each of the species. In general PGR include all species which contribute to people by providing food, medicine, feed, fiber etc. PGR is our heritage which needs conservation for prosperity.

Genetic resources management is a complex process starting from the identification of a target gene pool for conservation to the use of genetic resources. Many of these activities not only generate but also require geo-referenced data (Spatial data). Analysis of these data can lead to a greater understanding of the eco-geographic representation of existing collections, increasing the efficiency of conserving and managing genetic resources and broadening our understanding of the distribution of genetic diversity. Genetic resource databases could potentially serve as an information resource to a broader scientific community, providing relatively independent global data sets to meet the unique needs of researcher. GIS can merge genetic diversity information with other geo-referenced data such as population density, climate, topography and soils, adding value to genetic resources. GIS analysis can help develop conservation strategies, monitor genetic diversity, select potential collecting sites, design *in situ* reserves and enhance genetic resources use. Currently, GIS use is constrained by limited georeferenced information.

What is GIS?

A GIS defined as a database management system which can simultaneously handle spatial data in graphics form, i.e. maps, or the 'where', and related, logically attached, nonspatial, attribute data, i.e. the labels and descriptions of the different areas within a map, or the 'what'. GIS can be seen as a system of hardware, software and procedures designed to support the capture, management, manipulation, analysis, modeling and display of spatially-referenced data for solving complex planning and management problems. Although many other computer programs can use spatial data (e.g. AutoCAD and statistics packages), GISs include the additional ability to perform spatial operations

The main elements of a GIS are as follows

- Data input, verification and editing;
- Data storage, retrieval and management;
- Data manipulation and analysis; and
- Output.

Data input

Data can be entered into a GIS by digitizing paper maps and their associated attribute information *de novo* using a digitizing table or scanner, or by importing existing digital datasets, including remote sensing images. Genebank curators can import data from the database of their documentation system into a GIS, using the latitude and longitude fields in the passport data to provide the link to digitized maps and remote sensing images.

Data storage

There are two main types of GIS data: vector and raster. Vector files store geographic data as points, lines or polygons. Polygons represent areas of different sizes and shapes where a particular attribute is equal throughout. In contrast, in raster (or grid) data, an area is divided into an array of regularly shaped cells. Each individual cell is assigned a value for the variable being studied. Examples of vector data include maps of roads, or administrative zones. Typical raster datasets include satellite images, elevation (so called Digital Elevation Models) and interpolated climate data.

Data manipulation and analysis

The spatial processing system (to manipulate the 'where') and database management system (for the 'what') of a GIS allow the user to bring together diverse datasets, make them compatible, and combine and analyze them.

Output

GIS allows visualization of spatial data (maps) on a computer screen and the production of printed maps, allowing such manipulations as selecting areas or layers for output, and changing scale and colour. However, the outputs of GIS analyses are not just maps, but can also include tables, graphs and animations. Three-dimensional visualization is one of the key topics in current GIS development.

Why is GIS important?

GIS technology is to geographical analysis what the microscope, the telescope, and computers have been to other sciences. GIS integrates spatial and other kinds of information within a single system - it offers a consistent framework for analyzing geographical data. By putting maps and other kinds of spatial information into digital form, GIS allows us to manipulate and display geographical knowledge in new and exciting ways .GIS makes connections between activities based on geographic proximity.

Why is GIS so hot?

- high level of interest in new developments in computing
- GIS gives a "high tech" feel to geographic information
- maps are fascinating and so are maps in computers
- there is increasing interest in geography and geographic education

Process components of GIS to genetic resource management

Spatial analysis can be applied to locality, characterization and evaluation data in a GIS environment to enhance the efficiency of genetic resources management by following components of the process:

1. Ecogeographic surveying;
2. Field exploration;
3. Design, management and monitoring of in situ reserves;
4. Germplasm evaluation; and
5. Use of genetic resources.

1. Eco-geographic surveying

This is the process of collating information on the taxonomy, genetic diversity, geographic distribution, ecological adaptation and ethno-botany of a plant group, as well as on the geography, ecology, climate and the human setting of study regions. The sources of the information for such 'ecogeographic surveys' will include herbarium specimens, germplasm accession passport data, experts, the formal and grey literature, field notes and maps. It is only on the foundation of basic information such as (but not limited to) this that sensible conservation decisions can be made, for example regarding when, where and how to collect germplasm, and where genetic reserves might best be established and how they would need to be monitored and managed. One of the main objectives of ecogeographic surveys is thus the identification of those geographic areas which are:

- Likely to contain specific desired traits (adaptations), taxa or habitats of interest;
- Hhighly diverse (whether environmentally, taxonomically or genetically);
- Complementary to each other;
- Currently missing or under-represented in conservation efforts;
- Threatened with genetic erosion.

2. Field exploration

The ecogeographic study is essentially desk-based. It needs to be refined and expanded, and its results implemented, through fieldwork. GIS technology can contribute to this phase of genetic resources conservation through: the development of field aids; and providing information on the optimal timing of field visits.

(a) **Field aids:** Maps are crucial to fieldwork. Unfortunately, published maps showing a specific factor of interest to the conservationist. Mapping products develop by GIS can be used in the field. These maps allowed the identification of ecogeographic gradients in the field, the slope of which influenced decisions about sampling frequency. They were also used to monitor, during the course of the fieldwork, which combinations of environmental conditions were being adequately sampled and which were not.

(b) **Timing:** Rainfall can be unpredictable in both space and time, especially in the arid and semiarid tropics, making it difficult sometimes to time field exploration work to coincide with the most appropriate stage of vegetation development. Satellite imagery can provide data on the state of vegetation with relatively short lag-times. Use of such data could allow collectors to be more precise in timing their visit, doing greatly to the cost-effectiveness of collecting trips

3. Design, management and monitoring of *in situ* reserves

The design of protected areas, including questions of optimal size and shape, zonation and networking is a spatial problem, and spatial analysis in a GIS environment has been applied to them, although mainly in the context of ecosystem or animal conservation. Genetic resource manage within a GIS information on the cultural values of different features of the landscape, flora and fauna, with a view to developing resource and environmental management plans more in tune with the requirements of local people.

Given the rapid rate of environmental and socio-economic changes (including the threat of global climate change), it is important for reserve design to take into account possible future scenarios. GIS combined with computer modelling can be used to optimize reserve selection and management based not only on current climatic and vegetation patterns, but also under different assumptions of change. GIS used to document land use change, deforestation and habitat fragmentation in the Western Ghats of India, and investigate the socio-economic drivers of these processes to develop predictive models. These were then used to estimate the current and future effectiveness of the protected area network for biodiversity conservation. Once the *in situ* reserve has been established, some form of management intervention may need to be implemented, and its results on the frequency, abundance, demography and genetic diversity of target species monitored. GIS can assist in defining and implementing the management plan. Working in a GIS environment can ease the application of analytical techniques of plant demography as transition matrix sensitivity analysis , thus hopefully stimulating of plant genetic resources conservation.

4. Germplasm regeneration and evaluation

A key step in both germplasm multiplication and evaluation is to decide what sites are best for which accessions. The environment of the regeneration site must not only provide suitable conditions for reliable flowering and seed production, but also any necessary triggers for the different stages of plant development. Unless controlled pollination is practiced, pollinators must also be present. On the other hand, pests and diseases must not be prevalent and populations of species related to the one(s) being regenerated should not be found nearby. Climate data and distribution maps for pests, diseases, pollinators and wild relatives can be overlapped using GIS to identify potential sites for regeneration. If the choice of locations for regeneration is limited, such data can be used to estimate the relative suitability of different available sites for the regeneration of different accessions, by comparison with the environment at their site of collection (e.g. using FLORAMAP). They can also help to determine the unique needs of individual accessions so that growing procedures can be customized in sub-optimal sites. As for germplasm evaluation, both ecologically optimal and stress locations are usually necessary. Classification of testing locations using GIS-derived climatic data can be used to determine the suitability of different sites for the evaluation of specific traits, or combinations of traits, of interest. If a site has already been chosen, GIS can be used to determine the extent of the region where similar conditions apply, and hence the relevance and potential impact of the evaluation.

5. Use of genetic resources

The major aim of conserving germplasm is, of course, for it to be used. Adequate access policies, good linkages between genebanks and users and well-developed breeding and other use programmes are necessary conditions for genetic resources to be used to the full; but they are not sufficient. The lack of data on accessions and the large size of collections can still present significant bottlenecks to the use of conserved material. GIS can help to alleviate the problem in the first instance by improving the quality and quantity of the locality data associated with collections. Given adequate data on the origin of material, GIS can also help to identify particularly interesting germplasm and reveal the structure of diversity within collections. Finally, GIS can be used to assess the potential impact of the products of germplasm use for better locality data on accessions, germplasm targeting, structure of collections, assessing impact.

GIS as a Management Tool in Genetic Resource Management

The potential benefits of applying GIS tools to the conservation and management of genetic resources are providing strong incentives for curators and collection users to improve the quality and quantity of information. Potential GIS applications can be divided into four general areas:

(1) Monitoring of inter- and intraspecific biodiversity;

(2) Developing effective acquisition strategies;

(3) Increasing understanding of existing collections; and

(4) Utilizing existing germplasm collections.

GIS software used for vegetable genetic resource management

The recent development and dissemination of specialized software are

DIVA – Turning data into maps

The geographic distribution of diversity is an important consideration at all stages of genetic resource exploration and conservation, yet a difficult parameter to visualize and analyze. Diversity studies generally divide the target area into smaller areas of equal size (often squares in a grid pattern), for which a measure of diversity is calculated and compared to that of the other units.Off-the-shelf GIS software can carry out such an analysis but can be expensive and difficult to learn. In order to support national plant genetic resources programmes and regional networks assess their genetic diversity, Bioversity and the International Potato Center (CIP), with the financial support of CGIAR's System-Wide Genetic Resource Programme (SGRP) have developed DIVA - GIS, a GIS software tool specifically designed to calculate diversity indices on the bases of latitude, longitude and characterization data of a set of accessions. The software presents the results in a map that can be used for decision-making. It was developed specifically for use with genebank data available through national or international genebank documentation systems, such as SINGER, the genetic resources information exchange network of the CGIAR, and the European Plant Genetic Resources Catalogue (EURISCO). European Forest Genetic Resource Network (EUFORGEN) has used DIVA-GIS to plot distribution maps for many European tree species.

FloraMap – A tool for predicting the distribution of plants and other organisms in the wild FloraMap, the product of more than 20 years of research at the International Centre for Tropical Agriculture (CIAT), is a software that allows biodiversity specialists to map the likely distribution of species in nature, plan programmes to collect them and decide where to set up *in situ* conservation programs. FloraMap can also be used to study the taxonomic and genetic variation of particular species, and to map the distribution of crop pests and their natural enemies.In South America, Bioversity, in collaboration with CIAT, has used Floramap to predict where wild varieties of peanuts (*Arachis hypogaea)*could be found and help sound an alarm about the threats these varieties faced from climate change and development.

Prospects

The application of GIS to estimating the impact of improved germplasm opens the way for a fully integrated approach to the use of GIS in plant genetic resources management, in which GIS-based analysis guides and facilitates the process from start to finish and back again. First, the likely impact of a particular potential use product in a given area is assessed using GIS; for example a more frost-tolerant potato variety. Next, retro-classification of past collection sites is carried out with GIS to identify germplasm in existing national and international collections which could be useful in developing the proposed product. If sufficient suitable material is not available in genebanks, climatic data are used to identify areas where further germplasm collection should be done. This could be approached by superimposing maps of frost severity and potato cultivation, or by running the collecting localities of frost-tolerant accessions through FLORAMAP. Once promising material has been obtained, GIS is used to identify appropriate evaluation sites. The best material coming out of the evaluation programme is then used to develop the improved product, the actual performance of which is finally fed back into the impact model to start the cycle all over again. The potential is clearly there for GIS-based analysis of data on germplasm accessions to add greatly to the value of the data, and thus of the germplasm to which they pertain. It should significantly enhance the cost-effectiveness of conservation efforts, and facilitate use of germplasm by breeders and others. However, it must be admitted that GIS technology has not been taken up by plant genetic resources conservation programmes to the extent that one might have predicted on the basis of its potential. Part of the reason is that many such programmes, particularly in developing countries, have significant resource constraints, and GIS hardware, software and data are perceived as being expensive, difficult to obtain and complex to use. The perceived 'barriers to entry' may still be too high, but the current revolution in GIS technology is putting both the data and the analytical tools within the reach of many. However, if the potential of GIS in the field of plant genetic resources conservation and use is to be fulfilled, the plant genetic resources conservation community needs to take positive action itself at the international, regional and national levels, and not simply wait for the technology to come to its aid.

References

Bailey, T.C. 1994 A review of statistical spatial analysis in geographical information systems. In: Fotheringham, S. and Rogerson, P. (eds) *Spatial Analysis and GIS.* Taylor & Francis, London, pp. 13–44.

Guarino, L. 1995. Geographic information systems and remote sensing for the plant germplasm collector. In: Guarino,L., Ramanatha Rao, V. and Reid, R. (eds) *Collecting Plant Genetic Diversity. Technical Guidelines*. CAB International, Wallingford, UK, pp. 315–328.

Harmsworth, G. 1998. Indigenous values and GIS: a method and a framework. *Indigenous Knowledge and Development Monitor* **6:** 3–7

Huaccho, L. and Hijmans, R.J. 1999. A global geo-referenced database of potato production for 1995–1997 (GPOT97). *Production Systems and Natural Resource Management Department Working Paper* 1. CIP, Lima, Peru. www.cipotato.org/data/potato_atlas/gpot97.htm

Veitch, N., Web, N.R. and Wyatt, B.K. (1995) The application of geographic information systems and remotely sensed data to the conservation of heathland fragments. *Biological Conservation* **72:** 91–97.

Chapter - 26

Harvesting the Potential of Indigenous Genetic Resources for Improvement of Brinjal

A.S. Sidhu

Introduction

Eggplant or aubergine {*Solanum melongena* L. Solanaceae), is indigenous to a vast area stretching from northeast India and Burma, to Northern Thailand, Laos, Viet nam and Southwest China and wild plants can still be found in these locations. There is a wealth of eggplant common names. Eggplant is a major fruit vegetable with world production exceeding 31 million tonnes (Mt). Leading producers are China (17 Mt) and India (8 Mt), Egypt (1 Mt), Turkey (0.9 Mt), Japan and Italy (0.4 Mt). Eggplant is particularly favoured in Asia where it has been cultivated for millennia, and in India it is considered King of Vegetables. Brinjal, native to India, shares 31 % area (5.1×10^5 ha) and 28% production (82×10^5 t) of the world (Anon., 2003). It contributes 9% of the total vegetable production in the country. Three states/provinces *viz*. Orissa (1.41×10^5 ha), West Bengal (1.29×10^5 ha) and Bihar (0.72×10^5 ha) cover 67% of total brinjal area in India (Anon., 2006). It is widely adaptive and highly productive (17.78 t/ha) crop of tropical and sub-tropical regions. It is comparable with tomato for nutritional values like vitamin C, iron and fiber, thereby, can play an important role in combating malnutrition in under-nourished regions (Singh *et al*., 2001). However, genetic improvement, production technology, post harvest handling and plant protection measures are important for increasing the production, productivity and consumption of brinjal.

Genetic Improvement

The genetic diversity is the base of improvement and the range of variability for different traits of brinjal in India is given in Table 1 (Singh and Kalda, 2001; Som and Maity, 2002; Singh *et al.,* 2004; Rai and Yadav, 2005; Anon., 2006). The wider adaptability and often-cross pollinated nature of brinjal could be the reason for its broader base. The objectives of genetic improvement remains high yield, earliness, better quality and resistance to insect pests and diseases. However, consumer preference for shape, size and colour is also given due consideration. In North-Western India, purple and dark purple colour in South, it is green and striped and in East India, white-coloured fruits are preferred. Similarly, for *'bhartha'* (roast) making round and big fruits are used, for stuffing small round are ideal and for mix vegetable long fruits are cooked. So depending upon the preference, numerous cultivars have been identified in different parts of the country (Table 2). Beside these improved varieties and hybrids, many local genotypes are also popular in different parts of the country. Based upon shape of the fruit cultivars are divided into small-round, long, round, oblong and jumbo groups. The average fruit weight in these groups range between 30-60 g (small-round), 60-200 g (long and medium round), 200-300 g (round and oblong) and more than 300 g (jumbo). The fruit colour in all these groups can be light purple, purple, shining purple, black purple, green, white tipped.

Host plant resistance is economical, durable and hazard free. Many cultivars have been developed for resistance against bacterial wilt (Arka Nidhi, SM 6-6, IIHR-3, IIHR-7 Swarna Shyamli, Swama Pratiba, BB-46-13, BWR-12, BB-64, Pusa Purple cluster and WCGR 112-8), phomopsis blight (Pusa Bhairav), little leaf (PPL, PPC, PPR, Sel-212 Sel 252-1-1, Nurki, Hisar Shyamal, H-10), jassid (Majri Gota, Vaishali, Junagarh Sel-I, Aushey, R-3, H-4, T-3, Dorli, Bergona-1) and nematodes (Vijay, Banaras Giant, T-2): However, complete resistance against shoot and fruit borer *(Leucinodes orbonalis)* is not available in cultivated germplasm, but some genotypes like PPL, PPC, H-I28, H-129 Aushey, IHR-191, SM-202, S-519, S-4, H-4, S-34, Pb. Chamkila, Kusumkar, P-5-8, S~258, SM-62, Malkapuri, Rajangona, Tambalwadi, Gopuri, Jalgaon Local, Higna Doria, Gulabi Doria, Bhattu,

Bhagyamati, PBR-129-5 and SM-17-4 are reported to be tolerant (Som and Maity, 2002; Rai and Yadav, 2005). The field testing of transgenic brinjal carrying *cry* IAc gene has given encouraging results against shoot and fruit borer (Anon., 2006).

Table 1. Variability range of different characteristics in brinjal.

Character	Range	Character	Range
Growth:		**Floral:**	
Plant height	27.8- 167.7	Days to flowering	31.4-114.5
Terminal shoots/plant	9.9-86.3	No. of flowers/axil Style	1-37.4
Leaves/plant	53.5-405.7	length (mm)	1.5-15.5
Size of leaf (cm^2)	52.5-333.7	Anther length (mm)	2.3-8.1

Contd...

Primary branches/plant	2.1 - 17.4	Ovary diameter (mm)	1.26-5.1
Fruit:		**Biochemical:**	
Length (cm)	5.6-67.5	Moisture (%)	88.4-9.11
Girth (cm)	1.6-28.7	Crude protein (%)	1.2- 1.9
Weight	15.2-2116.4	Total sugars (%)	1.4-3.9
Fruits/plant	1.4- 101.8	Total phenol (%)	0.7-8.0
Fruit yield/plant	0.2 - 5.2	Anthocyanin (%)	0.2-4.6
Fruit colour	White-Purple-Black-Red-Pink-Green-Striped.	Ascorbic acid (%)	0.5- 1.1
		Titratable acidity (%)	0.1-0.5
		Total carbohydrates (%)	2.0-5.8

Table 2. Cultivated varieties and hybrids of brinjal in India.

Varieties

Long: Annamalai, Arka Keshav, Arka Kusumakar, Arka Nidhi. Arka Shccl, ARU-1, ARU-2C, Green Long, Gujrat Brinjal-6, Gulabi, Hisar Pragati, Junagarh Long, KS-331, Nurki Long, Pant Samrat, Pb. Chamikla, Sel-4, Pb. Barsati, Pb. Sadabahar, Pusa Purple Cluster, Pusa Purple Long, Rajendra Baingan, Swetha, Swarna Pratibha, Utkal Jyoti, Ulkal Keshri, Utkal Madhuri and Azad Kranti.

Round: Azad B-1, KS-224, Baramashi, CHBR-1, H-8, Nurki Round, Jamuni Gola, Pb. Neelam, MDU-1, Black Beauty, Pant Rituraj, P-8, Pb. Bahar, Pusa Purple Round, Pusa Upkar and Swarna Mani

Oblong: Pusa Kranti, Bhagyamati, CO-2, Hisar Jamuni, JB-15, Pusa Uttam, Surya, Swarna Shree, Suphal, Swarna Syamli, Utkal Tarini and Ram Nagar Giant

Small Round: Aruna, CO-1, DBSR-44, DBSR-91, KKM-1, PKM-1, PLR-1, Pb. Moti, BSR-11, Pusa Ankur and Pusa Bindu.

Hybrids

Long: Pusa Hybrid-5 and RHRBH-3, BSS-127, BSS-513, Navina, Long Purple, Mamta, PPL-74, HABH-18, Sachin, TSX-251, NBH-171, GS-82-2WA, ARBH-201

Round: Pusa Hybrid-6 and VRBHR-1, Utsav, BSS-540, Raveena, Surbhi, Neelma, Kuroi, HABH-17, Anurag, NDBH-1.

Oblong: Arka Navneet, BH-1, BH-2, Pusa Hybrid-9, Ravaiya, ARBH 201, Vijay, Suman, Utkal, NBH-202, KBHR-4.

Small Round: MHB-10, RHRBH-2, Mridula, BSS-426, BSS-468, ZEH-4078, ZEH-4011, EPH-139, NBH-627, Ajeet Virat. OS-84-2004. Phule Hybrid-2, MHB-39, ABH-1.

Production Technology

Adoption of suitable pr oduction technology can harvest full genetic potential of any genotype. Brinjal requires long and warm growing season with average temperature range of 21-27°C for growth, development and fruit setting. However, it can be grown throughout the year in India, but sowing and transplanting time differ as per agro-climatic conditions. In the northern plains of India, mainly two sowing are preferred for autumn-winter crop sowing is done in Jun-July for spring crop is October-November. The winter-sown nursery is protected from the frost in December and January. Sowing time in central and southern India is July-August and December-January. In hilly regions, seed is sown in March-April and seedlings are transplanted in May (Singh and Kalda, 2001).

For raising one hectare crop, 200 g seeds are required. The nursery is sown on raised beds of 1.25 m width and 20 cm height. Fumigation with 1.5% solution of formalin is found effective for controlling the mortality at seedling stage. Seed treatment with fungicide like thiram or captan @ 3g/kg of seed before sowing protects the seedlings from soil borne diseases. Seed priming with gibberellins improves the seed vigour, germination and uniformity of the nursery (Som and Maity, 2002). Seed treatment with IAA (30 ppm) and subsequent application of phorate and endosulfan showed beneficial effect on yield and control of shoot and fruit borer. The nursery raising in Pro-trays containing coco peat, vermiculite and perlite in 3:1:1 ratio v/v on volume basis is also gaining ground due to high cost of F_1 hybrid seeds. The planting distance depends upon the soil fertility, growing season and cultivar. Usually, 75 × 60 and 90 × 90 cm distance between rows and plants is given to vigorous growing, round and high yielding cultivars, 45 × 45 cm to early and dwarf type and 60 × 45 cm to semi-vigorous or mid season (Singh *et al.*, 1997; Singh and Kalda, 2001).

The fertilizer requirement of brinjal also depends upon soil type, crop rotation, season, genotype and the region of growing. It was observed that brinjal grown after leguminous crop or green manuring yield higher than following a non-leguminous or without green manuring. In a study using 'CO-2' cultivar, production of one tone brinjal requires 7.6 kg N, 1.4 kg P and 17.3 kg K (Singh and Kalda, 2001). The recommended doses of N, P and K fertilizers ranged between 50-160, 10-80 and 25-110 kg/ha in different states. The foliar spray of 1 % urea increased the size and number of fruits. The inorganic form of N is helpful for early flowering compared with that of organic forms. Fruit weight, length, diameter, volume and specific gravity showed linear relationship with increased nitrogen, however under stress, percentage of short style flowers increased and resulted in lower yield. Application of Neem cake at 5 t/ha along with 80:60:60 kg NPK gave beneficial effects on yield of brinjal. Poultry manure at 15 t/ha is also satisfactory for eggplant production. For more yield, sandy soils can be amended with liquid sewage and sludge to increase P, K, Ca and Mg content than that of NPK form of fertilizers only. The application of micronutrients alone or in combination with major nutrients not only increased the yield, but improved the quality also.

Although, vegetative growth is not affected, but Cu and Mn increase the number of flowers and fruits; Zn improves the fruit weight; B and Mn stimulated photosynthesis and increased the yield by 11 to 23%. Application of $ZnS0_4$ at 25 kg/ha along with 2 foliar sprays of 0.5% and CuS04 at 125 kg//ha along with 2 foliar sprays of 0.15 % at 30 and 60 days after transplanting on Zn and Cu deficient clay loam soils, improves the yield and vitamin C content of the brinjal. The leaf tip yellowing of distal ends of young leaves can be corrected with 200 mM B in Borax form. Use of bio-fertilizers like *Azospirillium* is beneficial for increasing the yield (Parsana and Rajan, 2001; Anburani and Manivannan, 2002; Som and Maity, 2002).

Brinjal prefers a balance soil moisture regime, not lower than 60% available water capacity but not exceeding field capacity. Flooding causes anoxic condition, which damage roots, affects root metabolism and ultimately uptake of nutrients. Drip system of irrigation effectively reduces the water use, controls the weeds and increases the yield. With drip system and white polythene mulching, irrigation requirements reduced by 29% and yield increased by 18%. The water use efficiency increased by 66% than the open plot. Trickle irrigation in brinjal requires 6000 cm3/ ha water compared with 11000 cm_3 in furrow system. The fertilizer use efficiency also increases with trickle irrigation, as N doze of 180 kg/ha gave same yield with drip system as that of 360 kg/ha with furrow.

The protected cultivation of brinjal under Agro net-house (40 mesh size net) effectively controlled the shoot and fruit borer with 100% increase in marketable yield. Using the vertical space of net-house, training with two parallel shoots gave 62% advantage in yield than the non-trained plants. Grafting of cultivated brinjal variety 'Punjab Sada Bahar' on *Solanum torvum* proved effective in increasing the yield, duration and nematode resistance (Dhatt, unpublished data). Grafting eggplant on tomato or *Solanum* species (e.g. *S. torvum* or *S. integrifolium)* rootstocks is often used in greenhouse production to overcome root diseases. Annual yields of 460 t/ha have been achieved in intensive greenhouse production in The Netherlands, but this is exceptional. There are several related cultivated *Solanum* species also referred to as eggplants, namely the African Gboma eggplant, S. *macrocarpon* (section *Melongena),* and the African scarlet eggplant, *S. aethiopicum* (section *Oligsnthes).*

Plant hormones provide growth regulation within the plants as directed by the genetic code within the DNA of the chromosomes. Along with naturally occurring hormones (auxins, gibberellins, cytokinins and abscisic acid), which are present in very low concentration, some natural and synthetic chemicals when applied exogenously gave hormonal effects and increased yield and improved the quality of brinjal (Som and Maity, 2002; Meena and Dhaka, 2003).

Conclusion

Brinjal *(Solanum melongena* L.) is an important and indigenous vegetable crop of India. It contributes 9% of the total vegetable production of the country. It is due to improvement in production technology, protection measures and the genetic improvement which has shown significant advancement in yield, quality, diseases and insect-pest resistance. The yield of long, round, oblong and small round varieties have reached at 50, 65, 60 and 40 t/ha, respectively, whereas, F_1 hybrids raised it to 62.5, 79, 75 and 50 t/ha in these respective groups. The cultivars of low glycoalkaloids content with different sizes, shapes and colour increased the market acceptability of the fruits. Resistance breeding for bacterial wilt sustained the cultivation of brinjal in sick soils. The development of Bt-transgenic will lower the shoot and fruit borer damage and ultimately the pesticide use in brinjal. The seed production practices enhanced the seed yield of varieties and hybrids. However, seedling health has been improved by precise nursery raising practices. Standardization of cultural practices, irrigation and nutritional requirements of different cultivars under different soils and climatic conditions helps in better crop stand. The use of herbicides and mulching practices suppressed the fast growth of weeds in brinjal. The water use efficiency has been increased by 66% using trickle irrigation. Protected cultivation makes the availability of brinjal during off-season and proved to be effective for borer free fruits. Use of growth hormones induced parthenocarpy, earliness and yield while different chemical formulations helped in checking of various insect-pest and diseases of brinjal. The grafting of brinjal cultivars on perennial and wild species increased the yield and availability period of the fruits.

References

Anburani, A. and Manivannan, K. 2002. Effect of integrated nutrient management on growth in brinjal *(Solanum melongena* L.) cv. Annamalai. *S. Indian Hort.*, **50:** 4-6.

Anonymous, 2003. FAO Production Year Book. **57:** 149. Food and Agriculture Organization of United Nations, Rome, Italy.

Anonymous. 2006. Horticulture Data Base. National Horticulture Board, Gurgaon, India. **12:** 195- 198.

Daunay, M. C., Janick, J. and Laterrot, H. 2007. Iconography of the Solanaceae from antiquity to the 17th century. A rich source of information on genetic diversity and uses. *Acta Hort.*, **745:** 59-88.

Meena, S. S. and Dhaka, R.S. 2003. Economics of plant growth regulators in brinjal *(Solanum melongena* L.) under semi-arid conditions of Rajasthan. *Ann. Agril. Res.*, **24:** 273-275.

Parsana, K. P. and Rajan, S. 2001. Effect of organic farming on storage life of brinjal fruit. *S. Indian Hort.*, **49:** 288-291.

Rai, N. and Yadav, D. S. 2005. Advances in Vegetable Production. Research Book Centre, New Delhi, India. p. 417-440.

Singh, H., Saimbi, M. S., Bal, S. S. and Singh, H. 1997. A note on the effect of plant population density on growth and yield of brinjal hybrids. *Veg. Sci.* **24:**164-166.

Singh, I., Kalloo, G. and Singh, K.P. 2001. Nutritive Value of Vegetables. Vegetable Crops: Nutritional Security. IIVR, Varanasi, India. p. 20.

Singh, N. P., Bhardwaj, A. K., Kumar, A. and Singh, K. M. 2004. Brinjal. In: Modern Technology on Vegetable Production, International Book Distributing Co, Lucknow, India. p 99-109.

Singh, N. and Kalda, T. S. 2001. Brinjal *(Solarium melongena* L.). *In: S. Thamburaj and N. Singh (eds.),* Text Book of Vegetables, Tuber crops and Spices, ICAR New Delhi, India. p. 29-49.

Som, M.G. and Maity, T.K. 2002. Brinjal. *In: T.K. Bose, J. Kabir, T.K. Maity, V.A. Parthasarthy and M.G. Som* Vegetable Crops, *(eds.),* Naya Prakash, Kolkata, India. p. 265-341.

Chapter – 27

Indigenous Genetic Resources of Underutilized Vegetable Crops

K.V. Peter, P.G. Sadhan Kumar and S. Nirmala Devi

Introduction

Indian subcontinent is one of the twelve mega diversity centres for cultivated plants and their wild relatives. More than 15000 species of flowering plants are indigenous to this region, which include 160 species of economic importance, 320 species of wild ancestral forms and approximately 800 species of ethnobotanical origin. Global diversity in vegetable crops is estimated around 400, of which 80 originated in India. These indigenous vegetables have immense potential in meeting the vegetable requirement of the country. They are rich in nutritional value especially in proteins, minerals and vitamins. Many of these vegetables have medicinal uses. Most of them are not used to their potential and their popularisation is needed in meeting the vegetable requirement of the country in coming days.

Ash gourd (*Benincasa hispida* Cong.)

Ash gourd also known as Wax gourd (*Benincasa hispida* Cong.) belonging to family Cucurbitaceae, is cultivated for its immature and mature fruits used as vegetable and in confectionery and ayurvedic medicines. A small fruited medicinal ash gourd is also grown in Kerala which is good for people suffering from nervousness. In areas where winter is mild, the crop can be grown throughout the year. The released varieties are KAU Local, Indu, Co.1, Co.2, APAU Shakthi, Karikumbala, Boodikumbala and IVAG -502. In addition, MAH-1 and MAH-2 are F1 hybrids released from private sector. The crop is raised by sowing seeds in pits at a spacing of 2.5 × 2.0 m in Tamil Nadu, 3.4 × 1.8-2.5 m in West Bengal and 4.5 × 2.0 m in Kerala. The seeds can also be sown in channels at a spacing of 2 × 0.5m. The crop is trailed on branches and twigs spread on the ground. The fruits are harvested at immature or fully ripe stage depending on demand. Immature fruits can be harvested 21 days after anthesis. Mature fruits for storage, long distance transport and seed extraction are harvested after full development of waxy coating on fruit surface. The average yield are 30-35 t/ha.

Ivy gourd (*Coccinia grandis* L.)

Ivy gourd also known as coccinia, kundru, little gourd or tondli is an under-exploited cucurbit vegetable originated in India. It is a dioecious perennial grown in Southern and Eastern states. Fruits are good for diabetic patients. The crop is raised by planting three nodded cuttings taken from female plants. Sulabha (CG 23), a high yielding variety, released by Kerala Agricultural University, produces light green, cylindrical, long fruits measuring 9.25 cm length and 6.70 cm girth. The average fruit weight is 18.4 g. Fruits have continuous striations on it. Plants come to flower in 37 days, fruits set partheocarpically and the first harvest can be made in 45 days after planting. It gives an average yield of 18.16 kg/fruit/plant/year. From IGKV, Raipur two high yielding genotypes were released Sarnaik *et al*., (2005). They are Indira Kunduru -5 which produces fruits of 4.30 cm length, 2.63 cm diameter and yield of 22.94 kg fruits/plant/year and Indira Kunduru-35 which produces fruits of 6.0 cm length, 2.43 cm diameter and yield of 21.08 kg fruits/plant/year.

Ridge gourd (*Luffa acutangula* Roxb.) and Smooth gourd (*L.cylidrica* Roem.)

Ridge gourd (*Luffa acutangula* Roxb.) and smooth gourd (*L.cylidrica* Roem.) are two important underexploited vegetables belonging to family cucrbitaceae. The genus name was derived from the product 'loofah' used as bathing sponge scrubber pads, door mats, pillows etc. Both crops are cultivated on commercial scale and in homesteads for its immature fruits, easily digestible and appetizing and prescribed for those who are suffering from malaria. The crop is raised by sowing seeds during Rabi and summer in pits or furrows. A row to row distance of 1.5 to 2.5 m and plant to plant distance of 0.60-1.20 m is required under bower and trellis system. The important varieties of *L.acutangula* are Deepthi, Arka Sumeet, Arka Sujat, Co.1, Co.2, Pusa Nasdar, Pant Torai-1, Swarna Manjari, Swarna Uphar, Konkan Harith and Punjab Sadabahar and that of *L.cylindrica* are Pusa Chikni, Pusa Supriya, Pusa Sneha, Phule Prajakta and Rajendra Nenua 1. The crop is ready for harvest in about 60 days after sowing. The fruits attain vegetable maturity 5-7 days after anthesis and picking is done at 3-4 days interval. The major pests infecting these crops are fruit fly, red pumpkin beetle and epilachna beetle. The average yield of the crop is 10-15 t/ha.

Snake gourd (*Trichosanthes anguina* L.)

Snake gourd is a vegetable, the fruits of which have unique smell, harvested at tender stage and used after cooking. It is a good source of vit. A and vit. B. The fruits have medicinal properties and are useful in treatment of blood pressure, heart disease and rheumatism. It improves appetite and acts as tonic and stomachic. Snake gourd is cultivated by trailing on bower. The important varieties are Kaumudi, Baby, Manusree, Co.1 and Co.2. The crop can be raised during September-December and January-April by sowing seeds in pits at a spacing of 2 x 2 m. It flowers 45-60 days after sowing depending on the variety. Fruits attain vegetable maturity in 7-10 days after flowering. In long and slender fruited varieties, there is a practice of hanging small weight at the growing tip of the fruit to get straight fruits. Harvesting is done twice a week and the average yield varies from 30-35 t/ha.

Spine gourd (*Momordica dioica* Roxb.)

Spine gourd or kartoli is another underutilized cucurbit vegetable grown for its tender fruits. In India it is grown in Assam, Meghalaya, West Bengal, Uttar Pradesh, Bihar, Maharashtra, Madya Pradesh, Gujarat and Andaman Islands. It is propagated by tuberous roots. Plants come to flower in 55-60 days after planting. Dubey *et al*. (2007) studied floral biology of spine gourd. They found that the male flowers appeared in the second week of August while the female flowers appeared in the third week of August. Maximum anthesis was noted at 8.30 pm at a temperature range of 26.8 to 27°C. The dehiscence started with appearance of a slit on the anther lobe which resulted in the liberation of pollen grains on the anther surface. Dehiscence started four and half hours before anthesis and individual flower took 45 minutes to complete the process. Maximum release of pollen grains was observed at 2.45 to 4.15 pm. The stigma was receptive upto 12 hours before and after anthesis. Stigma receptivity remained at a peak for 6-9 hours before and after anthesis. Fruits are ready to harvest in 15-20 days after set. It gives an average yield of 12-14 t/ha.

Kakrol (*Momordica cochinchinensis* Zour.)

Kakrol, Sweet gourd or Kheska is an underutilized indigenous cucurbit perennial vegetable grown for its nutritious fruits. It is grown in isolated pockets in Assam, Meghalaya, West Bengal, Uttar Pradesh, Bihar, Maharashtra, Madhya Pradesh, Gujarat and South India. The crop comes up well in hot humid climate. This is a dioecious crop. Plants are propagated by tuberous roots or seeds. Sprouted tubers collected from female plants are planted in pits at a distance of 2m during rainy season. It starts fruiting in 55-65 days after planting. Fruits can be harvested 11-12 days after anthesis. Fruits are harvested at tender stage. It gives an average yield of 7.5-12 t/ha.

Parwal (Pointed gourd) (*Trichosanthes dioica* Roxb.)

Pointed gourd is a cucurbit vegetable indigenous to India. It is grown in Bihar, Uttar Pradesh, West Bengal, Orissa, Assam, Madhya Pradesh, Gujarat, Andhra Pradesh and Tamil Nadu. It is a perennial trailing cucurbit, dioecious in nature. The crop performs the best in hot humid conditions. Well drained sandy loam rich in organic matter is suited for the crop. The plants are propagated by vine cuttings. Cuttings collected during October are planted in sand for root development and later transplanted to the main field. For every 20 females, one male plant is retained. It can also be propagated through root suckers. Vines start fruiting 80-90 days after planting. Fruits are harvested at immature stage. The important varieties are Rajendra parwal-1, Rajendra parwal-2, Faizabad parwal-1, Faizabad parwal-3, Faizabad parwal-4, Swarna Rekha and Swarna Alaukik.

Lablab bean (*Lablab purpureus* L. Sweet)

Lablab bean or hyacinth bean known as 'Sem" in Hindi is a very nutritive vegetable grown for consumption of its green pods, green seeds and dry seeds as pulse. It is a perennial herb, agriculturally treated as an annual, twining or bushy in growth with wide variation in color of stem, foliage, flowers, pods and seeds. Both short day and day neutral types exist in this crop. Twining types are mainly short day plants and bushy types are mostly photoinsensitive. Pusa Early Prolific, Pusa Sem 2, Pusa Sem 3, Hima, Grace, Rajni, Deepaliwal, Dasrawal, Co.1, Co.2, Co.3, Co.4 are photosensitive varieties and Arka Jay, Arka Vijay, Konkan Bhushan, Hebbal Averre-1, Hebbal Averre-3 and Hebal Averre-4 are bush types. The crop is raised during July-August with onset of monsoon. Seeds are sown in pits at a spacing of 1.0 x 0.75 m .The plants are trailed on the bower wherever necessary. The crop starts flowering by November-December. The bush varieties come to flower 50-60 days after sowing. The average yield varies from 14-17 t/ha.

Cluster bean (*Cyamopsis tetragonoloba* L.)

Cluster bean is a hardy crop. The tender pods, primarily used as a vegetable is a rich source of protein, iron, vit. 'A' and vit.'C'. The kharif crop is sown with onset of monsoon and summer season crop is raised during February-March. Pusa Mausmi, Pusa Sadabahar, Pusa Navbahar and P28-1 are some of the improved varieties. The crop is raised by sowing seeds in channels taken at a spacing of 45-60 x 20-30 cm. Harvesting starts 40 days after sowing and pods are harvested at tender stage. The average yield is 5-8 t/ha.

Amaranth (*Amaranthus* sp.)

Among tropical leaf vegetables, amaranth belonging to genus *Amaranthus* has attained commercial significance in India. It is often described as poor man's spinach. It is grown successfully in hot summer season and humid conditions and ensures food, nutritional and livelihood security for people inhabiting the marginal, hot and dry regions of the world. The common *Amaranthus* sp. present in India are *A.tricolor* (*A tristis, A gangeticus*), *A. viridis, A. spinosus, A dubius, A blitum (A. lividus), A hypochondriacus (A flavus), A cruentus (A. paniculatus) and A caudatus (A edulis). A. tricolor* has 5.2 g protein, 6.1 g carbohydrate, 0.5 g fat, 1 g fiber, 5520 μg carotene, 99 mg vitamin C, .03 mg thiamine, 0.3 mg riboflavin, 1.2 mg niacin 2.7 g minerals, 397 mg

Ca, 3.49 mg Fe,122 mg Mg, 230 mg Na and 341 mg. potassium /100 g edible portion. Presence of ant nutrient factors like oxalate and nitrate is a limiting factor in the large scale consumption of amaranth. About 40% of oxalates found in amaranth is in Free State and affect the calcium metabolism (Devadas and Mallika, 1991). Another anti nutrient factor present in amaranth is free nitrates. The oxalate content of various Amaranthus sp. varied from 0.77 to 4.3 mg and that of nitrate from 0.20 to2.47 mg/100 g. According to Krishnakumary (2000) accumulation of oxalates was higher in leaves than in stem and during summer than in rainy season.

A number of varieties in different edible species of amaranth are released from different research institutes.They are Arun, Mohini, Krishnasree, Renusree CO.1, CO.2 CO.3, CO.4, CO.5 Chloti chaulai Badi chaulai, Pusa Keerthi, Pusa Kiran, Pusa Lal Chaulai *(A.tricolor)* Arka Suguna and Arka Arunima. The crop is raised by direct seeding or by transplanting. For direct sowing, small plots of 3.0-3.6 m long and 1.0-1.5 m width are taken and seeds mixed with fine sand are uniformly broadcasted in it followed by light irrigation. The seedlings can be pulled out for marketing one month after sowing. The 20-25 days old seedlings can be transplanted in trenches at 20-25 x 10-15 cm spacing. The average yield varies from 20-25 t/ha.

Basella (*Basella alba/B.rubra*)

Basella also known as Malabar spinach, Ceylon spinach, Indian spinach, Vine spinach, Poi, Chinese spinach, Malabar climbing spinach and East Indian spinach is an underutilized leaf vegetable of Indian origin. It is grown in almost all parts of India for its fresh tender leaves and stems, consumed as leaf vegetable after cooking. The mucilaginous qualities of plant make it an excellent thickening agent in soups, stews etc. The juice of leaves is prescribed against constipation especially for children and pregnant women .A paste of root is applied to swellings and is also used as a rubefacient. The leaf juice is a demulcent, diuretic, febrifuge and laxative. A paste of leaves is applied externally to treat boils. The flowers are used as an antidote to poison. Basella belonging to family Basellaceae is a fleshy annual and biennial, twining, much branched herb with alternate, broadly ovate leaves which are pointed at the apex. There are three types *B. alba*, B. rubra and *B. cordifolia.*

Basella grows well in hot and humid climates. Although the crop is adapted to many soils, sandy loam is the most suitable. The soil should be moist, fertile and well supplied with organic matter. The crop comes up well when soil pH is 5.5 to 8.0. Basella is planted either by direct seeding or by transplanting. It can also be propagated through stem cuttings or by root cuttings. Crop raised from seeds will be ready for harvest with edible stems and leaves 8-10 weeks after sowing. The plants rose from root or stem cuttings will be ready for harvest in about 6 weeks after planting. Yield varies from 14-19.5 t/ha.

Chekkurmanis (*Sauropus androgynous* Merr.)

Chekkurmanis (*Sauropus androgynous* Merr.) is a perennial leaf vegetable of Indian origin. It also called as multi vitamin and multi mineral packed leaf vegetable. Its leaves are very rich in protein, minerals and vitamins A, B and C. The leaves can be cooked like other greens. The leaves are used to give a light green colour to pastry and to fermented rice in the Dutch East Indies. Leaves are used for preparation of soup. The plant is also useful for growing as a hedge around home gardens. It is often planted in live fences and in midst of garden beds to provide light shade for other vegetables planted in beds. Besides other uses, leaves are used as cattle and poultry feed in certain parts of the country. In some other places, plants are planted as a soil binder to prevent soil erosion.

Chekkurmanis has several medicinal properties. The juice of leaves pounded with roots of pomegranate (*Punica granatum*) and leaves of jasmine (*Jassminum sambac*) is used against eye troubles. A decoction of its roots is often recommended for fever in rural areas. Pounded roots and leaves are used as poultice for ulcers in the nose *S. androgynous* Merr. belong to family Euphorbiaceae. The genus S*auropus* consists of several other species of which S. *assimilis* Thw., *S. netroversus* Wight., *S. rigidus* Thw. and *S. quadrangularis* Muell are related to *S. androgynous* Merr.

The plant is a slow growing perennial shrub. It grows up to 2-3.5 m. in height. The branches are tesete and flaccid. The leaves are alternate entire, sessile ovate - oblong and 3.5 cm in length. Flowers small, greenish red, monoecious, minute, axillary, pedicelled and clustered. Fruits sessile, white or pinkish, 0.2 cm in diameter with a fleshy epicarp.

The plant grows well in all types of soil. A warm humid climate with good rainfall is the best suited for the luxuriant, succulent growth of leaves and twigs. It tolerates shade to some extent. The crop comes up well in mild humid locations also. It is propagated by stem cuttings which root easily and also through fresh seeds. After the onset of monsoon, it does not require much irrigation. The cuttings come up very well and would be ready for harvest within 3 - 4 months of planting. Apex of the plant is nipped off which enables plants in putting forth new branches. The tender shoots and leaves can be harvested intermittently for several subsequent years. Even though it withstands the hot dry weather for a long period, watering of plants in such condition is desirable for getting constant appearance and growth of new leaves.

Curry leaf (*Murraya koenigii* L.)

Curry leaf is indigenous to India and is popular in South India. Its leaves are used to flavour food. Ground curry leaf with mature coconut kernels and spices form a good preserve. The crop can be raised in a wide range of soils. It can tolerate a temperature up to 37°C. The plant is propagated through seeds or root suckers. Planting is done during rainy season. Pits of 30 cm x 30 cm are taken at a spacing of 4 m x 4 m. Farmyard manure is applied @ 10k g/pit. A full grown tree starts yielding 15 months after planting and gives an average yield of 100 kg leaves / plant/year.

Bread fruit (*Artocarpus altilis* Postperg)

Bread fruit (*Artocarpus altilis)* belonging to family Moraceae is grown throughout the tropics for its fruits. It can be grown in a variety of soils having good drainage. Bread fruit is propagated by root cuttings. Root suckers can also be used. Another method is by air layers made from offshoots. It is planted in pits of 60 x 60 x 60 cm taken at a spacing of 10-12 m. The tree comes to bearing in 3 years after planting. A fully grown tree can yield 1000-2000 fruits /year.

Drumstick (Moringa oleifera Lem.)

Drumstick is an indigenous, nutritious, vitamin rich tree vegetable even though it is very common in South India; it is not so popular in other parts of the country. The fruits, leaves and flowers are used in culinary preparations. Oil from seed is used as a lubricant in watch industry and in preparation of cosmetics. This is a tropical plant and grows well in the plains. It performs well in dry arid tracts. It grows well in almost all soil types except stiff clays. Sandy loam soil containing good amounts of lime is best for the crop.

The crop can be raised by seeds or limb cuttings. Seed propagation is adopted in annual moringa and limb cuttings are popular in other types. Limb cuttings of 1 to 1.35 m length and 14 to 16 cm in circumference are planted during rainy season in pits of size 60 × 60 × 60 cm at a spacing of 2.40 to 3.00 m in square system. The crop should be irrigated in rain free period till it establishes.

For raising seed moringa or annual drumstick, pits of size 45 × 45 × 45 cm are taken at a spacing of 2.5 × 2.50 m in square system. Farmyard manure or compost is applied @ 15 kg /pit. Sow one seed per pit .Gap filling should be done after one month. When the seedlings are 75 cm in height, tip should be pinched to facilitate branching. Three months after planting apply urea 100 g, super phosphate 100 g and muriate of potash 50 g/pit. Apply 100g urea/plant when the plants start flowering. Irrigate the crop depending upon moisture availability.

References

Devadas, V. S. and Mallika, V. K. 1991. Review of research on vegetables and tuber crops- Amaranthus. Kerala Agricultural University, Thrissur, pp 60.

Dubey, A.K., Srivastava, J.P and Singh, N. P. 2007. Studies on floral biology of Spine gourd (*Momordica dioica* Roxb.). *Acta Horticulture*, **752:** 453-458.

Krishnakumary, K. 2000. Genotypic and seasonal influence of leaf spot disease in Amaranth. Ph.D. thesis, Kerala Agricultural University, Thrissur.

Peter, K. V. ed 2007. Underutilized and Underexploited Horticultural Crops.Vol.1, 2, 3.New India Publishing Agency, New Delhi (www.bookfactoryindia.com).

Sarnaik, D.A., Sathish, K.V. and Sharma, G.L. 2005. Evaluation of Ivy gourd (*Coccinia grandis* L.) germplasm. Proceedings of National Seminar in Cucurbits Breeding and Production Technology, 22-23 September, 2005. G.B. Pant University of Agriculture and Technology, Pantnagar, pp 205-209

□□□

Chapter – 28

Genetic Resources for Pea Improvement

Y.V. Singh, Shri Dhar and K.B. Bhushen

Introduction

Peas (*Pisum sativum* L., 2n = 2x =14) are consumed as fresh vegetables or dry seeds throughout the world. In India, peas are grown as winter vegetable in plains and as summer vegetable in the hills. As field pea, it occupies about 0.45 m ha area in India, accounting for only about 2% of the total pulse area. About 90% of its area and production is limited to Uttar Pradesh alone. The area under vegetable peas is on increase. The acreage of vegetable pea in India is 2.72 lakhs ha with a productivity of 9.9 tons/ha. Pea is a rich source of protein (7.2% in green pea & 20 % in dry pea) and minerals (0.8%).

The geographical region comprising of Central Asia, the Near East, Abyssinia and the Mediterranean is considered as centre of origin based on genetic diversity. According to Blixt (1970), the Mediterranean is the primary center of diversity with secondary centers in Ethiopia and the Near East.

Garden pea, seeds have been observed in archeological excavations of Mohan Jodaro and Harappa, there are distinct hot weather tolerant and round-seeded varieties which carry high resistance to fusarium wilt and powdery mildew (*Erysiphe polygoni*). These varieties have a different use as a pulse crop for meeting protein requirements of common people. These are considered as intermediate stages of domestication of the present day vegetable (garden) pea, which is sweet and wrinkled-seeded. In fact, the quick spread of Arkel variety, an introduction from England in seventies, has displaced some of the native round-seeded varieties like Asauji, Hara Bonia, Hoshiarpuri, Kaparkheda, Kalanagni, Lucknow Bonia etc. there is need to conserve these valuable germplasm, before they are swept away by genetic erosion.

Pisum sativum L. is classified under the order Fabales, family Fabaceae which have more than 450 genera and over 12000 species. Approximately 90 % of legume species interact with Rhizobium to fix nitrogen. The six *Pisum* species reported are *P. sativum* L. (garden pea), *P. arvense* L. (field pea), *P. elatius* Stev. (Mediterranean pea), *P. abyssinicum* Braun (Abyssinian pea), *P. humile* Boiss and Noe (dwarf pea) and *P. fulvum* Sibth and Sm. (red yellow pea). These species can intercross freely with few sterility barriers but *P. fulvum* shows divergence with partial mating barriers. Interogressions between *P. sativum, P. humile* and *P. elatius* created variation in the cultivated *P. sativum*. The species *P. arvense, P. humile, P. elatius* and *P. abyssinicum* are considered now to be the forms or ecotypes of *P. sativum*. *P. elatius* found in moist habitats of the Mediterranean is a tall climber with long pods and small seeds. The species *P. humile* with a dwarf habit is widely distributed in the drier habitats of the near East.

Pea is an annual herbaceous plant. It has a tap root system. Stems are slender, usually single and upright in growth. Leaves are pinnately compound with two to several leaflets. The rachis terminates in a simple or branched tendril. There are large stipules at the base of leaf. The plant may be single stemmed or many axillary stems may originate at the cotyledonary node or any superior node, especially if the apical growing point is destroyed, leaflets of a pair are opposite or slightly alternate. The lower leaflets are larger than the upper leaflets. The margins of leaflets and stipules may be entire or serrated.

The inflorescence is raceme arising from the axil of a leaf. The lowest node at which flower initiation occurs is normally constant under a given set of conditions and is used in classifying the varieties into early and late types. Most early cultivars produce the first flower from nodes 5 to 11 and the late cultivars start flowering at about nôdes 13 to 15 (Gritton, 1986). Early cultivars are often single flowered or bear some single and double flowers. Late cultivars are usually double/triple flowered.

The flowers are typical papilionaceous with green calyx comprising of five united sepals, five petals (one standard, two wings and two keels). The stamens are in diadelphous (9 + 1) condition. Nine filaments are fused to form a staminal tube while the tenth is free throughout its length. The gynoecium is monocarpellary, with ovules (up to 13) alternately attached to the placentas. Style normally bends at right angle to the ovary. Stigma is sticky.

Pea is strictly self pollinated in nature. Stigma is receptive to pollen from 2-3 days prior to anthesis until 1 day or more after the flower wilts. Pollen is viable from the time anthers dehisce until several days thereafter.

The somatic chromosome number of peas is 14 (Yarnell, 1962). The translocations and other chromosome arrangements are commom. The seven characters studied by Mendal have been mapped as indicated below:

1. The shape of mature seeds, smooth/wrinkled (R/r), in chromosome 7
2. Seed colour, yellow/green (I/i), in chromosome 1

3. Flower, colour, purple/violet or white (A/a), in chromosome 1
4. Mature pods, smooth and expanded/wrinkled and indented (V/v) in chromosome 4
5. Colour of unripe pods, green/yellow (Gp/gp), in chromosome 5
6. Inflorescence, axillary/terminal (Fa/fa), in chromosome 4
7. Plant height, tall/dwarf (Le/le), in chromosome 4

Thus, Mendal probably delt with the genes a and i in chromosome 1, le, fa and v in chromosome 4, gp in chromosome 5, and r in chromosome 7. One may ask as to why he did not run into the complication of linkage while formulating the law of independent assortment. The answer is that with respect to a and *i* in chromosome 1 and fa in relation to le and v in chromosome 4, these genes are so distant, that linkage is not realized. The only two genes which could have shown linkage were le and v. as far as known, Mendal did not study the simultaneous segregation in these two (Blixt, 1974).

Genetics

Genetics of both qualitative and quantitative attributes in pea have been studied by several researches. Studies on linkages were also undertaken. The classical work of Mendal was on inheritance of some qualitative characters in pea. Besides the investigation on inheritance of various traits, the genes governing these were also assigned to different linkage groups on the chromosomes. The linkage map of pea prepared by using isozymes genetic marker and RFLP (restriction fragment length polymorphism) molecular marker has more than 200 loci. Blixt (1974) mentioned 324 genes for qualitative characters.

Augmentation of germplasm of pea

Exhaustive germplasms collections are also being maintained in USA, Germany, Italy and U. K. In India, germplasms of pea are being maintained at the National Bureau of Plant Genetic Resources (NBPGR), New Delhi and Indian Institute of Vegetable Research, Varanasi (UP.), State Agricultural University and some other institutes. Pea germplasm is also maintained at ICRISAT, Hyderabad, and Indian Institute of Pulse Research Station, Kanpur.

There are several local cultivars of pea grown in the country, such as, Boniya, Lucknow-Boniya, Local Yellow Batri, Hara Boniya, Asauji (Amritsar), Hoshiarpuri, Kap, Kanawari, Khaparkheda (Madhaya Pradesh), Kalanagini, Kinnauri and a few others. These cultivars possess useful traits which can be/have been transferred to improve the verities.

Major world resources of pea germplasm

Institute	Comments
Institute of Plant Introduction and Genetic Resources, Bulgaria	—
Zentralinstinstitut Fur Genetic Und Kulturpflanzenforschung, DDR	Old cvs (European); recent cvs, land-race populations and selections from these
National Bureau of Plant Genetic Resources, IARI Campus, New Delhi, India	—
Istituto de Germplasms del CNR, Bari, Itlay	Contains numerous accessions from USDA
Stacja Hodowli Roslin, Poland	Contains numerous accessions from Gatersleben (PIS) & Weibullsholm (WBH)
Weibullsholm Plant Breeding Institute, Sweden	World base collection. Contains numerous accessions from Gatersleben (PIS) and USDA (PI)
International Centre for Agricultural Research in the Dry Areas (ICARDA), Syria	—
John Innes Institute, UK	Contains numerous accessions from Weibullsholm (WBH) and USDA (PI).
N. I. Vavilov All Union Institute of Plant Industry, USSR	Contains duplicates of Gatersleben collection (PIS)
USDA-SEA North Central Regional Plant Introduction Stn, Iowa State University, Ames, USA	—
USDA-SEA National Seed Storage Laboratory, USA	Cultivars and breeding lines
USDA-SEA New York State Agriculture Experiment Station, USA	PI lines. Comprehensive collection including accessions from centre of diversity

Sources: Hebblethwaite *et al.* (1985).

Vegetable pea genetic resources at NBPGR and other institutions of India

Stations/Institutions

- NBPGR,Delhi (1738). Shimla (120), Bhowali (95)
- GBPUA&T, Pantnagar
- PAU, Ludhiana
- IIVR, Varanasi (NAGS)
- CCSHAU, Hissar
- CSKHPKVV, Palampur
- Y. S. Parmar Uni. of Hort. & Forestry, Nauni, Solan
- CSAUAT Kalyanpur, Kanpur
- BHU, Varanasi
- VPKAS, Almora
- JNKVV, Jabalpur

Characterization

The germplasms collected are characterized as per descriptors prepared by IBPGR Descriptors' List for characterization and evaluation of cowpea/other legume vegetables

1. Characterization traits

(1) Botanical Type (specify)

(2) Growth Habit
- Erect (bushy)
- Semi-erect
- Semi-spreading
- Spreading
- Trailing (prostrate)

(3) Growth Pattern
- Determinate
- Semi-determinate
- Indeterminate

(4) Basic plant vigour
- Highly vigorous
- Vigorous
- Average vigorous
- Poor
- Very poor

(5) Raceme position
- Mostly above canopy
- Upto upper 1/3rd of canopy
- Throughout the canopy

(6) Pod attachment to peduncle
- Pendent
- 30°-60° from erect
- Erect

(7) Utility type
- Vegetable type
- Dual-1 (vegetable + seed) type
- Seed type (pulse type)
- Dual-2 (seed + fodder) type
- Fodder type
- Dual-3 (fodder + vegetable) type

2. Flowering characteristics

(1) Days to first flowering
(2) Days to 50% flowering
(3) Duration of flowering
- Short
- Medium
- Long
- Number of flushing

3. Maturity traits

(1) Days to first pod picking (green stage)
(2) Days to first pod maturity
(3) Days to 50% maturity
(4) Days to complete maturity
(5) Senescence
- Uniform maturity
- Overlapping maturity (over 2 consecutive flushings)
- Highly variable maturity ranges

4. Plant canopy characteristics

(1) Plant height (cm)
(2) Number of primary branches
(3) Leafiness
(a) *Leaf size*
- Large
- Medium
- Small

(b) *Leaf density*
- Dense
- Medium
- Sparse

(c) *Leaf shape*
- Normal
- Moderately dissected
- Highly dissected

(d) Number of clusters per plant

5. Pod characteristics

(1) Number of pods per cluster
(2) Pod length (cm)
(3) Pod shape
- Straight
- Slightly curved
- Highly curved

(4) Pod fibrousness (At green picking stage)
- Very soft
- Average
- Highly fibrous

6. Seed characteristics

(1) Number of seeds per pod
(2) Seed size
- Big
- Medium
- Small

(3) Seed shape
- Round
- Oval
- Rhomboid
- Crowedge
- Other (specify)

(4) Seed colour Descriptive (specify)
(5) 100-seed weight (g)

7. Shattering

(1) Non Shattering
(2) Slightly Shattering
(3) Medium Shattering
(4) Shattering
(5) Highly Shattering

8. Yield traits

(1) Yielding ability aspect (yield potential) (visual score)

Evaluation

In pea, very rich germplasms of diverse types is available. A tremendous number of mutant genes have been investigated and most of them have been mapped on various chromosomes. Leaflets mutants and multilayered are classical examples of morphological variants in pea which are being used by a number of pea breeders. Similarly, sugar-podded varieties can be used as such for consumption. Accession resistant to powdery mildew, rust, mosaic and fusarium wilt have been identified by the vegetable pathologists and breeders. At Hisar, germplasms has been evaluated and accessions showing resistance to leaf **miner** and powdery mildew have been identified. There is no dearth of germplasm in pea having various economic attributes. The powdery mildew-resistant characteristic has been transferred from nonedible podded to Oregon Sugar pod. Varietal differences for the incidence of pea aphid (Acyrthosiphon pisum (Harr) have been reported by Bintcliffa and Wratten. Long-podded types (EC 95516, IC 2564), bold-seeded types (EC 4103 and EC 6185) and powdery mildew-resistant lines (PLP 75, PLP 91, EC 6209 and EC 15210) have been identified. Evaluation and characterization of germplasms of pea have been amply studied at Weibullsholm Plant Breeding Institute. Sweden, where characterization began in 1930 and presently, over 4000 traits have been studied and characterized in over 3000 accessions.

Molecular Characterization

Isozymes in genetic resources

The term isozyme was proposed by Markert and Moller as the multiple molecular forms of an enzyme. The different molecular forms of an enzyme are referred to as allozymes if their polypeptides are coded by different alleles at a single locus. This electrophoresis system can be used to study the diversity between, among and within the samples of genetic resources. It has been realized that this is the best currently available method for measuring genetic diversity of the germplasms. The genetic variation in a population can be measured in terms of the proportion of loci having more than a single allele and the proportion of genes which are under heterozygous in an individual. Isozyme analysis can be used for evaluation and characterization of germplasms which supplements the morphological procedures. Clearly it is used for more precise measurements of genetic variation within a single accession or several related accessions. The low degree of inter-accession differences can be studied by this technique. Isozyme electrophoresis is quick, relatively inexpensive and provides a rich source of genetic markers which are not influenced by environmental factors and which differentiate heterozygotes and homozygotes. Variation at the species level through isozymes have been studied in a number of crops. It shows the relationships between species which are progenitor and derivatives. Also, this can be used even to study the genetic relationship between parents and their hybrids

Since the development of the polymerase chain reaction (PCR) in generating random amplified polymorphic DNA in 1990, this technique has been found valuable in the construction of genetic maps in several species and in production of genetic markers linked to specific phenotypic traits in particular using bulked segregant pools. RAPD technique became popular because of its simplicity and ease of use. Laucou et al. (1998) constructed a genetic linkage map of *Pisum sativum* L. based primarily on RAPD markers that were carefully selected for their reproducibility and scored in a population of 139 recombinent inbred lines (RILs). The mapping population was derived from a cross between a protein rich dry-seed cultivar 'Terese' and an increased branching mutant (K 586) obtained from the pea cultivar 'Torsdag'. The map currently comprises nine linkage groups with two groups comprising only 6 markers (n = 7 in pea) and covers 1139 cM. This RAPD-based map has been aligned with the map based on the (J1281 x J1399) RILs population that includes 355 markers in seven linkage groups covering 1881 cM. for this alignment to the gene Uni were used as common markers and scored in both populations.

Genes for which linked markers have been reported in the pea are listed in Table (Myers *et al.*, 2001)

Molecular markers linked with certain genes in pea

S. N.	Trait	Gene	Marker
1	Cotyledon shape	*rb*	Ve-5 (RFLP)
2	Pea seed borne mosaic virus resistance	*sbm-1*	GS185 (RFLP)
3	Pea enation mosaic virus resistance	*en*	Adh-1 (isozyme)
4	Bean yellow mosaic virus resistance	*mo*	Pgm-p(isozyme) P252 (RFLP)
5	Fusarium wilt resistant	*Fw*	H19, Y14, Y15(RFLP); p254, p248, p277, p105 (RFLP)
6	Powdery mildew resistant	*er -1*	P236 (RFLP) PD10650 (RFLP converted to SCAR)
		er-1 *er-2*	Sc-OPO-181200' Sc-OPE-161600 (SCAR) 3 AFLP primers
7	Ascochyta blight resistant	*Several QTL*	Af & I (linkage group I); p227, p105 (RFLP, linkage group IV; p236 RFLP, linkage group VI)

McClendon *et al.* (2002) identified DNA markers linked to fusarium wilt race 1 resistance in pea. Eighty recombinant inbred lines (RILs) from the cross of Green Arrow (resistant) and PI 179449 (susceptible) were developed from single seed descend for disease reaction in race 1 infested field soil and the greenhouse using single-isolate inoculum. The RILs segregated 38 resistant and 42 susceptible fitting the expected 1:1 segregation ratio for a single dominant gene (÷2=0.200). Bulk segregant analysis (BSA) was used to screen 64 amplified fragment length polymorphism (AFLP) primer pairs and previously mapped random amplified polymorphic DNA (RAPD) primers to identify candidate markers. Eight AFLP primer pairs and 15 RAPD primers were used to screen the RIL mapping population and generate a linkage map. One AFLP marker, ACG: CAT_222 was within 1.4 cM of the *Fw gene*. Two other markers, AFLP marker ACC: CTG_159 at 2.6 cMlinked to the susceptible allele, and RAPD marker Y15_1050 at 4.6 cM linked to the resistant allele, were also identified. The probability of correctly identifying resistant lines to fusarium wilt race 1, with DNA marker ACG: CAT_222 is 96%. These markers will be useful for marker assisted breeding in applied pea breeding programs.

Conservation

Plant diversity can be conserved in natural habitat (*in situ*) or away from natural habitat in gene banks (ex situ). Seeds can be stored for long terms (25 – 100 years), medium term (2 - 25 years) or short term (6 - 24 months). Germplasms under long term conservation is designated as base collection. These are not disturbed except for regeneration. Long term storage is in cold storage modules at -18 to -20°C. The germplasms under medium term storage (2 - 5°C, 40% RH) is referred to as active collection available for multiplication and distribution. The short term collection (10 - 12°C, 40% RH) referred to as working collection is maintained by plant breeders/researchers for routine seed storage.

Utilization

Important donors for pea breeding programme

Singh (1991, 1995) has compiled extensive information on genetics and breeding of peas including listing of superior lines with multiple disease resistance in pulse crops. Kalloo (1993) and Narsinghani and Tewari (1993) have also given detailed accounts of pea breeding. A few examples on peas are as follows:

Earliness	:	Asauji, lucknow bonia, Hans, EC 3
More pods/plant	:	PLP 26, 50. 69, 179, 279, 496
Long pods	:	EC 109171, 109176, 109190, 109195
Bold pods	:	EC 4103, 6185, 95924
Powdery mildew	:	EC 326, 42959, 109190, 109196, T 10, P 19\85, P 288, PC 6578, 4048, P 6587, P 6588, BHU 159, EC 42959, IC 4604, JP 501, JP 179, VP 7906
Wilt	:	Early perfection, Bonneville, PL 43, 124, 6101, Glacier
Rust	:	PJ 207508, 222117, EC 109188, EC 42959, IC 4604, Pj
	:	207508, JP Batri Brown 3, JP Batri Brown 4,
Pea mosaic	:	American Wonder, Perfection Canner's Gem, Dwarf
	:	White Sugar, Little Marvel
Leaf miner	:	EC 16704, 21711, 25173, JP 179, JP 169-1, JP 77, LMR 4, LMR 10, LMR 20
Leaf weevil	:	JP 9, JP 179
Pea stem fly	:	Bonneville, Asaugi, Boach Sel., GC 141, IP 3 (Pant
(tolerant)	:	Uphar), Dwarf Gray Sugar, T 10, T 163
Abiotic stress	:	leafless pea
(tolerent)	:	
High protein and Sugar content	:	GC 195, Local cultivar Kinnauri
Processing quality		
Dehydration	:	Arkel,
Canning	:	T 19, Bonneviile

Local cultivars of pea- Boniya, Lucknow Boniya, Asauji (Amritsar), Hoshiarpur, Kap, Kanawari, Kinnauri, Lakanagini, Khaparkheda

Improved vareieties of pea

Several varieties were introduced by Plant Introduction Division of IARI, New Delhi in early seventies. After evaluation a few of them were recommended for large scale evaluation e.g. Bonneville, Arkel, Meteor, New Line Perfection, Thompson Laxton, Little Marvel, Kelevdon Wonder, Linclon, Gloriosa, Early Gaint, Sylvia etc.

Some of these introductions utilized in crossing programme to develop improved varieties of vegetable pea are given in following table:

Table: Improved varieties of pea

Early Maturing Cultivar	Parentage	Maturity (Number of days)	Yield (t/ha)	Source
Arkel	Introduction from England	55-60	10	IARI, New Delhi
Pusa Pragati	—	60-65	7	IARI, New Delhi
Jwahar Matar 4	T19 XEarly Badger	50	5	JNKVV, Jabalpur
Pant Matar2	Early Badger X IP3	55-60	6	GBPUAT, Pantnagar
Hisar Harit	—	60	10	HAU, Hisar
Ageta 6	—	50	6	PAU, Ludhiana

Mid season and late group

Bonneville	Introduction from USA	85	12	IARI, New Delhi
Lincoln	Introduction from USA	85-90	9-10	IARI, New Delhi
Jawahar Matar 1	T19XGreater progress	85-90	12	JNKVV, Jabalpur
Jawahar Matar 2	Russian 2X Greater progress	85-90	10	JNKVV, Jabalpur
Pant Uphar	—	85-90	10	GBPUAT, Pantnagar
Punjab 88	Pusa2 XMorassis 55	100	22.5	PAU, Ludhiana
VL3	Old Sugar X Early wrinkled Dwarf 2-2-a	85-90	9	VPKAS, Almora
Mith Phali	—	90	11-12	PAU, Ludhiana
JP 19	—	90	10-11	JNKVV, Jabalpur

Source: Vishnu, Swaroop. (2006) Vegetable Science and Technology in India.

Future Prospects

Collection, characterization and evaluation of pea germplasms using uniform list of descriptors for documentation and indexing for appropriate choice of initial breeding materials, provides the most optimum resource handling cum-plant improvement strategy for making per se selections for choice of parents in pea breeding programmes.

Literature Cited

Hebblethwaite, P. D. Heath, M.C. & Dawkins, T.C.K. (1985). The Pea crop. Butterworths

Kalloo, G. (1994). Vegetable breeding Vol. III. Panima Educational Book agency, New Delhi

Ram, H. H. (2005). Vegetable breeding Principles and Practices. Kalyani publishers. pp 247-255

Rana, R.S. Gupta, P.N. Rai Mathura & S. Kochhar. (1995).Genetic resources of vegetable crops. NBPGR, New Delhi.

Vishnu Swaroop. (2006). Vegetable Science and Technology in India. Kalyani publishers.

❑❑❑

Chapter – 29

Genetic Resources for Improvement of Radish

N. Ahmed, A.J. Gupta and K. Hussain

Introduction

Radish (*Raphanus sativus* L., 2n=2x=18) is a root cum leafy vegetable suitable for sub tropical and temperate climate. Radish probably originated in Europe and Asia. It has been under extensive cultivation in Egypt since long. It was introduced to England and France in the beginning of 16th century. In 1806, it was introduced to America. Radish does not exist in wild stack. It is believed to have originated from *R. raphanistrum* which is widely distributed as weed in Europe. Radish is primarily a winter crop but with the development and release of new varieties, it can be grown all the year round. The new varieties can withstand heat and does not bolt in spring. Radish is a favourite crop because of its quick growth and availability in 4-7 weeks.

Its fleshy roots are modified roots, developed from both the primary root and hypocotyl. The leaves and roots are consumed both as salad and cooked. The radish root is a good appetizer; its different preparations are useful in curing liver and gall bladder problems. Roots are also used in treating urinary complaints and piles. The juice of fresh leaves is useful as diuretic and laxative. Pungency in radish is due to volatile isothiocyanates while pink colour is due to pigment anthocyanin.

Radish is a good source of vitamin C containing 15-40 mg per 100g of edible portion. Roots also contain proteins, fat, minerals, fibre and carbohydrates, besides trace elements like aluminium, barium, lithium, manganese, silicon, titanium, fluorine and iodine. Pink skinned radish is generally richer in vitamin C than the white skinned type. Radish contains glucose as the major sugar and also contains fructose and sucrose in smaller quantities. Pectin and pentosans are also reported to be present. The leaves are good source of extraction of protein on a commercial scale and seeds are potential source for non-drying fatty oil suitable for soap making, illuminating and for the edible purposes.

Floral Biology

The inflorescence of radish is a typical raceme of Cruciferae. The flowers are small, white, rose or lilac in colour with purple vein in bractless racemes. Sepals are erect and petals clawed. Fruit indehiscent, 3-7 cm long, upto 1.5 cm in diameter with 6-12 seeds and long conical beak. Radish seeds are globose and about 3mm in diameter. The seed takes 7-14 days for germination under optimum environmental conditions and remain viable upto 5 years.

Radish is normally self-incompatible and entomophilous. The incompatibility is sporophytic in which the papillae on the stigma surface form the barrier to the pollen tube penetration. It is inferred that incompatibility is sporophytically determined by a multiple allelic series (S 1-56) with dominance relations. By disrupting the stigma surface or by applying the pollen immediately to the conducting tissue of the style, it is possible to get normal seed set through selfing. Bud pollination is effective for developing inbreds and the time immediately after flower opening is ideal for effective crossing.

Pseudo self-compatibility is also observed in radish. Pseudo self-compatibility represents a breakdown of the self incompatibility system, giving limited to full seed set after an incompatible pollination. Male sterility controlled by the interaction of recessive gene 'ms' and sterile cytoplasmic factor 'S' has been observed and exploited in radish.

Breeding Objectives

Breeding work in radish is mostly concentrated on productive and qualitative characteristics such as high yielding ability, early maturity, late bolting, edible quality (pungency), late pod formation, cold-hardiness, drought resistance, heat tolerance, wet tolerance, soil adaptability and cylindrical roots of appropriate length for mechanical harvesting, resistance to viral diseases, soft rot, downy mildew, grey leaf spot and other diseases.

In winter radish, breeding objectives are cold-hardiness, low temperature plumpness, delayed bolting and less pore formation. Summer radish is mainly utilized for eating fresh as grated radish and many varieties for this season were derived from 'Minowase' strains that had promising characteristic such as heat tolerance, drought resistance and virus disease resistance.

Genetic Resources

The variability existing among the cultivated forms of radish for morphology and ecology signifies the multi centre origin of this crop. A few important wild species available in Mediterranean region are considered to be the probable progenitors of European radish while Japanese type has originated from the wild species that remain in the coastal regions of Japan. *R. caudatus* L., the non-root forming rat-tail radish is closely related to the Indian type of radishes whose ancestor has been lost. The Indian type radishes got evolved in the near eastern and Indo-Burma centre of origin. The European types originated directly or through hybridization from a few wild species like *R. maritimus, R. landra* and *R. rostratus*. The Japanese types are derived from *R. sativus f. raphanistroides*.

The small cool season radish, large widely adapted radish and rat-tail fodder radish all grouped under cultivated radish and belong to species *Raphanus sativus* L. Radish is an annual or biennial herb with rosette of leaves, which may vary in size from 10-45 cm depending upon the variety. It is a cross-pollinated crop due to sporophytic type of self incompatibility and shows considerable inbreeding depression on selfing.

Cultivated radishes are put under three different species based on plant morphology and ecological requirements. The European types which are frost tolerant and strictly annuals are called *R. sativus L.* The Indian types which are heat tolerant, pungent and mostly used for pickle making are called *R. indicus*. The Japanese types which are intermediate to Indian and European types growing in the coastal regions of Japan and neighbouring regions and the turnip shaped giant radishes cultivated in the Sakurjuna Island of Japan are called *R. raphanistroides*.

Genetic divergence and variability

The genetics of yield and yield contributing traits is essential for formulating a suitable breeding approach to improve its yielding ability. Genetic variation in quantitative traits may arise from additive, dominance and epistatic gene effects. Not much information is available on the genetical behaviour of quantitative traits in radish.

The phenotypic and genotypic variability in 30 exotic and indigenous cultivars were evaluated by Arumugam and Muthukrishnan (1975) utilizing parameters like phenotypic and genotypic coefficient of variation, heritability and genetic advance. The study indicated the predominance of genetic effects and less of environmental influence. Lal and Srivastava (1975) crossed Kaliyanpur No.2 with large root and small fruit and rat-tail radish (*R. sativus* var. *caudatus*) with small root and long fruits and found that all characters except root thickness showed high estimates of heritability; high value for genetic advance were recorded for cortex thickness and fruit length. Improvement in root or fruit length can be achieved through selection breeding.

Nijjar *et al.* (1996) estimated genetic variability and heritability for ascorbic acid and isothiocyanate content in 30 cultigens of radish and reported highly significant differences among genotypes for both the characters. Kutty and Sirohi (2003) crossed eight divergent genotypes of radish in a diallel fashion excluding reciprocals and evaluated 28 F_1 hybrid including parents. The analysis of variance showed highly significant differences among the genotypes, parents, hybrids, and parents vs hybrids.

Correlation studies

Young *et al.* (1994) studied heritability and genetic correlation of three floral traits (corolla width, pollen production per flower and pod size) and reported that corolla width and pollen production showed significant heritabilities while pollen size variation appeared to be under little genetic control. Only corolla width and pollen production were significantly genetically correlated. Murali *et al.* (1998) studied correlation among yield and yield contributing traits in radish and reported that root yield had a significant positive correlation with root girth and root dry weight at the phenotypic and genotypic levels. Root girth had a positive correlation with root dry weight while the number of leaves had a significant negative correlation with root girth at the genotypic level.

Danu and Lal (1998) in a correlation study conducted with 16 cultivars of radish, revealed that root weight was significantly correlated both at genotypic and phenotypic levels with the root length, number of leaves per plant, average leaf length, average leaf size, leaf weight per plant and plant weight, showing these characters are to be more yield contributory and the selection on the basis of these characters will be effective. The results also indicated that there was a strong inheritance association between the characters, number of leaves per plant, average leaf size, leaf weight, total plant weight and root weight as they showed highly significant positive correlations amongst them.

Combining ability and gene action

Duplicate and complementary type of epistatic has been observed for root yield to the extent of 98.40 % and 92.92% over mid and better parent, respectively (Brar and Padda 1972, Chandel K.S. *et. al.* 1993, and Pandey *et al.* 1981). The complementary epistasis and high magnitude of non-additive genetic action and presence of over-dominance has been reported for root length in various crosses. The heterotic effects and the presence of over dominance for the trait like root girth has been reported by Chandel *et al.* (1993). Heterosis for root length and girth has also been reported by Pandey *et al.* (1981).

Kutty and Sirohi (2000) evaluated 8 radish cultivars and their 28 F_1 hybrids in diallel fashion and revealed partial dominance for leaves per plant, root length, root shoot ratio and days to harvest. For all these characters, additive component of genetic variance was greater than the dominance component of variance. High narrow sense heritability for these characters indicated the possibility of improving these traits by selection methods. Over dominance was observed for 5 characters namely root weight, shoot weight, leaf length, leaf width and yield per plant while dominance was observed for petiole length and root diameter. The predominance of non-additive genetic variance for majority of the characters including yield per plant may be exploited by developing F_1 hybrids.

Chandel *et al.* (2002) studied the effects of general and specific combining ability in radish using line x tester mating design involving 14 inbred lines and four cultivars as lines and testers respectively and reported that the hybrids recorded high specific combining ability for root yield and most of the yield components. Kumar and Sirohi (2003) crossed eight divergent genetics of radish in a diallel fashion excluding reciprocals and evaluated 28 F_1 hybrid including parents. The F_1 hybrids revealed wide range of heterosis over the better parent for all the eleven characters viz. number of days to harvest, root weight, shoot weight, number of leaves per plant, leaf length, leaf width, petiole length, root length, root diameter, root shoot ratio and yield per plant.

Kumar and Chandel (2003) studied genetics of root yield and quality traits (ascorbic acid, dry matter and nitrate contents) in a line x tester mating design including 14 lines (females) and 4 testers (males) of radish and reported that the estimates of specific combining ability were of higher magnitude than the general combining ability for root yield, dry matter, ascorbic acid and nitrate content which revealed the preponderance of non-additive gene action.

Breeding Methods

Radish is a highly cross pollinated crop and shows a lot of variability. Mainly selection method of crop improvement has been followed for isolation of desirable genotypes. Mass selection is done in open pollinated materials either from the crop raised in experimental farm or from farmers field where large non-pithy roots having other desirable trials are selected and seeds are produced in isolation *en masse*. Attempt has been made to combine sub-tropical type root qualities to temperate type through hybridization. Introduction has also been followed and number of varieties have introduced such as Japanese White Long, Scarlet Globe, Chinese Pink etc. which are still ruling in the country. Today many hybrids have been developed throughout the world keeping specific objectives in view by exploiting incompatibility and male sterility phenomena.

Mass Selection

This is practiced in land races/ cultivars collected from the farmers field. Roots are allowed to reach an over mature stage. They are dug-up and leaves removed (but not growing points). Bare roots after discarding the undesirable types are immersed in a container of water. Roots which float being pithy and full of air spaces are discarded and only the large sinking roots are retained for seed production in isolation *en masses* small sinkers are also rejected.

Pedigree Method

Variety 'Pusa Himani' has been developed as selection from cross between Black Radish and Japanese White. Pusa Rashmi is also such a selection from a cross Green Top x Desi. Singh *et al.* (1998) developed Punjab Pasand, a new variety of radish through repeated crossing of selected plants from a base population of cv. Pusa Chetki.

Polyploidy Breeding

Polyploid radishes with 2n=36 were produced. They have no distinct advantages over normal diploids. Two polyploid varieties Sofia Delicious (2n=36) and Semilong Red Gaint (2n=36) have yielded more than the diploids.

Interspecific Hybridization

Three types of radishes *R. indicus, R. sativus* and R. *raphanistroides* are completely cross-compatible. This is being done to develop heat tolerant lines, resistant to diseases and pests.

Intergeneric Hybridization

Radish crosses readily with cabbage and cauliflower (*Brassica oleracea*) when used as female parent. Varying degrees of sterility were observed when radish was crossed with Abyssinian mustard (*B. carinata*), Chinese cabbage (*B. chinensis*) and turnip (*B. rapa*).

Heterosis Breeding

Crosses between selfed lines have been studied as early as 1923. These crosses showed variations and usually exceeded the better parent in root size and all plant characters. At IARI, New Delhi, high yielding hybrids of radish were obtained which yielded 30 to 60 % more than the better parent (Pal and Sikka, 1956). It has been observed that in crop of any radish variety, the hybrid plants resulting from natural crossing were conspicuous by their vigour of foliage as well as the roots. Song and Wei (1996) used male-sterile radish (*Raphanus sativus*) line Zhe 3A in crosses as a maternal parent with radish lines differing in root shape and quality to study the inheritance of these traits. The length and diameter of edible roots, and weight and number of leaves per plant had varying degrees of heterosis over mid parent value. Heterosis over the better parent was observed for edible root yield and leaf length. Chandel *et al.* (1996) conducted genetic analysis for days to marketable maturity in radish (*Raphanus sativus*) using Hayman model, 1958 involving six inbred lines and reported that all hybrids showed negative heterosis over mid and better parent.

Kumar *et al.* (2002) studied heterosis in 56 crosses of radish for five important characters (root yield, weight, length, girth and diameter) in a line (14) x testers (4) mating design and reported heterosis for root yield. Heterosis in radish has been reported in our country which generally ranges from 30-60 per cent. Vinod and Chandel (2003) working on heterosis in radish, reported that the estimates of specific combining ability (sigma 2 sca) were of higher magnitude than the general combining ability (Sigma 2 gca) for root yield, dry matter, ascorbic acid and nitrate content, which revealed the preponderance of non-additive gene action. The cross combinations Mino Early White x Pusa Himani and Japanese White x DPR-1 exhibited maximum heterosis to the extent of 59.02 and 56.27 % over Japanese White (SC 1) and 63.11 and 69.29 % over Chinese Pink (SC 2) with significant sca effects. The gca effects showed that the lines Mino Early White, Japanese White and Chinese Pink and the testers Pusa Himani and DPR-1 were good general combiners for these traits. Kumar *et al.* (2003) reported that the cross combinations Mino Early White x Pusa Himani and Japanese White x DPR-1 exhibited maximum heterosis to the extent of 59.02 and 56.27 % over Japanese White (SC-1) and 63.11 and 69.29 % over Chinese Pink (SC-2) with significant SCA effects.

Different hybrids were found to possess heterosis for various characters viz., Qiufeng-2 developed by Li, *et al* (2005), Fengyu No. 1 developed by Wu, *et al.* (2005), Zheluo 1 developed by Wei and Song (1993), Man Tanng Hong developed by Lin and Li (1993), 95-2 developed by Sugahara, *et al.* (1997), and Jin Ruobo No.3 developed by Wu, *et al.* (2001). Lu Lyobu-8 developed by Li, *et al.* (1997) and Fuyuan No. 1 developed by Zhang, *et al.* (2003).

Breeding for Different Attributes

Breeding methods depend on the source of resistance and its inheritance. For simply inherited resistance, generally back cross method of breeding is commonly employed to transfer the resistance from the donor parent to commercial variety. However, in certain cases simple selection, pedigree methods and combination of backcross and pedigree method are employed in breeding. In polygenic control of resistance, mass selection, recurrent selection, controlled matting (among the resistant progeny) in the F_2 and succeeding generations and other breeding methods involving gene pyramiding are employed. Biotechnological approaches are being employed to overcome interspecific and even intergeneric. Generation of transgenic plants resistant to diseases have been mentioned later.

Breeding for Disease Resistance

Mosaic virus, yellows disease, soft rot and club root are the most important diseases of radish. The micro-organisms which attack the allied *Brassica* crops also attack radish although susceptibilities fluctuate according to the differentiation of new races. Variability in the level of disease resistance has been observed among radish cultivars and the genetic composition for this trait was introduced into commercial varieties by intervarietal crossing. In many radish diseases, resistance is under the control of a multiple genes. Accumulation of these minor genes has been accomplished in the course of selection. In recent years combined resistance is required for radish production. The resistance to mosaic virus disease in radish was studied and the resistant genes were found to be present in some Japanese varieties (Hida, 1983), Bansei mino (Minowase group) and Aki-wase and Takakua (Merima group) (Shimizu *et al.* 1963). Breeding for yellows disease resistance, well advanced in the U S A, resulting in release of a promising variety known as 'Red Prince, derived from cultivar Early Scarlet Globe by mass selection (Ponds, 1959). This disease is controlled by quantitative multiple genes without any dominant or recessive interrelationship (Ashizawa *et al.* 1979). Genetic material for soft rot resistance was also identified in some Japanese and North Chinese radish but further exploitation and search for resistance to this disease is in progress.

For the main radish diseases, it is possible to breed resistant varieties by intervarietal crossing. Genetic resources present in cultivated radish varieties could be utilized which are under control of multiple gene system. Resistant traits from related *Brassica* species can also be utilized to transfer the traits through bridge plants of amphidiploid intergeneric hybrids. The Variety Fuyuan No. 1 exhibited resistance to virus and black rot and the hybrid Zheluo (F_1 hybrid) exhibited resistance to turnip mosaic poty virus.

Resistance to Alternaria Blight

Kumar and Singh (2003) screened genotypes for resistance to *Alternaria* blight, caused by *A. brassicae* and reported seven genotypes (Accessions 7401, 6802, 8801, 8803, 8805, Pusa Desi and Jaunpuri) resistant to *Alternaria* blight and eight genotypes were moderately resistant viz; 7212, 7110, 7108, 7208, 7218, 8804, 8271 and IHR-1.

Resistance to Club root

Somatic hybrids between Japanese radish (*Raphanus sativus*) and Cauliflower (*Brassica Oleracea* var. *botrytis*) were obtained with clubroot (*Plasmodiophora brassica*) resistance introduced from the Japanese radish parent by selfing, progeny over three generations were obtained. Resistant plants were perfectly resistant, while susceptible plants were as susceptible as Kairan. Only two BC-2 plants were obtained from resistant BC 1 plants × Kairan, and both were resistant.

Resistance to Fusarium Wilt

Leeman *et al.* (1995) reported that in commercial green house trials, *P. fluorescens* strain WCS 374 suppressed *Fusarium* wilt and increased radish yield. The involvement of induced resistance in natural soil bioassays and commercial green house trials is suggested.

Resistance to Root knot nematodes

According to Bunte *et al.* (1997), the resistance of oil radish to root knot nematodes may be effective and may thus provide new possibilities for the management of *M. hapla* and *M. incognita*.

Table 1. Sources of resistance for various diseases.

Disease	Resistant Varieties
Wilt (*Fusarium oxysporum* sp. raphani)	Fix
Black rot (*Aphanomyces raphani)*	Summer Wonder, Regensburg Market
Downy mildew (*Albugo candida*)	China Rose White, Round Black Spanish
Leaf spot (*Alternaria* sp.)	Pujab Safed
Mosaic (Virus)	Country's Wealth, High Agriculture, Shogani, Early Large, Nerina, Bansei mino, Aki-wase, Takaua
Yellow disease	Red Prince

Breeding for insect resistance

Wild radish is an annual plant that exhibits broad spectrum resistance including to those of herbivores. Plant families had different levels of herbivory by rabbits and to oviposition by *P. rapae* but no to herbivory by flee beetles. Manual clipping is a poor inducer of plant responses. Induced responses in wild radish (*Raphanus raphanistrum* and *R. sativus*) can be double edged sword, increasing herbivory by some herbivores under certain conditions, while reducing herbivory by other herbivores (Agrawal and Sheriff, 2001).

Breeding for Herbicide Resistance

Yu-Qin *et al.* (2003) worked on biochemical and molecular basis of resistance to aceto lactate synthase (ALS)- inhibiting herbicides and noticed that the selection pressure from chlorosulfan on wild radish populations has resulted in target site mutation at the same proline residue in the ALS gene. They also reported that higher ALS activity also may play a role in resistance level. Walsh *et al.* (2004) carried out experiments on multiple herbicide resistance across four modes of action in wild radish and observed that there was multiple herbicide resistance across many herbicides. The multiple herbicide resistance status of wild radish populations developed from conventional herbicide usage in intensive cropping rotations indicates a dramatic change for the future control of wild radish.

Breeding for Quality Traits

Quality in radish is related to the amount of sugar, pungency, water content and pore extent. From Consumer and market point of view, external appearance, root length, shape, smoothness of skin, colour, colour of elongated roots are important. Most of the quality traits are heritable and therefore mass selection could be practiced for isolating genotype with rich quality attributes. The taste of radish for salad is one of the most important quality trait and consumers prefer mostly slightly pungent and sweet radish. Therefore, breeding for taste aimed at reducing the pungent content is required. But mode of inheritance of pungency is still yet to be ascertained. Pore development does much damage to the quality of radish by destroying its commercial value. Pores are formed by the collapse of parenchymatous cells in root tissue, caused by excessive root growth in comparison with the corresponding assimilation ability of leaf tissues. It was suggested that a variety with early root enlargement and late pore formation may be developed by crossing between varieties that had high assimilation ability and early growth (Hagiya, 1975). The breeding work to develop varieties with late pore development when they are sown in autumn and early spring is one of the important objective.

Capecka *et al.* (1998) investigated 6 radish cultivars and reported that the low tendency of pithiness of daikon cultivars allowed harvest date to be adapted to market demands. Further more, the higher concentration of the majority of organic compounds and macro elements in storage roots of winter cultivars made them suitable for long term storage. According to Nikornpun *et al.* (2004), hybrids 27-1-3 x 18-1, 27-1-3 x 18-2, 27-1-3 x 18-3 and 27-1-3 x 18 – 6 and some of hybrids 77-2 x 18-11 and 77-4 x 18-7 exhibited good horticultural characteristics which make them market acceptable. Qiu *et al.* (1993) studied 65 samples representing 7 varieties or hybrids of red radish and indicated that vitamin C (ascorbic acid) and anthocyanin contents were highly and significantly correlated ($r=0.37$, $P<0.01$).

Role of Biotechnology

The importance and application of cytoplasmic male sterility, (CMS) in crop improvement by breeding is well documented in different species. Giancola *et al.* (2003) discovered the radish Rfo gene which restores male fertility in radish plants carrying Ogura cytoplasmic male sterility and transferred to rapeseed for the production of F_1 hybrid seeds. A single radish nuclear gene, Rfo, restores Ogura (Ogu) cytoplasmic male sterility normal *Brassica napus*. A map-based cloning approach relying on synteny between radish and Arabidopsis was used to clone Rfo (Brown *et al.* 2003). Betlt and Lydiate (2004) reported three *Raphanus* populations (BC1, F2 and R8) each segregating for the restoration of Ogura CMS was used to map restorer loci. The three restorer loci, Rf-1, Rf -2 and Rf -3, each exhibited dominant restoring alleles and were each mutually epistatic. Hirata *et al.* (2004) obtained CMS lines by intergeneric protoplast fusion between radish (*R. sativus*) and cabbage (*B. oleracea*). Two types of different CMS lines stably maintain the radish CMS specific Qrf 138 and the surrounding structure in the mitochondria, although parental materials in both CMS lines are normal plants.

Evaluation and characterization of population are of prime importance in plant breeding. Such evaluation is conventionally done on the basis of morpho-physiological and reproductive characters. More efficient and direct study can be done by means of the analysis of biochemical markers because these markers are less affected by environmental factors and genetic control of the makers, particularly isozyme and DNA polymorphisms is easily understood. Co-segregation study of isozyme markers and male fertility restoration showed that homozygous and heterozygous resorted plants could be separated by specific PGI-2 phenotypes, (Delourme and Eber, 1992).

Thus, the Pgi-2 marker is now currently used in restorer breeding programmes. Pradhan *et al.* (2004) worked on molecular markers and identified several markers associated with high or low seed weight, germination proportion, seedling length and fresh weight. It is suggested that the method involving molecular markers for identifying potential RAPD, markers may be useful for maker assisted breeding and selection of improved radish cultivars. Identification of cultivars is extremely important both for cultivation and breeding of crop plants. Cultivar identification based on morphological characteristics can be difficult and complicated. Polymerase chain reaction technologies, such as random amplified polymorphic DNA (RAPD) analysis, can readily and quickly identify cultivars using seeds and young leaves. Pradhan *et al.* (2004) examined radish cultivars for RAPD marker polymorphism and reported that RAPD markers could be effectively used for the identification of radish cultivars. The intra-specific genetic relationships of several accessions based on RADP analysis studied by Rabbani *et al.* (1998) were related primarily to their collection sites rather than to their phenotypic affinities. The level of polymorphism exhibited by the various covariates could be exploited in genetic mapping populations to tag economically important traits.

PCR amplifications were performed using the genomic DNA isolated from S201-homozygous inbred line R 9301 (Niikura and Matsuura, 1997). Northern blot analysis confirmed that the major A9 transcript was detected in stigmas, but not in leaves or anthers, and appeared one day before anthesis prior to the onset of the self incompatibility response.

Pradhan *et al,* (2004) used RAPD analysis to study the genetic variation at a molecular level within and between cultivars; some polymorphisms were consistent within cultivars while others differed between genotypes. They also reported seed grading, improved genetic purity and maker assisted selection in future breeding and selection to reduce morphological variation and increase uniformity of radish.

Yamagishi (2004) determined the genetic relationship between wild and cultivated radish species and those among the cultivated species. He studied structural and sequential variation in the mitochondria Orf B gene region in one of the cultivated and two wild species of *Raphanus.* The comparison provided valuable information about the origin and differentiation of cultivated radish and the relationship between cultivated and wild radish. Huh and Ohnishi (2002) studied genetic diversity and genetic relationships of East Asian natural populations of wild radish and found that the AFLP variation showed a very close genetic relationship between *R. raphanistrum* and *R. sativus*, particularly Kazkhistan *R. sativus*, confirming the assumption that *R. raphanistrum* might be involved in the origin of *R. sativus.* Transgenic radish (*R. sativus* var. *longipinnatus* cv. Jin Ju Dae Pyong) plants were produced from the pungency of plants which were dipped in to a suspension of Agrobacterium carrying both beta–glucuronoidase (gus A) gene and a gene for resistance to herbicide Basta (glufosinate) (Curtis and Nam, 2001).

Development of Varieties

Variety/Hybrid	Organization	Pedigree	Brief Characteristics
Pusa Desi	IARI, New Delhi	Selection	Roots are 30-35cm long, tapering, pure white with green tops and pungent taste.
Purse Chetki	IARI, New Delhi	Line / Family breeding	Suitable for growing in hotter months when no other variety can be grown successfully. Roots are medium long, stumpy, pure white and mildly pungent with soft texture.
Japanese White	IARI Reg. Station, Katrain	Introduction from Japan	Roots are 25-30 cm long, 5 cm in diameter, cylindrical in shape with blunt tip, skin is pure white, smooth, flesh snow-white, crisp, solid, mildly pungent and leaves deeply cut.
Punjab Safed	PAU, Ludhiyana	Selection from cross White 5 x Japanese White	Roots 30-40 cm long, white, tapering, smooth, mild pungency and free from forking. Foliage light green, takes 50-60 days for root formation.
Co-1	TNAU, Coimbatore	Line / Family breeding	Roots milky white, less pungent, 23 cm long, cylindrical, suitable for growing in plains in all seasons.
Chinese Pink	Exotic	Introduction	Roots are 12-15 cm long with white towards tip, semi-stumpy to stumpy. It takes 50-55 days for root formation.
Pusa Himani	IARI, New Delhi	Selection from cross Radish Black x Japanese White	Suitable for Dec-Feb sowing when no other variety forms such good roots. It can be grown in hills through out the year barring three winter months.
Scarlet Globe	Exotic	Introduction	Roots are round, small, 2cm diameter, bright red in colour. The flesh is crisp and white.
Pusa Reshmi	IARI, New Delhi	Selection from Green Top x Desi Type (Asiatic)	Roots are 30-45cm long, white with green tinge on top, suitable for sowing in September. It takes 50-60 days for maturity.
Kaliyanpur No.1	CSAUAT, Kanpur	Selection	Roots are 22-23cm long, smooth, crispy, white with green shoulder, thick and tapering.
White Icicle	Exotic	Introduction	Roots are icicle-shaped, pure white, flesh icy white, crisp, juicy and sweet flavourd.
Arka Nishant	IIHR, Bangalore	Selection	Roots are medium sized, 25cm long 3-4cm in diameter, marble-white with crisp texture and mild pungency, resistant to pithiness, premature bolting, root branching and forking.
Punjab Pasand	PAU, Ludhiana	Repeated crossing of selected plants from Pusa Chetki.	Main Season quick growing variety, attaining marketable maturity within 40 days after sowing. Roots are long, pure white, semi-stumped and free from hair with mean root yield of 534.8 q/ha, 5.75 % dry matter, 3.4% total sugars, 0.67 % crude fibre and 16.48 mg /100 g ascorbic acid.

Future Prospects

Genetic analysis of each chromosome, mode of inheritance for various resistance genes, traits in cytoplasm and transfer of traits by wide crossing are important for radish breeding. Collection and conservation of genetic resources are fundamental to radish breeding programme. Development of excellent radish varieties adapted to tropical and the sub-tropical regions, the exploitation of varieties adapted to Europe and America and the improvement of varieties for salad and cooking qualities with increased sweetness and root colour as in the perpetual radishes of China should be our modern breeding objectives.

References

Agrawal, A.A. and Sheriff, M. F. 2001. Induced plant resistance and suceptibility to late season herbivores of wild radish. *Annals of the Entomological Society of America,* **94:** 1, 71-75.

Anai, T., Takai, R., Miyala, M., Uchida, H., Kosemura, S., Yammamura, S., Ishizaki, R. and Hasegawa, K. 1997. Isolation and characterization of an auxin binding protein gene from radish and its expression in insect cells. *Physiological Plantarum,* **101:** 3, 606-611.

Antonova, N. G. 1992. New radish varieties bred at the Institute of Vegetable Farming. *Kartofel-i-Ovoshchi,* **2:** 47.

Arumugam, R. and Muthukrishnan, C.R. 1975. Genetic variability in some quantitative characters of radish. *Indian Journal of Horticulture.* **32:** 179-181.

Ashizawa, M., Hida, K. and Yoshikawa, H. 1979. Studies on the breeding Fusarium resistance in radish . I. Screening of radish varieties for Fusarium resistance, *Bull. Veg. and Ornam. Crops Res. Sta.,* **6:** 39-43.

Bett, K.E. and Lydiate, D.J. 2004. Mapping and genetic characterization of loci controlling the restoration of male fertility in Ogura CMS radish. *Molecular breeding new strategies in plant improvement,* **13(2):**125-133.

Brar, J.S. and Padda, D.S. 1972. A study of heterosis for root and leaf characters in radish. *J. Res.* **9(1):** 27-31.

Brown, G. G., Formanova, N., Jin, H., Wargacchuk K, R, Dendy, C., Patil, P., Laforesh, M., Zhang, J., Cheung, W.Y. and Landy, B. S. 2003. *Plant Journal.* **35(2):** 262-272.

Bunte, R., Muller, J. and Friedt, W. 1997. Genetic variation and response to selection for resistance to root-knot nematodes in oil radish (*Raphanus sativus* ssp. oleiferus). *Plant Breeding,* **116(3):** 263-266.

Capecka, E., Libik, A., Thomas, G. and Monterio, A.A. 1998. Quality differences between radish (*Raphanus sativus* L.). Cultivars to determine the possibilities of their use. *Acta Horticulture,* **459:** 89-95.

Chandel, K.S., Sharma, V.K and Pathiang, N.K. 2002. Line x tester analysis for the study of combining ability in radish (*Raphanus sativus* L.). *Vegetable Science.,* **29:** 2, 110-112.

Chandel, K.S., Singh, A. K. and Hayman, B.I. 1996. Genetic analysis for days to marketable maturity in radish (*Raphanus sativus* L.). *Bhartiya-Krishi-Anusandhan-Patrika.,* **11:4,** 205-210.

Chandel, K.S., Singh, A.K. and Rathan, R. S. 1993. Genetics of quantitative traits in radish (*Raphanus sativus* L.). *Vegetable Science.,* **20(1):** 48-51.

Curtis, I.S. and Nam, H.G. 2001. Transgenic radish (*Raphanus sativus* L *longipinnatus* Bailey) by floral-dip method. Plant development and surfactant are important in optimizing transformation efficiency. *Transgenic Research* **10(4):** 363-371.

Danu, N.S. and Lal, S.D. 1998. Correlation studies in radish (*Raphanus sativus* L). *Progressive Horticulture.* **30(3-4):** 135-138.

Delourme, R. and Eber, F. 1992. Linkage between an isozyme marker and a restorer gene in radish cytoplasmic male sterility of rape seed (*Brassica napus* L.) *Theoretical and Applied Genetics*, **85(2-3):** 222-228.

Deng, D., Gong, Y., Wang, L. and Liu, L. 2004. Studies on some isozymes in cytoplasmic male sterility radish (*Raphanus sativus* L.). *South West China Journal of Agricultural Sciences*, **17(3):** 384-351.

Giancola, S., Morhadour, S., Desloire, S., Clouet, V., Falenlin G.H., Laloni, W., Falentin, C., Pelletier, G., Renard, M. and Bendahmane, A. 2003. Characterization of a radish introgression carrying the Ogura fertility restorer gene Rfo in rapeseed , using the Arabidopsis gene one sequence and radish genetic mapping. *Theo. Appl genet.*, **107(8):**1442-1451

Hagimori, M., Nishio, T. and Dore, C. 1995. Clubroot disease resistance and other characteristics of the progeny of somatic hybrids between Japanese radis and cauliflower. *Acta-Horticulture.* **392:** 77-80.

Hagiya, K. 1975. Phenomenon of 'pithy tissue' in radish , *Nhoko to Engei,* **9:** 42-47.

Hawlader, M. S. H. and Mian, M. A. K. 1997. Self-incompatibility studies in local cultivar of radish (*Raphanus sativus* L.) grown in Bangladesh. *Euphytica,* **96: 3**, 311-315.

Hawlader, M. S. H., Mian, M. A. K. and Ali, M. 1997. Identification of male sterility maintainer lines for Ogura radish (*Raphanus sativus* L).*Euphytica,* **96(2):** 297-300.

Hida, K. 1983. Testing and breeding on the mosaic disease resistance of radish. Data in Jpn. Soc. Seed Prod. Breed Engine of 1983. pp 49-55.

Hirata, Y., Motegi, T., Hogi, Y., Kurita, S. and Zhou, J. 2004. Induction mechanism of cytoplasmic male sterility by artificial chimera synthesis in *Brassica. Acta Horticulture,* **637:** 303-308

Huh, M.K. and Ohnishi, D. 2002. Genetic diversity and genetic relationships of East Asian natural populations of wild radish revealed by AFLP. *Breeding Science,***52:** 2, 79-88.

Kumar, P. and Singh, D.V. 2003. Screening of radish germplasm against Alternaria Blight. 2003. *Farm Science Journal*, **12(1):** 61-62.

Kumar, V. and Chandel, K.S. 2003. Genetics of root yield and quality traits in radish (*Raphanus sativus* L.) *Crop Improvement*, **30(1):** 33-38.

Kumar, V. , Chndel, K.S. and Pathania, N.K. 2002. Heterosis in Asiatic and European cultivars in radish (*Raphanus sativus* L.). *Vegetable Science*, **29: (1)**, 34-36.

Kutty, C. N and Sirohi, P. S. 2003. Exploitation of hybrid vigour in radish (*Raphanus sativus* L.). *Indian Journal of Horticulture*., **60(1):** 64-68.

Lal, S. and Srivastava, J. P. 1975. Heritability and genetic advance for root and fruit characters in radish. 1975. *Indian J. Hort.,* **32:** 91-93.

Leeman, M., Pelt, J.A.V., Ouden, F.M., Heinsbrock, M., Bakker, P. A. H. M., Schippers, B., Van, J. A. and Den, F. M. 1995. Induction of systematic resistance against *Fusarium* wilt of radish by lipopolysaccharides of *pseudomonas fluorescens*. *Phytopathology*, **85(9):** 1021-1027.

Li, M., Gai, S. Y., Li, M and Gai, Y.J. 1997. Selection of a new cultivar of radish-Lu Luobu 8. *China Vegetables,* **4:** 26-28.

Li, X., He, F., Li, C. and Yuan, H. 2005. A new radish cultivar-Qiufeng 2. *Acta Horticultural Science,* **32(6):** 1166.

Lin, X. L. and Li, Y. A. 1993. Preliminary report on breeding of the new radish cultivar Man Tang Hong. *Chinese Vegetables*, **6:** 8-9.

Mehta, U. C., Mahato, K. C. and Singh, J. N. 1995. Effect of parthenium extracts on pollen tetrad and pollen sterility in radish (*Raphanus sativus* L.). *Cruciferae Newsletter,* **17:** 48-49.

Murali, K., Reddy, N.S., Anjanappa, M. and Krihnappa, K.S. 1998. Correlation studies in radish (*Raphanus sativus* L.). *Karnataka Journal of Agricultural Sciences*, **11:** 4, 1140-1141.

Niikura, S. and Matsuura, S. 1997. Genomic sequence and expression of S-locus gene in radish (*Raphanus sativus* L.). *Sexual Plant Reproduction,* **10:** 4, 250-252.

Niikura, S., Matsuura, S., Dore, C., Dosba, F. and Baril, C. 2001. Genetic variation of the self-incompatibility alleles (S-alleles) in the cultivated radish (*Raphanus sativus* L.) by the PCR-RFLP method. *Acta Horticulture*, **546:** 359-365.

Nijjar, H.S., Singh, G. and Dhaliwal, M. S. 1996. Genetic variability for ascorbic acid and isothiocyanate content in radish (*Raphanus sativus* L.). *Advances in Horticultural Science*, **10(1):** 8-10.

Nikornpun, M., Putivoranat, M. and Zhong, L. 2004. Uses of self-incompatibility genes in Chinese radish. South West China Journal of Agricultural Sciences, **17(1):** 78-83.

Pal , B.P. and Sikka, S.M. 1956. *Indian J. Genet.* **16:** 98-104.

Pandey, S. C., Pandita, M.L., Dixit, J. 1981. Genetics of yield and yielding contributing traits in radish. *Haryana Agric. University. J. Res.*, **11:** 384-388.

Ponds , G.S. 1959. Red Prince in new radish, WISC. *Agr. Exp. Stn. Bull,.* 538, Port II. 93-94 .

Pradenan, A., Yan, G. and Peummer, J. A. 2004. Correlation of morphological traits with molecular markers in radish (*Raphanus sativus). Australian Journal of Experimental Agriculture,* 44**(8):** 813-819.

Pradhan , A., Plummer, J.A. and Guijun ,Y, 2004. Comparison of morphological and molecular variation in seeds and seedling of radish cultivars. *Acta Horticulture,* **637:** 263-270.

Pradhan, A., Yan, G. and Plummer, J. A. 2004. Development of DNA fingerprinting keys for the identification of radish cultivars. *Australian Journal of Experimental Agriculture,* **44(1):** 95-102.

Qiu, X.B., Song, F., Lin, X.L., Liu, Y. A and Chen, H. 1993. Analysis of correlation between content of vitamin C and anthocyanin in red radish. *Acta Agricultural Boreali-Sinka*, **8(1):** 1. 56-59.

Rabbani, M. A., Murakami, Y,. Kerginuki, Y. and Takayanagi, K. 1998. Genetic variation in radish (*Raphanus sativus* L.) germplasm from Pakistan using morphological traits and RAPDs. *Genetic Resources and Crop Evaluation*, **45(4):** 307-316.

Shen, X.Q.1993. Breeding of two spring radish male sterile lines. *Chinese Vegetables*, **6:** 15-16.

Shimizu, S., Kanazana, K. and Kono. H . 1963. Studies on the breeding of radish for resistance to virus diseases, *Bull. Hortic . Res. Sta.,* **2:** 83-87.

Singh, G., Kumar, J.C and Dhaliwal, M.S. 1998. Punjab Pasand: a new variety of radish (*Raphanus sativus* L.) *Journal of Research: Punjab Agricultural University,* **35:** 1-2:124.

Sirohi, P.S and Kutty, C.N. 2000. Genetic analysis of yield and its components in radish. *Vegetable Science.*, **27(2):** 142-144.

Song, M. X. and Wei, S. L. 1996. On the heterosis of hybrid radish and the strategy for parent choice. *Acta Agriculture-Zhejiangensis,* **8(1):** 46-49.

Sugahara, S., Aoyagi, M., Sakamori, M. and Ochiai, H. 1997. The breeding of a new radish cultivar. *Research Bulletin of the Aichi ken Agricultural Research Centre,* **29:** 97-101.

Vinod, K. and Chandel, K.S. 2003. Genetics of root yield and quality traits in radish (*Raphanus sativus* L.) *Crop Improvement*, **30(1):** 33-38.

Walsh, M.J., Powles, S. B., Beard, B. R., Parkin, B.T. and Porter, S. A. 2004. Multiple herbicide resistance across four modes of action in wild radish (*Raphanus raphanistrum*). *Weed Science*, **52(1):** 8-13.

Wei, S. L., and Song, M. X. 1993. Breeding the new radish cultivar Zheluo 1. *Chinese Vegetables*, **4:** 1-3.

Wu, L., Guo, S. and Hou, Z. 2005. Anew radish F_1 hybrid 'Fengyu No. 1'. *China Vegetables,* **10/11:** 103-104.

Wu, L., Hou, Z., Guo, S., Wu, L. X., Hou, Z. G. and Guo, S. Y. 2001. A new type of spring radish F_1 hybrid-'Jin Ruobo No. 3' *China Vegetables*, **3:** 22-23.

Yamagishi, H. 2004. Assessment of cytoplasm polymorphism by PCR-RFLP of the mitochondria orf B region in wild and cultivated radishes (*Raphanus*). *Plant Breeding,* **123(2):** 141-144.

Yamanaka, H., Kuginuki, Y., Kanno, T. and Nishio, T. 1992. Efficient production of somatic hybrids between *Raphanus sativus and Brassica oleracea. Japanese Journal of Breeding,,* **42(2):** 329-339.

Young, H.J., Stanton, M. L. and Ellstrand, N. C. and Clegg, J.M. 1994. Temporal and spatial variation in heritability and genetic correlations among floral traits in *Raphanus Sativus (*wild radish). *Heredity,* **73(3):** 298-308.

Yu, Q., Zhang, X., Abul, H., Walsh, M. J. and Powers, S. B. 2003. ALS gene proline (197) mutations confer ALS herbicide resistance in eight separated wild radish (*Raphanus raphanistrum*) populations. *Weed Science*, **51(6):** 831-838.

Zhang, L., Shen, X.Q. and Zhao, G.Y. 1999. Inheritance of male sterility in spring-summer radish. *Acta Horticulture-Sinica*, **26(4):** 238-243.

Zhang,. M., Li, Y. and Huo, Z. 2003. A new radish F_1 hybrid Fuynan No. 1. China Vegetables, **5:** 33-34.

❑❑❑

Chapter – 30

Genetic Resource Potential for Turnip Improvement

N. Ahmed, H. Choudary, A.J. Gupta and K. Hussain

Introduction

Turnip is an important root crop which is very popular particularly in northern India. It is a cool season crop mainly grown as a fall or winter crop in the hills and winter crop in the plains. Turnip (*Brassica rapa* L. syn. *Brassica compestris* var. *rapa*) belong to family *Brassicaceae*. It has a chromosome number of 2n=2x=20. Two main centers of origin have been indicated. The Mediterranean area is considered to be the primary centre of origin of European type/temperate types. While Eastern Afghanistan with adjoining areas of Pakistan is considered to be another primary centre with Asia Minor, Transcaucasus and Iran as secondary centre. The parents of cultivated turnip are found in the wild form in Russia, Siberia and Scandnavia. It was cultivated by the ancients and ranked next to grapes and cereals in Italy. It was introduced in England probably in 1590 and United States in 1606 (Yawalkar, 1980). Turnips are now cultivated throughout the world.

Turnips are botanically biennial herbs, but are cultivated as annual root crops for both animal and human consumption. The turnip root contains 41.6% moisture, 6.2 g carbohydrates, 0.5 g protein, 0.2 g fat, 0.04 mg thiamin, 0.04 mg riboflavin, 43 mg ascorbic acid, 30 mg Ca, 40 mg P and 0.4 mg Fe per 100 g of edible portion. The turnip green contains 15,669 IU vitamin A per 100 g (Aykroyd, 1963). Turnip is grown for its enlarged root as well as for its foliage. Extra seedlings from thinning are often used as greens. The fresh roots are consumed in salads or cooked as vegetable or used in pickles. The young leaves contain high amount of ascorbic acid and iron, rank second in vitamin A content and are cooked as greens.

The fleshy thickened underground portion of turnip is actually the hypocotyl. The colour and shape of this underground portion vary with the cultivars and the roots are flat globular to top-shaped and long. The colour of under ground portion may be white or yellow, while that of above ground portion may be red, purple, white, yellow or green. A distinct tap root and secondary roots arise from the lower part of the swollen hypocotyl. Thickening begins in the central part of the hypocotyle, followed by the upper and lower parts. Normally, the roots attain edible maturity in 40-80 days depending on cultivar and climatic conditions.

The leaves and petioles are hairy and coarse and yellowish-green in colour. The leaves are oblong to oval and may be entire, serrate or even pinnate, depending on cultivar and time of development. However, leaves formed early are less likely to be pinnate. The leaves developing on inflorescence are alternate, oblong or lancealate and entire or dentate. The inflorescence is a terminal raceme on the main stem.

Floral Biology and Pollination

The inflorescence of turnip is corymbose-raceme i.e. blend of corymb and raceme. The upper part is corymb, while the lower part is raceme. Flowers are very small with 4 sepals, 4 petals and outer stamens being shorter with curled filaments. Turnip is highly cross pollinated due to strong sporophytic system of self incompatibility. Pollination generally occurs through insects and honey bees which are the main pollinators. Self incompatibility is a common characteristic of plants. Because of this, there is no need for emasculation and crosses can be made by enclosing flower heads from two compatible plants in a muslin bag with flies as pollinator. Before enclosing two flower heads in one bag, open flowers should be removed. Selfing can be accomplished by application of fresh pollen on style after removing the stigma. Sodium chloride and carbon dioxide could also be used to overcome self incompatibility.

Tian *et al.* (1995) suggested that yield reductions in *Brassica rapa* cultivar Candle may occur because of the reduction of fertilization potential through rapid degeneration of embryo sacs following anthesis.

Genetic Resources

In addition to swede, *B. napus* exists as forage rape and as annual (spring) and biennial (winter) oilseed rape. There are also annual and biennial oilseed forms of *B. rapa*, as well as several important oriental vegetable crops including pak choi and Chinese cabbage (Toxopcus *et al*., 1984). All of these subspecies represent a wide gene pool and can be crossed easily with one another. *Brassica rapa* hybridizes readily with *B. napus* to produce triploid hybrids (3n=29) which are almost sterile (Frendsen, 1941). Both spontaneous and artificial amphidiploids (2n=58) have been reported and these might be used to introgress useful characteristics from one species to another. *Brassica oleracea* can be crossed with difficulty to both *B. napus* and *B. rapa*. Hybridization can be facilitated using embryo culture (Harberd, 1969). Artificial forms of *B. napus* can be synthesized from the diploid parent species and have been used to introgress disease resistance from *B. oleracea* to *B. rapa*.

Turnip belongs to the large *Brassica* family and bears similar yellow and occasionally white flowers. It does not cross readily with the *Olerecea* (cabbage types) and belongs to a separate species, *B. campestris*. Another crop swede (*B. napus* L. var. napobrassica) is directly related to turnip. The swede is an amphidiploid with 38 chromosomes and is known to be a natural hybrid between *B. compestris* (2n=20) and *B. oleracea* (2n=18). Neither of the ancestral types is known accurately, but it is thought that the cross occurred numerous times where the two species overlap in their natural habitat (from Western Europe to Eastern Asia). The interspecific class has been synthesized by several workers, but the result is not noticeably swede like. Cross-incompatibility between the parental species is quite marked and pollen tubes tend to become distorted in the style. Any successful fertilization seems to result in the production of a sterile hybrid.

Genetic Studies

Fukuoka (1986) carried out extensive biometrical studies on turnip. He calculated broad sense heritabilities and correlations between ten quantitative characters viz., root weight, leaf weight, total plant weight, root length, root diameter, root-form index, leaf length, leaf number, top/root ratio and dry-matter percentage. Although he was primarily concerned with breeding forage turnips, he did include cultivars of culinary turnip in his studies and his results are equally applicable to culinary turnips. In general, broad sense heritability of all ten characters was high, ranging from 0.89 for root weigh to 0.99 for root-form index and leaf number, and a negative correlation was found between total plant weight and root weight.

Nasrullah *et al.* (2003) reported that highly significant differences among genotypes were observed for all the characters. Heritability in broad sense was high for all the characters except root compactness. Genetic gain was maximum for root to top ratio followed by leaf weight and root weight. Root weight showed strong significant positive correlations with leaf number, leaf weight, root size, root to top ratio and root compactness; suggesting that selection based on these traits would be more effective for improving root yield. The quality traits namely; total dry matter, vitamin C and total soluble solids showed significant negative association with root weight and with most of the other quantitative traits.

Si and Thurling (2001) studied inheritance of the traits associated with higher-pre-anthesis growth at low temperature and to evaluate response to selection methods for those traits. The ability to use pollen selection in combination with pedigree selection for low potassium leakage should increase genetic gain of plant dry weight at low temperature in segregating populations.

Ahmed and Tanki (1988) reported that seed yield in turnip is positively correlated with growth and yield contributing attributes. Singh *et al.* (1989) observed that plant height, number of branches, number of siliquae, number of seeds/ siliquae and total yield were favourably influenced with larger root size stecklings. Root age enhanced bolting but did not influence various growths and yield attributes.

Pathania and Rattan (1992) derived information on combining ability on 6 yield and growth related traits in 7 inbred lines, derived from 7 varieties and their 21 F_1 hybrids. The highest GCA effect for seed yield, plant height and seed number/pod was recorded for Red-4. The highest SCA effects were determined in the hybrid White-4 x Golden Ball.

Wani and Zargar (1992) reported that epistasis was revealed for the characters like silique/plant, 1000-seed weight and seed yield/plant in 45 triple tests cross (TTC) families.

Ahmed and Tanki (1994) studied inheritance of root characteristics in the F_1, F_2 and back crossed (BC) generations of crosses between five turnip (*Brassica Campestris* var. rapa) varieties and reported purple root colour dominant over pink, partially purple and white; and was under the control of two genes, P and C. In the F_1 generation, round root shape was dominant over flat-round and was a monogenic trait.

Prasad and Shrivastava (1996) made segregation analysis of F_1 and F_2 progeny from crosses of the turnip (*Brassica rapa*) cultivar Purple Top White Globe (with a low mean number of chromocentres) with the cultivars Rose Red and Golden Ball (higher means for this trait) and reported that inheritance of this trait is controlled by polygenic effects.

Zhao *et al.* (1998) used isozyme analysis to compare the genetic variation of Chinese and European cultivars of *B. campestris* and *B. napus*. In *B. campestris*, the largest proportion (about 70%) of the total genetic diversity was due to the within-cultivar variation. In *B. napus*, the largest contribution to gene diversity was made by the variation among cultivars (about 80%) and concluded that exchange of breeding materials between China and Europe in *B. campestris* and between Chinese '+' and European '00' materials in *B. napus* is recommended to be an efficient approach for broadening the genetic base.

Lewis and Woods (2002) used seed (used as female parent) from an S_2 plant exhibiting self fertility (line 89-23-1) in backcrossing programmes with cultivar AC Sunshine to produce self-fertile summer turnip rape (*Brassica campestris* var. *rapa*). F_3 plant ratio from selfed F_2 plants did not differ significantly from dominant monohybrid inheritance for self fertility. Subsequent test crossing showed that self fertility was dominant and homozygosity was confirmed following backcrossing.

Cui *et al.* (2004) obtained Ogura cytoplasmic male sterile (CMS) lines of turnip (*B. campestris* subsp. rapa) by interspecific hybridization between Ogura CMS of *B. napus*, and turnip, followed by backcrosses with the turnip cultivars; Xinxuan Ogu28-4, Aijiaohuang Ogu33-14, Naibing Ogu 18-4 as male parents. The results indicate that the relation of nucleo-cytoplasmic interaction is steady and the cytoplasmic effects are hardly affected by the nucleus of the sterile lines.

Breeding Objectives

- Early days to attain marketable root size.
- Root colour as per consumers preference as white, purple types are more liked in India than golden ball types.

- Stump rooted varieties with thin tap root and non-branching habit.
- Slow bolting habit.
- Appropriate dry matter content (8-9%) in roots.
- Resistance to club root, powdery mildew, turnip mosaic virus, white rust, phyllody, cabbage root fly and turnip root fly.

Breeding Methods

No intensive breeding work has been done in turnip. A few Agricultural Universities and Institutes particularly in the north India are engaged in the improvement work on tropical and sub-tropical types, while improvement work for temperate types is being done at IARI Regional Station, Katrain. The major parameters on which the attention is being focused are:

1. To bring about more uniformity.
2. To evolve varieties with root quality of separate types and ability to produce the seed under tropical conditions.
3. Evolving varieties with varying maturity periods.
4. To develop disease resistant varieties particularly to white rust and phyllody of seed crop.

Improvement in uniformity is an important breeding objective in turnip. However, in turnip, being a self-incompatible crop, successful inbreeding is difficult to achieve, although not impossible. This difficulty can be over come by bud pollination.

Mass selection

This method of breeding has been commonly employed to breed several open-pollinated cultivars. Plants from selected roots are allowed mass-pollination in isolation. An increase of 25% over the base population was observed in mass selected progeny. Variety Improved Golden Ball has been selected through mass selection from a population of Golden Ball.

Zhang (1993) developed cultivar Mazhuang, which originated as a farmer's selection especially suitable for processing dry slivered turnip. The variety is resistant to the main diseases and has a growth period of 105 days.

Mutation breeding

A non-waxy mutant was developed using gamma rays on Pusa Sweti. The mutant showed resistance to aphids.

Polyploidy breeding

Polyploids have been developed through colchicine treatment. It was found that the first generation polyploids have low fertility which could be rectified by crossing and selection. The tetraploids had high root yield but reduced dry-matter content. A trisomic line (2n=21) has been isolated in turnip. It showed reduced fertility compared with diploid in open, cross and self pollination.

Crop	**Turnip**	**Swede**	**Chinese**	**Knoll-khol cabbage**	**Cabbage**
B. rapa (turnip)	SI	CC X	CC X	CC	CC
B. carinata (Swede 2n=34)	CC X	SI SC	CC	CC	CC
B. chinensis/ *B. Pekinensis* (2n=20)	CC (X)	CC	SI SC	CC	CC
B. Oleracea L. Var. Gongylodes (2n=18)	CC	CC	CC	SI SC	CC
B. oleracea var. capitata (2n=18)	CC	CC	CC	CC	SI CC

SI, Mainly self-incompatible; SC, self-compatible forms are also present; CC, cross-compatible; X, F_1 sterile; and (X) F_1 fertile.

Prasad and Jain (1998) developed highly fertile tetraploid line T76, derived from the diploid turnip (*Brassica campestris* var. *rapa*) variety Rose Red and crossed it with its parent to yield a triploid form. Triploids had the smallest seeds, but satisfactory seed germination. Numbers of chloroplasts per guard cell, numbers of leaves and branches and plant height increased with ploidy level.

Interspecific Hybridization

This is resorted to generate variability to effect useful selection.

Intergeneric Hybridization

A triploid (2n=28) was identified from a cross between a tetraploid fodder radish and a tetraploid stable turnip. The triploid had 9 bivalents and 10 univalents. Evidence of linkage between gene controlling leaf shape (C-) and root colour (R-P-) was found. Seed coat colour is controlled by pleiotropic effect of a gene pair pp which together with Cc also controls root colour. Brown seeded plants carry the allele Pp while creamy seeded plants are of pp genotypes.

Genetics of qualitative characters

Character	**Number of genes**	**Type of gene action**
Skin colour	Digenic (R-P-)	R-PP-Red rrP-Purple R-P-Deep red, rrPP-White
Root shape	Monogenic	Round incompletely dominant to flat root
Leaf type	Monogenic (CC)	Compound leaf dominant to simple leaf

Heterosis Breeding

There has been greater interest among breeders to produce hybrid cultivars of turnip utilizing self incompatibility. However, the same has not been achieved in case of horticultural types. Heterosis has been reported for several characters including the yield of root and leaf. Japanese breeders have been successful in developing several F_1 hybrid cultivars of white turnip. There has been some progress in breeding of F_1 hybrid turnip cultivars in the USA but the acceptance of these cultivars has been poor. However, considering the higher level of heterosis for root yield and availability of sporophytic system of self-incompatibity, it is suggested that hybrid cultivars of turnip should be developed following the procedures as outlined in case of cabbage.

Heterosis for dry matter yield has been demonstrated by Wit (1966). Synthetic cultivars, based on inbred lines of good combining ability, have been shown to out yield the best commercial cultivars in Holland. Brar and Gupta (1969) reported heterosis for yield and root length in turnip. Pathania *et al.* (1987) also studied heterosis and combining ability in half diallel of 6 parents. Most of the F_1 hybrids exhibited heterosis for root yield, root length, root diameter and number of leaves per plant.

Root yield was shown to be dependent upon plant size and foliage weight. Highly significant heterosis has been found for root length and diameter and fresh root weight (Grant *et al.* 1982). Since these three characters were found to be mainly under additive genetic control, the apparent heterosis was presumably due to accumulation of dispersed genes rather than over dominance.

Schuler *et al.* (1992) made hybrid between *Brassica rapa* var. *oleifera* cv. Tobin and 19 European and Canadian strains in order to evaluate potential for future heterosis breeding. Mid-parent heterosis of the F_1 strains averaged 18% for seed yield, 17% for yield, -0.7% for days to maturity and 7.0% for height. No heterosis was observed for days to flower or mean single seed weight. The level of heterosis observed may justify the development of commercial F_1 hybrids.

Ahmed *et al.* (1993) reported heterobeltiosis of the hybrids Pusa Chandrima x White Flat Round, Purple Top White Globe x Pusa Snow Ball, Pusa Snow Ball x Whte Flat Round , Pusa Snow Ball x Purple Globe , Pusa Snow Ball x White-4, Pusa Swarmina x Purple Globe, Pusa Swarmina x White-4, Pusa Sweti x Pink Flat Round, Pusa Chandrima x Pusa Swarnima and Pusa Chandrima x Pusa Sweti. These combinations recorded 23-69% higher root yield over better parent and hence may serve as potential hybrid in turnip.

Falk *et al.* (1994) conducted an experiment to evaluate performance of inter culture summer turnip rape hybrids and recorded an average of 13% heterosis for seed yield. Heterosis for seed yield was greatest in crosses between genetically diverse cultivars, which agree with classical theories on heterosis.

Bunin (1994) utilized various male sterility systems, nuclear and cytoplasmic, in breeding major crops in the Cruciferae including turnip with information on male-sterile forms obtained by genetic engineering, hybridization, tissue culture and treatment with some types of chemicals (mutagens, retardants) and reported better prospects for producing heterotic hybrids using these male-sterile forms.

Ahmed and Tanki (1999) reported heterosis over better parent for root yield in F_1, F_2 and F_3 generations. The results indicated presence of substantial amount of heterosis not only in F_1 (69.29%) but also in subsequent F_2 and F_3 generations to the extent of 31.91 and 24.70% respectively, suggesting that crosses (Pusa Chandrima x White Flat Round, Pusa Snow Ball x White Flat Round, Purple Top White Globe x Pusa Snow Ball and Pusa Chandrima x Pusa Sweti) with significant residual heterosis, could be utilized either in the synthesis or composites or directly as advance generations of single crosses. A method consisting of random selection of root in F_2 followed by their mating at random was also suggested to restore hybrid vigour in the later generations.

Breeding for disease resistance

Club root (*Plasmodiophora brassicae*) is one of the major diseases of turnip. The galls which form on the roots often making them unmarketable. The relationship between the variation found in *P. brassicae* and resistance in various *Brassica* crops including turnip, was reviewed by Crute *et al.* (1980). Resistance to different pathotypes of *P.brasicca* in turnip is apparently controlled by three independent dominant genes. Resistance is controlled by dominant pathotype-specific resistance genes. However, populations of *P. brassicae* exist which are pathogenic in all combinations of these resistance genes. Cytoplasm male sterile breeding lines have been derived from the turnip cultivar Tokyo Cross, with moderate resistance to clubroot and TuMV and a low level of resistance to downy mildew (*Peronospora parasitica*) (Leung and Williams, 1983).

Shattuck (1992) reported UG1 turnip germplasm resistance to the C3 strain of turnip mosaic potyvirus (TuMV-C3) derived from an intraspecific cross with TuMV-resistant Chinese cabbage (*Brassica rapa* subsp. *pekinensis*). The parents used in the initial hybridization were *B. rapa* subsp. *pekinensis* cv. Chen Chen and Purple Top White Globe. UG1 consisted of both self-fertile and self-incompatible plants.

Wang and Wang (1994) developed Weixian line through single plant selection. The new Weixian line was tolerant to heat and highly resistant to viruses and black rot (*Xanthomonas campestris*). Rajpurohit and Choudhary (1995) screened 100 biotypes of *B. tournefortii* for downy mildew and reported that SPS 9-4 and SPS 1-1 exhibited the least disease incidence.

Zhang *et al.* (1999) explored the TVCV subgroup from a TMV-like ancestor, the infection of turnip plants by TMV and by chimeras between TMV and TVCV and concluded that the TMV MP is not responsible for the limitation of TMV spread in turnips.

Woods *et al.* (2000) while working on resistance to brown girdling root rot in turnip screened 260 *Brassica rapa* lines and resulted in identification of 5 lines with partial resistance to brown girdling root rot (BGRR, caused by *Rhizoctonia solani*, *Fusarium* and *Pythium ultimum*). A method for simultaneously selecting canola quality traits and reduced disease susceptibility was developed and used to screen progeny of crosses to BGRR susceptible canola quality lines.

Klein and Woods (2004) examined the distribution of brown girdling root rot from spaced rows of *B. rapa* 'Tobin' (control) and revealed a minimal loss of precision in breeding programmes by reducing the frequency of control rows to one in four, which represents a considerable saving in cost compared with the current practice.

Pest resistance

The fodder cultivars 'Angus' and 'Melfort' bred at the Scottish Crop Research Institute possess good resistance to turnip root fly (*Delia floralis* L). Angus may also have some resistance to the cabbage root fly (*D. radicum*). Resistance to turnip root fly is a dominant character and is apparently correlated with increased dry-matter content. However, root resistance to larval feeding is not due to increased tissue hardness in cultivars with high dry-matter content, but is probably due to differences between cultivars in the chemical content of the outer root tissues which influence larval feeding and development.

In the USA cold-tolerance was combined with resistance to the turnip aphid (*Hydaphis erysimi*) derived from the cultivar Shogoin. Two cultivars were produced by recurrent selection among progeny of crosses between Shogoin, Raab Salad (cold-tolerant) and Purple Top White Globe (root quality). The cultivar Tina, bred in New Zealand, is reported to have moderate resistance to cabbage aphid (*Brevicoryne brassicae*) (Lammerink, 1985).

Recurrent selection in progeny of crosses involving cultivars such as Purple Top White Globe. Raab Salad and Shogoin turnip has resulted in considerable progress towards developing cold tolerant and aphid resistant turnips (Robbins, 1977). Clubroot resistant turnip cv. Manga has also been reported from New Zealand (Lanmerink *et al.* 1978) however, this cultivar is slightly low yielding and has a higher tendency to bolting when grown in spring. Nevertheless, this can be utilized in producing disease resistant cultivars.

Breeding for Quality Attributes

Gill *et al.* (1974) conducted an experiment on varietal variation in chemical constituents (ascorbic acid, total soluble solids and dry matter) in turnip greens. Significant varietal differences were reported in respect of all the characters studies. Ascorbic acid was retained to a great extent even after cooking. The loss in cooking was mostly proportional to the extent of vitamin C recorded in raw condition and vice-versa. Saimbhi and Singh (1987) evaluated the turnip varieties for dehydration and Purple Top White Globe was the best followed by Red-4 and White-4 for this purpose.

Role of Biotechnology

There has been extensive research on developing novel techniques for the improvement of oilseed forms of both *B. napus* and *B. rapa*. Most of this work is applicable to turnip and some techniques have already been utilized. Root protoplasts of turnip cultivar 'Snowball' regenerated roots via callus stage but failed to produce any shoots (Xu *et al.*, 1982). Whole turnip plants and protoplasts have been successfully transformed utilizing a modified cauliflower mosaic virus genome (Paszkowski *et al.* 1986). Techniques of DNA hybridization have been used to assess sequence homologies of turnip mt DNA with other plant species and between the nuclear genome of turnip and *Arabidopsis* and their somatic hybrid produced by protoplast fusion (Miroshnichenko, 1986). Restriction fragment length analysis was used to demonstrate that the male sterile cytoplasm of *B. napus* was same as that of *B. rapa* but different from that of *B. oleracea* (Ohkawa, 1985).

Tanhuanpaa *et al.* (1993) screened three Finnish spring turnip rape (*B. campestris* var. oleifera) cultivars (Valtti, Kova and Kulta) for differences in their random amplified polymorphic DNA (RAPD) patterns. The reproducibility of individual RAPD patterns was high when the DNA templates were of equal quality and quantity. It was possible to use both a rapid DNA extraction method and DNA extracted from a combined sample of ten individuals without losing any of the major fragments. When the cultivars were compared for RAPD pattern, most polymorphic loci exhibited differences only in allele frequency. The RAPD markers provided a fast and reliable method for analysing individuals and cultivars of turnip rape.

Mikkelsen (1996) analysed two back cross families using RAPD markers and reported two of the plants, as in *B. campestris*, had 20 chromosomes and a high pollen fertility, but also had some swede rape markers, indicating that rapid gene transfer of transgenes from swede rape to wild *B. campestris* might occur.

Fujimoto and Yamagishi (1996) applied RAPD analysis to strains of Sugukina and turnip varieties (*Brassica campestris*) to clarify the phylogenetic relationship between the two. The results indicated that Sugukina is a unique group of vegetables, containing large genetic variations and having different genetic characteristics from the turnip varieties growing in Japan.

Tanhuanpaa *et al.* (1996a) constructed a linkage map of spring turnip rape (*Brassica rapa* ssp. *Oleifera)* from an F_2 population of the cross Jo4002 X Sv3402. The map contained 22 RFLP loci, 144 RAPDs, one microsatellite and one morphological marker (seed colour). All ten linkage groups could be identified and the total map distance was 519 cM. A proportion of the markers (13%), most of which were located in two linkage groups, showed segregation distortion.

Tanhyanpaa *et al.* (1996b) used bulk segregant analysis to search for RAPD markers linked to gene(s) affecting oleic acid concentration in an F_2 population from the *Brassica rapa* subsp. *oleifera* cross Jo4002 X a high oleic acid individual from line Jo4072. The most suitable marker for oleic acid content is OPH-17, a codominant marker close (< 4 cM) to the QTL. The mean seed oleic acid content in the F_2 individuals carrying the larger allele of this marker was $80.14 \pm 9.76\%$, in individuals with the smaller allele, $54.53 \pm 6.83\%$; in the heterozygotes, $65.47 \pm 8.15\%$. To increase reproducibility, the RAPD marker was converted into a SCAR (sequence characterized amplified region) marker with specific primers. Marker OPH-17 can be used to select spring turnip rape individuals with the desired oleic acid content.

Tanhunpaa *et al.* (1998) used mapping and cloning technique for comparison of the wild-type and high-oleic-acid allele of the locus and revealed only one difference in their nucleic acid sequences leading to an amino acid change. This substitution of leucine by proline most likely affects the folding of the protein and thereby the activity of the enzyme. Allele-specific markers will be very effective in selection for plants with high-oleic-acid content derived from Jo4072 because they are located exactly at the locus and can differentiate between homo- and heterozygotes.

Tanhnanpaa and Vilkki (1999) studied the efficiency of marker-assisted selection for oleic acid content in an F2 population of 5 crosses of spring turnip rape (*Brassica rapa* subsp. *oleifera*) using a sequence characterized amplified region (SCAR) marker, the distance of which from the oleic acid gene is 11.5 cM, and allele-specific markers which are located at the gene locus. As expected, the allele-specific markers recognized the oleic acid genotypes more precisely than the SCAR marker.

Wu *et al.* (2004) reported that differential gene expression patterns at different stages of the inbred line Baimanjing 001-24 of turnip (*B. campestris* subsp. rapa) and their hybrids were analysed using cDNA-AFLP. Four types of differential gene expression patterns were detected between the hybrid and their parents.

Development of Varieties

There are two distinct groups of turnip (i) biennial or temperate or European type, (ii) annual or tropical or Asiatic type. The seed of former group can be produced in the hills and that of latter in plains. Temperate types undergo dormancy or rest period which can only be broken to initiate the development of seed stalk when the roots are subjected to low temperature i.e. less than 4-8 ^{0}C for a period of 4-6 weeks at any time either in storage or under field condition. It is for this reason the seeds can only be produced in places like Kashmir valley, Kullu, Kalpa, Katrain, Darjeeling and Nilgiri hills where winters are severe. Many good varieties differing in shape and colour are now available for cultivation in both the groups. Some of the commonly grown varieties of the two groups are:

Variety	Organization	Pedigree	Brief Characteristics
Temperate type			
Pusa Chandrima	IARI Reg. Station, Katrain	Hybridization between Asiatic type and European type	Its roots are medium to large, nearly flattened, globular, smooth, pure white skin with five grains. The flesh is sweet and tender. Top medium and leaves not so deeply cut. It is an early maturing (50-60 days), heavy cropper with an average yield of 400 q /ha. It is suitable for sowing from October to December in plains.
Pusa Swarnima	IARI, Reg. Station, Katrain	Hybridization between Asiatic type (Japanese White) and European type (Golden Ball)	The roots are flattish round with creamy-yellow skin and pale-umber coloured flesh of fine texture and flavour. Its top is medium, leaf blade is not so deeply cut. It is suitable for growing from June to October in hills and October to December in plains. It matures in 65-70 days.

Contd...

Nageen 1	SKUAST Srinagar	PTWG X PSB Line Breeding	Its roots are white, large, round compact & sweet. It can be used for cooking as well as salad purpose. It has good TSS and vitamin C, content. Its roots are free from pithiness
Purple Top White Globe	IARI, Reg. Station, Katrain	Introduction from Europe	It is a heavy-yielder and large-rooted variety. The roots are nearly rounded, upper part purplish, lower portion is creamy. The flesh is white firm crisp and mildly sweet flavoured. Top is small erected with cut leaves. It is suitable for growing during cooler months.
Golden Ball	IARI, Reg. Station, Katrain	Introduction from Europe	Its roots are perfectly globe shaped, medium sized and smooth. It has bright, creamy yellow skin and pale umber coloured flesh of fine textures and flavour. The top is small, erect with cut leaves.
Snow ball	IARI, Reg. Station, Katrain	Introduction from Europe	This variety is an early temperate type with medium-sized small top. Its leaves are erect, cut and medium green. The roots are round, smooth with pure white skin. The flesh is white, fine grained, sweet and tender.
Asiatic type			
Pusa Kanchan	IARI, New Delhi	Hybridization between Asiatic types (Local Red Round) and European type (Golden Ball)	It contains good qualities of both the parents. The roots look just like the Local Red Round. The skin is red; flesh is creamy yellow with excellent flavour and taste. The leaf top is shorter than the Local Red Round. It becomes ready for harvesting in about 10 days later than the local parent. Its roots can be kept for a longer time than Local Red Round in field with out becoming spongy.
Pusa Sweti	IARI, New Delhi	Line/Family breeding from Exotic line.	Attractive, white roots of Pusa Sweti, matures 45-50 days after sowing. A very early maturing variety, it is suitable for October sowing in plains.
Punjab safed-4	PAU. Ludhiana		An early maturing variety, commonly grown in Punjab and Haryana. The roots are pure white, round, medium sized with mild taste.
Early Milan Red Top	—	Introduced from Europe	Its roots are deep flat with purplish red top and white under earth. The flesh is pure white, well grained, crisp and mildly pungent. The top is very small with 4-6 sessile leaves. It is an extra early and very high yielding variety.

Future Prospects

The main objectives of turnip breeding include earliness, improving colour, enriching carotene content, increasing yield potential and incorporating resistance to pests and diseases. Wild forms of turnip are seen in Russia and Siberia, which have potential use in its improvement.

References

Ahmed, N. and Tanki, M. I. 1994. Inheritance of root characters in turnip. *Indian Journal of Genetics and Plant Breeding,* **54(3):** 247-252.

Ahmed, N. and Tanki, M. L. 1988. Effect of nitrogen and phosphorous on seed production of turnip (*Brassica rapa* L.). *Veg. Sci.* **15:** 1-7.

Ahmed, N. and Tanki, M.I. 1999. Exploitation of residual heterosis for improvement of root yield in turnip (*Brassica rapa* L.). *Applied Biological Research,* **1(1):** 71-74.

Ahmed, N., Tanki, M. I. and Khan, S. H. 1993. Heterosis and combining ability studies in turnip (*Brassica rapa* L.). *Vegetable Science,* **20(1):** 72-76.

Aykroyd, W. R. 1963. ICMR Special Report. Series **No. 42.**

Brar, J.S. and Gupta, V.P. 1969. A Genetic Study of yield and duration in *Brassica rapa* L. *Plant Science,* **1:** 24-29.

Bunin, M. S. 1994. Male sterility in crop plants of the family Brassicaceae L. and its use in breeding (review of the foreign literature). *Sel'skokhozyaistvennaya Biologiya,* **No.1**, 20-31.

Crute, I.R., Gray, A.R., Crisp, P. and Bruczacki, S.T. 1980. Variation in *Plasmodiophora brassicae* and resistance to clubroot diseasse in brassicas and allied crop-a critical review. *Plant Breeding Abst.*, **50:** 91.

Cui, H. M., Cao, J. S., Zhang, M. L., Yao, X. T. and Xiang, X. 2004. Production of the Ogura cytoplasmic male sterile (CMS) lines of Chinese cabbage-pak-choi (*Brassica campestris* L. ssp. chinensis var. Cmmunis and turnip (*B. compestris* L. ssp. *Rapifera*) and cytological observation of their sterile organs. *Acta Horticulturae Sintica,* **31(4):** 467-471.

Falk, K.C., Rakow, G., Dawney, R. K. and Spurr, D.T. 1994. Performance of inter-culture summer turnip rape hybrids in Saskatchewan. *Canadian Journal of Plant Sciences*, **74(3):** 441-445.

Frandsen, K.J. 1941. Contribution to the cytogenetics of *Brassica napus* L., *Brassica compestris* L., and their hybrid the amphidiploids *Brassica napocampestris* Arsskr. *K. Vet. - Landbohojsk,* **1941:** 59.

Fujimoto, T. and Yamagishi, H. 1996. Investigation on the phylogenetic relationship between 'Sugukina' and turnip (*Brassica campestris* L.) by RAPD analysis. *Journal of the Japanese Society for Horticultural Science.* **65(3):** 609-614.

Fukuoka, H. 1986. Biometrical breeding of forage turnip in Japan. *Jap. Agriric Res. Quart,* **19:** 281.

Gill., H.S., Tiwari, R. N. and Singh, R. 1974. Varietal variation in chemical constituents (ascorbic acid, total soluble solids and dry matter) in turnip greens. *Science and Cultivar,* **40:** 491-492.

Grant, I., Harney, P.M. and Christie, B.R. 1982. Inheritance of yield and other quantitative characters in *Brassica napus* var. *napobrassica. Can. J. Genet. Cytol,* **24:** 459.

Harberd, D.J., 1969. A simple effective embryo culture technique for Brassica. *Euphytica,* **18:** 425.

Klein, G. H.W. and Woods, D. L. 2004. Number of check rows required to select summer turnip rape for resistance to brown girdling root rot. *Canadian Journal of Plant Science,* **84(2):** 691-694.

Lammerink, J. and Hart, R.W. 1985. Tina, a new swede cultivar with resistance to dry rot and clubroot. *N.Z.J. Ext. Agric.*, **13:** 417.

Leung, H. and Williams, P.H. 1983. Cytoplasmic male sterile *Brassica campestris* breeding lines with resistance to clubroot, turnip mosaic and downy mildew. *Hort. Science*, **18:** 774.

Microshnichenko, G.P., Volkov, R.A. and Borisyuk, N.V. 1976. Structure of the genomes of folder turnip, *Arabidopsis* and their somatic hybrid. *Biokhimiya,* **51:** 84.

Mikkelsen, T. R. 1996. Inheritance of swede rape RAPD markers in the cross (swede rape X turnip rape) X turnip rape. *SP Rapport Statens Planteavlsforsog,* **4(2):** 28-29.

Nasrullah, Ahmed, N. and Wani, S. A. 2003. Genetic variability and character association in turnip (*Brassica rapa* L.). *SKUAST-J Res.,* **5:** 69-74.

Ohkawa, Y. 1985. Occurence of cytoplasmic male steritlity in *Brassica campestris* and comparison with that of *B. napus*. *Bull. Nat. Inst. Ag. Sci. Japan,*, **36:** 1.

Paszkowski, J., Pisan, B., Shillito, R.D., Hohn, T., Hohn, B. and Potrykus, I. 1986. Genetic transformation of *Brassica campestris* var. rapa protoplasts with engineered cauliflower mosaic virus genome. *Plant Molec. Biol.*, **6:** 303.

Pathania, N. K and Rattan, R. S. 1992. Combining ability studies in turnip. *South Indian Horticulture,* **40(5):** 266-268.

Pathania, N.K. Rattan, R.S. and Thakur, M.C. 1987. Heterosis and combining ability in turnip (*Brasica rapa* L.). *Veg. Sci.* **14:** 161-168.

Prasad, C. and Jain, S. C. 1998. The promising triploid of cultivated turnip. *Cruciferae Newsletter*, No. **20:** 25-26.

Prasad, C. and Shrivastava, M. 1996. Polygenic control of chromocentres in cultivated turnip. *Cruciferae-Newsletter,* No. **18:** 22.

Rajpurohit, T. S. and Choudhary, B. R. 1995. Downy mildew (*Peronospora parasitica*) of wild turnip (*Brassica tournefortii*), a new record from Rajasthan. *Indian Journal of Agricultural Sciences*, **65(5):** 377.

Robbins, M.L. 1977. *Cuciferae News letter,* **No 2.** p. 31.

Saimbi, M.S. and Singh, B. 1987. Evaluation of turnip (*Brassica rapa* L.) varieties for dehydration. *Veg. Sci.*, **14:** 203-205.

Schuler, T. J, Hutcheson, D. S. and Downey, R. K. 1992. Heterosis in intervarietal hybrids of summer turnip rape in western Canada. *Canadian Journal of Plant Science*, **72(1):** 127-136.

Shattuck, V. I. 1992. UG1 turnip germplasm possessing resistance to turnip mosaic virus. *Hort. Science,* **27(8):** 938-939.

Si, P. and Thurling, N. 2001. Genetic improvement of pre-anthesis growth of turnip rape (*Brassica rapa* L.) at low temperature. *Australian Journal of Agricultural Research*, **52(6):** 653-660.

Singh C. G., Pandita, M. L. and Khurana, S.C. 1989. Studies on effect of root age, size and spacing on seed yield of turnip (*Brassica campestris* var. rapa L.) *Veg. Sci.*, **16:** 119-124.

Tanhuanpaa, P. and Vilkki, J. 1999. Marker-assisted selection for oleic acid content in spring turnip rape. *Plant Breeding*, **118(6):** 568-570.

Tanhuanpaa, P. K., Vilkki, J. P. and Vilkki, H. J. 1996. A linkage map of spring turnip rape based on RFLP and RAPD markers. *Agricultural and Food Science in Finland,* **5(2):** 209-217.

Tanhuanpaa, P. Vilkki, J, Vilkki, J. and Pulli, S. 1993. Genetic polymorphism at RAPD loci in spring turnip rape (*Brassica rapa* ssp. oleifera). *Agricultural Science in Finland,* **2(4):** 303-310.

Tanhuanpaa, P., Vilkki, J. and Vihinen, M. 1998. Mapping and cloning of FAD2 gene to develop allele-specific PCR for oleic acid in spring turnip rape (*Brassica rapa* ssp. oleifera). *Molecular Breeding,* **4(6):** 543-550.

Tanhuanpaa, P.K. Vilkki, J.P. and Vilkki, H. J. 1996. Mapping of a QTL for oleic acid concentration in spring turnip rape (*Brassica rapa* spp. oleifera). *Theoretical and Applied Genetics,* **92(8):** 952-956.

Tian, X. Y., Caeseele, L. V., Sumner, M. J. Van, C. L. and Tian, X. Y. 1995. The possible role of embryo sac degeneration in yield reduction of summer turnip rape cultivar Candle. *Canadian Journal of Plant Science,* **75(3):** 595-598.

Toxopeus, H., Oost, E. and Reuling, G. 1984. Current aspects of the taxonomy of cultivated *Brassica* species. The use of *Brassica rapa* L. versus *B. campestris* L. and a proposal for a new intra specific classification of *B. rapa* L. *Cruciferae Newsletter*, **9:** 55.

Wang, H.X. and Wang, J. J. 1994. Newly selected Weixian turnip line with small plants and purple flowers. *China Vegetables*, No. **2:** 51-52.

Wani, S. A. and Zargar, G. H. 1992. Detection of epistatic, additive and dominance variation in turnip rape

(*Brassica compestris* L. var. brown sarson). *Crop Research Hisar,* **5(2):** 286-289.

Wit, F. 1966. *Qualitas Plant Mater. Veg,* **13:** 305-310.

Woods, D. L., Turkington, T.K., McLaren, D. and Davidson, J. G. N. 2000. Breeding summer turnip rape for resistance to brown girdling root rot. *Canadian Journal of Plant Science,* **80(1):** 199-202.

Wu, C. J. Cao, J. S. He, Y. and Dong, D. K. 2004. The relationship between differential gene expression patterns in rosette stages and heterosis in Chinese cabbage-pak-choi, turnip and their hybrids. *Acta Horticulturae Sinica,* **31(4):** 508-510.

Xu, Z.H., Davey, M.R. and Cocking, E.C. 1982. Plant regeneration from root protoplasts of *Brassica. Platn Sci. Lett.,* **24:** 117.

Yawalkar, K.S. 1980. Vegtable Crops of India. Agri-Horticultural Publishing House, Nagpure.

Zhang, X. C. 1993. Liuan Ma Zhuang turnip. *Chinese Vegetables,* **No. 3**, 48-49.

Zhang, Y. Lartey, R. T, Hartson, S. D. and Voss, T. C. and Melcher, U. 1999. Limitations to tobacco mosaic virus infection of turnip. *Archives of Virology,* **144(5):** 957-971.

Zhao, J. Y. Becker, H. C. and Zhao, J. Y. 1998. Genetic variation in Chinese and European oilseed rape (*B. napus*) and turnip rape (*B. campestris*) analysed with isozymes. *Acta Agronomica Sinica,* **24(2):** 213-220.

❑❑❑

Chapter – 31

Genetic Resource Potential for Carrot Improvement

N. Ahmed, H. Choudary, A.J. Gupta and K. Hussain

Introduction

Carrot (*Daucus carota* L., 2n=2x=18) is a cool-season root vegetable grown all over the world in temperate as well as sub-tropical climates. In India, it is cultivated over an area of 24,000 hectares with an annual production of 3,50,000 metric tones with a productivity much lower (14.58 t/ha) than the world average (22.17 t/ha) as per FAO, 2004. Carrots are used for human consumption as well as for animal feed. Carrot can thus be used for augmenting fresh fodder supply to milch animals during the lean winter in Kashmir (Hussan and Ahmed, 1999). Carrot has a significant place as an ingredient in soups and sauces and in dietary composition and also as a salad. A sweet preparation called *Gajar Halwa* is very famous dish in North India. Besides canning, it is also used in preparation of pickles. It is also exported in the form of fresh roots to the countries like Kuwait and Sharjah. Carrot is a rich source of á and â- carotene. The total carotenoid content in the edible portion of carrot roots ranges from 6.00 to 54.80 mg/100g. The yellow and orange coloured roots contain relatively more carotenoid. Carrot roots contain protein, fat, carbohydrate, minerals (especially Ca, Mg, Na, K, Cu, S etc.) besides vitamin A, B_1, B_2, nicotinic acid and vitamin C. In carrot roots, sucrose is the most abundant with endogenous sugar contents 10 times than those of glucose and fructose.

Asia minor, Afghanistan, North West India, Iran and Turkey are the centers of origin of carrot. As per Encyclopedia Britannica, carrots have originated from Afghanistan, Punjab and Kashmir with secondary centers in Ethiopia and North America. The European carotene carrot (syn. Western carotene carrot, temperate carrot) has been derived from the Asiatic anthocyanin containing forms of carrot (syn. Eastern anthocyanin carrot, tropical carrot). Afghanistan is the centre of diversity of the purple coloured carrot (anthocyanin carrot) and as such this country is suggested as the primary centre for origin. Integration between the Asiatic and European carrots has given the present day forms which are fleshy, smoother, less forked and better coloured.

Punjab and Kashmir are also considered as primary centre of origin with secondary centers in Ethopia and North America. It is still found in its wild form in Himachal Pradesh and Kashmir valley where wild animals especially brown bear feed on its roots. People still eat the wild carrots in some parts of Kashmir (Yawalkar, 1980). In north India highly coloured types of carrots are found which are not available in Europe. The colour of carrot ranges from absolute colourless to light lemon, light orange, orange and deep orange, light purple, deep purple and almost black. It is also an indication of this region being a primary centre of origin of carrot.

Carrot is an annual herb for root production and a biennial for flowering and seed set. The stem is erect, branched, 30-120 cm high arising from a thick fleshy taproot, which is 5-30 cm long. The root consists of outer core and inner core. The outer core consists of a thin periderm, a layer of cark cells and a relatively wide band of secondary phloem. In this region only the sugar is stored. The inner core has the secondary system and pith. The shape of the root can vary from tapering to long conical. The colour of flesh may be white, yellow, orange, red, purple or dark purple.

Floral Biology

The inflorescence of carrot is a compound umbel. A primary umbel can have over 1000 flowers at maturity, whereas secondary, tertiary and quaternary umbels bear fewer flowers (Peterson and Simon, 1986). Floral development is centripetal i.e. the flower to dehisce first are on the outer edges of the outer umbellate. It produces the first order umbel called king umbel and second and third order umbels are called as secondary and tertiary umbels etc. Carrot flowers are protandrous and stigma becomes receptive on third or fourth day after anthesis. A carrot plant takes about 4-6 weeks for complete blooming. In II, III and subsequent orders, there is a certain amount of embryoless seed. Stigma receptivity starts when dehiscence is completed in whole umbel. King umbel comes first in flower and gets pollen from secondary umbels and secondary umbels get pollens from tertiary umbels and so on.

Genetic Resources

Wild carrot (Queen Anne's lace) grows in temperate regions around the world but the extent of genetic variability in wild populations has not been evaluated in *D. carota* or related species. Constance, (1971), Heywood (1983) and Mathias (1971) have contributed much to our understanding of Umbelliferae systematics, but relatively little carrot germplasm has been collected for distribution. The other *Daucus* species, which mainly occur in the Mediterranean region, should also be collected more extensively.

Both the status of carrot germplasm resources and world breeding efforts would benefit from the incorporation of carrot into an international crops center and a greater emphasis on carrot by the International Board for Plant Genetic Resources (Bioversity International).

Genetic divergence and variability

Yielding ability increases with the root length but it should not exceed 25-30 cm because longer roots are liable to damage during harvesting and handling. Conical roots are reportedly less affected by unfavorable weather conditions (Bose and Som, 1986). Gill and Kataria (1974) studied the biochemical aspects of European and Asiatic varieties of carrot. A marked variation in dry matter, total soluble solids and carotene content amongst the varieties was found. European types of carrot were superior to Asiatic types in respect of all the attributes included in the study. The bigger size of core and high percentage of moisture was mainly responsible for poor nutritive value of Asiatic type of carrot.

Saini *et al.* (1981) studied the variability and genotypic environment interactions in carrot. Low genetic coefficient of variability, moderate and high heritability and low genetic advance as percentage to mean was observed for root diameter, whereas these estimates were comparatively high for all other attributes. Singh *et al.* (1987) worked out the genetic variability, heritability and genetic advance in 40 exotic and indigenous genotypes. It is reported that selection based on phenotypic values may prove quite effective because of high heritability estimates. Chandel *et al.* (1988) conducted studies to determine the role of epistasis and evaluated the importance of additive versus dominant effects for traits not influenced by epistasis. They observed that both additive and dominance variance were significant for other characteristics. Ahmed and Tanki (1992) reported that all the characters except root girth showed high heritability along with high genetic advance, indicating predominance of additive gene effects.

Nicolle *et al.* (2004) investigated the effect of carrot genetic back ground on the content of several micronutrients and found there were large differences among the cultivars in carotenoid content (0.32 to 17 mg/100g of fresh weight). In yellow and purple carrots, lutein represented nearly half of the total caroteniods. In orange carrots, beta-carotene represented the major carotenoid (65%). The concentration of vitamin E ranged from 191 to 703 micro g/100g of fresh weight, where as the concentration of ascorbic acid ranged from 1.4 to 5.8 mg/100g.

Ragheb *et al.* (1989) reported that wide range of phenotypic variation, with high values for the genetic coefficient of variation, heritability and genetic advance, were found for root fresh weight, top fresh weight, root/top ratio and root toughness. Taka (1996) reported that plant to plant phenotypic differentiation of skin quality, extent of splitting, colour of shoulders, petiole attachment and root colour were reduced to almost zero in dense stands, differentiation in the absence of competition increased many fold (from 1.19 to 41 times), allowing efficient selection, as evidenced by the high heritability values.

Tewatia and Dudi (1999) reported that root length and shoot length exhibited highest heritability, indicating the importance of genotypic variance. High heritability together with high genetic advance was recorded for root weight and shoot weight.

Hussain *et al.* (2005) reported wide range of variability for most of the characters studied. High phenotypic and genotypic coefficients of variation were observed for leaf number, shoot length, shoot weight, core diameter, top root ratio, root yield and total carotene. The traits namely leaf number, core diameter, root yield, top root ratio, shoot weight and total carotene recorded moderate to high genetic gain along with moderate to high heritability. Highly significant differences for all the characters under studies were observed by Verma and Gupta (2005) and Gupta *et al.* (2006).

Breeding Objectives

In carrot the following broad objectives should be considered for meeting consumer preferences and demand of processing industries.

1. Developing varieties with high yield, thick fleshy roots, broad shoulder, good root colour, thin self coloured core, attractive size, earliness, proper root to shoot ratio and resistance to cracking, forking and splitting.
2. Evolving cultivars with high carotene content as well as increasing the carotene content in Asiatic types by crossing with European types.
3. Cultivars adapted to abiotic and biotic stresses.
4. Production of F_1 hybrids using stable cytoplasmic male sterile lines for uniformity, high yield and quality.

Breeding Methods

Carrot is a natural out breeder, showing severe inbreeding depression. Its scientific breeding started only recently. The methods previously employed were introduction, mass selection, combined mass-pedigree selection, line breeding etc. where much attention was given to develop breeding techniques, selection for carotene content and inheritance of root colour. Selection of inbred lines and development for high degree of uniformity has been hindered by its out breeding character and consequent inbreeding depression. Uniformity is important for mechanical harvesting and handling which has now been exploited through F1 hybrid approach, based upon male sterile lines. Different breeding methods used in carrot improvement are as below:

Introduction: Introduction is an important method of breeding in carrot. The varieties successfully introduced to India are Imperator, Danvers, Nantes and Perfection.

Mass selection: Natural variability arises due to out crossing and spontaneous mutations coupled with selection have resulted in development of cultivated carrots. Mass selection from variable population for various metric traits including colour and carotene content resulted in the selection of high yielding lines with superior root colour such as Sel-29.

Bulk population method: This method has been successfully used to evolve Pusa Kesar from a cross between Nantes and Local Asiatic.

Line breeding: This method has been successfully exploited for isolation of a high yielding, orange coloured carotene rich variety Chamman from a heterogeneous local population.

Mutation breeding: Chemical mutagens (0.025-0.6 %) NEU (N-nitroso-N-ethyl urea) and 0.012% NMU (N-nitroso + N-M-methyl urea) were successfully used to develop male sterile lines in carrot.

Polyploidy breeding: Tetraploids (2n=36) and octaploids (2n=72) have been developed in carrot, but they have only limited utility in carrot.

Heterosis breeding: Heterosis has been reported for earliness, root length, root yield, carotene content, top weight, core diameter and root diameter. In F_2 considerable loss in vigour has also been observed. Bhagchandani and Choudhury (1971) studied the heterosis and inbreeding depression in Asiatic carrot. Timin and Vasilevskii (1995) considered the fertile lines 1163-2-915, 639-920-123, 521-528-587 and 639-656-569 and the CMS lines 1124-748-94, 1184-639-123, 1141-707-534 and 787-96-207 as promising for heterosis breeding. The hybrids so for produced with these lines out yield standard varieties by 30-50% and produced uniform roots with a high content of carotene.

Duan *et al.* (1996) observed positive heterosis for plot yield, single root weight, root length and root width. These traits also had high positive GCA values. Gauciene and Viskelis (1997) reported that F_1 hybrids out yielded the standard variety Satrija by 28-250% and 10-40% high for carotene content over Satrija. Suh *et al.* (1999) reported that F_1 hybrids showed strong hybrid vigour in quantitative characters such as leaf and root weight. The increase of root weight in F_1 hybrids was more influenced by increase in root diameter than that of root length. Verma *et al.* (2002) reported remarkable heterosis for length, girth, unmarketable and marketable yield of roots over the standard check and top parent. In order of merit the three best performing F_1 hybrids (240602 x N) x 1061, (28 x PY) x 1060 and (95 x N) x 1061 exhibited heterosis percentage of 59.79, 44.52 and 27.85, respectively for marketable yield over the standard check.

Desirable characters for selection of inbreds

1. Maturity: It can be judged by the number of days from sowing to the date when the roots have attained the maximum marketable size.

2. Shoot growth or top characteristic: At maturity the shoot or leafy growth generally known as top should be small or light. A heavy top should be avoided as far as possible.

3. Root characteristics:

- **Colour of skin:** In India, carrots of red or scarlet / orange are preferred. Roots for colour are selected during full day light when they are at fully grown marketable stage.

- **Flesh colour and thickness:** The colour of flesh in commercial varieties of carrot is generally red or scarlet or orange. The flesh must be thick in carrot.
- **Core colour and thickness:** A thin and self-coloured core is a better character for selection. Thick yellow core as occurring in most of the Asiatic carrots should be avoided.
- **Shape of root:** Selection should be made for a broad-shouldered, cylindrical, uniformly tapering or stump rooted carrot with non-branching or forking habit.
- **Size of root:** The size of root in carrot is much influenced by environmental conditions, particularly soil, temperature, moisture etc. The size is expressed in terms of length and girth of roots.
- **Carotene content:** Carotene is an important character for selection, particularly in the European varieties. The Indian varieties have generally *anthocyanin* pigmentation in roots. The carotene content is expressed as total *carotenoids* which comprises alpha and beta – carotenes up to 90% whereas *xanthophylls* and colourless polyenes constitute the remainder part of carotenoids. In selecting for carotene content, the stage of development and shape indicating the degree of maturity should be taken into account.
- **Sugar and dry matter content:** Sugar and dry matter contents are directly related to total soluble solids. With the help of a hand refrectometer total soluble solids (TSS) can be determined for this purpose.
- **Longitudinal cracking of roots:** Cracking of roots, particularly in the Nantes types is a serious problem. It is governed by a single dominant gene, but influenced by environment like soil moisture, root age, wider spacing and undecomposed FYM.

Breeding for quality improvement

Efforts to breed for improved quality have also progressed in recent years. Phenotypic recurrent selection has yielded carrot breeding stocks averaging 500 ppm carotene in contrast to 100-150ppm in widely grown cultivars (Simon *et.al.* 1989). The same breeding method has indicated that heritability of root sugar content is 0.40 to 0.45 (Stommel and Simon, 1989). Singh *et al.* (2002) studied root morphology and carotene content in Asiatic and European carrots and found maximum TSS content in the Asiatic type collections namely Local Rewari Black and Local Jaipur Black ranging from 9.0-9.1%. The European types had higher carotene content ranging from 2.039 to 14.166 mg/100g fresh weight to that of Asiatic type (0.594 to3.827 mg/100g). Similarly, vitamin A content (retinol) was also higher in European types, ranging from 3399 to 23610 IU/ 100g fresh weight whereas Asiatic types recorded only 990 to 6378 IU/100g fresh weight. The gene for the super sweet protein *thaumatin* II from *thaumatococeus daniellii* was used for flavour improvement in carrot by Dolgon *et.al.* (1999).

Further research on investigating the genetics and biochemistry of carrot quality is required as demand for more nutritious and appealing foods is increasing. The application of mutagenesis as a tool in generating new genetic variation has also been examined (Al-Safadi and Simon, 1990). Interestingly very few true breeding variants were identified in M_3 progeny.

Resistance Breeding

Breeding has played an important role in producing varieties resistant to diseases and pests (Table. 1). Resistance to downy mildew (*Erysiphe heraclii*) was found in *D. carota* sp. *dentatus* and gene Eh was identified.

The development of nematode resistance in carrots has received much attention. Huang *el al.* (1986) found resistance to *Meloidogyne incognita* in 'Brasilia' and partial resistance in 'Kuronan'. The heritability of resistance in half–sib progenies was estimated to be 0.16 to 0.48 for galling depending on variety. The inheritance of partial resistance to *Meloidogyne hapla* was polygenic (Wricke *et al.,* 1985) and was estimated from fibrous root weight (Frese and Weber, 1984). It is interesting to note that, wild carrot *D. carota* sub sp. hispanicus has the source of resistance (Frese, 1983) which can reduce pesticides to control nematodes.

Boiteux *et al.* (1993) reported that recurrent selection based on half sib progeny tests would be more effective than simple phenotypic recurrent selection in increasing the resistance level of Brasilia to *Alternaria dauci*. Villineuve *et al.* (1997) reported that Nandor appeared to be the most resistant and Tino the least resistant against *Pythium* disorders. However, the reaction of the varieties was strongly dependent on the *Pythium* species and isolates used. Simon and Strandberg (1998) reported that a five parent (B 3475, B 5280, B 5344, B 6253 and B 9695) diallel evaluation for resistance to *Alternaria* leaf blight suggested a preponderance of additive variation with some dominant gene action for resistance.

Simon *et al.* (2000) determined the inheritance of resistance to root-galling and reproduction by *M. javanica* from the cross BR-1252 x B-6274 (a susceptible inbred line). On the basis of galling and egg production on fibrous roots and the segregation pattern they reported simply inherited dominant resistance in selection BR-1252. For protecting against fungal and microbial attack, the plant defentin gene (PD) has been transferred to carrot by *Agrobacterium* mediatel transformation (Dolgor *et. al.* 1999).

Vieira *et al.* (2003) estimated heritability and gain from selection of traits associated with field resistance to multiple root knot nematode species in carrot. Brasilia was among the genotypes selected for the following cycles of recombination reinforcing the notion that this cultivar is one of the most promising sources of stable, wide spectrum field resistance to *Meloidogeyne* species in (*D. carota*).

Table 1. Source of resistance for diseases and pests

S. N.	Diseases / Pests	Resistant source
1.	*Cercospora carotae*	Inbred line WCRI
2.	Motley virus	Sweet Crop, Top Weight
3.	CMV	Virginia Savoy
4.	Pythium *spp.*	Nantes, Imperator, Danvers
5.	*Heterodera carotae*	Vilmorium 66
6.	Tip burn (physiological disorder)	Scarlet, Nantes
7.	*Meloidogyne incognita*	Brasilia,Kuronan
8.	*Meloidogyne javanica*	BR-1252.

Role of Biotechnology

Tremendous progress has been made in biotechnological field in carrot for improvement of the crop in respect of different desirable attributes. However, the techniques developed have not been fully applied to the improvement of commercial carrot varieties.

Carrot exhibits irregular seed germination and lack of genetic uniformity. Since carrot cell and somatic embryo can be grown in large number, they offer a great potential for large scale clonal propagation (Ammirato 1984). Matsubara *et al.* (1995) induced callus and adventitious embryoids development from carrot microspores after one month of anther culture. Callus regenerated from anthers of different stages on the media produced many adventitious buds and roots.

Transfer of cytoplasmic male sterility by means of protoplast fusion has been tried by several workers. Three potential novel CMS sources with different phenotypes have been identified in carrot based on the cytoplasm of *D. carota* sub sp. *gemmifera, maritimus* and *gadacie*. These new CMS sources are associated with brown anther and petaloid types of sterility (Nothnagel *et al.,* 1997) and could be used for production of parents for hybrid seed production.

A valuable male sterile line can be multiplied using clonal propagation for producing female plants for hybrid seed production. Application of haploid plant production offers tremendous potential for carrot breeding and crop improvement. Protoplasts derived from haploid plants can be very useful. This particular technology not only increases the range of variability in carrot plant but also generate hybrid hitherto impossible by means of conventional breeding. Through cell culture there is immense potential for the direct selection of specific traits such as disease resistance, herbicide resistance, cold tolerance and salt tolerance. It is also possible to find among regenerated plants variants with altered plant and taproot characteristics including better taste, colour, texture, shape, yield etc.

Gorecka and Krzyzanowska (2001) elaborated technology for producing homozygous lines of carrot using anther culture and regenerated carrot embryos on MS medium. The high frequency of CMS among regenerates from '*selendro*' carrot cell cultures provided an efficient method for developing sterile M tandem lines and corresponding new hybrid varieties (Wright *et al.*, 1996). Matsubayashi *et al.*, (2004) reported that chemical nursing using PSK showed promise as a tool for basic research in carrot biology and biotechnological applications. A differential RNA approach to trying to understand the nature of mechanisms causing variability in plant cell cultures was also reported (Marshall *et al.*, 2002).

The RAPD markers for genotype identification of carrot lines and F_1 hybrids have been reported to be useful as though it appeared to be limited for hybrid purity testing due to the high level of heterogeneity within the analysed hybrids. Nakajima *et al.* (1999) taxonomically identified mitochondrial and nuclear genomic diversities of 8 carrot varieties including 6 pure lines and 2 cytoplasmic male sterile (cms) lines using PCR with 19 RAPD primers.

Bradeen and Simon (1998) reported a Y2 locus controlling carotene accumulation in the root xylem core, and identified six AFLP fragments linked to the Y2 locus through a combination of F_2 mapping and bulked segregent analysis while Bradeen *et al.*, (2002) analyzed molecular diversity of cultivated carrot (*D. carota*) and wild *Daucus* populations and reported extensive variation within *D. carota*. Grzebelus *et.al.*, (2001) reported use of AFLP markers for the identification of carrot breeding lines and F_1 hybrids, and to examine their genetic relationship and produced markers useful for testing hybrid seed purity.

Genetic transformation in carrot has also been carried out employing various methods such as *Agrobacterium* mediated transformation with *Agrobacterium tumefaciens* and *A. rhizogenes*, using electrical field, electroporation, microinjection and protoplast transformation. Punja and Raharjo (1996) transformed carrot cultivars, Nanco and Golden state, with 2 chitinase genes from tobacco and found that the rate and extent of lesion development due to inoculation with *Botrytis linerea, Rhizoctonia solani* and *Sclerotium rolfsii* were significantly lower in the transgenic carrot than in non – transgenic carrot.

Punja and Chen (2004) further developed transgenic carrot plants expressing a *thaumatin like* protein (TLP) from rice with potential antifungal activity through *Agrobacterium tumefaciens* mediated genetic transformation. Two genes encoding *phosphinothricin* acetyltransferase (bar) and *hygromycin phosphotransferase* (htp) were used as selectable markers. The transgenic lines showed varying levels of tolerance to phosphinothricin (glufosinate), the active ingredient in several commercially available herbicides.

The advanced biotechnological techniques are in place for substantial crop improvement in carrot but they need to be integrated with conventional breeding programme.

Development of Varieties/ Hybrids

There are two distinct groups of carrot, viz. tropical or Asiatic and temperate or European. The temperate types are mostly orange and are rich in carotenes. They form roots both under tropical and temperate conditions but set seed only under temperate conditions as they need low temperature of 5-8°C for 40-60 days before bolting. The

Asiatic types are high yielding and produce seed under tropical conditions. But they are poor in carotene content and other quality attributes. The wild forms of carrot are annual but the cultivated types believed to have been derived from the wild types are biennial.

Table 2. List of improved varieties of carrot

Variety	Organization	Pedigree	Brief Characteristics
		Tropical or Asiatic varieties	
Pusa Kesar	IARI, New Delhi	Selection from cross Local Red x Nantes Half Long	Roots are scarlet in colour with sufficiently red coloured core and richer in carotene (38 mg/100g). Roots stay longer in field without bolting.
Pusa Meghali	IARI, New Delhi	Selection from cross Pusa Kesar x Nantes	Long orange roots tapering with self coloured core, richer in carotene (11571 IU/100g) and produces seeds in plains. It is suitable for early sowing and takes 110-120 days to attain harvest maturity.
Sel-233	PAU, Ludhiana	Selection from cross Nantes x No. 29	Roots are long, smooth, semi cylindrical, orange with light orange coloured core. It takes 90-100 days for root formation.
Sel-29	PAU, Ludhiana	Selection	Roots are long, tapering, thin and light red in colour.
		Temperate or European types	
Nantes Half Long	IARI, Regional Station, Katrain	Introduction	Suitable for sowing during winter months. Roots are cylindrical, stumpy, well shaped with abrupt tail, orange scarlet with self coloured core, flesh sweet with good flavour. It takes 110-120 days for root formation.
Zeno	State Dept. Hort., Tamil Nadue	Introduction	Roots are deep orange with self coloured core. It takes 110-120 days for root formation.
Pusa Yamdagni	IARI, Regional Station, Katrain	Selection from cross (EC 9981 x Nantes).	Roots are 15-16cm long, orange with self coloured core, slightly tapering with stumpy to semi-stumpy ending.
Chantenay	Exotic	Introduction	Roots are conical to near conical, thick, attractive orange, smooth, thick at the shoulders core indistinct, flesh tender, sweet and fine textures.
Early Nantes	Exotic	Introduction	Almost cylindrical roots terminating abruptly in small, thin, 12-15 cm long, fine texture, orange flesh.
Chamman	SKUAST-K, Shalimar	Line breeding	Roots are long, cylindrical, semi blunt with orange coloured core, tender, sweet fine textured with less cracking and forking. Its yield potential is 25 t/ha.
Shalimar carrot-1	SKUAST-K, Shalimar	Selection	High yielding (228.0 q/ha) temperate type with 21.75 percent yield advantage over best National check Early Nantes and Pusa Yamdagni, early maturing with high percentage of uniform cylindrical orange blunt roots with small self coloured core, roots are tender, sweet and fine textured, rich in carotenes (56.1 mg/100g), dry matter (12.4%) and TSS (10.5^{0}B).

Future Prospect

Carrot is an important source of carotene. There is need to develop heat tolerant carrot lines rich in carotene. Breeding varieties suitable for non- traditional areas like warm-humid tropics would be another important and useful step. Breeding for resistance also needs to be further intensified.

References

Allard, R.W., 1999. *Principles of Plant Breeding* (2nd Edn.). John Wiley and Sons, New York, USA.

Ahmed, N. and Tanki, M. I. (1992). Variability, heritability and genetic advance in carrot (*Daucus carota* L.)- A Note. *Haryana Journal of Hort. Sci.*, **21(3-4):** 311-313.

Al-Safadi, B. and Simon, P.W. 1990. The effects of gamma irradiation on the growth and cytology of carrot (*D. carota* L.) tissue culture. *Exp. Environ. Bot.*, **30:** 361-365.

Ammirato, P.V. 1984. In: Cell culture and somatic cell genetics of plants. Vol. I (*Ed.* I.K. Vasil). Academic Press, New Delhi, pp 139-151.

Bhagchandani, P. M and Choudhury, B. 1971. Heterosis and inbreeding depression in carrot (*Daucus carota* L.) *Progressive Hort.*, **2:** 65-75.

Boiteux, L. S., Della, V.P.T. and Reifschneider, F.J.B. 1993. Heritability estimate for resistance to *Alternaria douci* in carrot. Plant Breeding **110(2):** 165 – 167.

Bose, T. K. and Som, M. G. 1986. Vegetable Crops in India. Naya Pradash, Calcutta.

Bradeen, J.M. and Simon, P.W. 1998. Conversion of an AFLP fragment linked to the carrot Y2 locus to a simple. Co dominant, PCR bared form. *Theor. Appl. Genet.*, **97(5-6):** 960-967.

Bradeen, J.M., Bach, T.C., Briard, M., Le–Clere, V., Grzebelus, D., Senalik, D.A. and Simon, P.W. 2002. Molecular diversity analysis of cultivated carrot (*D. carota* L.) and wild *Daucus* populations reveals a genetically nonstructural composition. *J. Am. Soc. Hortic. Sci.,* **127(3):** 383-391.

Chandel, Singh, K. and Rattan, R.S. 1988. Epistasis, additive and dominance variation in carrot. *Veg. Sci.,* **15:** 31-32.

Dolgon, S.V., Lebeder, V.G., Anisimova, S.S., Lavrova, N., Serdobinskiy, L.A, Tjukarin, G.B. Shadenkov, S.A. and Lunin, V.G., 1999. Phytopathogen resistance improvement of horticultural crops by plant – defensin gene introduction. Genetics and breeding for crop quality and resistance. Proceeding of the XV EUCARPIA Congress. Viterbo, Italy. Pp 111-118.

Dolgon, S.V., Lebeder, V.G., Firsor, A.P., Taran, S.A. Tjukavin, G.B, Scarascia M.G.T., Poreceddu, E. and Pagenotta, M.A. 1999. Expression of thaumatin II gene in horticultural crops. Genetics and breeding of the EUCARPIA Congress, Viterbo, Italy. pp 165-172.

Duan, Y., Wong, Y., Ren, X.Y., and Du, G.Q. 1996. Analysis of heterosis and combining ability for main yield characteristics in carrot. *China Vegetables*. **2:** 13-15.

FAO STAT, 2004. Agriculture Data base, Rome, Italy.

Frese, L. 1983. Resistance of the wild carrot (*D. carota* spp. hispanicus) to the root knot nematode (*M. hapla*). Gartenbauwissenschak. **48:** 259-264.

Frese, L. and Weber, W.E. 1984. A method for the differentiation of breeding material of carrot with partial resistance to *M. hapla. Z. Pflanzenzucht.* **91:** 396-402.

Gauciene, O. and Viskelis, P. 1997. Breeding of carrot hybrids F_1 in Lithuania. Journal of *Applied Genetics.* **38(A):** 186-192.

Gill, H.S. and Kataria, A.S. 1974. Some biochemical studies in European and Asiatic varieties of carrot (*Daucus carota* L). *Current Science* **43:** 184-185.

Gorecka, K. and Krzyzanowska, D. 2001. The use of androgenesis in vitro to obtain homozygous lines of vegetable plants. *Vegetable Crops Research Bulletin*, **54(1):** 7-11.

Grzebelus, D., Senalik, D., Jagosz, B., Simon, P.W. and Michalik, B. 2001. The use of AFLP markers for the identification of carrot breeding lines and F_1 hybrids. *Plant breed. Berlin*: **120(6):** 526-528.

Gupta, A.J., Verma, T.S., Sethi, S. and Singh, G. 2006. Evaluation of European carrot genotypes including F1 hybrids for their roots quality, yield and nutritive characters. Indian Journal of Horticulture. **63(1):** 48-52.

Hansche, P.E. and Gobelman, W.H. 1963. Phenotypic stability of pollen sterile carrots (*D. carota* L.) *Proc. Am. Soc. Hort. Sci.* **82:** 341-350.

Huang, S.P. Della, V.P.T. and Ferreira, P.L. 1986. Varietal response and estimates of heritability of resistance to *Melidogyne Javanica* in carrots. *J. Nematol.* **18:** 496-501.

Hussain, K., Singh, D. K., Ahmed, N., Nazir, G. and Rasool, R. 2005. Genetic variability for qualitative and quantitative traits in carrot (*Daucus carota* L.) *Environment and Ecology,* **23(3):** 644 - 647.

Hussan, B. and Ahmed, N. 1999. Vegetables for augmenting green fodder resources during winters of temperate Kashmir. Indian Farming. March 1999: 17-23.

Marshall, G.B., Smith, M.A.L., Lee, C.K.C., Deroles, S.C. and Davies, K.M. 2002. Differential gene expression between pigmented and non- pigmented cell culture lines of *Daucus carota. Plant Cell, Tissue and Organ Culture.* **70(1):** 91-97.

Matsubara, D., Dohya, N. and Murakami, K. (1995) Acta Horticulture, **392:** 129-137.

Matsubayashi, Y., Goto, T. and Sakagami, Y. 2004. chemical nursing: phytosulfokine improves genetic transformation efficiency by promoting the proliferation of surviving cell on selective media. *Plant Cell Report.* **23(3):** 155-158.

Nakajima, Y., Yamamoto, T., Muranaka, T. and Oeda, K. (1999) Theor. Appl. Genet., **99:** 837-843.

Nicolle, C., Simon, G., Rock., E., Amouroux, P. and Remesy. C. 2004. Genetic variability influences carotenoid, vitamin, phenolic and mineral content in white, yellow, purple, orange and dark orange carrot cultivars. *Journal of the American Society for Horticultural Sciences.* **129(4):** 523-529.

Nothnagel, T., Straka, P. and Budhn, H. (1997) J. Appl Genet., **38A :** 172-177.

Park, S.H, Kim, C., Pike, L.M., Smith, R.H. and Hirschi, K.D. 2004. Increased calcium in carrots by expression of an Arabidopsis H^+ / Ca^{2+} transpoter. *Molecular Breeding.* **14(3):** 275-282.

Peterson, C.E. and Simon, P.W. 1986. Carrot breeding. In: Breeding Vegetable Crops (*Ed.* M.J. Bassett), AVI Publ. Co. pp 321-356.

Punja, Z.K and Chen, W.P. 2004. Transgenic carrots expressing enhanced tolerance to herbicide and fungal pathogen infection. *Acta-Horticulture*, **637:** 295-302.

Ragheb, W. S., Mahamedein, S. E. and El Raggal, T. A. 1989. A privinary study on phenotypic and genotypic variability in carrot (*Daucus carota* L.) *Assiut Jounal of Agricultural Sciences.* **20(4):** 129-139.

Saini, S. S., Karla, B. N. and Rastogi, K.B. 1981. Genetic variability and genotype environment interaction in carrot (*Daucus carota* L.) *Veg. Sci.* **8:** 93-99.

Simon, P.W. 1993. Carrot (*Daucus carota* L) In: G. Kalloo and B.O. Bergh (Eds.). Genetic Improvement of Vegetable Crops pp 479-484. Pergamon Press.

Simon, P.W. and Strandberg, J. O. 1998. Diallel analysis of resistance in carrot to *Alternaria* leaf blight. *Journal of the American Society for Horticultural Sciences* **123(3):** 412-415.

Simon, P.W. Wolff, X.Y. Peterson, C.F., Kamnalohr, D., Rubarzky, V., Strandberg. J., Barset, M. and White, T. 1989. High carotene mass carrot population. *Hort Science*, **24:**174-180.

Simon, P.W., Matthews, W.C. and Roberts, P.A. 2000. Evidence for simply inherited dominant resistance to *Meloidogyne javamica* in carrot. *Theoretical and Applied Genetics.* **100(5):** 735-742.

Singh, J., Sing, B., Kalloo, G. and Singh, J. 2002. Root morphology and carotene content in Asiatic and European carrot (*D. Carota* var. sativa). *Indian J. Agric. Sci.*, **72(4):** 225-227.

Singh, K., Chandel, Rattan, R. S. and Singh, A. K. 1992. Genetics of quality traits in carrot (*Daucus carota* L.). *Veg. Sci.* **19(2):** 181-185.

Singh, R., Sukhjia, B.S. and Hundal, T.S. 1987. Variability studies in carrot (*Daucus carota* L). *Veg. Sci.* **14:** 33-36.

Stommel, J.R. and Simon, P.W. 1989. Phenotypic recurrent selection and heritability estimates for total dissolved solids and sugar type in carrot. *J. Am. Soc. Hortic. Sci.* **114:** 695-702.

Suh, Y.K. Youn, G. H., Cho, Y.H. and Peak, K. Y. 1999. Expression of some quantitative and qualitative characters among breeding lines and their F_1 hybrids in carrot. *Journal of the Korean Society for Horticultural Science*. **40(6):** 697-701.

Taka, M.E., 1996. Effects of competition on phenotypic expression and differentiation of five quality traits of carot (*Daucus carota* L.) and their implications in breeding. *Scientia-Horticulture*, **65(4):** 335-340.

Tewatia, A.S. and Dudi, B. S. 1999. Genetic variability and heritability studies in carrot (*Daucus carota* L.). *Annals of Agri. Bio Research,* **4(2):** 213-214.

Thompson, D.J. 1961. *Proc. Am. Soc. Hortic. Sci.* **72.**332-338.

Timin, N. I. and Vasilevskii, V. A. 1995. Carrot lines for heterosis breeding using cytoplasmic male sterility (cms). *Kartofel-i-Ovoshchi*, **3:** 27-28.

Verma, T.S. and Gupta, A.J. 2005. Performance of temperate carrot *(Daucus carota L.)* genotypes including F1 hybrids and varieties in multi-location trials. Indian Journal of Agricultural Sciences. **75(4):** 298-300.

Verma, T.S., Sharma, S.C. and Lal, H. 2002. Heterosis in temperate carrot (*Daucus carrota* L.) *Vegetable Science,* **29(2):** 119-122.

Vieira, J. V., Charchar, J.M., Aragao, F.A.S. and Boitenx, L. S. 2003. Heritability and genetic gain from selection for field resistance against multiple root-knot nematode species (*M. incognita* race 1 and *M. Javahica*) in carrot. *Euphytia*, **130(1):** 11-16.

Villineuve, F., Bose, J. P., Rouxel, F. and Breton, D. 1997. Intra and inter specific variability of *Pythium* and possibility of varietal resistance improvement in carrot. *Journal of Applied Genetics*. **38(A):** 71-80.

Wricke, G. Frere, L., Sandmann, M. and Kraus, C, 1985. Genetic and epidemiological investigations on the resistance of carrot (*D. carota L.*) to the northern root – knot nematode (*M. hapla*). Deutschn Forssh. 1-5.

Wright, J., Reilley, A., Labriola, J., Kut, S. and Ortan, T. 1996. Petloid male sterile plants from carrot cell cultures. *Hort. Science,* **31(3):** 421-425.

Yawalkar, K.S. 1980. Vegetable Crops of India. Agri-Horticultural Publishing House, Nagpur.

❑❑❑

Chapter – 32

Marker Assisted Selection (MAS) for Genetic Improvement in Cucurbitaceous Vegetables

T.K. Behera

Introduction

The traditional approach to transferring genes from wild to cultivated species is based on interspecific hybridization followed by selection of hybrids that combine the trait with the cultivated genetic background. This breeding strategy is achieved by various backcross generations in which the selected hybrids at each generation are crossed back to the cultivated genotype with the aim of reducing the wild genome and its undesirable traits. The use of molecular markers has allowed this breeding approach to be greatly improved, since these markers directly reveal genetic variability through DNA analysis (Staub *et al.* 1996), and therefore their detection is not influenced by environmental effects. Since the development of numerous molecular markers for plant genometatic analysis, the possibility to select the genotype instead of the phenotype has been closely examined, leading to the concept of molecular marker-assisted selection (MAS; Paterson *et al.* 1991). The basic concept of MAS for crop improvement has been well discussed by Collard *et al.* (2005) which is most appropriate for this chapter. The most widely used markers suitable for MAS are RFLP, RAPD, AFLP, and SSR. Their common origin is point mutation or chromosome rearrangements that were accumulated during the evolution of the species without negatively influencing their survival and reproduction. The choice of the most suitable markers for MAS can differ and depends on the labor required for their detection, possibility of revealing single or multiple loci, dominant or co dominant nature and costs.

The application of molecular marker-assisted selection to the introgression of genes from one genotype (the dònor genotype) to another one (the recurrent genotype) through a backcross breeding scheme clearly points out the great advantages of the use of molecular markers for improving the cultivated varieties, as described by Tanksley *et al.* (1989). The theoretical model proposed that 99% of the cultivated genome in tomato can be recovered with only three backcross generations using MAS, instead of the six to seven generations required to recover the same percentage of genome without the use of molecular markers.

Molecular markers closely linked to the gene controlling the trait to be transferred allow precocious screening to be performed directly on DNA extracted from young leaves (positive selection) without waiting for the specific developmental stage at which the trait is expressed, which could also be flowering time or complete fruit ripeness. This leads to a reduction in both selection time and space, the advantage of which is clearly evident when the trait under selection requires the analysis of thousands of genotypes. Indeed, the concurrent analysis of more than one trait and the realization of more than one selection cycle per year are also possible. Without the use of molecular markers this screening at each backcross generation is based on morphological analysis and is laborious, especially for traits that are not easily scored. On the contrary, the availability of molecular markers specific for the wild donor species allows the genotypes that recovered the highest recurrent genome to be selected directly through DNA analysis. These genotypes serve as parents to obtain the next backcross generation.

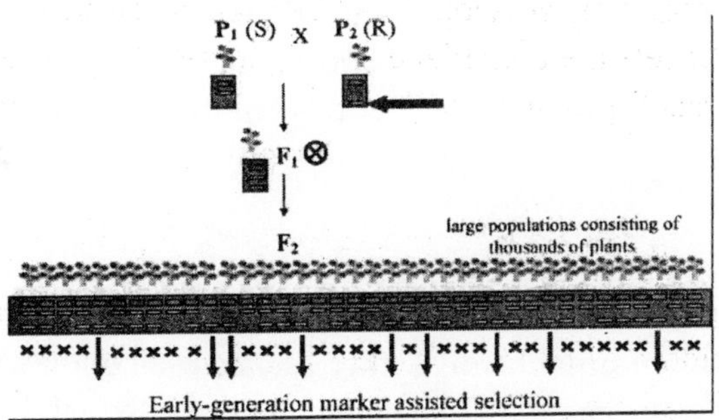

Selecting plants in a segregating progeny that contain appropriate combinations of genes is a critical component of plant breeding. Moreover, plant breeders typically work with hundreds or even thousands of populations, which often contain large numbers. MAS may greatly increase the efficiency and effectiveness in plant breeding compared to conventional breeding methods. Once markers that are tightly linked to genes or QTLs of interest have been identified, prior to field evaluation of large numbers of plants, breeders may use specific DNA marker alleles as a diagnostic tool to identify plants carrying the genes or QTLs (Michelmore, 1995; Ribaut *et al.*, 1997.

The Advantages of MAS include

- time saving from the substitution of complex field trials (that need to be conducted at particular times of year or at specific locations, or are technically complicated) with molecular tests;
- elimination of unreliable phenotypic evaluation associated with field trials due to environmental effects;
- selection of genotypes at seedling stage;
- gene 'pyramiding' or combining multiple genes simultaneously;
- Avoid the transfer of undesirable or deleterious genes ('linkage drag'; this is of particular relevance when the introgression of genes from wild species is involved).
- selecting for traits with low heritability
- Testing for specific traits where phenotypic evaluation is not feasible (e.g. quarantine restrictions may prevent exotic pathogens to be used for screening).

Cucurbits have small genomes; the genome size of melon is 4.5×10^8 (bp), cucumber (3.67×10^8 bp), watermelon (4.25×10^8 bp) and squash (5.2×10^8 bp). During the past decade, significant progress has been achieved in marker development, genetic mapping, and molecular breeding in cucurbits. Most progress has been made in melon and cucumber, which recently culminated in the cloning of the first cucurbit disease resistance genes in melon (Joobeur *et al*., 2004; Taler *et al*., 2004). Molecular markers are identified as potentially tool and subsequently developed into efficient and effective genotyping systems in cucurbits. These markers are then placed on a genetic map and associated with QTL through progeny analysis for their subsequent use in Marker assisted selection (MAS).

Steps in MAS

Marker-assisted selection (MAS) is a method whereby a phenotype is selected on the basis of genotype of a marker. However, the markers identified in preliminary genetic mapping studies are seldom suitable for marker-assisted selection without further testing and possibly further development. Markers that are not adequately tested before use in MAS programs may not be reliable for predicting phenotype, and will therefore be useless. Generally, the steps required for the development of markers for use in MAS includes: high resolution mapping, validation of markers and possibly marker conversion.

High-resolution Mapping of QTLs

The preliminary aim of QTL mapping is to produce a comprehensive 'framework' (also 'skeletal' or 'scaffold' linkage map) that covers all chromosomes evenly in order to identify markers flanking those QTLs that control traits of interest. There are several more steps required, because even the closest markers flanking a QTL may not be tightly linked to a gene of interest (Michelmore, 1995). This means that recombination can occur between a marker and QTL, thus reducing the reliability and usefulness of the marker. By using larger population sizes and a greater number of markers, more tightly-linked markers can be identified; this process is termed 'high-resolution mapping' (also 'fine mapping'). Therefore, high-resolution mapping of QTLs may be used to develop reliable markers for marker assisted selection (at least <5 cM but ideally <1 cM away from the gene) and also to discriminate between a single gene and several linked genes (Michelmore, 1995; Mohan *et al*., 1997).

There is no universal number for the appropriate population size required for high-resolution mapping. However, population sizes that have been used for high-resolution mapping have consisted of>1000 individuals to resolve QTLs to distances between flanking markers of <1 cM (Blair *et al*., 2003; Chunwongse *et al*., 1997; Li *et al*., 2003). Bulked-segregant analysis may also be used to identify additional markers linked to specific chromosomal regions (Campbell *et al*., 2001; Giovannoni *et al*., 1991). However, the extent to which framework maps can be saturated depends on the size of the population used to construct the map. In many cases, the sizes of segregating populations being used are too small to permit high resolution mapping, since smaller populations have fewer recombinants than larger populations (Tanksley, 1993). High-resolution maps of specific chromosomal regions may also be constructed by using NILs (Blair *et al*., 2003). Markers that are polymorphic between NILs and the recurrent parent should represent markers that are linked to the target gene and can be incorporated into a high-resolution map.

In cucurbits, linkage maps with molecular markers were first developed for cucumber by using intra- and interspecific (between C. *sativus* var. *sativus* and C. *sativus* var. *hardwickii)* F_2 populations (Kennard *et al.* 1994). Because of the small number of markers in these maps, the average distance between markers was significantly larger than in maps generated later (Park *et al.* 2000; Bradeen *et al.* 2001; Fazio *et al.* 2003b). The densest of these contained 347 markers and covered 816 cM, within a range of 750-1000 cM estimated by Staub and Meglic (1993). In squash, Brown and Myers (2002) created a RAPD map from BC_1 of *C. pepo* (A0449) × *C. moschata* (Nigerian Local) with 148 markers in 28 linkage groups covering 1954 cM.

Baudarcco-Arnas and Pitrat (1996) produced the first genetic map of melon with 102 RAPD and RFLP markers followed by Wang *et al.* (1997) using 188 predominantly AFLP markers. Perin *et al.* (2002) constructed a composite map consists of 668 AFLP, IMA, and phenotypic markers. Although not as high-throughput as AFLP, RFLP markers were the predominant markers used in the map by Oliver *et al.* (2001). Being co-dominant, RFLP is efficient in mapping F_2 populations and may also be useful in comparative mapping. Genome mapping efforts in watermelon are more recent, although the first map was developed by Hashizume *et al.* (1996) and they also released a high-density map of watermelon with 554 markers in 2003, most of which were RAPD markers. Since cucurbit breeding places great emphasis on disease resistance, mapping populations used often segregated for more than one disease resistance gene. In melon,

MR-1 used by Wang *et al.* (1997) was resistant to fusarium wilt (*Fom-l* and *Fom-2),* downy and powdery mildews (Baudarcco-Arnas and Pitrat 1996 and Perin *et al.* 1998, 2002).

Validation of markers

Generally, markers should be validated by testing their effectiveness in determining the target phenotype in independent populations and different genetic backgrounds, which is referred to as marker validation. In other words, marker validation involves testing the reliability of markers to predict phenotype. This indicates whether or not a marker could be used in routine screening for MAS. Markers should also be validated by testing for the presence of the marker on a range of cultivars and other important genotypes. Some studies have warned of the danger of assuming that marker-QTL linkages will remain in different genetic backgrounds or in different testing environments, especially for complex traits such as yield (Reyna and Sneller, 2001). Even when a single gene controls a particular trait, there is no guarantee that DNA markers identified in one population will be useful in different populations, especially when the populations originate from distantly related germplasm for markers to be most useful in breeding programs, they should reveal polymorphism in different populations derived from a wide range of different parental genotypes (Langridge *et al.*, 2001).

Marker conversion

There are two instances where markers may need to be converted into other types of markers: when there are problems of reproducibility (e.g. RAPDs) and when the marker technique is complicated, time-consuming or expensive (e.g. RFLPs or AFLPs). The problem of reproducibility may be overcome by the development of sequence characterised amplified regions (SCARs) or sequence-tagged sites (STSs) derived by cloning and sequencing specific RAPD markers (Jung *et al.*, 1999; Paran & Michelmore, 1993). Dominant markers (RAPD and AFLP) were useful initially in the development of moderately saturated maps (Serquen *et al.*, 1997; Bradeen *et al.*, 2001), but are not preferred in breeding programs. The mapped RAPD loci were, nevertheless, strategically important during early map construction (Serquen *et al.*,1997) in cucumber and were therefore, subjected to conversion to more preferable sequence amplified characterized region (SCAR) markers by silver staining-mediated sequencing (Horejsi *et al.*, 2000). Although 62 (83%) of the 75 RAPDs were successfully cloned, only 48 (64%) RAPD markers were successfully converted to SCARs markers and 11 (15%) of these reproduced the polymorphism observed with the original RAPD marker. The emergence of automated sequencing technologies made possible the development of codominant SSR and SNP technologies (Fazio *et al.*, 2003a) and the reassessment of RAPD to SCAR as well as SCAR to SNP marker conversion. A total of 39 new markers (SCAR and SNP) have recently been developed in cucumber, seven of which have proven effective in MAS.

Selection of QTLs for MAS

Theoretically, all markers that are tightly linked to QTLs could be used for MAS. However, due to the cost of utilizing several QTLs, only markers that are tightly linked to no more than three QTLs are typically used (Ribaut and Betran, 1999), although there have been reports of up to 5 QTLs being introgressed into tomato *via* MAS (Lecomte *et al*., 2004). Even selecting for a single QTL *via* MAS can be beneficial in plant breeding; such a QTL should account for the largest proportion of phenotypic variance for the trait (Tanksley, 1993). Furthermore, all QTLs selected for MAS should be stable across environments (Hittalmani *et al*., 2002).

Like that of other crops, several horticultural traits in cucurbits are also controlled by quantitative trait loci (QTLs). The goal of QTL mapping is to dissect the complex inheritance of quantitative traits into Mendelian-like factors amenable to selection through the analysis of the flanking molecular markers. These markers can then be used in molecular breeding and to clone the genes controlling the QTLs. Although any segregating population can be used for RIL mapping, use of RILs has certain advantages. RILs are near homozygous, which allows multiple replicates to assess phenotypic values, reducing the environmental effects and increasing the power and accuracy to detect QTL. Once QTLs are identified, they can be introgressed to elite germplasm through MAS, much like monogenic traits. In cucumber and melons a number of QTLs have been mapped and their use in MAS is in progress. Many horticultural traits, including yield are under polygenic control with considerable environmental influence and genotype by environment interaction on trait expression.

A number of horticultural traits have been mapped in cucumber. Bradeen *et al*. (2001) linked little leaf (ll) to RAPD marker BC551 at 0.6 cM and flanked determinate habit *(de)* by AFLP marker E14/M50-F137-P2 and RAPD marker L18_2 at 3.1 and 6.9 cM, respectively. One RFLP (CSP056/H3) and two AFLP markers (E14/M49-F-274-P1 and E14/M62-MO02) were found to co-segregate with F (gynoecy) (Bradeen *et al*. 2001). *F* was mapped by Fazio *et al*. (2003b) at 5.0 cM from RFLP marker CSWCT28. Trebitsh *et al*. (1997) found that *F* co-segregated with 1-aminocyclopropane-1-carboxylic acid (ACC) synthase gene when mapped with 73 F_2's from Gy14 x PI 183967. Mapping of quantitatively inherited traits in a narrow-based U.S. processing cucumber population (i.e., Gy-7 and H-19) led to the identification of QTL associated with yield components (Serquen *et al*., 1997; Fazio *et al*. 2003a) that were successfully used in the marker-assisted backcross introgression of one metric trait, multiple lateral branching (MLB; four QTL) over two cycles of selection (Fazio *et al*. 2003b). These were utilized for line extraction and population development in cucumber for improving plant architecture (MLB, GYN, L:D ratio) the strategic use of both PHE selection and MAS will likely enhance breeding strategies.

The cost of using MAS compared to conventional plant breeding varies considerably between studies. The cost-effectiveness needs to be considered on a case by case basis. Factors that influence the cost of utilizing markers include: inheritance of the trait, method of phenotypic evaluation, field/glasshouse and labour costs, and the cost of resources. In some cases, phenotypic screening is cheaper compared to marker-assisted selection (Bohn *et al*., 2001). However, in other cases, phenotypic screening may require time-consuming and expensive assays, and the use of markers will then be preferable. Some studies involving markers for disease resistance have shown that once markers have been developed for MAS, it is cheaper than conventional methods. In other situations, phenotypic evaluation may be time-consuming and/or difficult and therefore using markers may be cheaper and preferable.

MAS has not been widely used for the improvement of polygenic traits because QTL mapping techniques remain insufficiently precise in cucurbits and because the QTL information can not be easily extrapolated from mapping populations to other breeding populations. It is also hindered by numerous difficulties like genetic heterogeneity for polygenic traits; the expenses of applying high density marker assays across many populations; limited knowledge in linkage disequilibrium among species, populations and genomic regions; lack of specific statistical approaches for combining genotypic and phenotypic data and computational difficulties.

References

Baudracco-Arnas, S. and M. Pitrat. 1996. A genetic map of melon *(Cucumis melo* L.) with RFLP, RAPD, isozyme, disease resistance and morphological markers. *Theor. Appl. Genet.* **93**:57-64.

Blair, M., Garris, A., Iyer, A., Chapman,B., Kresovich, S. and Mc-Couch, S. 2003. High resolution genetic mapping and candidate gene identification at the *xa5* locus for bacterial blight resistance in rice (*Oryza sativa* L.). *Theor Appl Genet* **107**: 62–73.

Bohn, M., Groh, S., Khairallah, M.M., Hoisington, D.A., Utz, H.F. and Melchinger, A.E. 2001. Re-evaluation of the prospects of marker assisted selection for improving insect resistance against *Diatraea* spp. in tropical maize by cross validation and independent validation. *Theor Appl Genet* **103**: 1059–1067.

Bradeen, J. M., Staub, J. E., Wye, Antonise, C. R., and Peleman, J. 2001. Towards an expanded and integrated linkage map of cucumber *(Cucumis sativus* L.). *Genome* **44**:111-119.

Brown, R. N. and Myers, J. R. 2002. A genetic map of squash *(Cucurbita* sp.) with randomly amplified polymorphic DNA markers and morphological markers. *J. Am. Soc. Hort. Sci.* **127**:568-575.

Campbell, A.W., Daggard, G., Bekes, F., Pedler, A., Sutherland, M.W. and Appels, R. 2001. Targetting AFLP-DNA markers to specific traits and chromosome regions. *Aust J Agric Res* **52**: 1153–1160.

Chunwongse, J., Doganlar, S., Crossman, C., Jiang, J. and Tanksley, S.D. 1997. High-resolution genetic map the Lv resistance locus in tomato. *Theor Appl Genet* **95**: 220–223.

Collard, B.C.Y., Jahufer, M.Z.Z., Brouwer, J.B. and Pang, E.C.K. 2005. An introduction to markers, quantitative trait loci (QTL) mapping and marker-assisted selection for crop improvement: The basic concepts. *Euphytica* **142**: 169–196

Cultivar discrimination: a case study in cucumber. *HortTechnology* **3**:291-300.

Fazio, G., Staub, J. E. and Stevens, M. R. 2003b. Genetic mapping and QTL analysis of horticultural traits in cucumber *(Cucumis sativus* L.) Using recombinant inbred lines. *Theor. Appl. Genet.* **107**:864-874.

Fazio, G., Chung, S. M. and Staub. J. E. 2003a. Comparative analysis of response to phenotypic and marker-assisted selection for multiple lateral branching in cucumber *(Cucumis sativus* L.). *Theor. Appl. Genet.* **107**:875-883.

Giovannoni, J., Wing, R., Ganal, M. and Tanksley, S. 1991. Isolation of molecular markers from specific chromosomal intervals using DNApools from existing mapping populations. *Nucleic Acid Res.* **19**: 6553–6 558.

Hashizume, Shimamoto, T.,l., Harushima, Y., Yui, M., Sato, T., Imai, T. and Hirai, M. 1996. Construction of a linkage map for watermelon *(Citrullus lanatus* (Thumb.) Matsum & Nakai) using random amplified polymorphic DNA (RAPD). *Euphytica* **90:**265-273.

Hittalmani, S., Shashidhar, H.E., Bagali, P.G., Huang, N., Sidhu, J.S., Singh, V.P. and Khush, G.S. 2002. Molecular mapping of quantitative trait loci for plant growth, yield and yield related traits across three diverse locations in a doubled haploid rice population. *Euphytica* **125**: 207–214.

Horejsi, T., Staub, J. E. and Thomas, C. E. 2000. Linkage of random amplified polymorphic DNA markers to downy mildew resistance in cucumber *(Cucumis sativus* L.). *Euphytica* **115:**105-113.

Joobeur, King, T., J. J., Nolin, S. J., Thomas, C. E. and Dea, R. A. 2004. The fusarium wilt resistance locus *Fom-2* of melon contains a single resistance gene with complex features. *Plant J.* **39**:283-297.

Jung, G., Skroch, P.W., Nienhuis, J., Coyne, D.P., Arnaud-Santana, E. Ariyarathne H.M. & Marita, J.M. 1999. Confirmation of QTL associated with common bacterial blight resistance in four different genetic backgrounds in common bean. *Crop Sci.* **39**: 1448–1455.

Kennard, W. C., Poetter, K., Dijkhuizen, A., Meglic, V., Staub, J. E. and Havey, M. J. 1994. Linkages among RFLP, RAPD, isozyme, disease-resistance, and morphological markers in narrow and wide crosses of cucumber. *Theor. Appl. Genet.* **89:**42-48.

Langridge, P., Lagudah, E., Holton, T., Appels, R., Sharp, P. and Chalmers, K. 2001. Trends in genetic and genome analyses in wheat: A review. *Aust J Agric Res* **52**: 1043–1077.

Lecomte, L., Duffe, P., Buret, M., Servin, B., Hospital, F. and Causse, M. 2004. Marker-assisted introgression of five QTLs controlling fruit quality traits into three tomato lines revealed interactions between QTLs and genetic backgrounds. *Theor Appl Genet* **109**: 658–668.

Li, L., Lu, S., O'Halloran, D., Garvin, D.and Vrebalo, J. 2003. High resolution genetic and physical mapping of the cauliflower high beta-carotene gene Or (Orange). *Mol Genet Genomic* **270**: 132–138.

Markers in cucumber. *J. Am. Soc. Hort. Sci.* **127:**545-557.

Michelmore, R. 1995. Molecular approaches to manipulation of disease resistance genes. *Annu Rev Phytopathol* **33**: 393–427.

Mohan, M., Nair, S., Bhagwat, A., Krishna, T.G., Yano, M., Bhatia, C.R. and Sasaki, T. 1997. Genome mapping, molecular markers and marker-assisted selection in crop plants. *Mol Breed* **3**: 87–103.

Oliver, M., Garcia-Mas, J., Cardtis, M., Pueyo, N., L6pez-Sese, A. I., Arroyo, M. Gomez Paniagua, H., Artis, P. and de Vicente, M. C. 2001. Construction of a reference linkage map for melon. *Genome* **44:**836-845.

Paran, I. and Michelmore, R. 1993. Development of reliable PCR based markers linked to downy mildewresistance genes in lettuce. *Theor. Appl. Genet.* **85**: 985–993.

Park, Y. Sensoy. H. S., Wye. C., Antonise, R., Peleman. J. and Havey, M. J. 2000. A genetic map of cucumber composed of RAPDs, RFLPs, AFLPs, and loci conditioning resistance to papaya ringspot and zucchini yellow mosaic viruses. *Genome.* **43**:1003-1010.

Paterson, A., Tanksley, S. and Sorrels, M.E. 1991. DNA markers in plant improvement. *Adv Agron* **44**: 39–90.

Perin, C., Hagen, L. Dogimont, C., De Conto, V. and M. Pitrat, 1998. Construction of a genetic map of melon with molecular markers and horticultural traits. p. 370—376. In: J. D. McCreight (ed.). Cucurbitaceae '98. ASHS Press. Alexandria, VA.

Perin, C., Hagen, L., De Conto, V., Katzir, N., Danin-Poleg, Y., Portnoy, V., Baudracco-Arnas, S., Chadoeuf, J., Dogimont, C. and Pitrat, M. 2002. A reference map of *Cucumis mela* based on two recombinant inbred line populations. *Theor. Appl. Genet.* **104**:1017-1034.

Reyna, N. and Sneller, C.H. 2001. Evaluation of marker-assisted introgression of yield QTL alleles into adapted soybean. *Crop Sci* **41**: 1317–1321.

Ribaut, J.-M., Hu, X., Hoisington, D. and Gonzalez-De-Leon, D. 1997. Use of STSs and SSRs as rapid and reliable pre-selection tools in marker-assisted selection backcross scheme. *Plant Mol Biol Report* **15**: 156–164.

Serquen, F., Bacher, J. and Staub, J. 1997. Mapping and QTL analysis of horticultural traits in a narrow cross in cucumber (*Cucumis sativus* L.) using random-amplified polymorphic DNA markers. *Mol Breed* **3**: 257–268.

Staub, J. E., Serquen, F. and Gupta, M. 1996. Genetic markers, map construction and their application in Plant Breed. *HortScience* **31**: 729–741.

Taler, D., Galperin, M., Benjamin, I., Cohen, Y. and Kenigsbuch, D. 2004. Plant *eR* genes that encode photorespiratory enzymes confer resistance against disease. *Plant Cell* **16**:172-184.

Tanksley, S.D., 1993. Mapping polygenes. *Annu Rev Genet* **27**: 205–233.

Tanksley, S.D., Young, N.D., Paterson, A.H. and Bonierbale, M. 1989. RFLP mapping in plant breeding: New tools for an old science. *Biotechnology* **7**: 257–264.

Trebitsh, T., Staub, J. E. and O'Neill, S. D. 1997. Identification of a 1-aminocyclopropane 1-carboxylic acid synthase gene linked to the female (F) locus that enhances female sex expression in cucumber. *Plant Physiol.* **113**:987-995.

40. Wang, Y.-H., Thomas, C. E. and Dean, R. A. 1997. A genetic map of melon *(Cucumis melo* L.) based on amplified fragment length polymorphism (AFLP) markers. *Theor. Appl. Genet.* **95**:791-798.

❑❑❑

Chapter – 33

Exploitation of Indigenous Genetic Resources for Chilli Improvement

Durvesh Kumar Singh

Introduction

Chilli is one of the most important spice and condiment of daily use in all most all the Indian houses. The fresh and dry fruits are the rich source of Vitamin A (292 IU) and Vitamin C (111 mg/100g). Its pungency not only improves the taste of the different preparations but also improve their storability. The colour of the fruits is due to the presence of the pigment capsanthin and pungency and acridity is due to the presence of alkaloid oleoresins knows as capsaicin ($C_{18} H_{27} O_3 N$). The capsaicin has a very good medicinal importance and also having importance for the pharmaceuticals industries for preparation of the different types of the drugs, because of the presence of the antioxidants like Vitamin A, C and P (Rutin). The green and red sweet pepper, bell pepper, Pimiento (heart shape) and cherry pepper are used in the fresh market processing industries for stuffing pickling and dehydrated processed meat. It is used as green as well as dry powder, pickles, salad and for industry to prepare the medicines for external and internal uses.

Uses of Pepper

As spices, salad, vegetable, pickles, medicinal property

A. Internal: Stimulant, carminatives, digestion and despelfiatulence [gas]

B. External: Capsicum plasters, Balm, bend aids, antifungal, antibacterial, Anti radioactivity, pain killer, ointment [capsidol] and oil

Preparations - 3 Types

1. Oleoresins capsicum [3.9 -14%]
2. Oleoresins red pepper [0.6-3.9% of capsaicin]
3. Oleoresins paprika [less than 0.6%]

Chilli is good stimulant carminatives and has good digestion properties. It is also used in the preparations of balms, capsicum plasters, bend aids, pain killers etc.

Kind of pepper

1. Red pepper / Hot pepper [*Capsicum annuum* var *longum* or *Capsicum annuum* var *acuminatum*]
2. Sweet pepper/ capsicum [*Capsicum annuum* var *grossum*]
3. Paprika [*Capsicum annuum* var *annuum L.*]

Ornamental chilli

1. Christmas pepper– plant compact, cone shape fruit and green on maturity turn into purple than yellow & red
2. Himinishki– plant very compact, produce round, erect and purple fruit above the foliage
3. Holiday cheer– plant compact and have showy round
4. Red missile- it is a F_1 hybrid and 5cm long tape red fruits
5. Fiesta type-it is also a F_1 orange and light yellow fruits (Hosmani,1982)

Characters of Paprika

Paprika has brilliant red colour fruits with mild flavour used for garnishing, colouring material. It has high vitamin- C contents but still it is not much popular in the country.

Kind of Paprika

Hungarian, Bulgarian, Spanish, Yougoslavian, Romanian, Chilian, Czekoslovakian, Mordokan, Turkish, Portugese.

Table 1. Nutritive value of Hot pepper and Sweet Pepper (Dry fruits)

Hot pepper		Sweet Pepper	
Nutrient content	**Quantity**	**Nutrient content**	**Quantity**
Water (%)	13.0	Water (%)	92.4
Dry matter (%)	34.6	Energy (calorie)	29.0
Energy (kcal)	116	Protein (%)	1.2
Protein (%)	6.3	Calcium(mg.)	11.0

Contd...

Fiber (g)	15.0	Vitamin A (I.U.)	878
Calcium (mg)	86	Vitamin C (mg)	175.0
Fe (mg)	3.6	Thiamine (mg)	0.06
Phosphorus (mg)	80	Riboflavin(mg)	0.03
Potassium (mg)	217	Niacin(mg)	0.55
Carotene (mg)	6.6		
Thiamine (mg)	0.37		
Riboflavin (mg)	0.51		
Niacin (mg)	2.50		
Vitamin C (mg)	96.0		

India is one of the leading country in the world with respect to chilli growing area (0.95 m ha) and production of (0.82 mt) of dry chilli in the year 2005-06. Chilli occupies an area of 1956 thousand hectare world wide with an annual production of 22168 thousand tones and having productivity 13.88 t/ha. In India, crop covers 6000 ha land with an annual production of 50 thousand tones having productivity 9.18 t/ha (FAO, 2002). India is the largest exporter of chilli to the gulf countries, European countries and Asian countries followed by China, Japan, Indonesia, Mexico, Urgentina, Kenya, and Nigeria. Chilli is exported in the form of dry powder per dry fruits, but there is a great scope to export chilli products in the form of oleoresin, chilli paste, sauces, ketchup, pickles, oil (Capsicum, Chilli and Paprika) and Paprika fruits, etc. of high quality to complete in international market.

Table 2. Area and Production of dry chilli in India

Year	Area (000'ha)	Production (000' tonnes)	Productivity (Qt/ha Dry chilli)
1970	682	895	5.79
1980	826	511	6.18
1996	810.4	745.9	9.2
2005	945	496	5.11
2006	556	505	9.81
2006	654	1014.6	15.51
2007	758	1234.1	16.28

Table 3. Export of Chilli

Crop	Quantity Tones	Value Rs. (Corer)	Total Value (%)
Pepper			
1999-2000	42806	88488.1	36.3
2000-01	19250	32632.8	46.48
Chilli			
1999-2000	64776	2506589	12.32
2000-01	61000	195235	12.57
2005-06	98300(G) 121240(D)		
2006-07	375000	8.07	

Chilli fruits contain capsaicin, beta carotene capsorubin, zeoxanthin, neoxanthin, viloxanthin, and luetin pigments responsible for pericarp colour of the fruit (Hosmani, 1982). The fruit parts have different amount of capsaicin content.

Table 4. Capsaicin content in different part of the fruits

Part of fruit	Capsaicin content
Capsaicin	$C_{18}H_{27}O_3N$
Whole fruit (Av.)	0.6 (%)
Range	0.4 – 0.9 (%)
Placenta and septa	2.5 %
Seed	0.7 %
Pericarp	0.03 %

Pigments

Chilli fruit contains the following pigments:

1. Capsanthin (red colour)
2. Beta carotene (yellow colour)
3. Capsorubin
4. Zeoxanthin
5. Neoxanthin
6. Viloxanthin
7. Lutein

Chilli is grown all over the world from 45^0 N to 45.5^0 S of hemisphere from temperate to tropical regions. It is a frost susceptible crop and grows best at the temperature range between 20 to 35 ^{0}C and not below 10^0 C. It can be grown up to 2000 m altitude and requires 70 to 120 cm annually normal rainfalls and needs frost free area for better growth and production for 240 days. It is cultivated in Africa, Europe, North Central America, South America and Asia. In Asia, India, China, Korea, Pakistan, Sri Lanka, Turkey and Japan are the major chilli growing countries. The most popular and important chilli growing states in the country are A.P, Maharastra, Karnataka, Tamil Nadu and Orissa which together constitute nearly 75 % of the total area and production and rest of 25 % area covers in other states.

Distribution of Chilli

Chilli was carried out of the old world by early explorers being introduced into Spain by Columbus on his return trip in 1493 (Boswall, 1949). Cultivation spread from the Mediterranean area to England by 1548 and to the central Europe by the close of the sixteenth century. The Portuguese brought capsicum to India from Brazil prior to 1585 and cultivation was reported to China during the late 17th century (Sturtvant, 1885).

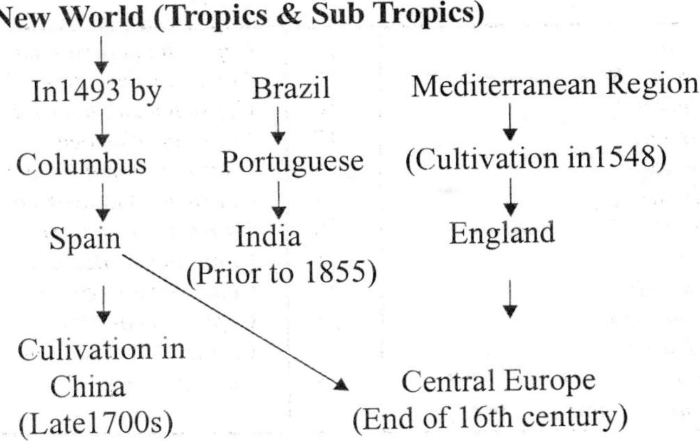

Almost all the capsicum species except *Capsicum annuum* (Japan) are native to tropical America particularly Peru and West Indies. Mexico in the primary centre of origin with secondary centre in GuataMala and Bulgaria (Safferd, 1986). In India it is known as hot pepper, chilli and sweet pepper. Chilli is also commercially cultivated in China, Korea, Indonesia, Pakistan, Sri Lanka, Turkey, Japan, Mexico, Ethiopea, Nigeria Uganda, Yugoslavia, Hungeri, Itally, Spain, Bulgaria and Romania. It has several vernacular names in different languages.

Table 5. Vernacular names of chilli

Languages	Vernacular Name	Language	Vernacular Name
English	Red Pepper, chillies, cayenne	Japanese	Jogarashi
Canada	Menasinkai	Portugese	Pimentaopiente
Chinese	Hung-fun-chiao	Russian	Stsuchkovy pyrets
Dutch	Spanse pepper	Spanish	Chile
French	Piment enrouge	Swedish	Spank pepper
Germen	Beissfee	Java and Malaya	Lombok
Italian	Peperance	Arabic	Filfil

Taxonomy

Chilli belongs to family solanaceae. The word capsicum has come out from a Greek word "Kapsa bit" and in Latin it is called as "Capsa" means box. The seeds are attached with the placenta under a box like structure and chilli word is of the Mexican origin. There are more than 25 species. Classified under the genus capsicum as follows:

Table 6. Cultivated and wild species of chilli.

S.N.	Species	S.N.	Species
1	*Capsicum annuum*	14	*Capsicum peruvianum*
2	*Capsicum frutescense*	15	*Capsicum cornutum*
3	*Capsicum pendulum*	16	*Capsicum galopogense*
4	*Capsicum pubescense*	17	*Capsicum chacoense*
5	*Capsicum chinense*	18	*Capsicum minutiflorum*
6	*Capsicum cardensi*	19	*Capsicum microcarpum*
7	*Capsicum eximium*	20	*Capsicum baccatum*
8	*Capsicum cornicum*	21	*Capsicum cardenassi*
9	*Capsicum geminifolium*	22	*Capsicum fastigiatum*
10	*Capsicum praetermissum*	23	*Capsicum cardiforme*
11	*Capsicum scolnikianum*	24	*Capsicum schottianum*
12	*Capsicum angulosum*	25	*Capsicum sinense*
13	*Capsicum anamalon (Japan)*		

Out of which only five are in cultivation such as *Capsicum annuum, Capsicum frutescense, Capsicum chinense, Capsicum pubescence, and Capsicum pendulum.*

Table 7. Cultivated species of chilli

S.N.	Species	Place of Origin
1.	*Capsicum annuum L.*	Mexico
2.	*Capsicum frutescens L.*	Amazonia
3.	*Capsicum chinense L.*	Amazonia
4.	*Capsicum pendulum L.*	Peru & Bolivia
5.	*Capsicum pubescens*	Peru & Bolivia

Table 8. Characteristics of cultivated species of chilli

Species	Calyx Teeth	Corolla colour	Spot on Corolla throat	Corolla shape	Anther colour	Seed colour	Flower/ node
Capsicum annuum L.	Present	White/ purplish	none	Rotate	Blue/ purple	Tangin	1
Capsicum frutescense L.	-	Greenish white	none	Rotate	Blue	Tangin	1-3
Capsicum chinense L.	Present	Greenish Yellow	none	Rotate	Blue	Tangin	1-5
Capsicum baccatum L.	Present	white	Greenish Yellow	*Rotate*	yellow	Tangin	1-2
Capsicum pubescense L	Present	Purple	none	Rotate	purple	Black	1

Species of *Capsicum annuum* L.

1. *Capsicum annuum* L.

   a. Capsicum annuum var. *annuum* L.-fruit thick, less pungent, fleshy, bulging base and blunt apex *b.* *Capsicum annuum* var. *longum* L. - fruit thin, very long, stout and broad base

c. *Capsicum annuum* var. *accuminatum L.*-fruit thin, long, pendulous and pungent

d. *Capsicum annuum* var. *cerasiforme*- fruit small, round highly pungent, erect cultivated in Salem & South Aerot distt. of Tamil Nadu.

e. *Capsicum annuum* var. *grossum*-fruit large, bell shape, mild pungent and all bell pepper belongs to this group.

f. *Capsicum annuum* var. *abbreviatum*- wrinkled fruit

g. *Capsicum annuum* var. *conoids*- cone type like tabasco

h. *Capsicum annuum* var. *fasciculatum*- cluster type very pungent

2. *Capsicum frutescense L.* – it is a perennial, but grown as annual, tall 2m in height, flower tiny, borne in cluster, 2-3 in axils, erect-sub erect, small fruit long stalk, highly pungent and calyx teeth is absent.
3. *Capsicum frutescense* var. *minima* –wild type and small fruit which is highly pungent, cvs. Lavangi, Seema mirapa & Golkonda mirpa.

Table 9. Characteristics of *C. annuum* and *C. fruitescense*

Character	C. annuum	C. fruitescense
Growth habit	Annual herbaceous	Shrubby perennial
Plant size	dwarf	Tall
Leaf	small	Broad
Flower	Small, solitary white/purple	Big, bear in cluster
Corolla	White	Greenish white
Fruit stalk	Shorter than fruit body	Longer than fruit body
Fruit size	Bigger, pendulous	Erect or semi erect, small
Pungency	Less	High
Ovary	2-4 celled	2 celled

Type of chilli on the basis of fruit shape and size

1. Tabasco group: bears small fruit, thin ,erect and highly pungent
2. Sianeys group: large fruit size10-30cm, base thicker, skin thicker in middle, red at ripe and pungency variable.
3. Cherry group: cherry type fruits, orange to red in colour and round to oval round shape highly pungent fruit.
4. Celestiol group: fruit erect, pointed, 1.8-3.2 cm long, yellow to purple in green stage and change in orange at maturity
5. Perfection group: fruit ovate with pointed end, 7.5-10cm long and commonly use for canning.

6. Tomato group: fruit round, flat ant, similar to tomato and 5-10 cm in diameter.
7. Bell group: fruit large, less pungent and used for salad,

Floral Biology

Chilli is a basically self pollinated crop but out crosses ranged between 35 to 95% due to the long styled flowers and generally feruled as often cross pollination and requires at least 200 to 400 m distance for breeder and foundation seed production, respectively.

Flower is solitary, axillary and sometime in pairs, bracteolate, pedicelate, bisexual and hypogynous. Calyx is companulate, spars usually five gamo sepalous and much shorter than fruit. Corolla is rotate, 5-6 lobed, twisted in the bud; lobes are thin, veined and in curved at tip.

Androecium is consists of five stamens which are epipetalous. Anthers dehisce, introse, non- counivalent in cone. Anther dehisce longitudinally bilateral sutures.

Gynoecium is superior, ovary is bicapillary with numerous ovules is each locule on exile of the placentation. Style is terminal and linar Stigma is sub capitate and faintly bifid.

Pollination in long styled flowers may takes place by bees (Honey Bees), Golden fly ants, thrips and flies like white fly. The out cross varied from genotype to genotype. Some cultivars have more number of short and medium styled flowers where as some has more number of long styles stigma exerted over the anthers. One line has identified from the germplasms (PCPGR 2064) which has more than 95% long styled flower and it require very care in seed production. In some of the genotypes, flowers remain open for 2-3 days. The percentage of fruit setting is ranged 40 to 50% (Purseglove, 1977). The flower opening and anther dehiscence to a large extent depends on the weather conditions. During cold as well as cloudy days, the opening is delayed.

Anthesis takes place during morning hours from 5.00 am to 11.30 am, it may vary due to climatic conditions. Anther dehiscence takes place after ½ hours of the anthesis and continues to the whole day. Stigma is receptive from the day of anthesis and continues for 2-3 days.

Germplasm of Chilli

Presently eleven centers have been identified for evaluation, characterization and maintenance of the germplasm of chilli *viz.* SAUs – Coimbatore, Jorhat, Dhanwad, Kalyani, IIVR, Varanasi, SKUAS&T, IIHR, Bangalore, HARP, Ranchi, CITH, Srinagar, UAS, Bangalore, IARI, New Delhi.

Pepper germplasms status in Long Term Storages (LT S) up to December 2006

Centre	Accession Nos.
NBPGR Delhi	2506
NBPGR, Bhowali	1208

PCPGR 620

AVRDC 6000

2500 stored for LTS and MTS

2800 (Primary characterization)

The International Board for Plant Genetic Resources (IBPGR) has made efforts to organize plant exploitation for both working and base global collection of capsicum genetic resources. During 1986, the Asian Vegetable Research and Development Centre (AVRDC) were invited to maintain a brake-up of global capsicum collections. The responsibility of the AVRDC and its genetic resources and seed unit is to systematize the multiplication, characterization, evaluation, documentation, preservation and distribution of capsicum genetic resources. Presently there are more then 6000 accessions in the collection and 2500 accessions multiplied and stored in gene bank for long term and medium term storage and about 2800 accessions completed for primary characterization by using 62 descriptors.

Table 10. Germplasms collection and maintenance at India & Abroad

Centre	Number of collection
IBPGR,Rome	>16000
AVRDC, Taiwan	6000
	2500 stored for LTS and MTS
	2800 (Primary characterization)
NBPGR,New Delhi	2506
NBPGR,Bhowali	1208 up to June2006
PCPGR, Pantnagar	0620

The exchange of material in South Asia is also, in progress through South Asia Vegetable Researching Network (SAVERNET) that is sponsored by Asian Development Bank and executed by AVRDC. There are some Identified centers (by ICAR) for maintenance and evaluation of the germplasms, are as follows:

1. NBPGR, Delhi,
2. NBPGR Regional Station, Bhowali
3. Regional Station, Akola.

Table 11. Germplasm collection, evaluation and maintenance centres in India

Crop	No. of centre	Name of centre
Chilli	11	Lam, Coimbatore, Jorhat, IIVR, IIHR, HARP-Ranchi, Dharwad, Kalyani, CITH-Srinagar, UAS Bangalore, SKUAS&T (K)
Capsicum	05	Katrain, Solan, IIHR, Palampur, SKUAS&T (K)
Paprika	05	Katrain, IIVR, IIHR, Dharwad, SKUAS&T (K)

AVRDC identified some of the lines namely AVRDC-13, AVRDC-16, CNPH-16, CNPH -2689 and CNPH-26 having resistance against the anthracnose disease and one line EC 334206 is reported to be resistant to the water logging areas and EC 32333 is reported for multiple diseases resistance by NBPGR. India is considered as secondary centre of capsicum diversity and has also initiated exchange of material with AVRDC through agreement with IBPGR and NBPGR, New Delhi. Apart from NBPGR during 1970-71, ICAR also selected three main centers, namely, Lam (A.P), Kovila Patti (T.N) and Kanpur (U.P) for collection, evaluation and maintenance of germplasm of chilli. Besides above centers, PCPGR, Pantnagar (U.K.) and other eleven centers were identified in the country for collection, evaluation, documentation and maintenance of germplasms in the country. An international hot pepper trial network (INTNOPE) was also started by AVRDC with objective of evaluation of hot pepper, land races and local as well as advance breeding lines for insect and disease resistance.

Chilli Improvement

Crop improvement work to evolve high yielding strains of chilli and package of practices were started at Lam, Guntur since 1928 and several new improved cultivars were developed. G-1 is the first released cultivar in 1937-38 and a local improved selection from the Nalla Padam. Later on another cultivar G-2 was isolated from a bulk of NP 46-A developed by Pusa which was superior than G-1 and recorded on an average 11% higher yield. Later on, G-2 has been replaced by G-3 which yields, 18 % greater yield and retains bright red colour. Later on, G-4 and G-5 ranked higher in AICVRP trial during 1972-73 and 1973-74, respectively and same was released as, Bhagyalakshmi (G-4) and Andhra Jyoti (G-5). Guntur developed a series of varieties of chilli. At the same time some other centres were also established for the vegetable improvement in the country such as Coimbatore (TN), Sabour (Bihar), and Kanpur (UP). These centres were also developed a good number of improved cultivars for the growers such as Co-1,Co-2, MDU-1, K-1, and K-2 from (TN), Kaliyanpur Red ,Kaliyanpur yellow, Kaliyanpur Chanchal from U.P and Puri Red, NP-30,and NP-46-A from Pusa, Bihar.

After, establishment of ICAR, during 1968, a large number of centers were established in the country under AICVRP for vegetable improvement. Presently, more than 15 improved cultivars were under commercial cultivation. These cultivars have their own identity for a specific character. Although there are large number of local cultivars, popular among the farmers in each state due to the regional preferences and better adaptation.

Breeding methods for improvement

1. Introduction
2. Selection
3. Hybridization
4. Heterosis

5. Inter species hybridization
6. Mutation
7. Male sterility

Table 12. Varieties developed by different methods

Method	Hot pepper	Sweet pepper
Introduction	Hot Portugal	California Wonder
	Albena	Yolo Wonder
	Floral Green	Bullnose
	Kalinkov	World Beater
	Albenak KAPIA	Ruby King
Selection	G-1, G-2, G-3	Arka Mohini
	G-4 (Bhagya Laxami)	Arka Gaurav
	G5 (Andhra Jyoti)	Arka Basant
	NP-46-A,	Pimiento
	HC-44, HC-28	
	CO-1,CO-2,K-1, CA-960,	
	Arka Jyoti, Musalbadi	

Foreign varieties grown in india

Hot Pepper: Tabasco, Maxican chilli / Hungarian, Anabein

Sweet Pepper: Rubyking ,Bullnose, Golden Green, Early Pimento, Sully book Spanish, Hungarian paprika ,Sweet Banana, King of North, Perfection, Haris early Giant

Paprika: Hungarian ,Bulgarian ,Romanian, Bydagi, Warrangal, Arka Abir, kt- 19 (Indian varieties.)

Table 13. Chilli varieties developed by Hybridization

Hot Pepper Variety	Crosses
Pusa Jwala	NP46-A x Purl Red
Andhra Jyoti	G-2 x Sabour local
Pant C-1	NP-46 Ax Local Kandhari
Pant C -2	NP-46Ax Local Kandhari
Pusa Sadabahar	Pusa Jwala x IC-31339
J-218	Kali Peeth x Pusa Jwala
X-235	G-4 x Anther Mutant
Punjab Lal	Perennial x Local Red
K-2	K-1 x Sattar Samba
LCA-206B	G-3 x Huntaka

Table 14. Varieties of Sweet Pepper:

Sweet Pepper Variety	Crosses
Spartan garnet	California Wonder x Pimento
Spartan	Marigold x California Wonder
Imerrold	Marigold x California wonder-
Sonnette	F-1 x Key stone Resistant Giant
Bharat-1	—
Pusa deepti (kt- 1)	—

By a large, vegetable research has been carried out in India by public institutions and SAUs. However, from 1969 onwards, there has been effort to start research and development activities by some private companies with foreign collaboration. Some private companies conducting research include M/S Mahyco. Jalna (Maharashtra), Nath Seed (Aurangabad), Sutton and Sons, Culcutta (W.B.) Beejo Sheetal Hybrid Seeds, Jalna (Maharashtra), Biogene (Bangalore), Nam Dhari Seeds Pvt. Ltd. (Bangalore), Novartis India Ltd. (Pune), Nunhamp Seed Co. etc. A number of open pollinated as well as hybrids of different crops have been developed by different private seed companies which have performed well under multilocation testing of All India Coordinated Vegetable Improvement Project. Almost all the companies are having development of hybrids and their interest in the seed production to sale them on higher prices to the farmers, but, ultimately, the main object is to increase the productivity of the crop to fulfill the demand of the peoples of the country.

To improve the production of chilli and other vegetables, the institutions and universities were also play the vital role in the development of the cultivars of the chilli and spices.Pantnagar, Kalyanpur, IIHR, Bangalore, IARI, New Delhi, Solan, J & K, Sobour, Ranchi, Udaipur, Hissar, PAU, Rahuri and Bhubneshwar etc. developed a good number of cultivars of chilli. Presently, 22 centers all over the country associated with AICRP for vegetable and eleven centers identified for collection, evaluation, characterization and maintenance of the chilli seeds in the country.

Table 15. Varieties develop by Mutation Breeding

Variety	Year	Mutagen	Character
Horgoska slatka x -3	1974 Yugo- salavia	Gama rays	CMV resistant
Albena	1974 Bulgara	Gama rays	Attractive fruits, good flavour
MDU-1	1974 Tamil.Nadu	Gama rays, K1	High yield & Capsaicin
Krichim Skinan	1976 Bulgaria	X-ray	High yield
Lyulin	1981	135 Gama rays	Early & High yield

Genetic variability in chilli/pepper

The success of heterosis breeding depends on the variability present in the material or genetic stock. In chilli a large amount of variability with respect to different quantitative as well as qualitative characters available and reported (Nandpuri *et al.*, 1971), Singh and Saini (1976), Awasthi *et al.* (1976), Gupta (2003) and Pramila (2005) indicated that there is great scope of improvement of this crop by selection and heterosis breeding. Des Pande (1933) and Verma *et al.* (2004) reported wide range of variability in different characters such as dry fruit yield per plant (7.8-48 g), number of fruits per plant (5.8-226.3) and stalk weight (19.0-58.6 %). Fruit number per plant in Punjab Local was noted up to 627.7 per plant (Singh & Kaur, 1990) and in Pant C-1 total number of fruits per plant was found 358 (Peter *et al.,* 1977). Fruit length ranged from 0.5 cm to 30 cm in chilli as reported by Hosmani (1982) and Verma *et al.* (2004). The variability also reported for fruit colour at maturity stage as well as green stage by large number of scientist (Ram and Lal, 1984 and Singh *et al.,* 2004). The colour varied from green to high green yellow, red, purple, violet etc. The colour differences due the presence of the pigments in the fruit are noted. In case of sweet pepper, there is lack in genetic variability. The breeder, therefore, face difficulty in the improvement because of low variability and another difficulty is the very narrow genetic base of the crop. This can be increased by crossing them with wild types as reported by Singh *et al.* (1993). Very less work has been conducted on the variability of sweet pepper. Arya and Saini (1977) found greater variability for rind thickness and fruit size. Singh *et al.* (2004) reported sufficient variability with respect to earliness, fruit weight, and number of fruits per/plant and weight of fruits per plant and total yield per plant later Pramila (2005) also recorded variability for fruit size, colour, weight of fruits per plant and total fruit yield.

Breeding for diseases and insect resistance

The main object of breeding is to develop the high yielding as well as biotic and a biotic resistant cultivars for the growers. It depends on the availability of the resistant source for a particular disease or insect and its nature of inheritance pattern, screening of germplasm lines, biochemical basis of resistance and coordination between the breeder, pathologist as well as entomologist for testing the material in the sick plot is essential for better result. In case of pepper or chilli, the breeding methods may be modified according to the resistant source, inheritance pattern, environment as well as parasitic relationship, In general, in case of disease or insect resistant, mostly back cross method is adopted to develop the resistant cultivar, if the character is grown by single or two genes. Besides this method, selection, pedigree method, combination of back cross and pedigree method or population improvements are followed when the resistance is grown by several genes or poly genes.

Table16. Diseases and their source of resistance and in heritance in chilli

Disease	Source of resistance	Inheritance
Fruit rot /Anthracnose	AVRDC-13, AVRDC-16, CNPH-2689, CNPH-26	Polygenic
Root rot (Phytoph thora sp.)	*C. chinense* PI 159236, C. *annuum* PL 201232 *C. chinense* PI 152225, Wax globe SCM 334	Monogenic dominant Monogenic recessive Monogenic
Anthracnose	—	Polygenic
Bacterial wilt	Pant C-1, White Kandhari, KAU Cluster	Monogenic
Bacterial leaf spot	*C. chacoence* (PI 260435)	Monogenic
Cercospora leaf spot	Punjab Lal, Perennial	(3 polygenic)
Damping off	—	Digenic
Soft rot	—	Olegogenic
Root knot Nematode	—	Monogenic and dominant
CMV	Roma IC 31339 (small fruited and perennial)	Monogenic and dominant
LCV	Pusa Sadabahar, Pant C-1, Punjab lal	Polygenic
Potato virus x	Delray Bell perennial SCM 334	Monogenic and recessive
TMV	*C. chacoence* PI 260429, PI 152225	Monogenic and dominant

Disease resistance has been reviewed by Green & Kim (1994). It was reported that a number of fungal and viral diseases damage the crop but among them anthracnose, cercospora leaf spot, phytophthora, root rot, bacterial wilt, TMV, CMV, PVX, PVY, CMR and tobacco leaf curl virus more dangerous and causes sever losses in the crop ranged from 80 to 100% have been in early infection of disease. The resistance source has been reported under abiotic and biotic diseases of the chilli. Many workers from India and abroad, exploited Indian hot pepper line which is small fruited and perennial as source of resistance to CMV and Poty virus. In India, very limited work has been done on disease resistance in chilli. Punjab Lal has been released as a multiple resistant chilli cultivar for growers in Punjab area. It has resistance against CMV, CMMV, wilt, and spedoptera leaf eating cater pillar. Other lines are Perennial, BG-1, Local, Taiwan, EC 323333 and Indonesia Selection. Pant C-1 was found to be resistant to mosaic and leaf curl virus. Pusa Sadabahar released by ICAR having resistance against CMV, TMV and leaf curl virus (Tewari, 1991). Pusa Jwala is reported to thrips resistant (Tewari and Ramanuj, 1974) Pant C-1 and Kandhari are reported resistant for bacterial wilt. New cultivars or varieties, Phule Sai, Musalwadi, Phule C-5, and Arka Lohit (Sel-1) were reported as resistant to fusarium wilt. Earlier it was reported that small fruited perennial wild type of Indian chilli is a source of resistant to CMV and poly virus and suggested a long term breeding programme to recover and stabilize resistant to viral diseases from wild type to other acceptable horticultural type for the growers.

Table 17. Some multiple resistance varieties of chilli.

Variety	Resistant against
Perennial	CMV, TMV, TLCV, Potato virus
Punjab Lal	CMV, CMMV, Bacterial wilt, Spedoptera leaf eating caterpiller
Pusa Sadabahar	CMV, TMV, LCV and Thrips
Pant C-1	Bacterial wilt, Phytophthora, Root rot, Root knot, nematode, CMV,TMV, TLCV
White Kandhari	Bacterial wilt, Phytophthora, Root rot, Root knot nematode
Tambeli and Tambeli	Tobacco ectch virus, TMV, potato virus
Myliddy-1, and Myliddy- 2	TLCV, Pepper mottle virus, Mites, thrips, aphids

Diseases of Chilli

Capsicums are affected by a number of diseases, pest and a biotic stresses. The crop requires continuous monitoring to assure that the crop should not be affected by the biotic stresses (i.e. diseases) if you can apply suitable or resistant variety or control measures to protect the crop. Some important diseases and resistance sources are given in the following table that can be utilized to develop the resistant varieties of the pepper.

Table 18. Source of resistance against disease of chilli.

Disease	Causal agent	Source of resistance
Anthracnose	*Colletotrichium capsici*	Chinese Giant, Yolo Wonder, Hungarian Yellow Wax, Spartan Emerald, Paprika
Leaf blight and fruit rot (hot pepper)	*Phytophthora capsici* *Phytophthora nicotianae* var. vicotianae	Hot pepper cultivar, Russian yellow, Solan yellow, Javiott Salimar, Hot Portugal, Perennial
Sweet pepper	-do-	PBG-631, Kauda Ghat Sel, PBG-631 and Waxy Globe (Single dominant gene)

Breeding method for disease resistance

The first step in resistance breeding programme is the identification of source of resistance which is available in primary and secondary centers of origin. Peru, AVRDC, (Taiwan), NBPGR (New Delhi), PCPGR (Pantnagar). Domestication of chilli for different purposes. Further, added to the variability in different countries of introduction, land races and resistant cultivars. The enrichment of germplasm is essential for identifying the reliable source of resistance. Field evaluation in hot spots with out actual inoculation of pathogen often fails to identify the true resistance for this standardization of inoculation techniques and scoring procedures are vital.

The major problems occur during the development of resistant varieties of chilli or peppers are the existence of variability among the pathogens. The resistance breaking down after the exposure of resistant cultivars to virulent pathogens and lack of complete knowledge of the inheritance of resistance. Before, adopting any breeding procedure to develop the varieties the knowledge of the mode of genetic control to a particular disease is very essential. Some diseases and their genetic resistance or mode of genetic control is presented in the following table.

Table 19. Genetics of resistance in chilli

Disease	Mode of genetic control
Anthracnose	Recessive, additive, dominant & epistatic effect are important
Cercospora leaf spot	Recessive, 3 complementary gene, additive and dominance
Damping off	Digenic
Bacterial leaf spot	Monogenic dominance
Soft rot	Oligogenic
Mosaic TEV	Monogenic and dominance
Mosaic CMV	Monogenic and recessive
Mosaic TMV	Monogenic and dominant
Root knot nematode	Monogenic and dominant

The conventional methods of breeding like simple selection mass selection, back crossing and pedigree methods are generally used for the development of new resistant cultivars. The gene pool (polygenic) for resistant genes is created by mutation breeding, Inter specific hybridization and disruptive selection method and somatic hybridization and now a days the genetic engineering (Transgenic) is the only method to incorporate the specific resistant gene into other variety for improvement.

Fungal diseases

The fungal diseases causes sever yield losses in bell pepper as compared to chilli. Nursery disease, damping off, fruit rot, and leaf blight, wilt and powdery mildew are the more destructive diseases on those areas where the plenty of rainfall takes place during the cropping period and due to the humid climate the fungal growth is higher and causes more than 70 per cent losses in the production. Therefore, control measures should be adopted to prevent the losses in yield. Anthracnose as well as die back is the harmful diseases of chilli seed production. Under suitable favorable environment these diseases caused 12 to 15% losses in the yield. Some important diseases and source of resistance is given in the following table.

Table 20. Resistance source in wild species against biotic & a biotic stresses in chilli

Species	Source of resistance
Capsicum chacoence	TMV, Anthracnose Bacterial spot, Poty virus, aphids, Thrips,
C. ciliaum	Root rot, water logging
C. eximium	Potato virus y, LCV drought, high temp.
C. flaxuasum	CMV, Potato virus x, soil salinity
C. cadenasii	Aphids, Thrips and low temp.
C. pubescens	Powdery mildew
C. microcarpum	Powdery mildew
C. baccatum var. pendulum	CMV, Potato Virus Y, Powdery mildew, drought, high temperature.

Bacterial diseases

Pepper or chilli crop is also affected by several bacterial diseases. Under humid conditions bacterial wilt caused by *Rolastoma solanacearum* in more harmful. Some time whole crop is damaged by this bacterial disease.These bacteria affect more than 200 plant species belonging to 33 families. Some of the resistant source is given in the following table.

Table 21. Bacterial diseases & source of resistance in pepper

Disease	Causal Agent	Resistance Source
Bacterial Leaf spot	*Xanthomonas vasicatoria*	Nilum, Jelapeno, Agronomica 8-2-17 I163192,260435,163189,163192,271322,322719
Bacterial Soft rot	*Erwinia carotovra*	Jalapeno
Bacterial wilt	*Ralostona solanacearum*	Kandari,Punjent pride, Cherry red, Pant C.-1

Insects resistance

In chilli, the important damaging insects are thrips, mites, pod borer, and aphids.

These insects are more dangerous in the tropical region of the country and causes about 25 to 50% yield losses. Some important lines having tolerance/resistance against the insects are as follows

The resistances require understanding of variability among the host and pest, source of resistance and mechanism of resistant. Resistance to thrips species *Fravliniella occidentaliza* was confirm in the five commercial varieties and was shown to be based on mechanism of tolerance to insect feeding rather than antibiosis or antixenosis .as reported by Frey & Seralk (1991).Very few reports are available on insect resistance, so the screening of germplasm is essential for a particular insect. Genetic of pest resistant and transfer of resistance to new pepper cultivars need further, investigation.

Table 22. Line tolerant to insects

Crosses	Initial crosses	$F_{1\,Seed}$	$F_{2\,Seed}$	Back Crosses
C. annuum x *C. frutescens*	+	+	+	+
C. annuum x *C. chinense*	++	++	++	++
C. annuum x *C. pubescens*	-	-	-	-
C. chinense x *C. frutescens*	+	+	+	+
C. annuum x *C. baccatum* var. pendulum	**E**	**E**	+	-
C. frutescens x *C. chinense*	+	+	+	+
C. frutescens x *C. baccatum* var. pendulum	++	++	+	+
C. chinense x *C. frutescens*	+	+		+
C. chinense x *C. annuum*	+	+	+	+
C. chinense x *C. pubescens*	**E**	**E**	-	-

Inter-specific hybridization and resistance breeding

There are more than 30 species of wild type out of which only 5 are cultivated and grown in different part of the world. *C. annuum* L. includes mainly 3 closely related species. As *C. anunum* L, *C. frutescense* & *C. chinense* are widely grown all over the world.

C. annuum includes most of the Mexican type chilli or pepper and majority of the hot pepper of Africa and Asia and some varieties of sweet pepper are grown in temperate climate of the country. *C. frutescense* and *C. Chinense* are well adapted to the humid, low land tropics of Latin America. C. frutescense is also cultivated in Africa and Asia as a spice crop where as *C. baccatum* and *C. pubescens* are mostly cultivated in the Latin America.

The crossibility of the different species and hybrid fertility 1st time reported by Smith and Heiser (1957) is presented in the table as follows.

Table 23. Inter species hybridization in chilli

Name of insect	Source of tolerance
Thrips	Calapin Red,Chamatkar,N.P.46A, X1068, X743,B. G.4, X230, X233, Pusa Jwala, and Pant C-1
Mite	LEC-1, X1068, Pant C-1, X204, Punjab Lal & Kalyanpur Red Goli
Aphids	LEC-28, LEC-30, LEC-1, X-1068, Pant C-1, Punjab Lal, Kalianpur Red, and Pusa Jwala

Some achievements in developing breeding lines or varieties of *C. annuum* by inter specific hybridization were made through transfer of B52 gene for bacterial spot resistance from *C. chacoense and* partial tolerance to CMV from *C. baccatum* var pendulum was TMV resistance from *C. chinense and C. chacoense.* Recessive poty virus resistance from C. *chinense* and C. *chacoense* as root rot resistance from *C. ciliatum.*

Viral diseases of chilli

Many virus diseases are reported in chilli. There are about 14 viruses were reported in chilli causes damage to the crop in early stages. The losses are greater in early infections, many viral diseases are reported in Brazil, USA, Columbia, India, Israel, Japan, Korea, Mexico, and Rico. Pepper is susceptible to over 50 viral diseases. Some are common as TMV, CMV, PVY, PMV, TEV, and TRSV (Tobacco ring spot virus) are causing moderate to heavy losses in most of the chilli growing areas. Punjab Lal a multiple virus resistant cultivar crossed with two susceptible line viz. Ludhiana Local Sel. and Hungarian Sweet Yellow to stay the genetic control of virus resistance against chilli mosaic. It was suggested that resistance is grown by monogenic recessive gene and advocated back cross breeding to recover resistant type. One line i.e. PI 342947 possesses resistance to Texas isolates of TMV, TEV and PVY. This line used in multiple virus resistance breeding programme with a vector (resistant to TEV and PVY) for evolving Hidalgo Sarvcuro pepper. It was found that Sel 38-2-1, 964-9-3, 101-2-3 from the cross of Pusa Jwala x Delhi local posses tolerance to TMV, CMV, PVY and TLCV (Peter and Pradeep Kumar, 1999).

In tropical regions, TLCV is an important virus causes sever losses in chilli or pepper. In association with sucking insects like thrips, aphids, flies, mites etc. TLCV causes heavy damage to the crop particularly in summer months. Puri Red and Puri Orange become susceptible now by the attack of TLCV which were reported resistant earlier. Pusa Jwala developed by crossing (NP 46A x Puri Red) found to be tolerant to TLCV and mosaic (Tewari and Ramanujam, 1974).

Later on it become susceptible under intensive cultivation. During 1980 a local chilli (IC 3133a), perennial with purple unripe fruits, small, erect was reported to be tolerant to TLCV. Pusa Jwala and a cross was made between Pusa Jwala x IC 31399 and selections were made in advanced generations for TLCV and mosaic resistance along with earliness with desirable fruit characters of Pusa Jwala. Thus, a new cultivar Pusa Sadar Bahar was released for commercial cultivation in India. by Tewari (1991).

Research in chilli

Research on chilli initially has focus on high yielding varieties. The chilli research work started during 1928 at Lam, (AP), combatore (TN), Kovil Patti (UP), ICAR (New Delhi, Sabour Bihar). Play a major role in the improvement of chilli cultivars. Lam K-1 (AP) has released 1st variety of chilli G_1 from a local material Naslla Padu in 1937. Later on G-2 was isolated from the bulk population of NP 46-A. Later on a large no. of varieties were developed by different Institute and universities for cultivation by adopting various methods .are given as follows

Table 24. Varieties of Hot pepper

Name of variety	Institute/ University
Bhagya Laxmi (G-4)	ARS, Lam
Andhra Jyoti (G-5)	-do-
CA-960 (Sindhur)	-do-
X-235	-do-
LCA-206	do- -
NP46-A	IARI, New Delhi
Pusa Jwala	-do-
Pusa Sada Bahar	-do-
Pant C-1	Pantnagar
Co-1	TNAU. Coimbarore
Co-2	-do-
MDUL-1	-do-
K-1	-do-
K-2	-do-

Contd..

Jabahar-218	JNKVV, Jabalpur
JCA-283	-do-
JCA-154	-do-
MUSALBADI	MPKVP, Rahuri
Phule sai	MPKVP, Rahuri
Phule Mukta	-do-
Phule jyoti	-do-
HC-28	CCHAU, Hjsar
HC-44	-do-
Punjab Lal	PAU, Ludhiana
PunjabGuchhdar	-do-
CH-1 ,CH-2	-do-
Kashi Anmol	IIVR,Varanasi
Solan Yellow	YSPUH&T, Solan

Table 25. Varieties of Sweet pepper

Name of variety	Institute/ University
Arka Lohit	IIHR,Bangalore
Arka Jyoti	-do-
Arka Basant	-do-
Arka Mohini	-do-
Arka Gaurav	-do-
Arka Abhir	-do-
Kalyanpur Red	CSAUA&T. Kanpur
Kalyanpur Yellow	CSAUA&T. Kanpur
Kalyanpur Chanchal	CSAUA&T. Kanpur

Table 26. Hybrids of Sweet pepper

F1 hybrids	Institute/University
Pusa deepti,(Kt-1)	IARI Katrain N. Delhi
DARL 202	DARL, Pithoragarh
Bhrat-1	Indo-Amarican
Hira	Nath
Indira, Lario	Sandoz
Early Bounty	Sutton
Nandani(6503), Alankar (6802), Harit (6703), Nutan(Jamini)	Nunhamps Proagro Seed

Multiple disease resistance

The breeding for multiple disease resistance is complicated due to variable nature of the insects and pathogens and bringing such a manifold number of resistant genes in a single plant. Generally resistance breeding is done for only disease and combined for many diseases in the advanced generations. In those cases back cross method is followed but when the nature of resistance is poly genic breeder may have to devise modifications as controlled mating. (Among the resistant progenies in the back cross, F_2 or succeeding generations). Some times, it may essential to go for a second crossing with resistant parent to bring in more resistant genes. For hot chilli, the Pant C-1 was resistant to four diseases and white Kandhari was found resistant to 3 diseases and Punjab Lal is also having multiple resistances.

During, 2002, (at Pantnagar) 837 lines were obtained from NBPGR Bhowali, screened for quantitative and qualitative characters during rainy season. Out of which 79 lines were found resistant to anthracnose, fruit rot, dieback, chilli mosaic virus, Y. L.C.V. under field conditions (Singh *et al.*, 2002).

Table 27. Multiple Resistance varieties of chilli

Variety	Resistance to
Pant C-1	Bacterial wilt, Phytophthera, Root rot, Root knot, Nematode, CMV, TMV, TLCV
Punjab Lal	CMV, CMMV, Bacterial Wilt, Spodoptera leaf eating catterpiller
Perinnial	CMV, TMV, TLCV, Potato virus
Pusa Sadabahar	CMV, TMV, LCV, Thrips
White Kandhari	Bacterial Wilt, Root knot, Nematode, phytophthera, Root rot
Tambeli-1	TMV, Potato virus, Tobacco etch virus
Tembeli-2	——— do ———
Myliddy-1	TLCV, Pepper mottle virus, Mites, Thrips, Aphids
Myliddy-2	——— do ———

Improvement through Heterosis

The term heterosis signifies increase or decrease vigor of F_1 hybrids over average performance of the parent or over better parent. Positive heterosis implies hybrid vigor and negative heterosis indicates heterotic depression of the F_1 hybrid as compared to the parent. Heterosis mainly applied for earliness, uniformity in shape and size, maturity and higher yield

The utilization of hybrid vigor is an important tool in plant breeding for the improvement of the crop. The magnitude of heterosis for maturity, plant height, fruit weight, fruit size and number of fruit per plant is considered importance. Heterosis in chilli was first reported by Desh Pande at IARI during (1933) and in okra at Coimbatore. Later on Pal (1945) also found that a hybrid between two genotypes of Pusa type gave higher yield, mature earlier and had thicker fruits than better parent. After that a number of scientist reported heterosis for different character in chilli (Singh and Singh 1978, Joshi and Singh, 1980, Hundal and Khurana, 1988 and Ram, 1982) got early yield with the cross of NP 46 x K. yellow and Pant C-1 x K yellow for number and weight of fruit (Tewari, 1984) and Ram and Lal (1989) and Singh (1998) observed heterosis for early yield between the cross Pant C-1 x BC-14. Singh *et al.* (1999) found early maturity and higher yield among the crosses between Pant C-1 x Pb. Surakh and Pant C-1 x JCA 283. The cross between perennial x Punjab Lal produced 235.5 % greater yield (Singh, 1997); Peter (1988) reported early yield in Rayboy x KAU. Singh (1987) found 70.9% higher yield in the crosses of MS 12 x S-27, MS-12 x LLS; and MS 12 x S -25.30 produced 70.0%, 81.9% and 93.0%, respectively. Tewari (2001) found resistance against anthracnose in PC 3 x Phule; Tewari (2001) recorded higher yield among the crosses of chilli (PS3-4 x AC-550) and Phule Sai x Pusa Sadabahar; Pusa Sadabahar x AC- 515 and PS-3-4 x AC-515 for higher and early yield of chilli.

Singh and Jain (1999) observed heterosis for yield when they have crossed Pant C-1 x BC -24, Pant C-1 x JCA- 283 and Pant C-1 x Pb Surkha. Tewari (2001) observed better crosses for higher yield as PS 3-4 x AC -515, Pusa Sadabahar x PS 3-4, Pusa Sadabahar x AC 515 better cross for more number of fruits per plant, yield per plant and diseases resistance. PC-3 x Phule Sai, PS B x Phule Sai, PS-3-4 x AC- 515 showed resistance/tolerance against anthracnose, fruit rot under field conditions. Pusa Sadabahar x Pant Sel -15 was found resistant to LCV, PC-3 (P. Sel 15 x AC- 515 for higher vitamin C content. By using genetic male sterile line, Hundal and Khurana (1988) were able to produce heterosis over Punjab Lal *as* compared to standard variety upto *235.7* % for green fruit yield and 138.00% for red ripe fruits.

In pepper commercial hybrid in red pepper have developed in Hungary during 1967. In India, 1st commercial hybrid Bharat-1 was released from IA.H.S.Co. Bangalore by Attawar and Bahtt (1973). Later on Singh and Singh (1977) reported another hybrid Kalyan for commercial cultivation and later on Pusa Deepti from Katrain, DARL -202, DARL-204 have been released for cultivation and Private sector have developed a number of hybrids in chilli and sweet pepper.

Improvement by using male sterile line

The use of hand emasculation and hand pollination is highly uneconomical method of seed production in chilli crop. Therefore, the use of male sterility for production of F_1 hybrids has been reported by various workers (Singh and Kaur, 1986). By using genetic male sterile line, Hundal and Khurana (1988) were able to produce the hybrids namely C H-1, (MS- 12 x LLS) and CH-3 (MS 12 x S-2530) and these hybrids were released for commercial cultivation in the country during 1992 and 2000, respectively by PAU Ludhiana.

The genetic male sterility is an important pollination control mechanism which is exploited commercially for hybrid seed production in chilli. The genetic male sterile line 1st time reported by Peterson (1958) in chilli later on male sterile plant has been observed in both sweet and hot pepper. Murthy and Laximi (1979) observed male sterility is either genetic or genetic cytoplasmic but not purly cytoplasmic. Genetic male sterility was found in bell pepper cv. 'All big' it was controlled by single recessive mutation designated genes as msms has been isolated from California Wonder which can be utilize to produce F_1 hybrid in Sweet pepper.

Later on a male sterile line in hot pepper in cv. G-2 has been recorded at A.P by Murthy and Laxmi in 1979. Later on PAU scientist were able to produce GMS and CMS line in the production of hybrids. The genetic male sterile line MS-12 was developed and utilized to develop the hybrids as CH-1, CH-2 and CH-3. Punjab Lal through back crossing (Singh and Kaur, 1986) by which this male sterile line (MS- 12). They have developed two F_1 hybrids as CH-1 (MS 12 x L.L.S.) and CH-3 (MS-12 x S 2530) for commercial seed production . Patel *et al.* (1998) also reported genetic male sterile line ACM 52, having monogenic recessive gene (cms_2) but no hybrids has been developed by using this male sterility.

Cytoplasmic genetic male sterility (C GMS) is also available in pepper it was 1st reported by Peteson in (1958) in an introduction of *C. annuum* from India (PI 164835). Most authors have reported that a single nuclear gene (rfl) interact with Scytoplasm to produce sterility and the restorer allele Rfl restorers fertility. Plants with N cytoplasm are fertile regardless of whether they have the Rfl or rfl allele. The cytoplasmic factor S interact with recessive nuclear gene ms and produce a male sterile line of the genotype. Smsms male sterile and known as line A in the hybrid seed production programme. The genotype with N cytoplasm is known as N ms ms and is male fertile. This is known as B line and is used as a maintainer for the male sterile line, S ms ms (A line) after repeated back crossing the male sterile A line and the maintainer line (Bihar) become almost isogenic. To produce hybrid seed, the line A is inter planted with the pollinator or C line having genotype N Ms Ms. The main advantage of the GCMS system over the GMS is that can get 100% male sterile plants for direct use as females. However, this type system (CGMS) was not exploited commercially to produce hybrid seed because of instability under fluctuating conditions such as temperatures and a low rate of natural cross pollination in cultivated peppers.

The male sterile line MS-12 was developed by transforming sterility gene from France (ms -509) into the cultivar Punjab Lal and by back crossing the sterile line was developed as MS-12 .which was used to develop the hybrids for commercial cultivation..

Identification of male fertile & sterile plants in female parent (MS 12)

Identification of male sterile and male fertile plants in female parent is done when plants are in blooming stage. Generally anthesis occurs between 5.0 to 7.0 A.M. and anthesis is followed by dehiscence. After dehiscence the plants can be checked for the presence or absence of pollen grains in the flower which can further verified on a black paper or cloth.

Evaluation of germplasm

Total 837 germplasm evaluated for different characters at NBPGR, Bhowali. The promising lines are as follows.

Table 28. Germplasm lines of chilli for disease resistance (Total no. 837, Resistantno.79)

PCPGR DOCUMENT F./NO.VI (Pantnagar, 2002)

Genotypes0.	Leaf spot	Fruit rot	Die back	C.M.V.	L.C.V.
EC 246019	R	R	R	R	R
EC 246024	R	R	R	R	R
EC 382079	R	R	R	R	R
EC 382175	R	R	R	R	R
EC 392679	R	R	R	R	R
IC 92124	R	R	R	R	R
IC 119276	R	R	R	R	R
IC 119277	R	R	R	R	R
IC 119285	R	R	R	R	R
IC 119382	R	R	R	R	R
IC 119389	R	R	R	R	R
NIC 20880	R	R	R	R	R
NIC 21577	R	R	R	R	R

1. At Pantnagar, 837 germplasm lines were evaluated for qualitative and quantitative characters. 79 genotypes were found resistant against leaf spot, fruit rot, die back chilli mosaic virus and LCV as reported by Singh *et al.*, 2002.
2. Regional Research Station, Hyderabad, 126 lines evaluated for 19 characters as reported by Yadav (2007). Line IC 4137 was early (39 days to flowering) and IC 446497, 446523, 4116215 were promising for yield per plant.
3. Kalyani BCKV evaluated 50 germplasm lines, the line BCC 32, 31, 30, 41, 27, 24 having high amount of ascorbic acid content than others and BCC 40, 30 and 31 were very early for 50% flowering.
4. Regional Research Station, Lam, (A.P.) registered 11 entries in NBPGR and obtained the IC number.
5. AVRDC-13, 16, CNPH-16, 2689, 26 were having resistance against anthracnose disease. EC 334206 found water logging tolerance, EC 32333 having multiple resis5tance . and CO 309, IC 383072, IC 364063 were found free from LCV under field conditions.
6. Highest capsaicin content was recorded in GP- 186 (0.99%) and lowest in 0.094 in GP -158. Capsanthin content was found greater in GA -159 (88450 E OA,).
7. Out of 300 lines, Oleoresin content was greater in GP -125 (16.72%) and lowest in GP- 152 (4.0%) only.
8. Drought resistant line NIC 23837, NIC 23797 were identified during 1999 and IC 119611 were resistant to mosaic virus.

9. PCWR-1, PCWR-2, PCWR-3 were reported as bacterial resistant line from Palam pur. EC- 580000 was found promising fruit yielder produced 1116.67 g per plant than rest seven other genotypes (Annual Report, AICVRP. 2007).
10. The scientist of PAU developed MS-12 line which cary genetic male sterility (GMS) controlled by recessive gene (msms). The male sterile line MS-12 was developed by transforming sterility gene from France (ms -509) into the cultivar Punjab Lal and by back crossing the sterile line was developed as MS-12.which was used to develop the hybrids for commercial cultivation..

Table 29. Promising entries registered in NBPGR & Obtained the IC NO.

Name of variety	National identity no.	Name of center
LCA 235	IC 548036	Lam
LCA 305	IC508037	AICVIP report
LCA 444	IC508038	May 2007
LCA436	IC508039	
LCA 424	IC508040	
LCA 353	IC508041	
G- 4	IC508042	
G -5	IC508043	
CA 960	IC508044	
LCA 315	IC508045	
LCA334	IC520856	

Future Thrust Areas

1. Collection evaluation, maintenance, documentation and exchange of germplasm
2. To develop high yielding varieties as well as hybrids in pepper
3. Development of varieties or hybrids resistance to biotic as well as a biotic stresses
4. Identification of germplasm for quality attributes such as capsaicin, oleoresin and capsanthin
5 Identification of genotypes to a specific character.
6 To develop varieties or hybrids having multiple disease resistance
7 Production of F_1 hybrid seed for commercial purpose to the growers
8 Development of varieties/hybrids having resistance against salinity, alkalinity high and low temperature and high moisture content (Water logging)

References

Anonymous, 2004, Area, production and Productivity of vegetables. Proceeding: Impact of vegetable Research in India ed. by S. Kumar Joshi, P.K. Suresh Pal, 2004, Page 70-71.

Arya, P.S. And Saini, S.S. 1977. Variability stud in pepper *cap. annuum* L. var. *Indian. J. Hort.*, **34.(4):** 425-32

Attavar, M. and Bhat, N. K.1973. High yield Indo American Capsicum hybrid. Lal Bagh.**18.(2):** 28-32.

Awasthi, D. N., Joshi, S. and Ghilayal, P.C. 1976. Studies in genetic variability, heritability and genetic advance in chilli. *C. annuum. Prog. Hort.,* **8:** 37-40.

Bhatt, J. P., Shah Deepak and Shah, D. 1996. Genetic Variability in hot pepper. *Recent hort.,* **3.(1):** 79-81.

Boswell, V. R 1949. Garden pepper, both a Veg. and a condiment.in our Veg. travelers. *Natl.Georg.,* **96.(1):** 145- 47

Chauhan, D.V.S. 1996. Chilli: In Vegetable Production in India 5th edition. Ram Prakash & Sons Agra. Page 356-366.

Cheema, G.S. Nagareth, B ana Dhareshwar, S.R. 1994.Improovment of chilli by selection in Bombey province. *Indian J. Hort.,* **2(1):** 49-61.

Despande, R.B. 1933. Studies in chilli 111. The inheritance of some charactes in *C. annuum* L. *Indian J. Agric. Sci.,* **3(1):** 219-300.

Dhanraj, K. S., Seth, M. L. and Basal, H. C. 1968. Reaction of certain chilli mutant and varieties to leaf curl virus Indian *Phyto pathology,* **21:** 342-43.

FAO, 2002. Chilli And Pepper green, FAO Production, Year book. **56:** 153-54

Green, S. K. and Kim, J. S. 1994. Source of resistance to virus of pepper (*Capsium annuum* L.). A catalogue AVRDC. Tech. Bulletin **20:** 72 page.

Gupta, M. K. 2003. Performance of chilli genotypes under Tarai conditions of Uttaranchal. M. Sc. Ag thesis GBPUA & T Pantnagar. 123p.

Hosmani, M. S. 1982. Chilli. University of Science, Dharwad, 206 p.

Hundal J. S. and Khurana, D. S. 1988. Heteerosis potential in chilli Proceeding National Seminar on Chilli. Ginger and Turmeric, held at 11-12 Janunary 1988 at Hyderabad, p. 33-57.

Hundal, J. S. and Dhak, R.K. 2006. Breeding for hybrid hot. Pepper. *In:* Hybrid Vegetable. development. (ed.) Singh, P. K., Das Gupta, and Tripathi, S.K. International Book Distribution Co. Lucknow. 1st pp 31-50.

Hundal, J. S. and Khurana, D.S. 2001. A new hybrid of chilli CH-3. *Suitable for processing Amer. Nat.,* **19:** 544-550.

Hundal, J. S. and Khurana, D.S., 1993. CH-1. A new hybrid of chilli Prog. Hort. 29-1: 11-13. *J. Res. PAU.,* **39.(2):** 326.

Joshi, G. C. and Choudhary, B. 1981. Sreening of Lycopersicon and Solanum species for resistance to leaf curl virus. *Vegetable Science.* **8(1):** 45-50.

Joshi, S. and Singh, B. 1980 A note on hybrid vigour in sweet pepper. *Haryana J. Hort. Sci.,***9.(1/2):** 90.92.

Kalloo, G. 1996. Solanacious Vegetables in 50 years of crop Science Res. In India ICAR Publication page 574-592.

Lal,G and Singh D.K.1997. Varietal trial on chilli. Annual Res. Rep. submitt. to DES, Pantnagar.

Mathai, P. J., Dubey, G. S., Peter, K.V., Saklani, U.D. and Singh, N. P. 1977. Pant C-1 and Pant C-2 two new promising selections of chilli (*C. annuum.* L.) South Indian Hort. **25(4):** 123-25.

Munshi, A. D. 1998. Dev of F_1 hybrid and tech. of hybrid seed production of chilli and sweet pepper. Summer school on hybrid and hybrid seed production technology of vegetable. Pub by Division of Veg. Crops IARI, New Delhi, (sponsored by ICAR) Page 117-124.

Munshi, A. D. 1999. Recent advances in breeding chillies for biotic and a biotic stresses. In Summer school on Advances in breeding of temperate and tropical vegetables for biotic and abiotic stresses. Pub by Division of Veg. Crops IARI, New D Delhi, (sponsored by ICAR) Page 133-140.

Murthy, S.R.N. and Laxmi, N.1979. Male sterile Mutant in cap. annuum. *Curr. Sci.,* **48(7):** 312-14.

Nandpuri, K.S., Gupta,V.P. and Thakur, P.C. 1971. variability studies in chillies *J. Res. P.A.U.* **8:**311-312

Pal, B. P. 1945. Studies in hybrid vigour. II note on manifestation of hybrid vigour in gram, Sesamum and chillies. *Indian J. Genet. Plant Breeding.* **5(1):** 106-14.

Pamila, 2005. Evaluation of chilli genotypes for qualitative and quantitative traits. M.Sc. (Ag) thesis G.B.P.U.A. & T., Pantnagar.

Patel, J. A. Shukla, M.R., Doshi, K.M. Patel, B.R. and Patel, S.A. 1998. Combining ability analysis for green fruit yield and yield components in chill. capsicum and egg plant *Newsletter*, **17.(1):** 34-37

Patel, J. A. Shukla, M. R. Doshi, K.M. Patel, S.B. and Patel, S.A. 1997 Hybrid vigour of quantitative traits in chilli. *Veg. Sci.*, **24.(2):** 107-10.

Peter K.V. 1997. Research advances in chilli breeding in summer school on Adv. Tech. in improvement of vegetable crops including Cole crops sponsored by ICAR held at Trissure w.e.f. May, 4th-24th at K.A.U. P, 1-12.

Peter K.V. and. Tewari P. K. 1999. Breeding for resistance to biotic and abiotic stresses in chilli. In Summer School on Advance in Breeding of Temperate and Tropical Vegetable Crops for Biotic and Abiotic stresses. By ICAR held at 17th to 6th June, 1999. Division of IARI, New Delhi Page 141-144.

Peter, K.V. 1998. Spices research and development. An update over view *Agro India* Aug. 16, page. 18

Peterson, P.A. 1958. Cytoplasmically inherited male sterility in capsicum. *Amer Naturalist* **92:** 111-119.

Purseglove, J. W. 1977. Tropical crops Dicotyledons I and II, Longman, London pp 524-25.

Ram B, and Lal, G. 1989. Heterosis analysis breeding depression in chilli *Prog. Hort.* **21(3-4):** 368-372.

Ram. B. 1982. Genetic variability in chilli. Ph. D. thesis GBPUAT, Pantnagar, page 94.

Ram. H.H. 1997. Bell pepper and chilli In. Veg. Breeding Principles and practices 1st Edition Kaly Pub. New Delhi, pp 195-20.

Safford, W. E. 1926. Our heritage from the American Indians Annu. Rep. Smith Sonian Inst. 1926: 405-10.

Saxena, A. Hundal, J.S. And Dhall, R.K. 2005 Fruit and seed setting studies on male sterile line of chilli. *Indian J. Hort.* **62(2):** 206-209

Sharma, O.P. and Singh, J. 1985 Reaction of different genotypes of pepper to cucumber mosaic and tobacco mosaic virus .*Capsicum new letter*. **4(1):** 47

Sharma, P.P. and Saini, S. S. 1977. Heterosis and combining ability for yield and agronomic characters in pepper. **4:** 43-48

Singh, A. and Singh H. N, Mittal, R.K.1973 Heterosis in chilli. Indian J. Genet. & Pl.Breeding. **33(3):** 398-400.

Singh, A. and Singh H.N. 1977. New hybrid of chilli. *Intensive Agriculture*.**15(5):** 15

Singh, A. and Singh H.N. 1978. Genetic divergence and heterosis in chilli *Plant Sci.* **10:** 17.

Singh, A. and Singh H.N. 1978. Combining ability in chilli. *Indian J. Agri. Sci.***48(1):**29.34.

Singh, D. K., Verma. S.K. and Jain, S. K. 2004 Performance of chilli genotypes. *Scientific Hort.***9(1):** 135-39.

Singh, D.K.and Jain, S.K. 1999 Development of new crosses in chilli. *Annual Res. Rep.Submitt.*

Singh, Durvesh., K., Akhatar, Jamil and H.H. Ram (2002). Evaluation of chilli germplasm from NBPFER R.S. Bhowali. PCPGR Document F. No. VI Prepared by PCPGR Pantnagar Aug. 2002.

Singh, J. and Kaur, S 1990. Development of multiple resistances in chilli. Malaysian plant protection Society (Malaysia). Proc. 3rd International Conference of Plant Protection in the Tropics.**V: 51**. 20-23 March.

Singh, R.V., Ram, H.H., Tweari, J.P. 2003. Role of indigenous germplasm in improvement of Hort. Crops. Proceeding of National Seminar organized by PCPGR wef. 24-25th June 2003.Pantnagar current status of Veg. Crops improvement using indigenous germplasm by H.H. Ram 2003. Page 1-23.

Singh, S. N., Srivastava, J. P. and S Ram. 1994. Natural out crossing in chilli *Veg.Sci.* **21(1):** 166-68

Smith P.G. and Heiser, C. B. 1957. Breeding behavior of cultivated pepper. *Proc.Amer. Soc. Hort.Sci.*, **70:** 286-90

Sturtevant, E. L. 1885. Kitchen garden esculents of American Origen. II pepper.

Tewari, P. K., 2001. Heterosis in chilli., M.Sc. Ag. Thesis submitted to GBPUAT, Pantnagar. p 85.

Tewari, R.S 1984. Trait wise selection efficiency and response of genotypes to application of NAA and fertility levels in chilli (*C. annuum*, L.) Ph. D. thesis GPPUA& Technology, Pantnagar. p.116

Tewari, V.P, 1987. Selection of promising line in perennial chilli (*Cap. Frutescense) Capsicum Newsletter.* **61:** 45-46

Tewari, V. P., 1991. A multi purpose perennial chilli. Pusa Sadabahar. *Indian Hort.* **35(4):** 29-31.

Tewari, V.P. and Viswanath, S. M. 1986. Breeding for multiple virus resistance in pepper. *Capsicum newsletter*. **51:** 49

Tewari, V.P. and Viswanath, S. M. 1988. Note on breeding for resistance in chillies today and tomorrow. *New Botanist.***15(23):** 185-86.

Tewari, V. P. and Ramanujam, S.1974. Grow Jwala A disease resistant high yielding chilli. *Indian Farming.* **24(1):** 20-21

Thakur, M. R. Singh, J., Singh, H. and Sooch, B. S. 1987. Punjab Lal a new multiple disease resistant chilli variety. *Prog. Hort.* **23.(9):** 11-16.

Thakur, P.C.1987. Gene action: an index for heterosis breeding in sweet pepper. *Capsicum News letter.* **6(1):** 41-42.

Thomas, P and Peter, K.V. 1988. Heterosis in inter varietal crosses of bell pepper (*C.annuum* var. grossum) and hot chilli (*C.annuum* var. fasciculatum). *Indian J. Agri. Sci.* **58(10):** 747-56.

Verma, A. K., Sharma, H. R. and Rattan R.S.1995. Heterosis studies in Bell pepper. National Sym. on Recent Dev. in Veg. Imp. Abstract, P-9, Indian Soc. of Veg. Sci. ISVS, Varanasi. IGKVV, Raipur 58p.

Verma, S.K., Singh R. K.and Arya, R.R. 2004 Genetic variability and correlation Studies in chillies.*Prog.Hort.* **36(1):** 113-17

Yadav, S.K. 2007. Collection, Evaluation and Conservation of germplasm. Annual Report of AICVRP 25th Group Meeting held at CCS HAU, Hissar. 30 May 2007.

❑❑❑

Chapter – 34

Genetic Resources for Potato Improvement in India

Dhirendra Singh and K.P. Singh

Introduction

Potato is a wonder crop that can be grown under diverse range of agro-climatic conditions. Short duration; high yield per unit area and time and wide flexibility in planting and harvesting time are important virtues of potato that enable its inclusion in intensive cropping systems.

Potato is not an Indian crop but is a native of high Andean region of South America. It was introduced in India perhaps in 16th or early 17th century by either the Portuguese traders or the British missionaries. The present day cultivated potatoes (tetraploid, 2n=4x=48) in most of the world represent *Solanum tuberosum* spp. *tuberosum* and *S. tuberosum* spp. *andigena*, although it's numerous diploid (2n=2x=24) relatives are still under cultivation in and around its primary and secondary centers of origin in South America (Pandey and Sarkar, 2005). The great Irish famine in the 19th century, caused by late blight disease, nearly wiped out all the earliest potato introductions in the Europe and dramatically changed the potato cultivation and production scenario in rest of the world.

Unlike in European Countries 80% of the potato is grown in India under short day conditions as a short duration crop i.e. 70-90 days. The tubers are exposed to high temperatures during planting (September-October) and harvesting (February-May). The crop is irrigated and suffers mid day water stress. The potato varieties introduced from Europe were unsuitable for growing under the above conditions. Systematic potato research in India spanned over the last half-a-century of the 20th century in the past millennium. The Central Potato Research Institute (CPRI), which was established in 1949, has bean mainly responsible for tremendous growth of potato production in the country. The indigenous technological advances made it possible to extend potato cultivation in almost all part of the country. Now in India, compared to the production, area and yield of potato in 1949-50, the increase over the same period was 1513%, 559% and 270% respectively (Naik and Thakur, 2007). According to the FAO, potato production worldwide stands at 329 million tones and covers more then 19 million ha.

Origin of Cultivated Potato

Hawks (1990) concluded that potato may have been domesticated in what in new the Lake Titicaca to take Pope Region of North Bolivia and it originated from the wild diploid species *S. leptophyes* some 10,000-7,000 years ago.

1. The first domesticated species was *Solanum stenotomum.*
2. *S. tuberosum* is a straight tetraploid of *S. stenotomum* but there are stronger evidences in support of the allotetraploid origin of *S. tuberosum* by hybridization between *S. stenotomum* and *S. sparsipilim.*
3. A diploid cultivated species, *S. phureja* evolved from *S. stenotomum* by human selection exercised for rapid maturity and lack of tuber dormancy to develop varieties to be grown 2-3 times in a year in the lower, Eastern frost free Andean valleys (Hawes 1990).
4. During the course of evolution diploid species *S. megistacrolobum* and the tetraploid spices *S. acacule* contributed frost resistance.

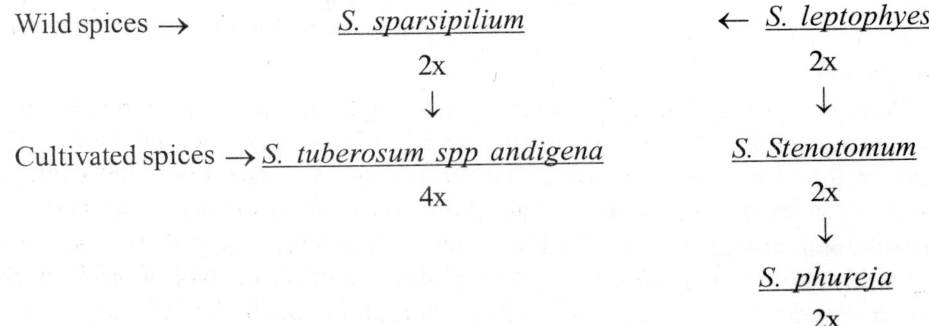

Figure 1. Evolution of cultivated potatoes.

Cultivated Potato Species

1. Diploid species (2n=2x=24)

Solanum stenotonum (Juz. and Buk.): The most primitive in appearance and cultivated in high mountainous regions of Southern Peru and Central Bolivia. Some forms are resistant to frost.

Solanum phureja (Juz. and Buk.): It is cultivated mainly in hot climates without frosts or drought, on the slopes of the Andean Mountains in Ecuador, Venezuela, Columbia, Peru and Northern Bolivia. The tubers of this species never remain dormant.

Solanum ajanhuiri (Juz. and Buk.): It is resistant to frost and cultivated in Northern Bolivia. It develops fine blue blossoms.

Solanum goniocalyx (Juz. and Buk.): It is cultivated in Peru. Flesh colour is intense yellow.

2. Triploid species (2n =3x=36)

Solanum chaucha (Juz. and Buk.): It is cultivated in central Bolivia and Peru.

Solanum juzepezukii (Buk.): It is resistant to frost. It is cultivated in high mountains of Peru and southern Bolivia. It develops fine blue flowers.

3. Tetraploid/autotetraploid species (2n=4x=48)

Solanum tuberosum subspecies *andigena*: It is cultivated in South and Central America

Solanum tuberosum subspecies *tuberosum*: It is cultivated all over the world.

4. Pentaploid species (2n= 5x=60)

Solanum curtilobum (Juz. and Buk.): Resistant to frost. It is cultivated in high mountains of Peru and Southern Bolivia. It develops large violet flowers.

Organization of Potato Research and Development in India

The beginning of potato research and development efforts by these Governments Continued up to 1935 when potato research was organized at national level by the then Imperial Council of Agriculture Research (New Indian Council of Agricultural Research). IARI opened a research station for potato breeding in Shimla hills and later at Kufri and Bhowali in 1945. To strengthen the potato improvement programme in India Central Potato Research Institute was established in August 1949 at Patna. After establishment of the Institute, it was soon realized that Patna was not an ideal place for carrying out research on potato because of the following constraints:

1. Prevalence of high aphid population did not allow healthy seed production as envisaged at the time of establishment of the Institute at Patna.
2. Rapid degeneration of the stocks due to high incidence of viruses did not allow maintenance of germplasm and segregating hybrid progenies in healthy condition.
3. Most of the genotypes did not flower under short photo-period prevalent in plains hence a wide spectrum of valuable germplasm could not be exploited in the breeding programme.

As a result, significant advancement in potato research could not be made at Patna during 1949 to 1956. The headquarters of the Institute were therefore shifted to Shimla in high hills of Himachal Pradesh where the above problems did not exist. Patna was retained as a regional station for Eastern plains. The Institute comes under ICAR in the year 1964. In 1971, All India Co-ordinated Potato Improvement Project was started by ICAR with Headquarter at Shimla with a chain of centers at State Agriculture Universities (SAUs) and in are AICRP (Potato) Centers located in 17 states of these, 7 centers are located at Regional Research Stations of CPRI and remaining 15 centers are based at the SAUs. The AICRP (Potato) network covers all distinct agro-climatic zones in the country. The project has been instrumental in coordinating and monitoring multilocational trials with improved production vis-à-vis identification of remunerative potato based cropping systems, plant protection measures at increasing production, productivity and utilization of potato in the country.

Genetic research and varietal development of potato in India

As potato is not native to India and so not much indigenous variability is available. Originally introduced potato type was Andigena which lost their identity and come to be known by local names viz., Phulwa, Satha, Gola and DRR. From as early as 1824 to 1939 several *Tuberosum* and *Andigena* varieties were introduced by various state/ provincial governments particularly, Bombay, Madras, Assam, Uttar Pradesh and Punjab. But yield of the introduced varieties were not consistent. The main reason was degeneration due to virus infection. Limited efforts made to evolve new varieties through crossing in Bombay, Assam, Punjab, Madras and Mysore, were not successful probably because of lack of information on combining ability of the parents, techniques to raise seedlings and handling and storing of seedling tubers.

During 1935 to 1955, nearly 400 genotypes grown in the country under different names were collected and studied. This led to identification which were being grown under different names. The study revealed that native varieties like Phulwa, Satha, Gola and Surkha which were of Andigena type were popularly in the plains and the European cultivars like Great Scot, Craig's Defiance, Magnum Bonum and Up-to-Date and Craig's Defiance (CD) in the North Indian hills and plains, and President and Great Scot in southern hills remained under cultivation till recently. The varieties collected in the survey formed the basic collection to which new exotic varieties were added to build up a modest germplasm collection. This collection was assessed for direct introduction and as parental lines for use in the breeding programme crosses were also made between widely grown native varieties and promising exotic introductions to breed cultivars suitable for sub tropical conditions in the plains and the temperate conditions in the hills. These efforts were not very successful mainly because of degeneration of breeding material during the period of testing in the plains. Attempts were made to overcome this problem by developing regional pattern of breeding wherein the initial breeding material was developed and maintained at Kufri, Shimla and part of the material was sent for testing in the regional stations located in the plains.

The major break-through in the potato improvement programme came with the development of seed plot technique (Puskarnath, 1967) which made it possible to raise, select and multiply breeding material under disease free condition in the plains. Starting from 1958 to date, a total of 41 high yielding varieties have been released for different agro climatic situations in the country but at present 23 hybrids are notified (Table.1).

Table1. List of varieties released by CPRI and AICRIP

Year of Release			Variety Salient features and Adaptability
1.	1958	**Kufri Kisan**	Late-maturing. North Indian plains
2.	1958	**Kufri Kuber**	Medium-maturing. Bihar and Maharashtra
3.	1958	**Kufri Kumar**	Late-maturing and moderately resistant to late blight. North Indian hills
4.	1958	**Kufri Kundan**	Medium-maturing, moderately resistant to late blight and good keeping quality. Himachal Pradesh and hills of Uttar Pradesh
5.	1958	**Kufri Red**	Medium-maturing and good keeping quality. Plains of Bihar and West Bengal
6.	1958	**Kufri Safed**	Late-maturing and good keeping quality. North Indian Plains
7.	1963	**Kufri Neela**	Late-maturing and moderately resistant to late blight. Nilgiri Hills
8.	1967	***Kufri Sindhuri**	Late-maturing, essentially short day adapted variety with red tuber. Heavy yielder even on low inputs. North Indian plains
9.	1968	**Kufri Alankar**	Medium-maturing. North Indian plains
10.	1968	**Kufri Chamatkar**	Late-maturing and resistant to early blight
11.	1968	***Kufri Chandramukhi**	Early-maturing and good for processing. North Indian plains and plateau region of peninsular India
12.	1968	**Kufri Jeevan**	Late-maturing. Himachal Pradesh
13.	1968	***Kufri Jyoti**	Medium-maturing, good for processing, field resistant to late and early blights and immune to wart, and tolerant to viruses. Wide adaptablity
14.	1968	**Kufri Khasigaro**	Late maturing and resistant to both late and early blight. Hills of Meghalaya
15.	1968	**Kufri Naveen**	Late maturing and resistant to late blight and immune to wart. Northern hills of West Bengal and Meghalaya
16.	1968	**Kufri Neelamani**	Late maturing and resistant to late blight. Nilgiri hills
17.	1968	**Kufri Sheetman**	Medium to late-maturing and resistant to frost. North Indian plains and tarai area of Uttar Pradesh
18.	1971	**Kufri Muthu**	Medium-maturing and resistant to late blight. Nilgiri hills
19.	1972	***Kufri Lauvkar**	Early-maturing and rapid bulking under warmer conditions, suitable for processing. Plateau region of peninsular India
20.	1973	**Kufri Dewa**	Medium maturing, good keeping quality and resistant to frost. Tarai area of western Uttar Pradesh

Contd..

21.	1979	***Kufri Badshah**	Medium- maturing, resistant to both late and early blights and PVX. North Indian plains and plateau region of peninsular India
22.	1980	***Kufri Bahar**	Medium-maturing and heavy yielder. North Indian plains
23.	1982	***Kufri Lalilma**	Medium-maturing with red tuber and resistant to virus 'X' North Indian plains
24.	1983	**Kufri Sherpa**	Medium-maturing, resistant to late blight and immune to wart. hills of west Bengal
25.	1985	***Kufri Swarna**	Medium-maturing, resistant to late blight and cyst nematode. Nilgiri hills
26.	1989	**Kufri Megha**	Medium-maturing, late blight resistant. Hills of Meghalaya
27.	1996	***Kufri Ashoka**	Short duration (75days). Plains of central and eastern Uttar Pradesh, Bihar and West Bengal
28.	1996	***Kufri Jawahar**	Medium-maturing, resistant to late blight and ideal for intercropping. Punjab, Haryana and the plateau regions of Madhya Pradesh, Gujarat and Karnataka
29.	1996	***Kufri Sutlej**	Medium-maturing and resistant to late blight. Western and central Indo-Gangetic plains
30.	1997	***Kufri Pukhraj**	Medium-maturing and resistant to late blight. Northern plains and plateau region
31.	1997	***Kufri Chipsona-1**	Medium-maturing and resistant to late blight. Excellent for chip making. Indo-Gangetic plains
32.	1997	***Kufri Chipsona-2**	Medium-maturing and resistant to late blight. Excellent for Chipping. Indo-Gangetic plains
33.	1997	***Kufri Giriraj**	Medium to late-maturing and resistant to late blight. North western hills
34.	1998	***Kufri Anand**	Medium maturing and resistant to late blight. Heavy yielder. Northern plains
35.	1999	***Kufri Kanchan**	For North Bengal hills
36.	2005	***Kufri Surya**	For North Traditional warmer area of the country and both in cooler north western and warmer central plain
37.	2005	***Kufri Shailja**	For Indian hills
38.	2005	***Kufri Pushkar**	For Indian plains and plateau region
39.	2005	***Kufri Arjun**	For Eastern plains
40.	2005	***Kufri Chipsona-3**	High dry matter and excellent chip colour and suitable for Indo-Gangetic plains
41.	2005	***Kufri Himalini**	for Indian hills

*notified varieties

Some varieties have been released for India, are very popular in near by country of India. (Table.2)

Table 2. Indian varieties/hybrids adopted elsewhere

Country	Indian varieties/hybrids
Afghanistan	K. Chandramukhi
Bangladesh	K. Sindhuri, K. Lalima
Bolivia	I-1039 (India)
Madagascar	I-1035 (Mailaka)
Mexico	I-654 (CCM 69.1)
Nepal	K. Jyoti, K. Sindhuri
Philippines	I-1035 (Montanosa), I-1085 (BSUP-04)
Sri Lanka	Kufri Jyoti, I-822(Krushi), I-1085 (Sita)
Vietnam	I-1039 (Red Skin)

Breeding Approaches to Utilize Genetic Resources for Improvement of Potato

Breeding procedure for improvement of any crop may be based on breeding objective. Following few objectives may be considered for potato improvement in Indian conditions.

Breeding objectives

1. To develop early bulking varieties that can fit well in intensive cropping systems.
2. To develop high yielding and photo-insensitive varieties having wide adaptability.
3. To develop varieties having good quality tubers i.e. medium sized, oblong tubers with shallow eyes, yellow fleshed, uniform distribution of dry matter, high vitamin C and protein content, high specific gravity, low sugar content less black spot bruising and glycoalkaloids, long shelf life and firm cooking quality.
4. To develop transgenic potato varieties having novel traits of nutritional superiority, high starch content, resistance to abiotic and biotic stresses etc.
5. To develop varieties and agronomic practices suited for organic cultivation and production of 'Baby potatoes', which are in great demand in the international markets.
6. To develop varieties suited for processing with high specific gravity (dry matter content) and suitable for French fries, chips and dehydrated products, low sugar content for chips and French fries to avoid browning; Cold Chippers i.e. the varieties that do not accumulate reducing sugars when stored at low temperatures.
7. To develop varieties with field resistance/tolerance to important biotic stresses: late blight, early blight, charcoal rot, wart, common scab, bacterial wilt, soft rot,

viral diseases (potato virus X, potato virus Y), cyst nematode (*Globodera rustochinensis*, *G. pallida*), root not nematode (*Meloidogyne incognita*), resistance/tolerance to aphids (*Myzus persicae*), potato tuber moth (*Phythorimaea operculella*).

8. To develop varieties resistant to abiotic stresses – heat, drought, frost, soil salinity.
9. To develop varieties tolerant against shrinkage, rottage, accumulation of sugars especially reducing sugar, having reasonable dormancy.

Following approaches can be adopted to improve potato crop.

(A) Conventional breeding approaches

Potatoes are propagated vegetatively and therefore their breeding is considered to be easier than that of crops with sexual reproduction. It is an important aspect as any selection will keep true to type, expect for rare mutations, as long as vegetative reproduction is allowed. New varieties may be selected as plants from parental crosses in the F_1 generation. F_1 generations if self provides enough variability due to heterozygosis of parental cultivars and involvement of autotetraploidy.

However, selection of promising parental cultivars for production of F_1 progeny in which desired characters are to be combined is difficult task. Four main types of parents are available to the potato breeder. They are:-

1. *Tuberousum* varieties
2. *Andigena* varieties
3. Varieties of cultivated diploid species
4. Wild species

It is now generally agreed that for most parts of the world *tuberosum* variety able to produce a high yield under long day conditions, should always be one parent of a cross and where a desired character can not be found in *tuberosum* varieties, the next source should be an *andigena* variety, cultivated diploid potato and wild species in this order less farourable characters are difficult to be completely avoided. In order to avoid the recurrence of wild characters in the progeny, potato breeders usually do not use unimproved materials particularly the wild species in the crossing programme. Some of genotypes have been identified as donor for important traits in *tuberosun*, *andigena* and other species (Table 3 and 4). It is safer to use commercial cultivars. While planning a crossing programme, it should be kept in mind that many varieties are male sterile and hence selection of male parent is troublesome. In India crosses are normally made in open fields at Kufri in Shimla hills although Indoor hybridization is also possible.

(B) Use of dihaploids in potato breeding

The traditional breeding procedure in potato is to intercross superior autotetraploid parents and select F_1 individuals. Such selection is laborious and the chances of finding a superior new recombinant are remote as an meiosis the optimally balanced, highly heterozygous genotypes of each parent divides into numerous and diverse male and female gametes. In a cross combination these are fertilized by a completely random process. Efforts to induce desirable changes in useful varieties by mutagenic treatment, leaving the overall genotype largely intact, have not been encouraging. Further narrowness of the genetic base of the commercial *tuberosum* potatoes is another reason for relatively slow progress in conventional potato breeding.

Following potato breeding steps can be suggested to overcome the constraints in conventional breeding methods:

1. Raising plants from gametes (dihaploids) of *tuberosum* and *andigena* potatoes.
2. Vegetative propagation of the dihaploids to enble a reliable evaluation of these genotypes for various kinds of resistance, yielding ability and quality characters.
3. Hybridization of dihaploids with different semi-cultivated and wild diploid species to incorporate the desirable traits such as resistance to biotic stresses, genetic diversity and quality traits.
4. Sexual terrapins dization using 2n gametes formation in dihaploid species hybrid through unilateral (4x-2x or 2x-4x) and bilateral (2x-2x) matings.

There are three essential components to this breeding method:

1. Wild and cultivated relatives provide genetic diversity.
2. Dihaploids of *tuberosum* and *andigena* effectively capture this genetic diversity and put it into usable far.
3. 2n gametes effectively and efficiently transmit this diversity to cultivated 4x potatoes. These 4x genotypes are allotertraploids and are called meiotic tetraploids, to distinguish these from autotetraploids produced from dihaploids through mitotic manipulation. The practical utility of this non conventional method of potato breeding is getting to be realized on meaningful scale.

(C) Use of genetic engineering for utilization of potato genetic resources to improve potato crop

In the production of potato, late blight, bacterial wilt, viruses, the potato tuber moth and sensitivity to high temperatures are the major constraints which can be well addressed by development of transgenic potatoes. Genetically engineered potato can express high levels of pokeweed antiviral protein (PAV) which conferred resistance to PVX and PVY when mechanically inoculated. PAV is an example of a gene encoding antiviral protein also known as ribosome inactivating protein (RIP) which can modify ribosomal RNA and interfere with translation.Genetically engineered potato with bacteriophage T4 Lysozyme to potato cell wall can reduce the extent of tissue maceration when plants were challenged with *Erwinia carotovera.*

The expression of genes encoding the *Bacillus thuringensis* (Bt) is well known to protect potato against insect such as tuber moth and Colorado beetle. *Helicoverpa armigera* in one of the important insect pest adversely affecting the yield of potatoes in India. A synthetic gene encoding the insecticidal crystal protein (Cry1ab) of *Bacillus thuringensis* has been found to show the resistance against this pest.

There is also a vast scope of developing transgenic potato against cold stress, photo-oxidative stress etc. Potato transgenic with tolerance to higher temperatures would enable the extension of potato cultivation into the warmer regions of the country.

Dormancy plays an important role for storage of potato tuber. A huge amount of potato is loosed due to its short dormancy. For increasing the length of dormancy genetic engineering plays an important role. When potato plants were transformed with the antisense genes G1-1 showed a significance increase in length of dormancy.

The transgenic potato with altered carbohydrate metabolism for the industrial production of cyclodextrin, fructan and quality starch would overcome certain utilization constraints. The transgenic potato can be utilized to produce vaccines for major livestock diseases such as Rinderpest and Foot and Mouth Disease of cattle, and New Castle disease of poultry.

The Central Potato Research Institute (CPRI), Shimla in collaboration with the National Research Center on Plant Biotechnology (NRCPB) and National Center for Plant Genome Research (NCPGR), has developed and is developing, transgenic potatoes which will possess resistance/tolerance to late blight, resistance/tolerance to potato tuber moth and superior nutritional (protein) qualities. At CPRI, controlled research trials have been conducted with Cry1Ab transgenic lines of Indian potato cultivars and an application has been presented to seek permission from the RCGM for open field trials in hot spot areas. The transgenic lines developed using the other two genes *viz*. Ama1 gene cloned from *Amaranthus hypochondriacus* having high protein content and Tobacco osmotin gene for control of late blight, are under testing.

Table 3. Useful traits found in *Solanum tuberosum* ssp. tuberosum and ssp. andigena

Traits	Donors
(A) Adaptation to different agro-climates	
Temperate long days	CP 1412 (Herkol),CP 2099 (MS 91-18), CP 2130 (Achirana Inta), CP 2383 (AGB-69-1)
Sub-tropical, Short days	CP 1475 (Goya), CP 1515 (Black's 3392 (1)), CP 1987 (Rossis 64. 953/74), CP 2023 (Universal)
Sub-tropical, near equinox conditions	CP 1798 (Rowels 316.1), CP 1824 (Rossis 62.66(1), CP 1832 (Ica cuantiva), CP 2003 (1-1150)
(B) Resistance to biotic stresses	
Late blight(*Phytophthera infestans*)	K. Jyoti, K. Alankar, K. Jawahar, CP 3094 (Hybrid-14), CP 2385 (AND-69-1), CP 2415 (MEX. 750821)
Wart (*Synchytrium endobioticum*)	CP 2291 (P-2), CP 2307 (SA 1310), CP 2308 (SPC-58), CP2417 (MEX.780838)
Common Scab (*Streptomyes scabies*)	CP 2213 to CP 2263
Potato tuber moth *Phthorimaca operculella*)	CP 1824 (62.66(1), CP 1832(Ica cuantiva), CP 2003 (1-1150), CP 2058 (CIP 379386), CP2335 (ka Sirena)

Contd..

(C) Tolerance to abotic stresses	
Heat	CP 2108 (LT-1), CP 2109 (LT-2), CP 2118 (Desiree) CP 2150 (K. Lourkar)
Frost	Kufri Deva, Kufri Sheetman
Mineral Nutrition	JEx/ANos.: 17,40,79,163,183,196,421, 459, 600, 665, 674, K. Sindhuri, K. Lalima, K. Jyoti.
Soil salinity	CP 1532 (Black's 2600 b (29) CP 2061 (CIP 379391)
(D) Tuber quality traits	
Dry matter	CP 1456 (Febricia), CP 1929 (BR 5948-1), CP 2142 (Kufri Dewa), CP 2417 (MEX. 7508838)
Protein	CP 1404 (Burmania), CP 2141 (K. Chandramukhi) CP 2158 (K. Sinduri)
Vitamin C	JEX/A Nos. 63, 68,208,226,275,426,479,493

Table 4. Sources of Resistance in wild and other spices

Traits	Donors
Resistance to late blight (*Phytophthera infestans*)	*S. demissum, S. bulbocastanum, S. polyadenium, S. pinnatisectum, S. stoloniferum, S. verrucosum, S. tuberosum subsp. andigena, S. Phureja, S. microdontum,S. berthaulti, S. tarijense, S. circacifolium, S. vernei.*
Resistance to wart (*Synchytrium endobioticum*)	*S. tuberosum* (both species) also to R_2 and R_3 races in a range of wild species from Bolivia including *S. sparsipilum S. acaule* and S. spegazzinii from Argentina
Resistance to common Scals (*Streptomyces scabies*)	*S. chacoense, S. commersonii, S. yungasense*
Resistance to Bactrial wilt (*Pseudomonas solanacearum*)	*S. chacoense, S. sparsipilum, S. phureja, S. stenotomun and S. microdontum.*
Resistance to Soft Rot *(Erwinia carotorora)*	*S. bulbocastanum, S. chacoense, S. demissum, S. hjetingii, S. leptophyes, S. microdontum,S. megistacrolobum, S. phureja, S. pinnatisectum, S. tuberosum*, subsp andigena.
Resistance to virus X	*S. acule, S. chacoense, S. curtilobum,S. phureja, S. sucrense, S. tarijense,S. sparsipilum, S.tuberosum, subsp andigena,S. microdontum*
Resistane to virus Y	*S. chacoense, S. stoloniferum, S. phureja, S. demissum, S. tuberosum subsp. Andigena, S. microdonum.*
Resistance to potato leaf roll virus (PLRV)	*S. brevidens, S. etuberosum, S. acaule, S. raphanifolium*
Resistance to spindle tuber viroid	*S. acaule, S. berthanltii, S. guerreroense*
Resistance to Colorado beetle (*Leptinotarsa decomlineata*)	*S. chacoense, S. demissum, S. commersonii, S. berthaultii, S. tarijense, S. polyadenium*
Resistance to Aphids (*Myzu persical*) *Macrosiphum euphorbiae*	*S. berthanltii, S. stoloniferum, S. multidissectum, S. medions, S. marinasense, S. lignicattle, S. infundibuliforme, S. chomatophilum, S. bulbocastanum, S. bukasovii, S. berthaultii*
Resistance to potato cyst nematode (*Globodera rostochiensis* & *G. Pallida*)	*S. acaule, S. spegazzinii, S. vernii, S. gourlagi, S. capsicibaccatum, S. boliviense, S. bulbocastanum, S. cardiophyllum, S. oplocense, S. sparsipilum, S. sucrense.*
Resistance to rootknot nematode (*Meloidgyne incognita)*	*S. chacoense, S. microdontum, S. phureja, S. sparsipilum, S. tuberosum, subsp andigena, S. curtilobum, S. spegazzinii.*
Resistance to frost	*S. acaule, S. ajanhuiri, S. vernei*
Resistance to heat and drought tuber blackening	*S. acaule, S. bulbocastanum, S. chacoense*

Major Constrains in Potato Improvement

1. It is difficult to produce wide genotypic variation, wide favourable genes for yield, quality characters and resistance to pests and diseases.
2. Inadvertent selection of genotypes susceptible to environmental influences can result in wasted efforts since most of the economic characters do not stabilize in early generations.
3. The propagation rate is low.
4. Breeding material and selections are liable to become infected with viruses, hence, degeneration set in.
5. For planning a programme for genetic improvement of a crop plant, knowledge of interaction of an economic trait is vital.
6. But, owing to a highly heterozygous autotetraploid nature of potato and the possibility of identification of a potential genotype at the F_1 and its further multiplication without the hazards of segregation, have deterred potato geneticists, in elucidating the genetics of important economic traits.

Exploitation of heterosis in potato

The cultivated autotetraploid potato is considered to be an out breeder species which suffers from in breeding depression and naturally express heterocyst upon crossing of suitable parents. Mendoza and Haynes (1974) have proposed a model of over dominant gene action to explain heterocyst for yield in the autotetraploid potato. Loci with multiple alleles and a maximum heterotic value for quadrigenic genotype structures have been postulated. Various experimental results have been analyzed on the basis of such a model in contrast with a dominance situation. The analysis suggests a close positive correlation between heterozygosity and yield. The implication of the proposed over dominance model to potato breeding would be that substantial genetic advance in yield should be made upon increasing the genetic diversity of the parental clones. However, the alien sources of germplasm should undergo some sity previous selection for adaptation. As proper balance between heterozygosity and adaptation, mainly to photo period, should maximize the hetrosis for yield in potato.

The potato varieties released in India and abroad have been using mainly the variability available in *Solanum tuberosum ssp tuberosum* which represents only a fraction of the variability available in the related short day adopted ssp. *andigena*. Since ssp. *andigena* are a rich source of resistance to several diseases, good keeping quality and several other derivable attributed and also provide crosses, it is proposed that the thrust in near future should be on exploitation of this subspecies in our breeding programmes.

True potato seed production

Potato flowers in long day condition which are prevalent in hills. The application of gibberellic acid (GA_3, 50 ppm) on young leaves, when floral initials have just formed result in flower induction in the shy flowering varieties even in short day conditions. Flowering may be induced by providing extra illumination (4-5 hours) with a sodium vapour lamp of 150 watts during the flowering season (December to January) in short day conditions of northern plains. One bulb is adequate for 100 m^2. Open pollinated as well as hybrid seed can be produced.

Hybrid TPS production

Male and female parents are selected on the basis of general combining ability especially for yield. Both male and female parents should have high pollen fertility. The selection of parental lines should be made on the basis of their yield potential, resistance to major diseases, ability to bloom under long and short day conditions. Set berries should have a high number of bold 'A' type seeds. Their progeny should have attributes of high germination percentage, seedling vigour, withstanding transplanting shock, early maturity, high yield potential and homogeneity for plant and tuber characteristics. Most of the economic characters are governed by both additive as well as non-additive gene actions. Male and female parents are planted in different blocks.

Staggered planting of male parent is done at 60 × 20 cm inter and intra row spacing about a week earlier than the female parent planted at 50 × 15 cm spacing. This allows sustained supply of pollen when female parent starts blooming. Planting of single eye pieces of parental lines in cross block gives single thick stems which do not require staking. The trimming of flower bunches (retaining 6 buds) influence synchronous flowering and increases bud and flower size. It also gives higher berry set, increased average berry weight and higher proportion of bolder and A type seeds. The collection of flowers at right stage and their overnight storage at room temperature gives a high recovery of pollen. The flowers with anthers at the verge of dehiscence and corolla fully opened produces highest quantity of pollen. Motorized minimixer has been found more efficient in milking the anthers for pollen extraction.

However, Application of pollen thrice at 8 hour interval on receptive stigma without emasculation ensures maximum production of TPS in the unit area and reduces cost of seed production. The optimum temperature for pollen germination is reported to be about 20°C. The berries are picked up 6-7 weeks after pollination and allowed to ripen at room temperature for about 35-40 days before seed extraction. Berries may be macerated by hand operated reverse screw juice extractor (100-150 kg berries can be macerated by 2 persons per day) and debris is removed by sieving through 3.13 mm nylon net. The process may be hastened by treating seed and pulp mass with 10 per cent HCl with continuous stirring for 20 minutes which helps in digestion of mucilage and obtaining clean seed. The seed is cleaned by 3-4 washing. The cleaned seed is spread under shed for about 72 hours for drying. The shade dried seeds are exposed to warm sun for half an hour to reduce the moisture content to a safe level of 5-7 per cent. The seed could be packed in polythene bags, sealed and stored in desiccators over $CaCl_2$ as desiccant. The desiccator may be kept in at 6-10°C or at room temperature. Seed from premature harvested berries are less dormant and less viable than seeds extracted from fully developed berries. Freshly harvested seeds remain dormant for 5 to 6 months. For immediate sowing dormancy is broken by treating it with 0.1 to 0.15 per cent gibberellic acid (GA) solution for 24 hours. However, more than 6 months old TPS is desirable for quick germination and vigorous seedling growth.

TPS varieties

Both open pollinated (OP) and hybrid varieties (family/population) have been developed. Among OP, TPS-C3, TPS-C-17, 97-PT-14 and among hybrids, HPS 1/13, HPS 2/13 have been developed. Besides, 83-P-47 × Agro-96, JX-214 × TPS/D-150 are also promising hybrids. Hybrids are higher yielder than open pollinated (OP) varieties.

Potato for processing

Indigenously developed potato varieties Kufri Chipsona-1 and Kufri Chipsona – 2 are presently the choice of processing sector for the production of quality potato chips and French fries and a recently released processing variety Kufri Chipsona – 3 is expected to augment the industry further. Emerging potato based processing sector may reduce the price uncertainly by accelerating the trade across countries and the continents. Besides, it can connect the poor small and marginal farmers to the markets. Enhanced direct marketing accessibility may prove to be beneficial to these millions of poor farmers struggling to survive in complex socio-economic back grounds. In India, potato processing industry mainly comprises four segments: potato chips (wafers) French fries, potato flakes/powder and other processed products. However, potato chips (wafers) still continue to be the most common and popular processed product. As a consequence, there is a steady increase in production of potato chips in the country.

Storage

Germplasm has to be identified for longer keeping quality as in India we do not have strong storage infrastructure. Potatoes can be stored under ordinary conditions particularly in sub-tropical plains where high temperatures and dry weather prevails soon after its harvest. Therefore, to sustain increased potato production, proper storage facilities are essential. Earlier, some non refrigerated storage structures were developed which included sandpits, diffused light storage rooms, thatched mud wall rooms etc. These structures are suitable to hold potatoes for 3-4 months. The storage of potato is most appropriate for holding potato tubers for longer periods. Cold stored potatoes become sweet due to accumulation of sugars hence unsuitable for processing and also not liked by the consumers for table purposes. Attempts are being made in India to store potatoes at 10-12 °C at which sugar accumulation is reduced and the tubers are suitable for table and processing purposes. But at this temperature potato tubers sprout quickly depending on their dormancy. Therefore, they need to be treated with sprout suppressant like CIPC (Chlorpropham) at least once or even twice after 8-12 weeks of initial treatment. Germplasm with red skinned is stay for loner time in open condition.

References

Hawks J.G. 1990. The potato: Evolution, Biodiversity and Genetic resources. Belhaven Press, London. pp259.

Mendoza, H. A. and Haynes, F.L. 1973. 1974. Some aspects of breeding and inbreeding in potatoes. *Am. Pot. J,* **50**, 216-222.

Naik, P.S. and Thakur, K.C. 2007. 35 years of AICRP (Potato). AICRP(Potato) Bulletin No.1.CPRI, Shimla.

Pandey, S.K. and Sarkar, D. 2005. Potato in India: Emerging trends and challenges in the new millennium. *Potato J.,* **32(3-4),** 93-104

Puskarnath, 1967. Potato in India. Indian Council of Agricultural Research, New Delhi.

❑❑❑

Chapter – 35

Breeding Potential of Indigenous Germplasm of Seed Spices Crops

S.K. Malhotra

Introduction

Spices have played a very important role in shaping the history of human culture and civilization. Spices condiments and aromatic plants were the first articles traded by ancient people. These prompted the foreigners to the land of spices. Among spices, seed spices is a group well distributed over different agro-climatic regions in India covering major part in semi-arid to arid region largely in Rajasthan and Gujarat. The other states where one or more of seed spices grown are Haryana, Punjab, M.P., Bihar, U.P., West Bengal, Orissa, Tamil Nadu and Karnataka. The seed spices are grown in different parts of the world covering mainly Mediterranean region, South Europe and Asia. These crops were introduced to India from Mediterranean and Central Asian region, perhaps a long time ago. Today India enjoys the position of largest producer and exporter in the world for these crops. The country is bestowed with immensely rich land races diversity in seed spices crops. Much of the country's agro-biodiversity is in the custody of farming community who followed age old farming systems. However the land races are being lost due to expansion of agriculture production in the frontier areas or replacement with the improved cultivars. Hence, scientific management of these valuable resources on different dimensions of activities in India *viz.* survey and collection, characterization, evaluation, documentation and conservation has assumed prime importance today. Out of the total 20 seed spices grown in India the ten are prominent. These important seed spices are divided into major and minor group of seed spices. The crops covered as major seed spices are coriander, cumin, fennel and fenugreek, whereas, ajowan, dill, nigella, celery and anise constitute minor group. The semi-arid to arid region is considerably rich in plant biodiversity of seed spices and it is required to conserve the natural wealth of these crops. The increasing loss of plant diversity owing to numerous natural and manmade causes calls for immediate efforts for its collection and safeguard.

Table 1. List of Seed spices crops, botanical names, family and their centre of origin

Seed Spices	Botanical name	Family	Centre of origin
Ajowan	*Trachyspermum ammi* Sprague	Apiaceae	India and Egypt
Anise	*Pimpinella anisum* L.	Apiaceae	Eastern Mediterranean region
Caraway	*Carum carvi L.*	Apiaceae	Mediterranean region
Celery	*Apium graveolens* L.	Apiaceae	Mediterranean region
Coriander	*Coriandrum sativum* L.	Apiaceae	Mediterranean region
Cumin	*Cuminum cyminum* L.	Apiaceae	Mediterranean region
Dill	*Anethum graveolens* L. *Anethum sowa* Kurz.	Apiaceae	Europe, Africa and Asia India for sowa
Fennel	*Foeniculum vulgare* Mill.	Apiaceae	South Europe and Mediterranean region
Fenugreek	*Trigonella foenumgraecum* *Trigonella corniculata* L.	Fabaceae	South East Europe and West Asia
Nigella	*Nigella sativa* L.	Ranunculaceae	Eastern Mediterranean region

Taxonomical classification

All of the seed spices listed above (Table 1) are dicotyledon and most of them belong to family Apiaceae except nigella which is from family Ranunculaceae. The most of seed spices crops have their centre of origin primarily in Mediterranean region and South Europe except Ajowan and Indian Dill (sowa) with nativity from India (Malhotra, 2003).

Ajowan : The ajowan (*Trachyspermum ammi* L.) belongs to the family Apiaceae and is classified as aromatic and herb spice, mostly dried fruits are used as spices. The common synonyms are *Trachyspermum copticum* link, *Carum copticum* Benth. The somatic chromosome numbers are 2n=18. The flowers are terminal and compound umbels. The flowers are white in colour, very small, crowded and all bisexual. The minute greyish white fruits are ovoid in shape.

Anise: The common cultivated anise is *Pimpinella anisum* and belongs to family Apiaceae. It is classified as aromatic and herbs spice, mostly dried fruits and leaves are used. The somatic chromosome numbers are 2n=18. The flowers are small and white, arranged in large compound umbels. A single umbel produces one to six clusters, each bearing 6 to 10 seeds. The two carpels containing the seed are hairy, aromatic, greenish brown to greyish brown in colour.

Caraway: Caraway (*Carum carvi* L.) belongs to family Apiaceae, is classified as aromatic spice whose dried fruits are mostly used as spice. The somatic chromosome numbers are 2n=20. Flowers are minute borne in terminally or axillary compound umbels producing clusters of white flowers. The fruits contain warm, sweet, slightly sharp taste and flavour.

Celery: Celery (*Apium graveolens* L.) belongs to family Apiaceae and is classified as aromatic and herbs spice, mostly dried fruits and leaves are used. There are three known horticultural types of celery i.e. *Apium graveolens* var. *dulce* – blanched celery, *A. graveolens* var. *rapaceum* – edible rooted celery and *Apium graveolens* var. *secalinum* – leafy type and seed type (smallage type). In India, *Apium graveolens* var. *secalinum* is cultivated for seeds produced as spices. The somatic chromosome numbers are 2n=22. The flowers are small, white in colour and inflorescence is a compound umbel.

Coriander: The genus *Coriandrum* from family Apiaceae has been reported to have two species but *C. sativum* L. is cultivated type. The small fruited variety is referred as *Coriandrum sativum* var. *microcarpum* DC is cultivated in temperate countries like Russia, Bulgaria and Western Europe and has higher oil content of 1.5-2.5%. Whereas the large fruited coriander are designated as *Coriandrum sativum* var. *vulgare* Alef. It is mostly cultivated in Morocco, India and Pakistan but possess lower essential oil content 0.3-0.4% and is considered suitable for ground spices. The chromosome number of 2n=22 has been reported for coriander. The small, white or pink flowers are borne in compound umbles about 4 cm across. Fruit is schizocarpic, consisting of two halves, the single seeded mericarps.

Cumin: The cumin (*Cuminum cyminum* L.) belongs to family Apiaceae and is known to occur only one species *cyminum*. There is no report of existence of any other species. The cumin has 2n=14 chromosomes which show regular bivalent formation and synchronous segregation during meiosis. The flowers are regular and bisexual. Flowers are white or pink in colour and distributed in small compound umbel of 2-3 cm in diameter. One compound umbel contains about 4 -7 umbellates.

Dill: Dill belongs to the family Apiaceae and comes under genera *Anethum*. There are two species under cultivation i.e. European dill (*Anethum graveolens* L. Syn. *Peucedanum graveolens* Benth & Hook.) and another closely related is Indian dill (*Anethum sowa* Roxb.). The somatic chromosome numbers are 2n=20. It is classified as aromatic and herbs spice, mostly dried fruits and leaves are used. The plant bears flat, terminal compound umbels, with numerous yellow flowers, whose small petals are rolled inward. The flowers are regular, bisexual, pentamerous and dichogamous.

Fennel: Fennel (*Foeniculum vulgare* Mill. Syn. F. *officinale*) from family Apiaceae is the widely cultivated species in India. Based on the appearance of the plant, seed and the compositional data of the volatiles from the fruit *F. vulgare* is distinguished into two subspecies and further classified into numerous varieties. The two sub speices are *F. vulgare* sub sp. *pipertum* (Holmboe) wild fennel, also called as Carosella fennel and the other *F. vulgare* sub sp. *capillaceum* (Gilib) is known for its high essential oil and further classifed into 3 varieties, *vulgare* (Miller) Thellung -bitter fennel; var. *dulce* (Miller) Thellung -sweet fennel and var. *ozoricum* (Miller) Thellung -culinary florence fennel. The Indian fennel is regarded as a distinct variety *panmorium*, the fruits of which are mostly chewed with betels. The genus *Foeniculum* has been reported to possess 2n=22 chromosomes. The flowers of fennel are small, yellow, arranged in compound terminal umbels and fruit cremocarp breaks up into 2 mericarps on little pressure.

Fenugreek: The fenugreek an annual herb belonging to subfamily Papilonaceae of the family Leguminosae called Fabaceae with new name. There are two species of the genus *Trigonella* which are of economic importance *viz. T. foenum graecum* L., the common methi with larger seed size and *T. corniculata*, the *kasuri methi* with smaller size of seeds. These two types differ in their growth habit and yield. The later one is slow growing type and remains in rosette form during most of its vegetative growth period. The fenugreek is diploid with 2n=16 chromosomes. The fenugreek, large seeded bears white coloured flowers with straight pods, whereas, kasuri methi, small seeded bears yellow to orange colour flowers with sickle shaped pods.

Nigella: Nigella (*Nigella sativa*) belongs to family Ranunculaceae and is classified as aromatic spice. The somatic chromosome numbers are 2n=12. The flowers are white; seeds trigonous, regulose, angular, small funnel shaped 0.2 cm long and 0.1 cm wide, black in colour, odour slightly aromatic and bitter in taste. Flowers solitary and terminal, beautiful, stamens numerous, carpels 5 partially united and the fruit is botanically a capsule containing seeds.

Management of Genetic Resources of Seed Spices

A germplasm collection with good variability for the desirable characters is the basic requirement of any successful crop improvement programme. Moreover, germplasm is the natural wealth of nation and is required to be safeguard. The germplasm so collected must be scientifically maintained so that the variability is fully conserved against the forces like genetic drift and selection. A systematic collection, evaluation of germplasm and its cataloguing is necessary for identification of material for specific breeding programme and hence must be carefully carried out.

Exploration and Collection of Germplasm

A number of collections of seed spices crops of indigenous and exotic origin were made and evaluated. Germplasm collection activities were also started initially from Rajasthan and Gujarat, the major area for cultivation of coriander, cumin, fennel, fenugreek, ajowan and dill. The earliest report for collection and conservation of the genetic resources of umbelliferous seed spices was in 1957 at IARI, New Delhi by Joshi *et al.*, 1963. However, most of the collections were lost due to lack of proper maintenance or follow-up. Systematic genetic resource activities for different crop species with long term objectives started after 1976 with the establishment of NBPGR in general and in the early nineties by four centres of AICRP (Spices) one each at Jobner (Rajasthan), Jagudan (Gujarat), Coimbatore (T.N.) and Lam (A.P.) in particular. Later, few more research stations were added as AICRP centres. Survey and collection work on four major seed spices namely coriander, cumin, fennel and fenugreek have been given importance only. The germplasm collections maintained at different centres are listed in Table 2. Besides AICRP Centres, few more SAU's, CSIR and their research centres are also maintaining little collections of various seed spices.

Since after the establishment of NRC on Seed Spices at Ajmer by ICAR during April, 2000, the survey and collection of plant genetic resources was given priority. In addition to major seed spices crops, the NRC on Seed Spices is maintaining collection of ajowan, dill, nigella, celery and anise also. Efforts are being made in introduction of caraway, sweet fennel and few others from the semi-arid agro climatic conditions. The total germplasm collection being maintained at NRCSS is 1357, which comprises 1124 from major group (Table 2) and 233 from minor group (Table 3). The germplasm of a crop species includes land races, primitive cultivars, released varieties, parental lines of released hybrids, genetic stocks with known desirable attributes and wild and weedy relatives of the cultivated species. None of the seed spices is native to India, but continuous natural selection and creative human selection in a wide range of agro-climatic regions and different cultures have led to enormous morphological and genetic diversity in India. The development of improved cultivars has improved the production and productivity of seed spices to some extent, which has endangered the accumulated diversity among the traditional cultivars of these crop species. Therefore, the conservation of germplasm of the seed spices is of profound importance for sustaining their production. If not collected and conserved these valuable genetic resource may be lost forever. The NRCSS is giving prime importance in collection and conservation of the biological diversity of spices from all over the country and abroad.

India is a country rich in spices diversity. In order to conserve the valuable germplasm attempts have been made to conserve them in various institutes. The list of germplasm collection available at different institutes show a total of about 5000 accessions available at different centres (Table 2 and 3) which counts duplicates also, if duplicates are removed the number shall go to 2750 accesions. But the fact appeared at recently organised XX AICRP Spices Workshop that germplasm being maintained at different centres is liberally exchanged with each other and thus equal chances of duplicates prevail. Hence it was suggested to remove such confusions by registering the germplasm at NBPGR and use of the unique IC (Indigenous Collection) number commonly in reports, publications etc.

Table 2. Germplasm collection of major seed spices at different centres

S. N.	AICRP Centres	Coriander		Cumin		Fennel		Fenugreek	
		I	E	I	E	I	E	I	E
1	RAU, Jobner	649	112	313	10	261	20	366	12
2.	GAU, Jagudan	70	18	173	7	115	4	76	-
3.	TNAU, Coimbatore	205	-	-	-	-	-	262	-
4.	APAU, Guntur	230	-	-	-	-	-	70	-
5.	RPAU, Dholi	102	-	-	40	171			
6.	IGAU, Raigarh	20	-	-	13				
7.	NDAU, Kumarganj	29	19	15	82				
8.	CCSHAU, Hisar	276	-	122	12	295			
9.	NRCSS, Ajmer	345	21	209	8	260	22	252	7

I -Indigenous collections
E- Exotic collections

Table 3. Germplasm collection of minor seed spices at different centres

S. N.	Centres	Ajowan		Dill		Nigella		Celery		Anise		Caraway	
		I	E	I	E	I	E	I	E	I	E	I	E
1	GAU, Jagudan	85	-	34	-	-		-		-		-	
2.	APAU, Guntur	46	-	-	-	-		-		-		-	
3.	MPUA&T, Pratapgarh	20	-	20	-	-		-		-		-	
4.	NRCSS, Ajmer	87	-	86	4	20	-	11	2	9	3	8	3

I- Indigenous collections
E- Exotic collections

Germplasm Maintenance

Proper maintenance of the germplasm should take care of maintaining the genetic structure of the original sampled population. In order to achieve this objective, two criterions need to be fulfilled. First, the breeding system must be controlled, which involves the prevention of out crossing with other entries and the second, to reduce the effect of natural selection in an environment other than the original one (Frankel and Soule, 1981). The mode of reproduction determines the genetic structure as well as the pattern of ecotypic differentiation and the appropriate maintenance procedure (Singh *et al.*, 2004). In the self-pollinated crops the precise gene and genotypic composition of the population is preserved in the absence of selection, whereas the cross-pollinated crops mandates controlled pollination techniques for their maintenance. The reproduction behavior of the umbelliferous seed spices have been studied and elaborated (Ramanujam *et al.*, 1964, Hore, 1989 and Peter *et al.*, 1994).

Coriander is an andromonoecious crop and is a highly cross-pollinated crop. Even though it is cross-pollinated crop, there is no inbreeding depression observed even after three generations of selfing (Romanenko *et al.*, 1992). Upto 60% cross-pollination has been observed in closely spaced plants (Dimri *et al.*, 1976). Therefore, the genebank accessions need to be isolated from each other for genetically identical reproduction of an accession. If sufficient field area is available, then isolation can be achieved by providing 80 m isolation distance between the accessions. However, most of the cases the number of accessions to be maintained is large and therefore, the isolation of accessions have to be achieved by technical tools such as per gamin bags. However, it may reduce fruit setting considerably. Alternatively if there are only few seeds, then isolation cabins can be used for successful reproduction of populations. Cumin and fennel are also cross-pollinated crops, bees being the major pollinator. The maintenance of these crop germplasm can be achieved by barriers, which can prevent bee movement. Fenugreek is a self-pollinated crop and hence its maintenance is easy compared to other seed spices. The accessions can be maintained by selfing using appropriate sized butter paper bags and avoiding contamination during threshing of the individual accessions.

The seeds should be harvested at maturity from the plants in order to ensure optimum viability. The seed spices have remarkable resistance to high temperature, but for long-term storage the drying temperature should not exceed 40° C. Therefore, the seeds harvested from the plants have to be dried, preferably under shade, to appropriate moisture depending on the storage condition. Care should be taken to ensure that the seeds are free from weeds, foreign particles, pests and diseases during processing.

Germplasm Conservation

Loss of diversity at intra-specific level with valuable genes for important quality and resistant characters, an offshoot of scientific breeding of new varieties, is dubbed as Breeders paradox. The spread of high yielding varieties though very important, adds to the erosion of landraces. Currently ex situ genebank, in vitro repositories, cryopreservation and in situ conservation strategies are adopted to conserve the biodiversity of spices.

The seed spices are annual in nature and are propagated through seeds. Thus in-situ preservation is the economical method. Under normal storage conditions the viability of seed spices seeds remain satisfactorily for two years. Fresh seeds therefore, have to be produced every third year. All of the seed spices except fenugreek are cross pollinated and pollination is brought about mainly through bees. Thus, 1 to 1.5 m long musline cloth cages are used as bee barriers. In each cage, the number of plants caged are 5 in fennel, dill and ajowain, whereas, 7-10 plants in cumin and coriander. The number is certainly small to regenerate the full spectrum of genetic variability of individual accession but is enough for maintenance of the large germplasm by overcoming the problem of genetic drift.

An alternative strategy is to form gene pools based on agro-morphological traits of the accessions. A limited number of non-competitive gene pools can be formed by grouping the germplasm accessions based on important traits such as maturity, habit, etc. These gene pools can then be maintained as composite populations by random mating among the accessions of each group in polycross blocks maintaining proper isolation. This approach has been successfully adopted in coriander where four composites have been developed based on maturity and habit of the accessions (Sharma, 1994).

In situ conservation refers to the conservation of the genetic resources within their natural habitat. It can be of two types namely conservation in genetic reserve area and on farm conservation. Among seed spices, the vegetatively propagated species of black cumin (through underground bulbs) that grows in wild state in high altitudes of Kashmir, Himachal Pradesh and Uttaranchal, the conservation through formation of genetic reserve areas could be a viable option. The on farm conservation refers to "continuous cultivation and management of diverse set of crop populations by farmers in a particular ecosystem where a crop has evolved" (Bhatt and Singh, 2004). This is a dynamic method of conservation followed by farmers for maintaining traditional land races and also involves the evolution of the cultivars in due course.

Ex situ conservation refers to conservation of genetic resources outside their natural habitat. The *ex situ* conservation methods includes seed gene bank, cryopreservation, *invitro* storage, DNA storage, pollen storage, field gene bank and botanical gardens (Tao, 1985). The seeds of seed spices such as coriander, cumin, fennel and fenugreek are orthodox seeds i.e., they can withstand desiccation without any significant loss in viability. Therefore, the conservation of seed spices germplasm can be accomplished through seed gene bank.

Conservation in the form of seeds is both the safest and cheapest method of conservation. Due to the orthodox nature of seeds, the seeds of the seed spices can be stored under sub-zero conditions by reducing the moisture content to minimum. Based on the storage conditions, seed storage can be classified into short, medium and long term storage (Table 4). The type of storage to be adopted will also be dependent on the purpose of storage.

Table 4. Types of seed storage based on storage conditions

Storage Type	Type of collection	Storage parameters		Period of storage
		Temperature (°C)	R.H. (%)	(years)
Short term	Seed before processing	15 to 20	30 – 40	2
Medium term	Active collection	0 to 10	20 – 30	5 - 25
Long term	Base collection	-18 to - 20	No control	> 25*

* Till the viability of above 85% is maintained during quick viability test (Ellis *et al.*, 1980).

Under room temperature, the viability of the seeds of seed spices remains satisfactory for 2 years. The advantage of long-term conservation over alternative procedure of regular regeneration at interval of 2 years (in case of seed spices) due to the avoidance/ reduction of natural selection, genetic drift in small populations, natural hybridization and destruction by parasites/ climatic rigors/ loss through human error (Frankel and

Soule,1981). Further, for the conservation of seeds without loosing viability and maintaining the genetic integrity of the accession, standards have been established for gene bank by IPGRI (Tao, 1985). Based on the gene bank standards, the implied seed standard for storage of seed spices germplasm is presented in Table 5.

Table 5. Seed standard for storage of seed spices.

Parameters	Preferable standards
Seed moisture content	3 - 7 % (Base collection) and 8 - 10 % (Active collection)
Sample size/ accession	3000 - 4000 seeds (fenugreek)
	4000 - 12000 seeds (all seed spices except fenugreek)
Viability monitoring interval	10 years (Base collection) and 5 years (Active collection)
Regeneration	Minimum population of 100 plants/ accession

The other option available for conservation of seed spices is cryopreservation of the seeds. Cryopreservation is the storage of the genetic material at ultra-low temperature, usually in liquid nitrogen at –196°C. Although, this technique has shown enormous potential of storage for prolonging the life span of stored material for indefinite period, it is regarded as a tool mainly for conservation of recalcitrant seeds and vegetatively propagated plant species. Technologies such as DNA storage and pollen storage are still in its infancy and could be a potential tool for conservation of crop species in the near future. Since, each method of conservation in the two basic approaches *in situ* and *ex situ* conservation has its own objectives, advantages and disadvantages, these methods should not be viewed as alternatives or confirming each other but rather as being complimentary. The biodiversity of a crop could be conserved through a "complementary conservation strategy" involving more than one method, which can be complimentary to each other (Tao, 1985).

Germplasm characterization:

The systematic germplasm characterisation provides information about the identity / uniqueness of accessions and basic information to help and promote its utilization in plant improvement. The seed spices being relatively less common and under utilised, were neglected from the characterisation point of view. Presently no any centre working on seed spices has given attention on characterisation work. The NRC on Seed Spices at Ajmer has initiated its exercise on characterisation aspect. The descriptor list on major and minor seed spices has been prepared and published at NRCSS for use by research workers (Malhotra and Vashishtha, 2006).

Germplasm evaluation

In order to facilitate effective utilization of Plant Genetic Resources, it is important that these are evaluated for productivity and its components, crop duration, resistance to biotic and abiotic stresses and quality of produce. In seed spices the quality characters need special attention on account of their export value. The germplasm material available at different centres has been evaluated and utilized for crop improvement. Research work carried out under the AICRP network on the conventional breeding work through selection from the germplasm resulted in the release of **74** improved varieties in seed spices coriander (25), cumin (8), fennel (14), fenugreek (20), Ajowan (2), Dill (2), Kalongi (1). The evaluation status of existing genetic resources of seed spices in India as per the major thrust areas is given in Table 9. The existing germplasm collection of seed spices at various centres in the country possess limited variability particularly in ajowain, fennel and cumin. There is lack of genetic resources with high volatile oil content in seeds and resistance to biotic and abiotic stresses. The promising genetic stock with potential genes or rare gene combinations are needed to be isolated / identified for effective utilisations in crop improvement programme. At the same time there is strong need to validate the identity of accessions at biochemical and molecular levels using RFLP, AFLP and RAPD.

The evaluation and characterisation information in case of seed spices is required to be documented and catalogued crop-wise for use by the breeders, geneticists, biotechnologists and other users. Documentation of germplasm is necessary for its sustainable use of genetic resources of seed spices. So far no serious attempt has been made for documentation of information on seed spices.

A wide range of variability was observed for different traits. The extent of variability available in the germplasm collection in seed spices such as coriander, cumin, fennel and fenugreek for some of the important agronomic traits has been discussed by Sharma (1994). The biotic (Table 6) and abiotic stresses (Table 7) and quality characters (Table 8) in seed spices crops are given below, for which the germplasm is required to be evaluated for identification and isolation of desired types for further utilization in the crop improvement programme.

1. Major yield reducing factors (biotic and abiotic stresses) :

Biotic and abiotic stresses are the major yield reducing factors and are given below crop wise.

Table 6. Major diseases and insect pests in seed spices

S.N.	Seed Spice Crop	Major Diseases and Insect Pests
1.	Coriander	Stem gall (*Protomyces macrosporus*), Powdery Mildew (*Erisiphe polygoni*), Wilt (*Fusarium oxysporum* f.sp. *corianderi*), Stem rot (*Rhizoctonia solani*), Root rot (*Curvularia pallescens*), Aphids, Mites
2.	Cumin	Alternaria Blight (*Alternaria burnsii*) Powdery Mildew (*Erisiphe polygoni*), Wilt (*Fusarium oxysporum f.sp. cumini*), Aphids
3.	Fennel	Alternaria Blight (*Alternaria tenuis*), Ramularia Blight (*Ramularia foeniculi*), Gall (*Protomyces macrosporus*), Powdery Mildew (*Leveillula taurica*), Wilt (*Fusarium oxysporum f.sp. corianderi*), Bacterial soft rot (*Erwinia carotovora*), Aphids, Mites
4.	Fenugreek	Leaf Spot (*Cercospora traversiana*), Powdery Mildew (*Erisiphe polygoni, Leveillula taurica*), Downey Mildew (*Peronospora trigonellae)*Collar rot (*Rhizoctonia solani*), Root rot (*Alternaria alternata*), Aphids,Mites

Contd..

5.	Ajowan	Collar and root rot (*Scleroticum rolfsii*), Powdery Mildew (*Erisiphe polygoni*), Aphids
6.	Dill	Root rot (*Fusarium oxysporum*), Powdery Mildew (*Erisiphe polygoni*), Aphids, Cut worms
7.	Nigella	Root rot (*Fusarium oxysporum*), Cut worms
8.	Anise	Alternaria leaf Blight (*Alternaria solani*), Powdery Mildew (*Erisiphe polygoni*), Aphids, Semilooper
9.	Caraway	Alternaria leaf Blight (*Alternaria solani*), Powdery Mildew (*Erisiphe polygoni*), Aphids,
10.	Celery	Celery leaf mosaic virus

(**Source:** Malhotra and Vahsishtha,2005)

Table 7. Abiotic stresses in seed spices

S.N.	Seed Spice crop	Abiotic stresses
1.	Coriander	Frost, drought, rainfed cultivation, salinity
2.	Cumin	Frost, drought, limited water conditions, salinity
3.	Fennel	Limited water conditions, salinity
4.	Fenugreek	Salinity, limited water conditions,
5.	Ajowan	Frost, drought, rainfed cultivation, salinity
6.	Dill	Frost, drought, rainfed cultivation, salinity
7.	Nigella	Limited water conditions, salinity
8.	Anise	Limited water conditions, salinity

2. Major nutritional and quality characters desired for quality improvement

The seed spices crops possess little nutritive value and are valued for peculiar aroma and medicinal value. Thus the important quality characters are presence of high volatile oil and other flavour contributing components.

Table 8. Quality characters in seed spices crops

S.N.	Seed Spice crop	Quality characters in seed spices
1.	Coriander	High volatile oil, high linalool content, dual purpose use both as seed and leaves
2.	Cumin	High volatile oil, high cuminaldehyde content
3.	Fennel	High volatile oil, high anethol content
4.	Fenugreek	High/low trigonellone content, high diosgenin content, dual purpose use both as seed and leaves
5.	Ajowan	High volatile oil, high thymol content
6.	Dill	High volatile oil, high carvone content, dillapiol toxin less, splitting/non-splitting type
7.	Nigella	High volatile oil, high nigellone content
8.	Anise	High volatile oil, high anethol content
9.	Caraway	High volatile oil, high carvone and limonene content
10.	Celery	High volatile oil, high limonene content

3. Sources identified for various biotic and abiotic stresses and for other desirable traits

In order to facilitate effective utilization of plant genetic resources, the germplasm continuously evaluated and the sources identified for resistance to biotic-abiotic stresses of other valuable characters are given in Table 6 and 7.

Table 9. Sources utilized

S.N.	Crops	Biotic and Abiotic Stresses (sources utilized)
1.	Coriander	Wilt (ACr -01-250), stem gall (ACr -01-250), Dual Purpose (Leaf & seed) (Acr-01-256),
2.	Cumin	Wilt resistance (AC-01-3, AC-01-167), Drought tolerance (AC-01-3), High oil content (AC-01-167),
3.	Fennel	Ramularia Blight (Sel 01-87), High oil content (Sel 01-119), *early* season (Sel 01-119),
4.	Fenugreek	Downy Mildew (AM-01-10), Root rot (AM-01-10), Early & Large podded (AM-01-35), Dual Purpose (Leaf & seed) (AM-01-35)
5.	Ajowan	Powdery Mildew (AA-01-19, AA-01-61), High oil content in seed (AA-01-19), Drought tolerance (AA-01-19, AA-01-61), Early & bold seeded (AA-01-61) (Both varieties identified for release as NRCSS-AA-1, NRCSS-AA-2 by AICRP)
6.	Dill	Powdery Mildew (AD-01- 32), High oil content in seed (AD-01- 32), Drought tolerance (AD-01- 6) Dillapiole toxin less (AD-01-43) (Both varieties identified for release as NRCSS-AD-1, NRCSS-AD-2 by AICRP)
7.	Nigella	High yielding (AN-01-1) variety identified for release as NRCSS-AN-1 by AICRP

4. Sources utilized by the national crop improvement programme

The germplasm material available at different centres has been evaluated and utilized for crop improvement. Research work carried out under the AICRP network on the conventional breeding work through selection from the germplasm resulted in the release of **74** improved varieties in seed spices (coriander-25, cumin-8, fennel-14, fenugreek-20, Ajowan (2), Dill (2), Kalongi (1) (Table 10). The suitable selection under crop improvement programme have been made by utilization of identified sources are given in Table 6.

Table 10. Seed spices varieties released through AICRPS and NRCSS

Crop	Variety Released
Coriander (25)	G Cr 1, G Cr 2, Co1 , Co2,Co3, CS-287, RCr 41, Sadhana, Swathi, Sindhu, RCr 20, RCr 435, RCr 436, RCr 684, Hisar Anand, Rajendra Swathi, Hisar Sugandh, Hisar Surbhi, NRCSS ACr1, LCC128, RCr480,Rcr 728, LCC 170, DH242
Cumin (8)	G C 1, G C 2, G C 3, GC 4, RZ 19, RZ 223, RZ341, RZ345
Fennel (14)	PF-35, G F1, G F 2, Co 1,RF 101, UF-125, Rajendra Sourabha, Hisar Swarup, GF11, RF 143, RF178, RF 205, NRCSS AF1, JF444-1
Fenugreek (20)	Co1, Co2, Rajendra Kanti, Lam sel.1, RMt 1, RMt 303, RMt 143, Hisar Sonali ,Guj. Methi-1, Rajendra Khushba, Hisar Suvarna, Hisar Mukta, Hisar Madhvi, RMt 305, GM2, GM351, RMt 361, NRCSS AM1, NRCSS AM2, LFC 84
Ajowan (2)	NRCSS-AA-1, NRCSS-AA-2
Dill (2)	NRCSS-AD-1, NRCSS-AD-2
Kalongi (1)	NRCSS-AN-1
Anise (1)	NRCSS-Ani 1
Celery (1)	NRCSS-Cel 1

Anthesis and Crossing Techniques

The anthesis and crossing techniques including emasculation and pollination methods are mentioned here (Malhotra, 2003).

Apiaceae crops: The flower anthesis proceeds from the periphery inwards both within the compound umbel and within the umbellate. Depending upon the temperature, the flower opening and dehiscence of pollens start between 6-8 a.m. and reaches to maximum between 12 to 2 p.m. The dehiscence of anthers is spread over a period of 8-10 hours at an interval of about 2-3 hours. The peripheral flowers in the umbel open first. It takes about 7-10 days for anthesis of all the flowers of each umbel. The anthesis of a compound umbel starts from the outer whorl of umbellate and move towards inner whorl of umbellate. The stigma is reported to be receptive for 5 days and the first 3 days give the highest seed set. Pollens remain viable for a maximum of 24 hours. The flowers are protandrous, with the anthers maturing before the stigmas become receptive. The flowers are self fertile. The hermaphrodite and staminate flowers occur within the umbel. Honeybees are chief agents leading to genetic contamination between different cultivars. Owing to small size of flowers, the crossing success is less. For emasculation and hand pollination the four days old buds are selected. The pollination around 11-12 a.m. favours more seed setting.

Fenugreek: Fenugreek is leguminous crop and is highly self pollinated in nature. The flowers are white or yellow in colour. The crop is a typical self pollinated, in which double fertilization occur within the unopened flower buds. Pollen fertility range form 95 to 99% in the unopened flower buds and 67 to 80% in the opened flower buds. The flowers open between 9.00 a.m. and 6.00p.m. with peak at 11.30 a.m. The anthesis takes place between 10.30 a.m. and 5.30 a.m. with peaks between 11.30 a.m and 12.30 p.m. The stigma becomes receptive 12 hours before flower opening and remains receptive for about 10 hours after the opening of the flowers.

Nigella: The flowers open between 9.00am and 6.00pm with peak between 11am to 12 noon. The flowers are protandrous and cross pollination occurs through insects. The flowers are self fertile. The stigma remains receptive for 2-5 days depending upon clomate. The flowers are solitary and conspicuous, therefore emasculation and pollination can be done easily.

Breeding Methods

Among the seed spice crops, coriander, fennel, ajowan, dill and nigella are cross pollinated, cumin often cross pollinated and fenugreek predominantly self pollinated crop. The breeding methodologies depend upon the pollination behaviour of the crops. Consequently, different breeding methods are used. Though the crops are self fertile but the pollination in crops from family Apicaceae, occur primarily by insects. The typical flower morphology is amenable mainly to selection. The other breeding methods reported are hybridisation, mutation breeding and biotechnological approaches. The varieties release through AICRP Spices and their development through different methods are described here.

1. Selection

The most of the recommended varieties of seed spices in India have been developed through selection. Different methods of selection have been followed for crops improvement work in coriander, fennel, cumin and fenugreek.

Coriander: In coriander, the selection through introduction was used earlier during 1969 from Bulgarian and Russian lines. Most of the varieties developed in coriander are selection from the local germplasm. The varieties, Gujarat Co1 by GAU, Jagudan Gujarat, Co1 by TNAU, Coimbatore, Rajendra Swathi by RAU Dholi, were developed through selection from local germplasm. Some varieties were developed through reselection from varieties viz. Co 2 from culture P2, Gujarat Co 2 from Co 2, CS 287 from Guntur collection and Co 3 from ACC 695. The other commonly used selection methods were mass selection and recurrent selection. The varieties developed through mass selection were Sadhana, Swathi and Sindhu at APAU, Lam and Hisar Anand at HAU Hisar. Whereas varieties RCr 41, RCr 20 at RAU Jobner and Acr-1-256, Acr-1-250 at NRCSS Ajmer were developed through recurrent selection. Here in coriander, family breeding and pedigree selection methods can also be used for crop improvement.

Fennel: Under the AICRP on Spices all of the varieties released so far have been developed through selection. The first released variety PF-35, was developed through selection from germplasm maintained at Spices Research Centre Jagudan during 1973, later Co1 was developed by making reselection from PF-35 in 1985 at TNAU, Coimbatore. Besides, pure line selection, pedigree selection and recurrent selection were also used for development of improved cutivars. The variety Gujarat Fennel 1 was developed by the pure line selection and Gujarat Fennel 2 by using pedigree selection method at GAU, Jagudan. The variety RF 101, from RAU, Jobner was developed through recurrent selection from a local collection. Recently AF-1-119 has been developed through recurrent selection based on individual plant progeny performance and is suitable for growing both as early and rabi crops. The family selection method can also be used.

Cumin: Out of five varieties released so far, four have been developed through selection. The first improved variety MC-43 was released during 1970 and was developed through selection from local germ plasm at GAU, Jagudan. Similarly, another selection from local germplasm was released as Gujarat Cumin1 during 1983. After the gap of 5 years, another variety RZ-19 was released in 1988 and was developed through recurrent selection from UC-19. Recently, Gujarat Cumin-3, developed through recurrent selection from a West Germany introduction (EC 232689) released during 1999 as a first Fusarium wilt resistant cultivar. Through cumin is often cross pollinated crop, but the typical flower morphology favours mainly the selection procedures particularly, the recurrent selection method.

Fenugreek: The research work on improvement of fenugreek, taken up at different research Centres under AICRP spices, revealed that selection is the common practice being followed by breeders. Being highly self pollinated crops, mostly pure line selection method has been used. The variety Co 1 was developed as reselection from TG 2356 at

TNAU, Coimbatore. Similarly Lam Sel 1 at APAU, Lam, Guntur and CO 2 at TNAU Coimbatore, were developed as the selections from local germ plasm. The pure line selection method was the commonly used method for improvement in fenugreek. The Rajendra Kanti at RAU Dholi was developed as pure line selection from Reghunathpur, RMt-1 at RAU Jobner as from Nagur local and Hisar Sonali, at HAU Hisar from local germplasm The Gurarat Methi-1 has been reported to be developed through recurrent selection from germplasm based on the individual pure line selection from JF 102. The NRCSS has developed AM-1-35 from local germplasm through pure line selection method.

Ajowan: The first report of ajowan selections was from IARI in 1963, these selections were S25, S95, S47, S 84 and IC3743. Later in the year, 1986, GA 1, selection from local germplasm from Gujarat ; Lam Sel 1 and 2 from AP.; RA1-80 and RA-19-80 from Bihar were reported in the early ninties. All of the above selections were developed from the local germplasm. Recently two selections, AA-01-19 and AA-01-61 from NRC Seed Spices have been developed through recurrent selection method through local germplasm.

Dill: The local selections Mehasana local, Ruby local of Gujarat and Pratapgarh local of Rajasthan are grown by farmers in the specific areas so for no improved cultivar of dill is available. Recently three selections including two Indian dill type and one European dill type have been developed. The Indian dill types are AD-1-6 (recurrent selection based on individual plant progeny from Nagaur local) and AD-1-32 (from Gujarat local), whereas, AD-1-43, European type developed through single plant selection.

Celery: The local selections Amritsari, Gurdaspuri are being grown by farmers in Punjab. An improved line RR 851 was developed through mass selection, which yields high essential oil content. The selections EC 99249-1 and PRL-85-1 were developed in past. The efforts are being made at NRCSS for acclimatization of celery for seed setting under semi-arid conditions using recurrent selection method.

Anise: So far no serious attention has been paid for development of improved varieties. A selection from EC 22091 is being acclimatized for seed setting under semi-arid conditions using recurrent selection method.

Caraway: A selection from Germany exotic line is being acclimatized for seed setting under semi-arid conditions using recurrent selection method.

Nigella: There is lack of superior varieties of Nigella. So far Azad Kalongi from Kanpur and AN 1 from NRCSS have been developed through selection method. AN 1 is known to possess 2.5% essential oil.

2. Hybridizations

The hybridisation is used to combine the superior characteristics of two or more genotypes. The manifestation of heterosis and gene action in yield of coriander, fennel, cumin and fenugreek have been studied by very few workers. The explanation of heterosis in fennel was attempted earlier by Ramanujam and Tiwari (1970) and reported F_1 hybrids yielding as high as 110% more yield than the best pure line. The sterility has been reported as cytoplasmic male sterility. The fenugreek is a highly self-pollinated crop and hybridisation has not been reported so far. But hybridisation can be used in crossing the diverse parents for combining the desired parents in hybrids and further selection through hybridisation can be practiced using pedigree selection method.

3. Mutation breeding

The mutation breeding has been used in seed spices viz. coriander cumin, fennel and fenugreek. The ionising radiations through X-rays and cobalt –60 gamma rays and chemicals such as Ethyne Methone Sulphonate (EMS) are successful in inducing mutation. The promising strains of seed spices developed possessed economic traits, more seed yield and oil content. In cumin one variety Gujarat Cumin 2 has been developed through induced mutation of MC-43, similarly another variety RCr-684 in coriander has been developed through mutation from RCr-20 by RAU, Jobner. The new fenugreek variety RMT 303 has been developed through mutation breeding from RMt-1 by RAU, Jobner. In case of fennel, Tiwari (1968) observed Mutant-1, Mutant 2 and Mutant-3 and indicated the possibility of obtaining success through mutation breeding.

4. Biotechnology

The use of biotechnological tools viz. micropropagation, development of soma clonal variations, haploids, di-haploids, secondary metabolites from cell cultures, *in vitro* conservation and molecular characteization of genetic resources and development of novel transgenics have opened opportunities for enhancing, stabilizing yield and improving quality. The successful exploitation of plant biotechnology depends on the establishment of efficient regeneration system. The micro propagation protocols for many seed spices are available (Nirmal Babu *et. al.,* (1997).

Breeding Objectives

The major breeding objectives related to seed spices crops are mentioned below.

Increased seed yield: The average seed yield of seed spices in India is very low in comparison to other countries. There is lack of sufficient number of improved varieties to suit different agroclimatic situations. There is lack of high yielding fertilzer responsive varieties also.

Plant types: The medium tall to short varieties in coriander, fennel, dill, ajowan with more branching are required to be developed. Such type of varieties resist lodging and more branching contribute positively to the higher seed yield and also assume significance for different cropping systems to get enhanced productivity.

Dual purpose varieties: Coriander and fenugreek, the important seed spices crops are used both for green leaves as vegetable and seeds as spices. Similarly, ajowan, dill, fennel possess both herbal value and seeds as spices value. Therefore there is need to develop dual purpose varieties.

Early maturity: Some of the seed spices are long duration in nature and take about 4-6 months for seed maturity. Among the seed spices fennel and ajowan takes about 6 months for seed maturity. Thus there is a need to develop early varieties. Suitable varieties for late, mid and early seasons should be developed for using in multiple cropping systems.

Disease resistance: The Seed Spices crops suffer from mainly wilts, blights, powdery mildew and root rots. Wilt in coriander and cumin; blights in cumin and fennel; powdery mildew in cumin, coriander, fenugreek; root rot in fenugreek has caused havoc in successful production of these crops. There is an urgent need to develop resistant varieties for such diseases.

Insect resistance: Aphid is the serious pest in all most all seed spice crops and cause severe losses to the crops viz. coriander, fennel, cumin, fenugreek and deteriorates the quality of seed/fruits. The progress in breeding for insect resistance in comparison to disease resistance is negligible due to lack of effective insect rearing methods, germplasm screening technologies and resistant sources.

Abiotic stress tolerance: The seed spices crops are facing drought, salinity and low temperature stresses. Comparatively the seed spices crops need lesser irrigations in comparison to other crops. Among seed spices crops coriander is successfully grown by farmers under rainfed conditions and cumin requires lesser irrigation. There is need to develop efficient drought hardy genotypes of seed spices. The cumin and coriander are vulnerable to frost mostly at flowering stage, thereby resulting in poor seed setting and yield. The frost tolerance in these crops is an important objective for breeding and improvement. The seed spices crops are mostly grown under semi arid to arid conditions with salinity in the soil. Thus salinity tolerant variety of seed spices crops are needed to be developed.

Breeding for quality: Among the quality characters the high essential oil content is the important criteria. Higher the oil content, the better is the quality. The oil content in coriander ranged from 0.17 to 0.26 % which is very less in Indian varieties whereas Russian lines contain more essential oil (0.7 %). The essential oil content in cumin seed varied from 3 to 6 %. The Indian varieties contain mostly 3% of the oil content whereas one exotic line EC 130202 has been reported to contain upto 6% of essential oil. The fruit of fennel yield a volatile oil of which the major constituent is anethole. The oil is used as flavouring agent. In different group of fennel, the volatile oil content ranged from 0.7 to 6.0 %. The Indian varieties of fennel contain very low range of oil content. In fenugreek from quality point of view, large seed size with attractive yellow colour and content of diosgenin a steroid are important criteria. The local fenugreek contain 0.4 to 1.26 % of diosgenin. The essential oil content ranged from 2 to 4 % and 26% fatty oils in ajowan; 1.5 to 4 % in dill; 1.5 to 3.0 % in celery; 1.9 to 3.1% in anise; 1.5 to 4.1% in Nigella.

Constraints in utilization of sources, if any

It is estimated that the country would need to produce 15-17 lakh tons of seed spices by 2020 AD to meet the domestic as well as export demand in the world market. This could be done only through increasing the productivity of seed spices crops in the country, which is at present very low. The crop improvement and breeding approach could be the best solution for resolving the constraints and thereby improving the productivity. In this context the constraints are to be looked for working strategies. The NRC on Seed Spices is striving hard to collect and conserve the valuable gene pool on different seed spices as a national repository for further utilization in crop improvement. The constraints in effective utilization of genetic sources are given below.

1. The inherent nature of slow germination of seed spices crops, lack of complete information on pollination behaviour, high diseases and insect-pests pressure viz, wilts, blights, rots, powdery mildews, aphids and store grain pests, quality degradation due to microbial load during storage, lack of information on postharvest techniques particularly storage methods for seed spice crops are the major challenges posing problem in effective utilization of seed spices germplasm.
2. Most of the available germplasm exists in the form of traditional / local varieties and have been subjected to natural selection for local adaptation. Thus the germplasm exit in the form of complex gene mixtures. Proper sampling as well as regeneration is essential to recover and maintain the full range of genetic variability of these crops for further utilization in seed spices crop improvement programme.
3. Present germplasm collection possesses limited variability. Strong and stable resistant sources are not available for biotic and abiotic stresses. The value added germplasm need to be identified from the collections regarding spices oil and oleoresins content, suitability to rainfed conditions, tolerance to frost, resistance to diseases and insect-pests. Dual purpose lines of coriander and fenugreek i.e. both for leaves and seeds. There is a need of collaborative explorations from Mediterranean regions (Morocco, Algeria, Libya, Egypt, Tunisia) and adjoining parts of Mediterranean sea, Iran, Iraq, Russia and Europe.
4. The collections made are random, without proper documentation both at field level and the maintenance site. The collections have been made by breeders whose prime objective was to develop improved varieties. Biasness for superior types might have occurred in such collections.
5. Serious attention has not been paid regarding survey and germplasm collection of ajowan, dill, celery, nigella, aniseed and caraway, the minor but important group of seed spices. The plant biodiversity related to these crops needed to be collected, conserved and utilized, as they are considered as under-exploited crops.

6. The plant biodiversity of seed spices required to be evaluated for different morpho-agronomic attributes, resistance against biotic and abiotic stresses, chemical and biochemical traits and promising stocks with potential genes are rare gene combinations are isolated. The characterisation information in case of seed spices is required to be documented and catalogued crop-wise for use by the breeders, geneticists, biotechnologists and other users. Documentation of germplasm is necessary for its sustainable use of genetic resources of seed spices. So far no serious attempt has been made for documentation of information on seed spices.

Future Perspective to Enhance Use of Genetic Resources in Crop Improvement

1. Most of the varieties released so far have been developed through selection. To solve the specific problems of biotic and abiotic stress resistance, modern techniques using biotechnology involving molecular assisted plant breeding techniques are required to be used.
2. In order to effectively counter any attempt of biopiracy of valuable germplasm of seed spices from India, the first and foremost step needed is stock taking our assets and there accurate documentation. In this regard, there is need to develop adequate research base for characterisation of accessions at biochemical and molecular level employing techniques like isozyme profiling, RFLP, AFLP and RAPD. Molecular characterization of important accessions of seed spices crops is required to validate the identity.
3. Conservation of seed spices germplasm in the National Gene Bank especially with reference to National Active Germplasm Site draws our attention as high priority action.
4. The seed spices are still considered as under-utilised crops and serious attention has not been paid regarding systematic germplasm collection, evaluation and conservation. The prevailing diversity of major and minor seed spice crops in remote and tribal pockets need to be revisited. There is tremendous scope for collection of valuable land races of such seed spices crops.
5. The varied agro-climatic conditions offer a good scope of promotion of non-traditional seed spices (anise, caraway, celery, Parsley, black caraway, to make India a broaden supply base in the world market. The systematic germplasm collection, evaluation and conservation of such spices have better future perspectives of crop improvement.

References

Bhatt, K.C. and Singh, R.K. 2004. On-farm *in situ* management of crop genetic resources. pp. 297-317. In: Plant Genetic Resource Management (eds. B.S. Dhillon, R.K. Tyagi, Arjun Lal and S. Saxena). Narosa Publishing House, New Delhi.

Dimri, B.P., Khan, M.N.A. and Narayana, M.R. 1976. Some promising selections of Bulgarian corinader (*Coriandrum sativum* Linn.) for seed and essential oil with a note on cultivation and distillation of oil. *Indian Perfumer*, **20:** 13-21.

Ellis, R.H., Roberts, E.H. and Whitehead, J. 1980. A new, more economic and accurate approach to monitoring the viability of accessions during storage in seed banks. *PGR Newsl.*, **41:** 3-18.

Frankel, O.H. and Soule, M.E. 1981. The conservation of plants used by man. pp. 225-251.In: *Conservation and Evolution* (eds. O.H. Frankel and Michael, E. Soule). Cambridge Universdity Press, Cambridge, U.K.

Hore, A. 1989. Improvement of minor (Umbelliferous) spices in India. *Economic Botany*, **33(3):** 290-297.

Joshi, B.S., Ramanujam, S. and Saxena, M.B.L. 1963. Improvement of some oil bnearing spices plants. *Bull. Reg. Res. Lab*. Jammu, **1:** 94-100.

Malhotra, S.K. and Vashishtha, B.B. (2005) Seed Spices. *Peter K.V. and J. Abrahm.* Biodiversity of Spices and Aromatic Crops. In Ed. Daya Publishing House, New Delhi.

Malhotra, S.K. and Vashishtha, B.B. (2006) A Manual on minimal Descriptors of Seed Spices, NRCSS Ajmer, p 1-51.

Malhotra, S.K. and Vijay,O.P.2003. Plant genetic resources of seed spices in India. Seed Spices Newsletter, **3(1):** 1-4.

Malhotra, S.K. 2003. Crop Improvement in seed spices crops. Ed. Sharma, AK., Breeding Field Crops. Academic Publishers, Bikaner.

Malhotra, S.K. 2002, Status paper on Plant Genetic Resources of Seed Spices in India. Concept paper presented in National Workshop on "Germplasm Management of Horticulture and Agro forestry crops for sustainable utililzation at NBPGR, New Delhi, held from 27-28, February, 2002.

Malhotra, S.K. 2004. Status paper on utilization of plant genetic resources of seed spices crops in India. Concept paper presented in National Workshop on Utilization of plant genetic resources at NBPGR, New Delhi held from 5-6, October, 2004.

Nirmal Babu,K.,Ravindran,P.N. and Peter,K.V.1997. Protocols for micropropagation of spices and aromatic crops. IISR, Calicut. p35.

Mahajan, R.K., Sapra, R.L., Umesh Srivastava, Mahendra Singh and Sharma, G.D. 2000. Minimal descriptors (For characterisation and evaluation) of agri-horticultural crops (Part I). pp. 217-224. National Bureau of Plant Genetic Resources, New Delhi, India.

Nair, M.K. and Ravindran, P.N. 1988. Genetic resources in spices – their diversity and utilisation in India. pp. 419-428. In: *Plant Genetic resources – Indian Perspective* (eds. R.S. Paroda, R.K. Arora and K.P.S. Chandel). National Bureau of Plant Genetic Resources, New Delhi, India.

Peter, G.B., Vedamuthu, MD. Abdul Khader and F. Salal Rajan. 1994. Improvement of seed spices. pp 345-374. In: *Advances in Horticulture vol. 9* (eds. Chadha, K.L. and P.Rethinam). Malhotra Publishing House, New Delhi, India.

Peter, K.V. and Nirmal Babu. 2005. Genetic resources of seed spices. *Indian J. Pl. Genet. Resour.*, **18(1):** 19.

Ramanujam, S., Joshi, B.S. and Saxena, M.B.L. 1964. Extent and randomness of cross-pollination in some umbelliferous spices of India. *Indian J. Genet.* **24(1):** 62-67.

Ramanujam S and Tiwari,V.P. 1970 Heterosis in fennel. Indian J. Genet. Plant Bread. **30(3):**732-37.

Romanenko., L.G., Nevkrytaja, N.V. and Kuznecova, E.J.U. 1992. Self fertility in coriander (in Russian). *Sel. Semenovod.* (Moskva) **1992:** 25-28.

Sharma, R.K. 1994. Genetic resources of seed spices. pp 193-205. In: In: *Advances in Horticulture vol. 9* (eds. Chadha, K.L. and P.Rethinam). Malhotra Publishing House, New Delhi, India.

Singh, M., Sharma, A.K, Shahana, S. and Dhillon, B.S. 2004. Plant genetic resource management in relation to mode of sexual reproduction. pp. 297-317. In: *Plant Genetic resources – Indian Perspective* (eds. R.S. Paroda, R.K. Arora and K.P.S. Chandel). National Bureau of Plant Genetic Resources, New Delhi, India.

Srivastava, Umesh, Mahajan, R.K., Gangopadhyay, K.K., Mahendra Singh and Dhillon, B.S. 2001. Minimal descriptors of agri-horticultural crops (Part II). pp. 221-243. National Bureau of Plant Genetic Resources, New Delhi, India.

Tao, K. L. 1985. Standards for genebanks. *PGR Newsl.*, **62:** 36-41.

Tiwari, VP 1968. Some floral mutants in fennel and sowa. Indian Agriculture **16(4):** 92-93.

❑❑❑

Chapter – 36

Biodiversitry and Utilization of Medicinal Mushrooms with Particular Reference to *Ganoderma lucidum* and *Cordyceps sinensis*

R.P. Singh & K.K. Mishra

Introduction

In 1990, the magnitude of fungal diversity, that is, the actual number of species worldwide, was estimated conservatively to be at least 1.5 million. Of the 1.5 million estimated fungi, it has been estimated that 14,000 species produce fruiting bodies of sufficient size and suitable structure to be considered macro-fungi which can be called mushrooms. Of these, about 50 % or 7,000 species are considered to possess varying degrees of edibility and more than 3,000 species from 31 genera are regarded as prime edible mushrooms. To date, only 200 of them are experimentally grown, 100 economically cultivated, approximately 60 commercially cultivated and about 10 have reached an industrial scale of production in many countries (Hawksworth, 2001). Of the 14,000 species of mushrooms in the world, around 700 have known for medicinal properties. Thus, mushrooms have vast prospects as sources of medicines. The early herbalists were more interested in the medicinal properties of mushrooms than in their basic value as a source of food. Humankind has constantly searched for new substances that can improve biological functions and thereby make people fitter and healthier. About 3.5 billion people worldwide, all over half of the world populations rely on plant-based medicines and dietary supplements for their primary health care. Of the plant materials involved in medicines or health tonics, quite a few are fungi. The practice of using fungi as herbal medicines can be traced in different early records of the 'materia medica'. The earliest book on medicinal materials the "Shen Nongs Herbal" recorded the medicinal effects of several fungi such as *Ganoderma lucidum, Poria cocos, Tremella fuciformis, Polyporus umbillatus* and other unidentified fungi. The most famous of all work on the traditional medicines "Pen Ts'ao Kang Mu" which was compiled by Li Shi-Zhen of the Ming Dynasty recorded the medicinal fungi totaling more than twenty species, including *Ganoderma lucidum, Poria cocos, Polyporus umbellatus, Lentinula edodes, Termitomyces albuminosus, Auricularia auricula, Pleurotus ostreatus, Armillaria mellea* etc. A very unique insect-infecting fungus, *Cordyceps sinensis*, was for the first time taken as a medicinal fungus in the book "Essentials of Materia Medica". There have been new discoveries and developments in the field of medicinal fungi, which have greatly enriched the treasure house of our traditional medical and medicinal sciences. As a result of large number of scientific studies on medicinal mushrooms in the past three decades, the traditional uses of many mushrooms have been confirmed and new wider uses found. Some traditionally important and leading medicinal fungi in the Oriental medicines are presented below:

1. *Ganoderma lucidum:* The leader of medicinal mushrooms

Ganoderma lucidum or Reishi is a Basidiomycete, lamella less fungus belonging to the phylum: Basidiomycota, order: Aphyllophorales and family Ganodermataceae (Alexopoulous *et al.,* 1996). In nature, it grows in densely wooded mountains of high humidity and dim lighting. It is rarely found since it flourishes mainly on the dried trunks of dead plum, *Quercus serrata* or Pasonia trees. Out of 10,000 such aged trees, perhaps only 2 or 3 will have Reishi growth, therefore it is very scarce. *Ganoderma* is found flourishing well in many high altitude regions of Uttarakhand in India (Singh *et al.*, 2007). Relatively rare and undiscovered in the West, Reishi has been revered as an herbal medicine for thousands of years in Japan and China. Emperors of the great Chinese dynasties and Japanese royalty drank teas and concoctions of the mushroom for vitality and long life. Reishi has long been known to extend life span, increase youthful vigor and vitality. It also promotes blood circulation by eliminating thrombi in the blood streams. As a result, the person feels renewed vitality. Deterioration of mind and body is arrested. Reishi is indeed an herb with multiple applications.

In ancient times, Reishi or *Ganoderma lucidum* has been considered as a very auspicious herbal medicine and its efficacy has been attested to the oldest Chinese Medical Text Book which is presumed to be about 2,000 years old. This book is known to the Japanese as 'Shinnoh Honsohkyo' and is accepted as being the original textbook of Oriental Medical Science. In it, around 365 different kinds of medicines are classified and explained. The medicines are basically of three main categories: 120 of them are regarded as "superior" medicines; another 120 are classified as "average" medicines while the remaining 125 are placed in the fair category. Of the superior medicines listed in the text, Reishi was rated as number one.

Morphological Characteristics

The size of the fruiting bodies is usually varying in nature. They may be stipitate, dimidiate or reniform. In some cases, they may also be of sub orbicular shape. They are thick, corky and yellowish along the growing margins and then turn brown in the matured part. They have a shining surface. The margins of fruiting bodies are usually thin and often slightly incurved. The stipe is lateral, sometimes eccentric. It is thick and dark in colour but later turns to purple-brown. Initially the pores are white. Browning appears at a later stage. Basidiospores are brown, ovate with a rounded base and truncate to narrowly rounded apex. The wall is complex and is composed of several layers (Furtado, 1962; Peglar and Young, 1973; Mishra, 2007). The outermost wall is convected to the inner wall by interwall pillars. The spore is the most characteristic and distinguishing feature of this mushroom family (Furtado, 1962; Donk, 1964). The epispores are smooth while the endospores are rough with a large central gutta. Nuss (1982) suggested that *G. lucidum* may produce two types of basidiospores. One type is produced in the early part of the season and is said to germinate only after insect ingestion and probably is dispersed in this manner. The second type of basidiospore is the common one produced throughout the rest of the growing season. This type of basidiospore is widely dispersed by air and germinates readily on agar without any special treatment.

Growth Parameters

Ganoderma lucidum, commonly known as a wood decaying fungus, causes white rot of a wide variety of trees and is phytopathogenic in nature. It is found more frequently in subtropical regions than in the temperate zones. It is an annual mushroom, growing on a wide variety of dead and dying trees. The growth parameters required for its optimum growth are summarized as follows:

(a) Temperature: Temperature for mycelial growth ranges from 15 to 35° C and the optimum temperature is 24 to 25° C; for primordial initiation it is from 18 to 25° C and for fruiting body development the range is 20 to 25° C.

(b) Water content in substrate: Should be maintained at 60 to 65 %.

(c) Relative humidity: for mycelial running in the range of 60 to 70 %, primordia initiation, 85 to 90 % fruiting development, 70 to 85 %.

(d) Air: During the fruitification period, good ventilation is necessary.

(e) pH value: The optimum for mycelial running is 5.0 to 5.5.

(f) Light: During primordia initiation, light is required at about 500 to 1000 lux and for fruiting body development, 750 to 1500 lux.

Artificial Cultivation of *Ganoderma lucidum*

Many developed and developing countries have developed cultivation technology of this mushroom and further advancement is required. Although the medicinal value of *G. lucidum* has been treasured in China for more than 2000 years, the mushroom was found infrequently in nature. This lack of availability was largely responsible for the mushroom being so highly cherished and expensive. During ancient times in China, any person who picked the mushroom from the natural environment and presented it to a high-ranking official was usually well rewarded. Even in the early 1950s, this custom continued and the mushroom was presented to Chinese leaders, both in Mainland China and in Taiwan, following its occasional discovery in the wild. As the supply of the wild mushroom is limited and as there is great difficulty to properly control the quality of its fruiting bodies in nature, it has become increasingly popular to use Ling Zhi that has been grown in a controlled environment.

Artificial culture and cultivation of this mushroom was initially attempted in 1937. Its mass production was first achieved in 1971 by cultivating the fungus in pots containing sawdust. After that, the cultivation method using bed logs or sawdust had been established through studies by many peoples and farmers. In 1991, Triratana and Chaiprasert tried to cultivate it on sawdust bags. The substrate was a mixture of pararubber (*Hevea brasiliensis*), sawdust, rice bran (5 %), gypsum (1 %) and magnesium sulfate (0.2 %). The substrate was packed in 7" X 8" polypropylene bags of only 300 g each. The bags were then incubated at 27-32°C. After the spawn run through the bottom, the bags were placed on the shelves in the mushroom house for fruiting. The culture bags were exposed to 80 % humidity and natural indoor daylight of about 150 lux. They found that average yield taken only once from the first harvest varied from 6.7-16.9 g/bag. This cultivation is still practiced in many parts of the world with variation in the quantity and quality of substrate basically the sawdust from different trees including *Hevea brasiliensis, Dipterocarpus alatus, Tectona grandis* etc. mixed with various supplements such as varying levels of rice bran, magnesium sulfate and calcium sulfate. Other methods of cultivation of this mushroom has been tried out during the times using different substrate in different proportions and combinations in different conditions of light, temperature and humidity (Chiu *et al.,* 2000; Rai , 2003; Wagner *et al.,* 2003; Dadwal and Jamaluddin, 2004; Mishra and Singh, 2006,. Veena and Pandey, 2006).

Artificial cultivation of this valuable mushroom was successfully achieved in the early 1970s and since 1980, particularly in China, production of *G. lucidum* has developed rapidly. Similarly, as for other cultivated edible mushrooms, the process for producing *G. lucidum* fruiting bodies can be divided into two major stages. The first stage involves the preparation of the fruiting culture, stock culture, mother spawn and planting spawn and the second stage entails the preparation of the growth substrate for mushroom cultivation. Currently, the methods most widely adopted for commercial production are the wood log, short wood segment, tree stump, sawdust bag and bottle procedures. Examples of cultivation substrate, using plastic bags or bottle as containers, include the following:

(a) Saw dust, wheat bran, gypsum and soybean powder.

(b) Bagasse, wheat bran, cane sugar, gypsum and soybean powder.

(c) Cottonseed hull, wheat bran, cane sugar and gypsum.

(d) Sawdust, corncob powder, wheat bran, gypsum and cereal straw ash.

(e) Corncob powder, wheat/rice bran, gypsum and straw ash.

After sterilization, the plastic bags can be laid horizontally on beds or on the ground for fruiting. Log cultivation methods include the use of natural logs and tree stumps, which are inoculated with spawn directly under natural conditions.

Chemical Composition and Therapeutic Potential of *Ganoderma lucidum*

Ganoderma is correctly called the "mushroom of immortality". In recent years, its active ingredients have been a subject of extensive research regarding their apparent ability to help, prevent or treat certain types of cancers, liver diseases, HIV infection, acute or recurrent hepatic infections, high blood pressure, chronic bronchitis, allergies asthma and to favorably modulate the immune functions (Chang, 1993, 1996b and 1996c). It has been reported to have multiple and remarkable beneficial medicinal effects on various diseases. This diversity in the beneficial and medicinal effects is attributed to the fact that the fruiting bodies are composed of a vast number of bioactive compounds. The major compounds with significant pharmacological activities appear to be the Triterpenes, especially the lanostane type derivatives (Luo *et al.*, 2002 and Ma *et al.*, 2002) and polysaccharides although bioactive proteins, nucleic acids, and other substances have also been identified (Kawagishi, 1997 and Zhu, 1999).

The most important active ingredients of this mushroom are as follows:

1. Polysaccharides: More than 100 types of polysaccharides have been isolated from the fruiting body and mycelia and most of them have a molecular weight in the range of 4×10^5 to 1×10^6 (Liu, 1996 and Su *et al.,* 1997) These components are the one that are responsible for the immunomodulatory effects of *Ganoderma*. Evidence from extensive research has indicated that these polysaccharides have significant biological activities, including immunomodulating, chemo- and radio-preventive, anti-ageing, hepatoprotective, anti-tumorogenic, anti-microgenic effects, antimicrobial, etc.(Wang, 1997 and Wasser and Weis, 1999)
2. Heteropolysaccharides and Glycoproteins: In addition to water soluble â-D-glucans, â-D-glucans also exist with heteropolysaccharide chain of xylose, mannose, galactose, uronic acid, ulucidenic acids, etc. (Wasser and Weis, 1999).
3. Triterpenoids: Currently, about 119 triterpenoids have been isolated from the fruiting body and mycelia, which include highly oxidized lanostane-type triterpenoids, such as ganoderic acid and lucidenic acid, ganodermic acids,

ganoderemic acids, lucidone, ganoderols and ganoderal etc (Kim and Kim, 1999). The structures of these triterpenoids have a lanostane skeleton and they are classified into 10 groups based on their numbers of carbons and state of oxidation (Komoda *et al.,* 1985). Most of these triterpenoids exhibit a wide spectrum of biological activities (Kim and Kim, 1999).

4. Proteins: Biologically active proteins have been isolated from *G. lucidum* (Tanaka *et al.,* 1989 and Vander Hem *et al.,* 1995). LZ-8 protein isolated from *G. lucidum* has similar variable regions of immunoglobulin heavy chains both in its sequence and its predictory secondary structure by various sequencing studies (Tanaka *et al.,* 1989).
5. Other components: These include small amounts of polyphenols, steroids, lignin, ganomycins, vitamins, lectin, nucleosides, nucleotides, alkaloids, amino acids and organic germanium Rosecke and Konig, 2000). The alkaloids (ganoine, ganodine and ã-butyrobetaine and their derivatives) from this mushroom have showed some anti-inflammatory activity in animal studies (Yang and Yu, 1990). These substances may also have some hepato-protective activity.

Medicinal Uses of *Ganoderma lucidum*

Mushrooms have been considered as ultimate health foods (King, 1993, Chilton, 1993) and have been used since ages. Ancient literatures and timely research investigations have proved the observations of the oriental herbalists that certain mushrooms possess very useful medicinal attributes (Pai *et al.,* 1990; Chang, 1996a; Ikekawa, 1969; Ito, 1972). In the recent past, a variety of medicinal preparations in the form of tablets, capsules and extracts from mushrooms for the treatment of various kinds of ailments, diseases and disorders have been produced and have a worldwide market and their usage has shown a tremendous increase since last few years (Vinning, 2003 and 2004; Mishra *et al.,* 2004; Singh *et al.,* 2006). In 1991, the value of world mushroom crop was estimated to be around 8.5 billion dollars and in the same year 1.2 billion dollars were estimated to have been generated from medicinal products from various medicinal mushrooms. Herbal medicinal preparations from different mushrooms have become a growing business in various parts of the world and have tremendous therapeutic potential (Wasser, 2002). Since ages, *Ganoderma lucidum* has been regarded as panacea of life, imparting youth, vigor and longevity. Comprehensive reviews on *Ganoderma lucidum* oriented towards modern medical science elaborate the use of this fungus, along with other mushrooms, in various treatments (Szedlay, 2002; Singh and Mishra, 2004). There are data of clinical trials that support the efficacy of *Ganoderma lucidum* as a medicinal herb, especially for disorders related to the liver, kidney and immune system. A number of studies indicate that *Ganoderma lucidum* possess certain anti-cancer, anti-allergic, antibiotic and antimicrobial activities along with definite immuno-stimulating properties. It is also reported to have anti-oxidant activity (Yamaguchi *et al.,* 2000; Li *et al.,* 2001b). The fruit bodies and mycelial culture of *Ganoderma lucidum* have been used time immemorial and are still used for strengthening the immune system and for many types of heart, kidney, liver, circulatory and for treating TB, asthma, back pain, reproductive disorders, cancer, etc.

2. *Cordyceps sinensis*

Cordyceps is a fungus of subphylum Ascomycotina, class Pyrenomycetes, order Clavicipitales and family Clavicipitaceae and includes more than 300 species worldwide (Mains, 1940; Saccardo, 1883; Massee, 1895). A new classification of *Cordyceps* species has been suggested on the basis of chemo-taxonomy of partial nucleotide sequence of 18 S rDNA obtained from four different species (Chen *et al.*, 2002a). *Cordyceps* species are parasitic, mainly on insects and other arthropods (Chen *et al.*, 2002b). Some of these are also parasitic on other fungi like the subterranean, truffle-like *Elaphomyces* and also on spiders. The mycelium invades and eventually replaces the host tissue, while the elongated fruiting body (stroma) may be cylindrical, branched or of complex shape. The genus has a worldwide distribution and most species have been described from Asia (notably China, Japan, Korea and Thailand). *Cordyceps* species are particularly abundant and diverse in humid temperate and tropical forests. Some *Cordyceps* species are sources of important biochemical substances like cordycepin which has very high medicinal properties. Charles (1941) has included 39 species of *Cordyceps* in a check list of entomogenous fungi found exclusively in North America. Of the 137 species, 34 are reported for North America alone. There have been some differences in generic limitations. Their identification as different species is done on differences in various characters as color, length and shape of the perithecium and the asci, nature of host etc. The species *Cordyceps sinensis*, that parasitizes the vegetable caterpillar, is the most famous amongst all the species of *Cordyceps* and has been considered a precious ingredient of high medicinal importance. Chen *et al.* (1999) also studied the genetic diversity and taxonomic status of *Cordyceps sinensis* using RAPD markers.

The mushroom and its host

C. sinensis is an entomophagous fungus of the family Clavicipitaceae (Charles, 1941). It parasitizes a range of grass root boring caterpillars, most commonly the Thitarodes (*Hepialus armoricanus*, family Hepialidae). In all, around 40 species of *Hepialus* moth have been recognized in the Tibetan Plateau region and around 30 of these species can be infected by *C. sinensis*. The mycelium of the fungus grows in soil and colonizes the buried larvae (caterpillar) of this moth. The caterpillar becomes mummified by the growth of the mycelium (Arif and Kumar, 2003) and hence is given the name, "caterpillar fungus" (Peglar *et al*, 1994). It has been reported that *C. sinensis* has evolved and developed a special adaptation to improve chances of reproductive success (Li *et al.*, 1999). Reproduction is highly host-specific (Nikoh and Fukatsu, 2000). Every single spore fragments into around 32 million propagules. These tiny propagules get attached to the larval stage of the insect. The larvae is then forced to move closer to the surface of ground (non-infected larvae will not hibernate close to the ground surface). The mycelium, which is composed of white thread like structures called hyphae, grows inside the body of the insect in the form of a cottony mesh. The hyphae fill the interior of the entire caterpillar and mummifies it, leaving behind the larval exoskeleton filled with only the white mycelium of fungus. When alpine grasses start sprouting, a fruiting body develops which, surprisingly, always emerges out from the head of the caterpillar (larvae). This fruiting body is usually 5-10 cm long, brown colored and club-shaped. The propagules present on the fruiting body are dispersed by the wind and can attach to new host insects. The fruiting body resembles grass sprouting but the difference is the colour which is dark blue to black.

Native occurrence of this entomophagus fungus is mostly confined to the high Himalayan Mountains in Tibet, Nepal and India, at an altitude ranging from 3000 to 5000 metres. The most common occurrence of this fungus is between 3500 and 4500 metres elevation in cold and arid environment. *C. sinensis* is endemic to the Tibetan Plateau including the adjoining high altitude areas of the Central and Eastern Himalayan range (covering areas of Nepal, Bhutan and Uttarakhand, Sikkim, Himachal Pradesh and Arunachal Pradesh in India). It is found in the high altitudes of Pithoragarh, Uttarakhand and other provinces at locations above 7000 feet. It is also common in the grasslands and shrub lands of the Tibetan Plateau including west Sichuan, North Yunnan and major areas of Qinghai and West Gansu. The distribution of *C. sinensis* is limited to those areas where the average annual precipitation is above the range 350-400 mm.

This caterpillar fungus thrives very well in sub-alpine and alpine grasslands or meadows and in open dwarf shrublands. In Lithang, it can be found on the north-facing slopes. In Machen and other regions of South Qinghai Province, it can be found on well-drained sunny slopes with good and rich grass vegetation. In Nagchu (Tibet), it is almost absent in the marginal areas but is present in ample amounts in the rich pasture areas. Extensive research has been conducted on the ecology, collection, utilization, trade route, management and significance and species diversity of *C. sinensis* in various parts of India, Tibet, Michigan, Bhutan, China and Korea (Sung, 2004; Zang and Kinjo, 1998; Jones and Yusipang, 2001; Garbyal, 2001).

Collection and trade of *C. sinensis*

Years back this fungus was collected and traded from Tibet to China in exchange for tea and other commercial goods like silk, grains, etc (Namgyel, 2003). The most appropriate time for harvesting of this fungus starts with the arrival of the spring season, in about starting of May. Local villagers and nomads search for the fungus in the grasslands and shrublands. But the harvesting is slightly a difficult process as only the stroma or the grass-like part of the fungus is visible over the surface which too is quite short, not longer than 2-5cm and has to be lifted out with the help of a sharp knife. Extreme care has to be taken while pulling out the fungus from the ground surface because if the stroma breaks off from the head of the larvae, it directly affects the commercial value of the mushroom resulting in a decline in its market rate.

During the harvesting season, all other activities come to a standstill as every one is focused on gathering more and more of the fruiting bodies. This has often led to many blooded wars in these areas. The Ministry of Population and Environment, Govt. of Nepal has banned its collection, trade and transportation. However, in Tibet it is an open trade. In some areas of Tibet, even the schools announce vacations for 15-20 day in late May so that the students can also help in the collection and harvesting of this precious mushroom. The daily collection may vary from 250 g to as much as 5 kg.

During recent years, caterpillar mushroom has emerged out be an important cash crop traded on a large scale and a new source of income in the rural areas in the higher altitude regions usually above 3000-5000 meters (Boesi, 2003). In the river valley of Gori Ganga alone, the number of fungus gatherers at alpine habitats has increased about four-fold since the year 2000. During 2002, nearly 900 persons went to seven different alpine habitats in search of the fungus (average 128 persons per habitat) and collected about 200 g of the fungus worth Rs. 8600, according to the purchase price in the year 2002. This was the price if the material was sold immediately in the alpine habitats. Carrying the collected fungus material a few kilometers down to the local market, the selling price increases to about Rs. 7000 per kg. Since the year 2000, purchase price at the field site and in the local market has increased tremendously and so also the income of a gatherer. Between the year 2000 and 2002, the income of a wild material gatherer had increased by 3.7 times and above four-fold if the material was sold at the field or to the agents in the local market respectively. Factors such as high price of the fungus, very often make the transaction secretive in the local market while due to cross-boundary trade between two countries, the rest of the trade is under the surface (Sharma, 2004). Cost of one kg of the fungus at the final destination (brokers in national and international markets) was much higher than the price paid to the field gatherers in the year 2002 and ranged between Rs. 68000 to 80,000 in Tibet, while in Nepal a slight increase was noted (Rs. 80,000 - 90,000). However in the Indian market, the material was sold at the rate of Rs. 125,000 to 130,000 per kg. Fresh fungus is sold for 9,000 - 15,000 RMB/Jin (1 Jin=1 metric pound=500 g) in Lithang. It is believed that in the International market the fungus may fetch a price between one and two million Rupees per kg (US$ 20,000 - 40,000). The amount paid varies among the trade channels which start from the wild material gatherers in the field, then to the brokers and agents who collect the dried material from the various locations and sell it at a higher price. However, rapid and immediate marketing of this fungus is not required as the fungus is usually sold and consumed in a dried form. Its small size and easy storage conditions make the transportation much easier.

Medicinal Uses of *Cordyceps sinensis*

Besides the attributes that make mushrooms the ultimate health foods (King, 1993), timely research investigations have proved the empirical observations of the oriental herbalists that certain mushrooms possess very useful medicinal attributes (Pai *et al.*, 1990; Chang, 1996b). In the recent past, a variety of medicinal preparations in the form of tablets, capsules and extracts from mushrooms for the treatment of various kinds of ailments, diseases and disorders have been produced and marketed (Chang and Buswell, 1996; Mizuno, 1996). In 1991, the value of world mushroom crop was estimated to be around 8.5 billion dollars and in the same year 1.2 billion dollars were estimated to have been generated from medicinal products from various medicinal mushrooms. Herbal medicinal preparations from different mushrooms have become a growing business in various parts of the world as for instance, Bhutan is an emerging market for *Cordyceps* and its usage has shown a tremendous increase since last few years (Vinning, 2003 and 2004). Comprehensive reviews on *C. sinensis* oriented towards modern medical science elaborate the use of this fungus, along with other mushrooms, in various treatments (Jason, 2005). There are data of clinical trials that support the efficacy of *C. sinensis* as a medicinal herb, especially for disorders related to the liver, kidney and immune system. A number of studies indicate that *C. sinensis* (and also its mycelial extract) possess certain anti-cancer, anti-metastatic and immuno-stimulating properties. It is also reported to have anti-oxidant activity (Li *et al.*, 2001a; Yamaguchi *et al.*, 2000). Medicinal use of *C. sinensis* by Tibetans has been documented for over 500 years. The fruit bodies and mycelial culture of *C.* sinensis has been and are still used for strengthening the immune system and for many types of heart, kidney, liver, circulatory and for treating TB, asthma, back pain, reproductive disorders, cancer, etc.(Manabe *et al.*, 2000). Since ages, *C. sinensis* has been regarded as panacea of life, imparting youth, vigor and longevity.

Artificial and Semi-artificial Cultivation of *C. sinensis*

Chinese laboratories pioneered the artificial cultivation of *C. sinensis* derived organisms in the early 1980s. The resulting asexual organisms are distinct from the original fungus and since the fungus is not parasitizing *Thitarodes* caterpillars, but feeding on other substrate, the organism develops differently and is thus, regarded as a different species. There are many new species known as agamotypes, i.e. *Paecilomyces hepialid, Hirsutella sinensis* and *Cephalosporium dongchong xiacae*. In China, liquid culture or fermentation is wide spread. In liquid culture or fermentation, the organism is introduced into a tank of sterilized liquid medium, which has been formulated to provide all of the necessary nutritional components for rapid growth of the mycelium. Sterilized silkworm residue is commonly used as a base for this medium. After growth in the liquid medium, the mycelium is harvested by straining it out of the liquid broth and drying, after which it can be used as it is or can be processed further. Generally, in this method, the extra-cellular compounds which were exuded by the fungus during the growth cycle are discarded with the spent broth. This represents a major loss of bioactive compounds as many of the active ingredients are extra-cellular in nature and are found only in small concentrations in the mycelium. The substrate of choice for most Chinese growers in a liquid media, as stated earlier, is based upon silkworm residue, with added carbohydrates and minerals. This seems to be a logical choice, as this mushroom is found in nature growing on insects (caterpillars). Dried silkworm bodies are the by-product of silk industry and have little other use. Therefore they are readily available and are cheaper than other substrate.

This silkworm based substrate seems to yield a relatively high quality product. But the major problem with silkworm residue based substrate is that in the United States, the FDA guidelines for mycelial products advocate them to be produced on a normally consumed human food source. Silkworms do not fit into that category. They are also not available as a raw material source to most of the worlds *Cordyceps* cultivators.

The second cultivation method is the solid-substrate method followed by most growers in Japan and America. In this cultivation system the mycelium is grown in plastic bags or glass jars full of sterilized medium, which is usually some type of cereal grain. This grain is usually rice, wheat or rye although many different types of grain have been used. The most usual substrate for the Japanese and American growers is rice. But it has been experimentally determined that rice is not a very suitable substrate for *Cordyceps* production if the target medicinal compounds are considered on an economic basis. Rice does not allow the full range of secondary metabolites to be expressed by the fungus and the quality of rice grown *Cordyceps* is inferior to other artificially produced mycelia. There is rarely any appreciable amount of Adenosine or Cordycepin present in rice-grown *Cordyceps*. Furthermore, there are growth-stunting metabolites which build up in the substrate when *Cordyceps* is grown on rice, limiting the growth stage to only about 22-24 days and allowing no more than about 40% of the rice to be converted into mycelial mass. This figure of 40% represents the high end of conversion and is usually around 25-30%. This means that when *Cordyceps* is grown on rice, dried and powdered, the resultant product is actually about 60-75% rice flour.

Rye grain is another substrate often used for solid culture and it yields a higher quality product than rice, as long as a source of vegetable oil as an amendment is added to the growth medium at the time of substrate makeup. The oil provides necessary nutrients, which the organism utilizes for bioactive compound production. But even rye has disadvantages too. The compounds in rye, which give it that characteristic rye smell and taste, are not broken down by the *Cordyceps* and they concentrate in the final product. This rye taste and smell overcomes the characteristic *Cordyceps* taste and smell, and even though the resultant product is of better quality than the rice grown mycelium, there are certain perceptual problems that needs to be overcome by the buyer to make this an economical alternative. Rice-grown *Cordyceps* may seem like a better product to the average buyer because the rice does not mask the characteristic *Cordyceps* smell and taste. Most buyers in the health supplements industry tend to purchase bulk products on perception and faith rather than requiring an independent analysis. Rye also has growth-limiting factors, which causes the *Cordyceps* growth to stunt at about 28-30 days, although this can be overcome to a slight degree with the addition of about 1% ground oyster shell buffer to the medium at time of make-up.

Millet is another very good choice of substrate when it is available. It has no strong taste or smell of its own and does not stunt the growth to any significant degree; instead it allows the full expression of the secondary metabolites by the organism. But even millet has a problem, which is the high ratio of chitinous outer husky layer to starch. This outer husk is not broken down and represents a large portion of the final product weight, about 15%. The chitinous husk cannot be removed from the grain ahead of time, since doing so allows the grain to become too sticky during sterilization and a high degree of anaerobic contamination follows. The husk can be removed from the final product through mechanical means or the product can be used for hot water extractions or other processing. *Cordyceps* does not grow as fast on millet as it does on other grains, but the quality of the end product is higher.

White Milo grain, also known as white kaffir corn or white sorghum is an excellent choice of substrate. The red variety of Milo does not work nearly as well as the white variety as a substrate. White Milo has all of the best characteristics; it is cheap, it has a high starch/husk ratio, it does not stunt the growth, allows full expression of bioactive compounds and has no strong odor or taste of its own to compete with the taste and smell of the resultant *Cordyceps* product. White Milo when used alone however lacks some essential ingredients required for optimum growth by the *Cordyceps*. The addition of some portion of millet to the white Milo speeds up the growth by a factor of 6 times. The millet to Milo ratio is optimum at 1 part millet to 4 parts white Milo.

Whatsoever the grain is used, after some period of growth, the mycelium is harvested along with the residual grain. While this is an easily mastered and low capital investment cultivation technique, the most striking disadvantage of this method is that the grain content is usually greater than the mycelium content (but it can be overcome by using grain of better choice as stated above). However, the advantage of this method is that extra-cellular compounds are harvested along with the substrate and mycelium. In America and Japan, unconverted grain substrates are frequently used in solid culture (Holliday *et al*., 2004, Zhang *et al*., 1992).

The method adopted for the culturing of *C. sinensis* mycelia has an effect on the quality of the resultant product. Apart from the methodology itself, the next most important factor in the production of particular secondary metabolites (or target medicinal compounds) is the nature and composition of the substrate itself. (Zhang *et al.,* 1992). While it seems that a substrate that favors rapid and strong growth of the mycelium would be an ideal substrate to use, this is not necessarily the case. Substrates are chosen on their availability and price, on historical usage or preference in handling. In fact, the only way to determine whether the substrate being used is the best choice or not, is to compare the resultant product after growth on that particular substrate with some standard. If the end goal of production is cordycepin or didioxyadenosine (or some other specific compound, then the analysis is fairly straightforward. The amount of cordycepin or didioxyadenosine present is analyzed and the method is adopted on industrial basis.

Cultivators like to advertise their product derived from the asexual propagation as superior, since it is asexually reproduced under controlled laboratory conditions and thus its potency is supposed to be guaranteed, but this does not assure that these products contain the same range of active ingredients as the natural product (Li *et al*., 2001a). Many of these do not even contain Cordycepin and Cordycepic acid, the two most important active ingredients of *C. sinensis* (Holliday *et al*., 2004 and Hsu *et al*., 2002).

Wild fungi, as long as they are available will not be replaced by the laboratory produced fungal strains, but prices might be negatively affected. Artificially grown *C. sinensis* mycelium so far has not impacted the market of wild Cordyceps in Asia at all, as clearly documented in recent price increases.

References

Alexopoulous, C. J., Mims, C. W. and Blackwell, M. 1996. Introductory Mycology. 4th Ed. John Wiley and Sons, Inc. New York, 869p.

Arif, M. and Kumar N. 2003. Medicinal insects and insect-fungus relationship in high altitude areas of Kumaon Hills in Central Himalayas. *Journal of Experimental Zoology India*, **6.1:** 45-55.

Boesi, A. 2003. The Dbyar rtswa dgun 'bu *(Cordyceps sinensis* Berk*):* An important trade item for the Tibetan population of the Lithang county. Sichuan Province. China. *The Tibet Journal* **28(3):** 29-42.

Chang, R. 1993. Limitations and Potential applications of *Ganoderma* and related fungal polyglycans in clinical ontology. *First International Conference on Mushroom Biology and Mushroom products*. pp. 96

Chang R. 1996a. Functional properties of edible mushrooms. *Nutrition Review* **54:** S91-S93

Chang, R. 1996c. Potential Application of *Ganoderma* Polysaccharides in the Immune Surveillance and Chemoprevention of Cancer. 153-160 *In:* Royse, D.J. (ed). 1996. Mushroom Biology and Mushroom Products. Proceedings of the Second International Congress.

Chang, R. 1996b. The Central Importance of the beta-glucan receptor as the basis of immunologic bioactivity of *Ganoderma* polysaccharides *In*: Reishi, Mizuno T, Kim B.K. (eds), II Yang Press, Seoul, pp.177-179.

Chang, S.T. and Buswell, J.A., 1996 Mushroom nutriceuticals. *World J. Microbiol. Biotechn.*, **12:**473-476.

Charles, V.K. 1941. A preliminary check list of the entomogenous fungi of North America. Insect and Pest Survey Bulletin Bureau of Entomology and Plant Quarantine. U.S. Dept. Agri. **21(9):** 707-785.

Chen, S.J., Yin, D.H., Zhong, G.Y. and Huang, T.F. 2002b. Study on the biology of adult parasite of *Cordyceps sinensis, Hepialus biruensis. Zhongguo Zhong Yao Za Zhi,* **27(12):** 893-895.

Chen, Y., Zhang, Y.P., Yang, Y. and Yang, D. 1999. Genetic diversity and taxonomic implication of *Cordyceps sinensis* as revealed by RAPD markers. *Biochem. Genet.*, **37:** 201-213.

Chen, Y.Q., Wang, N., Zhou, H. and Qu, L.H. 2002a. Differentiation of medicinal *Cordyceps* species by rDNA 18S sequence analysis. *Planta Medica ,* **68(7):** 635-639.

Chilton, J. 1993. What are the health benefits of mushrooms? Let's Live, Dec., pp. 24-29.

Chiu, S.W., Cheng, K.W., Chu, W.L., Law, W.M., Yeung, H.W. Au and Ma, S.Y. 2000. Developmental plasticity of Hong Kong Ling Zhi as a response to the environment. Science and Cultivation of edible fungi. Van Gnensven (ed.) Balkema, Rotterdam.

Dadwal, V.S. and Jamaluddin 2004. Cultivation of *Ganoderma lucidum* (Fr.) Karst. *Indian Forester.* April : 435-440.

Donk, M.A. 1964. A conspectus of the families of Aphyllophorales. *Persoonia.* **3:** 199-324.

Furtado, J.S. 1962. Structure of the spore of the Ganodermoideae Donk. *Rickia.* **1:** 227-241.

Garbyal, S.S. 2001. Occurrence of *Cordyceps sinensis* in upper Himalaya, Dharchula sub-division, Pithoragarh District, Uttaranchal, India. *Indian Forester.* **127(11):** 1229-1231.

Hawksworth, D.L., 2001. The magnitude of fungal diversity: The 1.5 million species estimate revisited. *Mycol. Res.*, **105:** 1422-1432.

Holliday, J.C.; Cleaver, B S; Loomis-Powers, M. and Patel, D.2004. The hybridization of *Cordyceps sinensis* strains and the modifications of their culture parameters, in order to optimize the production of target medicinal compounds. *http://www.nwbotanicals.org/nwb/lexicon /hybridcordyceps.htm.*

Hsu, M.J., Lee, S.S. and Lin, W.W. 2002. Polysaccharide purified from *Ganoderma lucidum* inhibits spontaneous and Fas-mediated apoptosis in human neutrophils through activation of the phosphatidylinositol 3 kinase/Akt signaling pathway. *J Leukoc Biol.*, **72(1):** 207-16.

Ikekawa, T. 1969. Antitumor activity of aqueous extracts of edible mushrooms. *Cancer Research.*, **29(3):** 734-735.

Ito, H. 1972. Antitumor activity of Basidiomycetes. *Nippon Yakurigaku Zasshi* **68(4):** 429-44.

Jason E. Barker 2005. Medicinal mushrooms. *Alternative and complementary Therapies*. Vol. **11(3):** 141-145.

Jones, N.H. and Yusipang. 2001. *Cordyceps sinensis* in Bhutan. A concept note on the sustainable collection and utilization RNR-RC. *Yusipang*. Thumpu, Bhutan.

Kawagishi, H. 1997. A lectin from mycelia of the fungus *Ganoderma lucidum. Phytochemistry,* **44(1):** 7-10

Kim, H.W. and Kim, B.K. 1999. Biomedicinal triterpenoids of *Ganoderma lucidum* (Curt.: Fr.) P. Karst. (Aphyllophoromycetidae). *International Journal of Medicinal Mushrooms,* **1:** 121-138.

King, T.A. 1993. Mushrooms, the ultimate health food but little research in US to prove it. *Mushroom News,* **41(2):** 29-46.

Komoda, Y., Nakamura, H. and Ishihara, S. 1985. Structure of new terpenoids constituents of *Ganoderma lucidum* (Fr.) Karst. (Polyporaceae). *Chem. Pharmaceut. Bulletin.*, **33:** 4829-4835.

Li, Q.S., Zeng, W., Yin D.H.and Hyang T.F. 1999. A preliminary study on alternation of generation of *Cordyceps sinensis.* In :*Zhonggero Zhong Yao Za Xhi, (China Jour. Chin,. Mat. Med.),* **23(4):** 210-218.

Li, S. P., Li, P., Ji, H., Zhang, P., Dong, T.T.X. and Tsim, K.W.K. 2001a. The contents and their change of nucleosides from natural *Cordyceps sinensis* and cultured *Cordyceps* mycelia. *Acta. Pharmaceutica. Sinica.,* **36(6):** 436-439.

Li., S. P. Li. P. Dong T.T. and Tsin K.W. 2001b. Antioxidation activity of different types of natural *Cordyceps sinensis* and cultured *Cordyceps* mycelia. *Phytomedicine.*, **8:** 207-212.

Liu, Z. B. 1996. Modern research of *Ganoderma. Beijing Medical University Peking Union Medical College Press, Beijing.*

Luo J., Zhao Y. Y. and Li, Z. B. 2002. A new lanostane-type triterpene from the fruiting bodies of *Ganoderma lucidum. Journal of Asian Natural Products Research,* **4(2):** 129-34.

Ma, J., Ye, Q., Hua, Y., Zhang, D., Cooper, R., Chang, M.N., Chang, J.Y. and Sun, H.H. 2002. New lanostanoids from the mushroom *Ganoderma lucidum. Journal of Natural Products.*, **65(1):** 72-5.

Mains, E. B. 1940. *Cordyceps* species from Michigan. *Papers Michigan Academy of Science,* **25:** 79-84.

Manabe, N., Azuma, Y., Sugimoto, M., Uchio, K., Miyomoto, M., Taketomo, N., Tsuchita, H. and Miyomoto, H. 2000. Effect of mycelial extract of cultured *Cordyceps sinensis* on in vivo hepatic energy metabolism and blood flow in dietary hypoferric anemic mice. *British Journal of Nutrition.*, **83:** 197-204.

Massee, G. 1895. A revision of the genus *Cordyceps. Nn. Bot.*, **9:** 1-44.

Mishra, K. K. 2007. Characterization and commercialization of *Ganoderma lucidum*-A Medicinal Mushroom. Paper presented at International Conference on Mushroom Biology and Biotechnology, National Research Centre for Mushroom, Solan, Himachal Pradesh.

Mishra, K. K. and Singh, R.P. 2006. Exploitation of indigenous *Ganoderma lucidum* for yield on different substrates. *Journal of Mycology and Plant Pathology.* **36 (2):** 130-133.

Mishra, K. K., Kushwaha, K. P. S. and Sigh, R.P. 2004. Mushrooms of medicinal importance. *Indian Farmer's Digest* **37 (12):** 30-32.

Mizuno, T. 1996. A development on anti tumor polysaccharides from mushroom fungi. *FFI Journal.* **167:** 69-85.

Namgyel, P. 2003. Household income, Property rights and Sustainable Use of NTFP in Subsistence Mountain Economy: The case of case of *Cordyceps* and *Matsutake* in Bhutan Himalayas. Paper presented at the Regional CBNRM Workshop, Nov. 2003, 1-23.

Nikoh, N. and Fukatsu, T. 2000. Interkingdom host jumping underground: phylogenetic analysis of entomoparasitic fungi of the genus *Cordyceps*. *Mol. Biol. Evol.*, **17 (4):** 629-638.

Nuss, I. 1982. Die Bedeutung der Prroterosporen : Schlussfolgerungen ans Untersuchungen an *Ganoderma* (Basidiomycetes). *Plant Syst. Evol.*, **141:** 53-70.

Pai, S.H., Jong, S.C. and Low, D.W., 1990. Usages of Mushroom. *Bioindustry.* **1:** 126-131.

Peglar, D.N. and Young, T.W.K. 1973. Basidiospore form in the British species of *Ganoderma* Karst. *Kew Bulletin.*, **28:** 351-369.

Peglar, D.N. Yao, Y.J. and Li, Y. 1994. The Chinese "Caterpillar fungus" *Mycologist,* **8:** 3-5.

Rai, R. D. 2003. Successful cultivation of the medicinal mushroom Reishi, *Ganoderma lucidum* in India. *Mushroom Research,* **12(2):** 87-91.

Rosecke, J. and Konig, W. A. 2000. Constituents of various wood-rotting basidiomycetes. *Phytochemistr.*, **54:** 603-610.

Saccardo, P. A. 1883. *Cordyceps. Syll. Fungi.*, **2:** 566-578.

Singh, R. P. and Mishra, K. K. 2004. The medicinal mushroom: *Ganoderma. Indian Farmer's Digest.* **37 (12):** 28-29.

Singh, R. P., Mishra, K. K. and Singh, M. 2006. *Biodiversity* and Utilization of medicinal mushrooms. *Journal of Mycology and Plant Pathology.*, **36 (3):** 446-448.

Singh, R. P., Verma, R. C., Arora, R. K., Mishra, K. K., Bhanu, C. and Singh, M. 2007. Medicinal Mushrooms of Uttaranchal with particular reference to *Ganoderma, Auricularia* and *Cordyceps sinensis.* Paper presented at *International Conference on Mushroom Biology and Biotechnology*, National Research Centre for Mushroom, Solan, Himachal Pradesh.

Su, C. H., Sun, C. S. and Juan, S.W. 1997. Fungal mycelia as the source of chitin and polysaccharides and their applications and as skin substitutes. *Biomaterials.*, **18 :** 1169-1174.

Sharma S. S. 2004. Trade of *Cordyceps sinensis* from high altitudes pf the Indian Himalaya: conservation and biotechnological priorities. *Current Science.* **86:** 12.

Sung Jao-Mo. 2004. *Cordyceps* diversity and its preservation in Korea. *Mycologia.*, **55 (4):** 1-3.

Szedlay, G. 2002. Is the widely used medicinal fungus the *Ganoderma lucidum* (Fr.) Karst. sensu stricto? (A short review). *Acta Microbiol Immunol Hung.* **49 (2-3):** 235-43. Review.

Tanaka, S., Ko, K., Kino, K., Tsuchiya, K., Yamashita, A., Murasugi, A., Sakuma, S. and Tsunoo, H. 1989. Complete amino acid sequence of an immuno-modulatory protein, Ling Zhi-8 (LZ-8). An immunomodulator from a fungus, *Ganoderma lucidum,* having similarity to immunoglobulin variable regions. *Journal of Biol. Chem.*, **264:** 16372-16377.

Triratana, S. and Chaiprasert, A. 1991. Sexuality of *Ganoderma lucidum. Science and Cultivation of edible Fungi.* Maher (ed.) Balkema, Rotterdam.

Van der Hem, L. G., Van der Vliet, J. A., Bocken, C. F., Kino, K., Hoitsma, A. J. and Tax, W. J. 1995. Ling Zhi-8: Studies of a new immunomodulating agent. *Transplantation*, **60:** 438-443.

Veena, S. S. and Pandey, M. 2006. Evaluation of locally available substrates for the cultivation of indigenous *Ganoderma lucidum* isolate. *Journal of Mycology and Plant Pathology*, **36(3):** 434-438.

Vinning, G. S. 2003. Hong Kong: Some marketing issues for Bhutanese natural products. *Rural Enterprise Development Programme. Ministry of Agriculture, Thimpu.*

Vinning, G. S. 2004. Herbal medicines: Some notes from the International Trade Centre, Geneva. *Agricultural Marketing Services, Ministry of Agriculture, Thimpu.*

Wagner, R., Mitchell, D. A., Sassaki, G. L., Amazonas, M. A. L. de A. and Berovic, M. 2003. Current techniques for the cultivation of *Ganoderma lucidum* for the production of biomass, ganoderic acid and polysaccharides. *Food Technology and Biotechnology.*, **41 (4):** 371-382.

Wang, S. Y., 1997. The anti-tumor effect of *Ganoderma lucidum* is mediated by cytokines released from activated macrophages and T lymphocytes. *International Journal of Cancer.* **70(6):** 699-705.

Wasser, S. P. 2002. Medicinal mushrooms as a source of antitumor and immunomodulating polysaccharides. *Appl Microbiol Biotechnol.* **60(3):** 258-74.

Wasser, S. P. and Weis, A. L. 1999. Therapeutic effects of substances occurring in higher basidiomycetes mushrooms : modern perspective. *Crit. Rev. Immunol.*, **19:** 65-96.

Yamaguchi, Y., Kagota, S. Nakamura, K. Shinozuka, K. and Kunitomo, M. 2000. Antioxidant activity of the extracts from fruiting bodies of cultured *Cordyceps sinensis*. *Phytother. Res.*, **14:** 647-649.

Yang, J. J. and Yu, D. Q. 1990. Synthesis of *Ganoderma* alkaloid A and B. *Yaoxue Xuebao. Acta Pharm Sin.*, **25:** 555-559.

Zang, M. and Kinjo, N. 1998. Notes on the Alpine *Cordyceps* of China and nearby nations *Mycotaxon.*, **66:** 215-229.

Zhang, C.; Zhou, Y.; Wu, Z. and Bai, Y. 1992. Nourishment of *Cordyceps sinensis* mycelium. *Weishengwuxue Tongbao*, **19(3):** 129-133.

Zhu, M. 1999. Triterpene antioxidants from *Ganoderma lucidum. Phytother Res.*, **13(6):** 529-31.

❑❑❑

About the Editors

Dr. D. K. Singh (Dinesh Kumar Singh) is an Eminent Young Vegetable Scientist and Professor, Department of Vegetable Science, G. B. Pant University of Agriculture & Technology, Pantnagar, India with research experience of over 17 years. He has published more than 105 research papers and popular articles. He has three manuals, four book chapters and three books to his credit. He has attended more than 60 national and international seminars, symposia, conferences and workshops. He has been invited speaker in various national and international seminars and conferences. Dr. Singh has developed 18 varieties of vegetable crops including 5 hybrids, 12 (OPV) and 1 pure line, successfully completed six research projects and currently handling four research projects funded by various funding agencies (Like DBT, DST, HTM, NAIP - ICAR etc.). He has also introduced polyhouse tomato breeding and parthenocarpic cucumber breeding programmes first time in India and has released 4 varieties in Parthenocarpic Cucumber and Plolyhouse Bred Tomato (Pant Parthenocarpic Cucumber-2, Pant Parthenocarpic Cucumber-3, and Pant Polyhouse Bred Tomaro-2, Pant Polyhouse Bred Hybrid Tomato-1).

He has guided eleven M. Sc. (Ag) and six Ph.D. students. He has also established vegetable research centre and molecular and seed technology lab in the department. He has been awarded with gold medal for securing top position in Masters Degree programme, and also received gold medal for his meritorious service, outstanding performance and remarkable contribution in his professional field by "Indian International friends Society". Dr. Singh has also recipient of best paper presentation award during his visit in China for attending 4th International Cucurbits Symposium at Changsa.

He has visited in Ethiopia under ATVET World Bank programme and done Post Graduate course in constraints Horticulture from University of Jerusalem Rehovot, Israel and Romania for crop biodiversity study of Romanian agricultural crops and also visited Bangkok, Thailand for attending the APSA group discussion. The editor is member of several professional societies and alumni of Shallom Society Israel. He has been an invited speaker in several national and international seminars and conferences and has also delivered lectures, held discussions at Universities and Agriculture Research Centers. Dr Singh also visited Romania under International Cooperation for study of Romanian Vegetable Crop Biodiversity. He has also selected by U.N.O.-FAO as a consultant for the improvement of vegetable research and development programme for Libya Republic.